Partial Differential Equations: Theory, Control and Approximation

Jacques-Louis Lions in 1998

Philippe G. Ciarlet • Tatsien Li • Yvon Maday
Editors

Partial Differential Equations: Theory, Control and Approximation

In Honor of the Scientific Heritage of Jacques-Louis Lions

Springer

Editors
Philippe G. Ciarlet
Dept. Mathematics
City University of Hong Kong
Hong Kong, People's Republic of China

Yvon Maday
Laboratoire Jacques-Louis Lions
Université Pierre et Marie
Paris, France

Tatsien Li
School of Mathematical Sciences
Fudan University
Shanghai, People's Republic of China

ISBN 978-3-662-51434-4 ISBN 978-3-642-41401-5 (eBook)
DOI 10.1007/978-3-642-41401-5
Springer Dordrecht Heidelberg New York London

Mathematics Subject Classification: 35-XX, 49-XX, 65-XX

Softcover reprint of the hardcover 1st edition 2014

Printed on acid-free paper

Springer is part of Springer Science+Business Media (www.springer.com)

Preface

Jacques-Louis Lions (1928–2001) was an exceptional mathematician, whose lasting influence is still deeply felt all over the world.

He was a universally recognized and admired expert in partial differential equations, to the study of which he has made outstanding contributions regarding not only the theoretical aspects such as existence and uniqueness of partial differential equations, regularity of the solutions, homogenization, and control, but also their numerical analysis and applications to fluid and solid mechanics, oceanography, climatology, etc.

Together with Enrico Magenes, he first produced an exhaustive analysis of linear boundary value problems posed in Sobolev spaces, which includes, in particular, a remarkably elegant proof of Korn's inequality. He then developed with Guido Stampacchia the theory of variational inequalities, visco-elasticity, or plasticity. But he is perhaps even more remembered for the manifold landmark contributions he made to the research of nonlinear partial differential equations, notably by recognizing the efficiency of compactness, monotony, regularization, and penalty methods for their analysis.

With an incredible intuition, Jacques-Louis Lions foresaw very early the advantage of Galerkin methods, for instance, how the finite element method exceeds the more traditional finite-difference methods. In so doing, he was highly instrumental in the creation of a very powerful school of numerical analysts "without frontiers" (across national boundaries), who made many extraordinary breakthroughs to the theoretical understanding as well as to the practical implementation of a wide array of methods for approximating the solutions of partial differential equations. He also made pioneering contributions to the analysis of problems with small parameters and, more generally, of singular perturbation problems.

But his ever-favorite subject was control theory, where, as far back as in 1958, he made milestone advances in the extension of optimal control to systems governed by partial differential equations. One highlight of his contributions to this field was the prestigious "John von Neumann Lecture" that he gave at the SIAM Congress in Boston in 1986, where he laid the foundations of his well-known "HUM method".

One can only be impressed by his immense works, for the quality, diversity, or novelty of the mathematics used, and for his permanent quest for new applications that had previously been believed to be inaccessible.

Jacques-Louis Lions was a visionary, who quickly understood that the availability of ever-increasing computational power would revolutionize the modeling of numerous phenomena, provided however that the required mathematics were simultaneously created and developed. This is the essence of his immense scientific heritage.

Jacques-Louis Lions justly received numerous honors. In particular, he was a member of twenty-two academies, which included the most prestigious ones, such as the Royal Society, the USSR Academy of Sciences, the National Academy of Sciences of the USA, the French Academy of Sciences, the Third World Academy of Sciences, the Accademia Nazionale dei Lincei, and the Chinese Academy of Sciences. He was also awarded such highly prestigious prizes as the John von Neumann Prize, the Lagrange Prize of the ICIAM, and the Japan Prize.

It is to honor the scientific heritage of Jacques-Louis Lions that an "International Conference on Partial Differential Equations: Theory, Control and Approximation" was organized and held at Fudan University in Shanghai from May 28th to June 1st, 2012. This conference brought together experts from all over the world, whose talks covered the fields of research that Jacques-Louis Lions created or contributed so much to create. This book gathers some of the most representative contributions to the Conference, which have been and will be separately published in Chinese Annals of Mathematics in 2013 and 2014. We thank Ms. Wei Wu of the Editorial Board Office of Chinese Annals of Mathematics for her enthusiastic and effective work in editing this collection of papers.

All those who approached Jacques-Louis Lions will cherish the memory of his warm personality, the vision that he so well conveyed, and his profound intelligence.

Philippe G. Ciarlet
Tatsien Li
Yvon Maday

Contents

Control and Nash Games with Mean Field Effect

Alain Bensoussan and Jens Frehse

Abstract Mean field theory has raised a lot of interest in the recent years (see in particular the results of Lasry-Lions in 2006 and 2007, of Gueant-Lasry-Lions in 2011, of Huang-Caines-Malham in 2007 and many others). There are a lot of applications. In general, the applications concern approximating an infinite number of players with common behavior by a representative agent. This agent has to solve a control problem perturbed by a field equation, representing in some way the behavior of the average infinite number of agents. This approach does not lead easily to the problems of Nash equilibrium for a finite number of players, perturbed by field equations, unless one considers averaging within different groups, which has not been done in the literature, and seems quite challenging. In this paper, the authors approach similar problems with a different motivation which makes sense for control and also for differential games. Thus the systems of nonlinear partial differential equations with mean field terms, which have not been addressed in the literature so far, are considered here.

Keywords Mean field · Dynamic programming · Nash games · Equilibrium · Calculus of variations

Mathematics Subject Classification 49L20

Project supported by the WCU World Class University program through the National Research Foundation of Korea funded by the Ministry of Education, Science and Technology (No. R31-20007) and the Research Grants Council of HKSAR (No. PolyU 5001/11P).

A. Bensoussan (✉)
International Center for Decision and Risk Analysis, School of Management, University of Texas-Dallas, Richardson, TX, USA
e-mail: alain.bensoussan@utdallas.edu

A. Bensoussan
School of Business, The Hong Kong Polytechnic University, Hong Kong, China

A. Bensoussan
Graduate Department of Financial Engineering, Ajou University, Suwon, Korea

J. Frehse
Institute for Applied Mathematics, University of Bonn, Bonn, Germany

P.G. Ciarlet et al. (eds.), *Partial Differential Equations: Theory, Control and Approximation*, DOI 10.1007/978-3-642-41401-5_1,

1 Introduction

In this paper, we study the systems of nonlinear partial differential equations (or PDE for short) with mean field coupling. This extends the usual theory of a single PDE with mean field coupling. This extension has not been considered in the literature, probably because the motivation of mean field theory is precisely to eliminate the game aspect, by an averaging consideration. In fact, the starting point is a Nash equilibrium for an infinite number of players, with similar behavior. The averaging concept reduces this infinite number to a representative agent, who has a control problem to solve, with an external effect, representing the averaged impact of the infinite number of players. Of course, this framework relies on the assumption that the players behave in a similar way. Nevertheless, it eliminates the situation of a remaining Nash equilibrium for a finite number of players, with mean field terms. One may imagine groups with non-homogeneous behavior, in which case it is likely that one may recover systems of nonlinear PDE with mean field coupling. Although interesting, this extension has not been considered in the literature, and seems quite challenging. This is why we develop here a different motivation, which has interest in itself. It makes sense for control problems as well as for differential games. The mean field coupling term in our case has a different interpretation. Another interesting feature of our approach is that we do not need to consider an ergodic situation, as it is the case in the standard approach of mean field theory. In fact, considering strictly positive discounts is quite meaningful in our applications. This leads to systems of nonlinear PDE with mean field coupling terms, that we can study with a minimum set of assumptions. This is the objective of this paper. The ergodic case, when the discount vanishes, requires much stringent assumptions, as is already the case when there is no mean field terms. This case will be dealt with in a following article. We refer to [2, 5–7] for the situation without mean field term. Basically, our set of assumptions remains valid, and we have to incorporate additional assumptions to deal with the mean field terms. Moreover, some related results can be found in [9–11, 13–15].

2 Control Framework

2.1 Bellman Equation

We consider a classical control problem here. We treat an infinite horizon problem, with stationary evolution of the state. In order to remain within a bounded domain, we assume that the state evolution, modelled as a diffusion, is reflected on the boundary of the domain. More precisely, we define a probability space Ω, with $\mathcal{A}$, P equipped with a filtration $\mathcal{F}^t$ and a standard n-dimensional $\mathcal{F}^t$ Wiener process $w(t)$. Let $\mathcal{O}$ be a smooth bounded domain of $\mathbb{R}^n$. We set $\Gamma = \partial\mathcal{O}$. We denote by $\nu = \nu(x)$ the outward unit normal on a point x of Γ. Let $g(x, v)$ be a continuously differentiable function from $\mathbb{R}^n \times \mathbb{R}^m \to \mathbb{R}^n$. The second argument represents the control. To simplify, we omit to consider constraints on the control. Let $v(t)$ be a stochastic process adapted to the filtration $\mathcal{F}^t$. A controlled diffusion reflected at the boundary Γ with initial state $x \in \mathcal{O}$ is a pair of processes $y(t)$, $\xi(t)$, such that $y(t)$ is continuous adapted, $y(t) \in \overline{\mathcal{O}}$, and $\xi(t)$ is continuous adapted scalar increasing

$$\begin{aligned} \mathrm{d}y(t) &= g\big(y(t), v(t)\big)\mathrm{d}t + \sqrt{2}\mathrm{d}w(t) - \nu\big(y(t)\big)\mathbb{1}_{y(t)\in\Gamma}\mathrm{d}\xi(t), \\ y(0) &= x. \end{aligned} \tag{2.1}$$

Next, let $f(x, v)$ be a scalar function on $\mathbb{R}^n \times \mathbb{R}^m$, which is continuous and continuously differentiable in v. We assume also that $f(x, v)$ is bounded below. We define the payoff

$$J_\alpha\big(x, v(\cdot)\big) = E\int_0^{+\infty} \exp(-\alpha t f\big(y(t), v(t)\big))\mathrm{d}t. \tag{2.2}$$

We define the value function

$$u_\alpha(x) = \inf_{v(\cdot)} J_\alpha\big(x, v(\cdot)\big). \tag{2.3}$$

It is a fundamental result of dynamic programming that the value function is the solution to a partial differential equation, the Hamilton-Jacobi-Bellman equation

$$\begin{aligned} &-\Delta u_\alpha(x) + \alpha u_\alpha(x) = H\big(x, Du_\alpha(x)\big), \quad x \in \mathcal{O}, \\ &\left.\frac{\partial u_\alpha}{\partial \nu}\right|_\Gamma = 0 \end{aligned} \tag{2.4}$$

with the following notations:

$$\begin{aligned} &H(x, q) : \mathbb{R}^n \times \mathbb{R}^n \to \mathbb{R}, \\ &H(x, q) = \inf_v L(x, v, q), \\ &L(x, v, q) = f(x, v) + q \cdot g(x, v). \end{aligned} \tag{2.5}$$

An essential question becomes solving the PDE (2.4), and finding a sufficiently smooth solution. In the interpretation, which we shall discuss, we assume the regularity allowing to perform the calculations that we describe (in particular, taking derivatives).

The function H is called the Hamiltonian, and the function L is called the Lagrangian. Since the function L is continuously differentiable in v, and the infimum is attained at points, such that

$$\frac{\partial L}{\partial v}(x, v, q) = 0. \tag{2.6}$$

We shall assume that we can find a measurable map $\widehat{v}(x, q)$, which satisfies (2.6) and achieves the infimum in (2.5). We then have

$$H(x, q) = f\big(x, \widehat{v}(x, q)\big) + q \cdot g\big(x, \widehat{v}(x, q)\big). \tag{2.7}$$

It is also convenient to write

$$G(x,q) = g\big(x, \widehat{v}(x,q)\big). \tag{2.8}$$

With this notation, we can write Bellman equation as follows:

$$\begin{aligned} &-\Delta u_\alpha(x) - G\big(x, Du_\alpha(x)\big) + \alpha u_\alpha(x) = f\big(x, \widehat{v}\big(x, Du_\alpha(x)\big)\big), \\ &\left.\frac{\partial u_\alpha}{\partial \nu}\right|_\Gamma = 0. \end{aligned} \tag{2.9}$$

The main result of dynamic programming is that the infimum in (2.3) is attained for the control

$$\widehat{v}(t) = \widehat{v}\big(\widehat{y}(t), Du_\alpha\big(\widehat{y}(t)\big)\big), \tag{2.10}$$

where the process $\widehat{y}(t)$, i.e., the optimal trajectory, together with an increasing process $\widehat{\xi}(t)$, is the solution to

$$\begin{aligned} &d\widehat{y}(t) = G\big(\widehat{y}(t), Du_\alpha\big(\widehat{y}(t)\big)\big)dt + \sqrt{2}dw(t) - \nu\big(\widehat{y}(t)\big)\mathbb{1}_{\widehat{y}(t)\in\Gamma}d\widehat{\xi}(t), \\ &\widehat{y}(0) = x, \quad \widehat{y}(t) \in \overline{\mathcal{O}}. \end{aligned} \tag{2.11}$$

The main feature is that the optimal control is obtained through a feedback $\widehat{v}(x, Du_\alpha(x))$. We note $\widehat{v}_\alpha(x) = \widehat{v}(x, Du_\alpha(x))$.

2.2 *Revisiting Bellman Equation*

The fundamental result of Dynamic Programming motivates the following approach. Suppose that we restrict ourselves to the controls defined through feedbacks. A feedback is simply a measurable map $v(x)$. In fact, x can be restricted to $\overline{\mathcal{O}}$. To each feedback, we associate the function $u_{v(\cdot),\alpha}(x)$ as a solution to

$$\begin{aligned} &-\Delta u_{v(\cdot),\alpha}(x) - g\big(x, v(x)\big)\cdot Du_{v(\cdot),\alpha}(x) + \alpha u_{v(\cdot),\alpha}(x) = f\big(x, v(x)\big), \\ &\left.\frac{\partial u_{v(\cdot),\alpha}}{\partial \nu}\right|_\Gamma = 0. \end{aligned} \tag{2.12}$$

In fact, a feedback defines a particular case of control. We define the trajectory related to the feedback $v(\cdot)$ by considering the reflected diffusion

$$\begin{aligned} &dy(t) = g\big(y(t), v\big(y(t)\big)\big)dt + \sqrt{2}dw(t) - \nu\big(y(t)\big)\mathbb{1}_{y(t)\in\Gamma}d\xi(t), \\ &y(0) = x. \end{aligned} \tag{2.13}$$

To save notation, we omit to write that the trajectory depends on the feedback. The control corresponding to $v(\cdot)$ is $v(y(t))$. The corresponding payoff (see (2.2)) is thus

$$E\int_0^{+\infty} \exp\bigl(-\alpha t f\bigl(y(t), v\bigl(y(t)\bigr)\bigr)\bigr)\mathrm{d}t. \tag{2.14}$$

We shall also write it as $J_\alpha(x, v(\cdot))$ to avoid redundant notation. However, here $v(\cdot)$ refers to the feedback. It is easy to check that

$$u_{v(\cdot),\alpha}(x) = J_\alpha\bigl(x, v(\cdot)\bigr). \tag{2.15}$$

If we take $v(\cdot) = \widehat{v}_\alpha(\cdot)$, then $u_{\widehat{v}_\alpha(\cdot),\alpha}(x) = u_\alpha(x)$, $\forall x$, where $u_\alpha(x)$ is the solution to Bellman equations (2.4) and (2.9). From maximum principle considerations, we can assert that

$$u_\alpha(x) \le u_{v(\cdot),\alpha}(x), \quad \forall v(\cdot),\ \forall x \in \overline{\mathcal{O}}. \tag{2.16}$$

We recover that $\widehat{v}_\alpha(\cdot)$ is an optimal feedback.

2.3 *Calculus of Variations Approach*

To avoid confusion of notation, we shall consider the process defined by (2.13), with an initial condition x_0. The corresponding process $y(t)$ is a Markov process, whose probability distribution has a density denoted by $p_{v(\cdot)}(x,t)$ to emphasize the dependence on the feedback $v(\cdot)$, which is the solution to the Chapman-Kolmogorov equation

$$\begin{aligned}
&\frac{\partial p}{\partial t} - \Delta p + \operatorname{div}\bigl(g\bigl(x, v(x)\bigr)p\bigr) = 0, \quad x \in \mathcal{O},\\
&\frac{\partial p}{\partial \nu} - g\bigl(x, v(x)\bigr)\cdot \nu(x) p = 0, \quad x \in \Gamma,\\
&p(x,0) = \delta_{x_0}(x).
\end{aligned} \tag{2.17}$$

By the smoothing effect of diffusions, $p_{v(\cdot)}(x,t)$ is a function and not a distribution for any positive t. Moreover, $p_{v(\cdot)}(x,t)$ is, for any t, a density probability on $\mathcal{O}$. Now we can express

$$\begin{aligned}
u_{v(\cdot),\alpha}(x_0) &= E\int_0^{+\infty} \exp\bigl(-\alpha t f\bigl(y(t), v\bigl(y(t)\bigr)\bigr)\bigr)\mathrm{d}t\\
&= \int_0^{+\infty} \exp\Bigl(-\alpha t\Bigl(\int_{\mathcal{O}} p_{v(\cdot)}(x,t) f\bigl(x, v(x)\bigr)\mathrm{d}x\Bigr)\Bigr)\mathrm{d}t.
\end{aligned} \tag{2.18}$$

Let us define

$$p_{v(\cdot),\alpha}(x) = \alpha\int_0^{+\infty} \exp\bigl(-\alpha t p_{v(\cdot)}(x,t)\bigr)\mathrm{d}t, \tag{2.19}$$

which is the solution to

$$\begin{aligned} & -\Delta p_\alpha + \operatorname{div}\big(g\big(x, v(x)\big) p_\alpha\big) + \alpha p_\alpha = \alpha \delta_{x_0}, \quad x \in \mathcal{O}, \\ & \frac{\partial p_\alpha}{\partial \nu} - g\big(x, v(x)\big) \cdot \nu(x) p_\alpha = 0, \quad x \in \Gamma. \end{aligned} \tag{2.20}$$

We then get the formula

$$\alpha u_{v(\cdot),\alpha}(x_0) = \int_{\mathcal{O}} p_{v(\cdot),\alpha}(x) f\big(x, v(x)\big) \mathrm{d}x. \tag{2.21}$$

We can then state the lemma as follows.

Lemma 2.1 *The functional $u_{v(\cdot),\alpha}(x_0)$ is Frechet differentiable in $v(\cdot)$ with the formula*

$$\alpha \frac{\mathrm{d}}{\mathrm{d}\theta} u_{v(\cdot)+\theta\widetilde{v}(\cdot),\alpha}(x_0)|_{\theta=0} = \int_{\mathcal{O}} p_{v(\cdot),\alpha}(x) \frac{\partial L}{\partial v}\big(x, v(x), Du_{v(\cdot),\alpha}(x)\big) \widetilde{v}(x) \mathrm{d}x. \tag{2.22}$$

Proof We first show that $p_{v(\cdot),\alpha}(x)$ is Frechet-differentiable in $v(\cdot)$ for fixed x. Indeed, by direct differentiation, we check that

$$\widetilde{p}_\alpha(x) = \frac{\mathrm{d}}{\mathrm{d}\theta} p_{v(\cdot)+\theta\widetilde{v}(\cdot),\alpha}(x)|_{\theta=0}$$

is the solution to

$$\begin{aligned} & -\Delta \widetilde{p}_\alpha + \operatorname{div}\big(g\big(x, v(x)\big) \widetilde{p}_\alpha\big) + \alpha \widetilde{p}_\alpha + \operatorname{div}\big(g_v\big(x, v(x)\big) \widetilde{v}(x) p_{v(\cdot),\alpha}(x)\big) = 0, \\ & x \in \mathcal{O}, \\ & \frac{\partial \widetilde{p}_\alpha}{\partial \nu} - g\big(x, v(x)\big) \cdot \nu(x) \widetilde{p}_\alpha - g_v\big(x, v(x)\big) \widetilde{v}(x) \cdot \nu(x) p_{v(\cdot),\alpha} = 0, \quad x \in \Gamma, \end{aligned} \tag{2.23}$$

in which

$$g_v(x, v) = \frac{\partial g}{\partial v}(x, v).$$

Therefore,

$$\begin{aligned} \alpha \frac{\mathrm{d}}{\mathrm{d}\theta} u_{v(\cdot)+\theta\widetilde{v}(\cdot),\alpha}(x_0)|_{\theta=0} &= \int_{\mathcal{O}} \widetilde{p}_\alpha(x) f\big(x, v(x)\big) \mathrm{d}x \\ &\quad + \int_{\mathcal{O}} p_{v(\cdot),\alpha}(x) \frac{\partial f}{\partial v}\big(x, v(x)\big) \widetilde{v}(x) \mathrm{d}x. \end{aligned}$$

But

$$\int_{\mathcal{O}} \widetilde{p}_\alpha(x) f\big(x, v(x)\big) \mathrm{d}x = \int_{\mathcal{O}} Du_{v(\cdot),\alpha}(x) \cdot g_v\big(x, v(x)\big) \widetilde{v}(x) \mathrm{d}x,$$

and the result follows immediately. □

Corollary 2.1 *A feedback $\widehat{v}_\alpha(\cdot)$, which minimizes $u_{v(\cdot),\alpha}(x_0)$, satisfies*

$$\widehat{v}_\alpha(x) = \widehat{v}\big(x, Du_\alpha(x)\big),$$

where $u_\alpha(x)$ is the solution to the Bellman equation (2.4).

Proof The Frechet derivative of $u_{v(\cdot),\alpha}(x_0)$ at $\widehat{v}_\alpha(\cdot)$ must vanish. From formula (2.22), we deduce

$$\frac{\partial L}{\partial v}\big(x, \widehat{v}_\alpha(x), Du_{\widehat{v}_\alpha(\cdot),\alpha}(x)\big) = 0.$$

But then $u_{\widehat{v}_\alpha(\cdot),\alpha}(x) = u_\alpha(x)$ and the result follows. □

Remark 2.1 We note that the feedback $\widehat{v}_\alpha(x)$ is optimal for any value of x_0. In this approach, Bellman equation appears in expressing a necessary condition of optimality for a calculus of variations problem. This is not at all the traditional way, in which Bellman equation is introduced as a sufficient condition of optimality for the original stochastic control problem (2.3). This calculus of variations approach is rather superfluous for the standard stochastic control problem, since it leads to weaker results. In particular, we need to restrict the class of controls to the feedback controls, whereas we know that the optimality of the feedback controls holds against any non-anticipative controls. However, the calculus of variations approach can be extended to more general classes of control problems, as considered in this work, whereas the traditional approach can not.

3 More General Control Problems

3.1 Motivation

We consider the same objective function as before, but we would also like to control a functional of the path. As an example, we want to minimize the modified functional

$$\begin{aligned} J_\alpha\big(x_0, v(\cdot)\big) = {} & E\int_0^{+\infty} \exp\big(-\alpha t f\big(y(t), v\big(y(t)\big)\big)\big)dt \\ & + \frac{\gamma}{2}\left(E\int_0^{+\infty} \exp\big(-\alpha t h\big(y(t)\big)\big)dt - M\right)^2, \end{aligned} \tag{3.1}$$

where $h(x)$ is continuous. We can regard the second term as transforming a constraint into a penalty term in the cost functional.

We restrict ourselves to the feedbacks $v(\cdot)$ and $y(0) = x_0$. Clearly, the dynamic programming approach fails for this problem, since $J_\alpha(x, v(\cdot))$ is not a solution to a PDE. However, we can extend the calculus of variations approach. Indeed, considering the probability $p_{v(\cdot),\alpha}(x)$ as a solution to (2.20), we can write $J_\alpha(x_0, v(\cdot))$ as

$$
\begin{aligned}
J_\alpha\big(x_0, v(\cdot)\big) &= \frac{1}{\alpha}\int_{\mathcal{O}} p_{v(\cdot),\alpha}(x) f\big(x, v(x)\big)\mathrm{d}x + \frac{\gamma}{2}\bigg(\frac{1}{\alpha}\int p_{v(\cdot),\alpha}(x) h(x)\mathrm{d}x - M\bigg)^2 \\
&= \frac{1}{\alpha}\int_{\mathcal{O}} p_{v(\cdot),\alpha}(x) f\big(x, v(x)\big)\mathrm{d}x + \frac{1}{\alpha}\Phi_\alpha(p_{v(\cdot),\alpha}),
\end{aligned}
\tag{3.2}
$$

where

$$
\Phi_\alpha(m) = \frac{\gamma}{2\alpha}\bigg(\int_{\mathcal{O}} m(x)h(x)\mathrm{d}x - \alpha M\bigg)^2 \tag{3.3}
$$

is a functional on the set $L^1(\mathcal{O})$.

3.2 *Calculus of Variations Problem*

To avoid Dirac measures on the right-hand side, we shall consider the state equation $p_{v(\cdot),\alpha}(\cdot)$ as a solution to

$$
\begin{aligned}
&-\Delta p_\alpha + \operatorname{div}\big(g\big(x, v(x)\big)p_\alpha\big) + \alpha p_\alpha = \alpha m_0, \quad x \in \mathcal{O}, \\
&\frac{\partial p_\alpha}{\partial \nu} - g\big(x, v(x)\big)\cdot \nu(x) p_\alpha = 0, \quad x \in \Gamma,
\end{aligned}
\tag{3.4}
$$

in which m_0 is a probability density on $\mathcal{O}$. It corresponds clearly to Eqs. (2.17)–(2.19) with initial condition m_0 instead of δ_{x_0}. It means that, going back to the reflected diffusion (2.13), we can not observe the initial state. However, since we apply a feedback on the state, we still consider that we can observe the state at any time strictly positive. We choose the feedback $v(\cdot)$ in order to minimize the payoff

$$
\alpha J_\alpha\big(v(\cdot)\big) = \int_{\mathcal{O}} p_{v(\cdot),\alpha}(x) f\big(x, v(x)\big)\mathrm{d}x + \Phi_\alpha(p_{v(\cdot),\alpha}). \tag{3.5}
$$

The functional $\Phi_\alpha(m)$ is defined on $L^1(\mathcal{O})$, and we assume that it is Frechet-differentiable, with derivative in $L^\infty(\mathcal{O})$. Namely

$$
\frac{\mathrm{d}\Phi_\alpha(m + \theta\widetilde{m})}{\mathrm{d}\theta}\bigg|_{\theta=0} = \int_{\mathcal{O}} V_{m,\alpha}(x)\widetilde{m}(x)\mathrm{d}x, \tag{3.6}
$$

where $V_{m,\alpha}(\cdot)$ is in $L^\infty(\mathcal{O})$. In the example (3.3), we simply have

$$
V_{m,\alpha}(x) = \frac{\gamma}{\alpha}h(x)\bigg(\int_{\mathcal{O}} m(\xi)h(\xi)\mathrm{d}\xi - \alpha M\bigg). \tag{3.7}
$$

Our problem is to minimize the functional $\alpha J_\alpha(v(\cdot))$. In fact, since there are no constraints on the feedback control, we will write a necessary condition of optimality for an optimal feedback.

3.3 Euler Condition of Optimality

We just check that the functional $\alpha J_\alpha(v(\cdot))$ has a Frechet derivative. We associate to a feedback $v(\cdot)$, i.e., the PDE with mean field term

$$\begin{aligned}
&-\Delta u_{v(\cdot),\alpha}(x)-g\big(x,v(x)\big)\cdot Du_{v(\cdot),\alpha}(x)+\alpha u_{v(\cdot),\alpha}(x)\\
&\quad=f\big(x,v(x)\big)+V_{p_{v(\cdot),\alpha},\alpha}(x),\\
&\frac{\partial u_{v(\cdot),\alpha}}{\partial\nu}\bigg|_\Gamma=0.
\end{aligned}\tag{3.8}$$

We see that, conversely to the case (2.12), the PDE depends explicitly on $p_{v(\cdot),\alpha}$.

Lemma 3.1 *The functional $\alpha J_\alpha(v(\cdot))$ has a Frechet differential given by*

$$\alpha\frac{\mathrm{d}}{\mathrm{d}\theta}J_\alpha\big(v(\cdot)+\theta\widetilde{v}(\cdot)\big)\big|_{\theta=0}=\int_{\mathcal{O}}p_{v(\cdot),\alpha}(x)\frac{\partial L}{\partial v}\big(x,v(x),Du_{v(\cdot),\alpha}(x)\big)\widetilde{v}(x)\mathrm{d}x.\tag{3.9}$$

Proof The proof is similar to that of Lemma 2.1. The Lagrangian $L(x,v,q)$ is defined in (2.5). □

Then we can give a necessary condition of optimality for a feedback $\widehat{v}_\alpha(\cdot)$. We recall the notations (2.6)–(2.8). We consider the system

$$\begin{aligned}
&-\Delta u_\alpha+\alpha u_\alpha=H(x,Du_\alpha)+V_{m_\alpha,\alpha}(x),\quad x\in\mathcal{O},\\
&\frac{\partial u_\alpha}{\partial\nu}\bigg|_\Gamma=0,\\
&-\Delta m_\alpha+\operatorname{div}\big(G(x,Du_\alpha)m_\alpha\big)+\alpha m_\alpha=\alpha m_0,\quad x\in\mathcal{O},\\
&\frac{\partial m_\alpha}{\partial\nu}-G(x,Du_\alpha)\cdot\nu(x)m_\alpha=0,\quad x\in\Gamma.
\end{aligned}\tag{3.10}$$

We then write

$$\widehat{v}_\alpha(x)=\widehat{v}\big(x,Du_\alpha(x)\big),\tag{3.11}$$

where we recall the definition of $\widehat{v}(x,q)$ as the solution to (2.6). $H(x,q)$ and $G(x,q)$ have been defined in (2.7) and (2.8), respectively.

We can state as follows.

Proposition 3.1 *For a feedback* $\widehat{v}_\alpha(\cdot)$ *to be optimal for the functional* (3.5), *it is necessary that Eqs.* (3.10) *and* (4.14) *hold.*

Proof From the expression (3.9) of the Frechet derivative, one must have

$$\frac{\partial L}{\partial v}\big(x, \widehat{v}_\alpha(x), Du_{\widehat{v}_\alpha(\cdot),\alpha}(x)\big) = 0.$$

Hence

$$\widehat{v}_\alpha(x) = \widehat{v}\big(x, Du_{\widehat{v}_\alpha(\cdot),\alpha}(x)\big).$$

If we set

$$u_\alpha(x) = u_{\widehat{v}_\alpha(\cdot),\alpha}(x), \qquad m_\alpha(x) = p_{\widehat{v}_\alpha(\cdot),\alpha}(x),$$

and from Eqs. (3.4) and (3.8), it is clear that $(u_\alpha(\cdot),\ m_\alpha(\cdot))$ is a solution to the system (3.10). This completes the proof. □

Remark 3.1 As mentioned in the case of standard dynamic programming, showing that the system (3.10) has a solution becomes a problem itself. The claim that it has a solution, as a consequence of necessary conditions of optimality, lies on the assumption that an optimal feedback for the control problem (3.2) exists, and is not fully rigorous. We will address this problem in the analytic part.

4 Nash Equilibrium

4.1 Definition of the Problem

To avoid redundant notation, we will not write explicitly the index α. We will generalize the calculus of variations problem described in Sect. 3.2, and then provide applications and examples. We consider N players, which decide on feedbacks $v^i(x)$ $(i = 1, \dots, N, x \in \mathbb{R}^n)$. We shall use the notation

$$v = \big(v^1, \dots, v^N\big) = \big(v^i, \overline{v}^i\big).$$

The second notation means that we emphasize the case of player i, so we indicate his decision v^i, and denote by $\overline{v}^i$ the vector of decisions of all other players. The decision v^i belongs to an Euclidean space $\mathbb{R}^{d_i}$. We next consider continuous functions $f^i(x, v) \in \mathbb{R}$ and $g^i(x, v) \in \mathbb{R}^n$.

An important difference from the case of a single player is that the decision of player i is not just the feedback $v^i(x)$, a measurable function from $\mathbb{R}^n$ to $\mathbb{R}^{d_i}$, but also the state $p^i(x)$, a probability density on $\mathcal{O}$ which is a continuous function. So player i chooses the pair $v^i(\cdot)$, $p^i(\cdot)$. We will require some constraints between these two decisions, but it is important to proceed in this way, for the reasons which will be explained below. In a way similar to v, we shall use the notation

$$p = (p^1, \ldots, p^N) = (p^i, \overline{p}^i)$$

to refer to the vector of states.

Each player wants to minimize his payoff

$$J^i\big(v(\cdot); p(\cdot)\big) = \int_{\mathcal{O}} p^i(x) f^i\big(x, v(x)\big)\mathrm{d}x + \Phi^i(p). \tag{4.1}$$

The functionals $\Phi^i(p)$ are defined on $(L^1(\mathcal{O}))^N$. The functionals have partial Frechet derivatives. More precisely, by our convention $\Phi^i(m) = \Phi^i(m^i, \overline{m}^i)$, we assume that

$$\left.\frac{\mathrm{d}\Phi^i(m^i + \theta\widetilde{m}^i, \overline{m}^i)}{\mathrm{d}\theta}\right|_{\theta=0} = \int_{\mathcal{O}} V^i_{[m]}(x)\widetilde{m}^i(x)\mathrm{d}x, \tag{4.2}$$

and the functions $V^i_{[m]}(x)$ are in $L^\infty(\mathcal{O})$.

Our concept of Nash equilibrium is as follows. A pair $(\widehat{v}(\cdot),\ \widehat{p}(\cdot))$ is a Nash equilibrium, if the following conditions are satisfied. Let $v^i(\cdot)$ be any feedback for player i. Define $p^i_{v^i(\cdot),\widehat{\overline{v}}(\cdot)^i}(x)$ as the solution to

$$\begin{aligned} & -\Delta p^i + \mathrm{div}\big(g^i\big(x, v^i(x), \widehat{\overline{v}}(x)^i\big)p^i\big) + \alpha p^i = \alpha m^i_0, \quad x \in \mathcal{O}, \\ & \frac{\partial p^i}{\partial \nu} - g^i\big(x, v^i(x), \widehat{\overline{v}}(x)^i\big)\cdot \nu(x)p^i = 0, \quad x \in \Gamma. \end{aligned} \tag{4.3}$$

We note that the feedbacks of all players except i are frozen at the values $\widehat{v}^j(\cdot)$ $(j \neq i)$. The player i can choose his own feedback $v^i(\cdot)$. His decision $p^i(\cdot)$ is not decided independent of $v^i(\cdot)$ and of the vector of other players' decisions $\widehat{\overline{v}}(\cdot)^i$. It is $p^i_{v^i(\cdot),\widehat{\overline{v}}(\cdot)^i}(\cdot)$. However, he considers the decisions of the other players as $\widehat{\overline{v}}(\cdot)^i,\ \widehat{\overline{p}}(\cdot)^i$. The important thing to notice is that, he can not influence either $\widehat{\overline{v}}(\cdot)^i$ as expected, or $\widehat{\overline{p}}(\cdot)^i$. Therefore, the first condition is

$$\widehat{p}^i(x) = p^i_{\widehat{v}^i(\cdot),\widehat{\overline{v}}(\cdot)^i}(x) = p^i_{\widehat{v}(\cdot)}(x). \tag{4.4}$$

The second condition is that

$$J^i\big(\widehat{v}(\cdot); \widehat{p}(\cdot)\big) \leq J^i\big(v^i(\cdot), \widehat{\overline{v}}(\cdot)^i; p^i_{v^i(\cdot),\widehat{\overline{v}}(\cdot)^i}(\cdot), \widehat{\overline{p}}(\cdot)^i\big). \tag{4.5}$$

This condition explains why the $p^i(\cdot)$ is also considered as a decision variable. If only the feedbacks $v^i(\cdot)$ (and not the pair $(v^i(\cdot), p^i(\cdot))$) were decision variables, we would have in (4.5), the vector of functions $p^j_{v^i(\cdot),\widehat{\overline{v}}(\cdot)^i}(x)$, $j \neq i$, instead of $\widehat{\overline{p}}(\cdot)^i$. We do not know how to solve this problem. The difficulty arises from the fact that the functional $\Phi^i(m)$ depends on all the functions. If it were dependent on m^i only, it would not be necessary to make the difference. This occurs, in particular, in the case of the control problem, when there is only one player.

4.2 Necessary Conditions for a Nash Equilibrium

Let $v(\cdot)=(v^i(\cdot),\overline{v}^i(\cdot))$ and $p(\cdot)=(p^i(\cdot),\overline{p}^i(\cdot))$ be a pair of vector feedbacks and probabilities. We associate probabilities $p^i_{v^i(\cdot),\overline{v}(\cdot)^i}(\cdot)$ as a solution to

$$\begin{aligned}
&-\Delta p^i+\operatorname{div}\big(g^i\big(x,v^i(x),\overline{v}(x)^i\big)p^i\big)+\alpha p^i=\alpha m_0^i, \quad x\in\mathcal{O},\\
&\frac{\partial p^i}{\partial \nu}-g^i\big(x,v^i(x),\overline{v}(x)^i\big)\cdot \nu(x)p^i=0, \quad x\in\Gamma.
\end{aligned} \tag{4.6}$$

We furthermore define functions $u^i_{v^i(\cdot),\overline{v}(\cdot)^i;\overline{p}^i(\cdot)}(x)$ by

$$\begin{aligned}
&-\Delta u^i-g^i\big(x,v^i(x),\overline{v}(x)^i\big)\cdot Du^i+\alpha u^i\\
&\quad=f^i\big(x,v^i(x),\overline{v}(x)^i\big)+V^i_{[p^i_{v^i(\cdot),\overline{v}(\cdot)^i}(\cdot),\overline{p}^i(\cdot)]}(x), \quad x\in\mathcal{O},\\
&\left.\frac{\partial u^i}{\partial \nu}\right|_{\Gamma}=0.
\end{aligned} \tag{4.7}$$

We then claim the following result.

Lemma 4.1 *The functional $J^i(v(\cdot);p(\cdot))$ satisfies*

$$\begin{aligned}
&\frac{\mathrm{d}}{\mathrm{d}\theta}J^i\big(v^i(\cdot)+\theta\widetilde{v}^i(\cdot),\overline{v}(\cdot)^i;p^i_{v^i(\cdot)+\theta\widetilde{v}^i(\cdot),\overline{v}(\cdot)^i}(\cdot),\overline{p}^i(\cdot)\big)|_{\theta=0}\\
&\quad=\int_{\mathcal{O}}p^i_{v^i(\cdot),\overline{v}(\cdot)^i}(x)\frac{\partial L^i}{\partial v^i}\big(x,v^i(x),\overline{v}(x)^i,Du^i_{v^i(\cdot),\overline{v}(\cdot)^i;\overline{p}^i(\cdot)}(x)\big)\mathrm{d}x
\end{aligned} \tag{4.8}$$

with

$$L^i\big(x,v,q^i\big)=f^i(x,v)+q^i\cdot g^i(x,v). \tag{4.9}$$

Proof The proof is similar to the case of a single player, since the vectors $\overline{v}(\cdot)^i$, $\overline{p}^i(\cdot)$ are fixed in the functional $J^i(v^i(\cdot),\overline{v}(\cdot)^i;p^i_{v^i(\cdot),\overline{v}(\cdot)^i}(\cdot),\overline{p}^i(\cdot))$. □

We will now state necessary conditions for a pair $\widehat{v}(\cdot)$, $\widehat{p}(\cdot)$ to be a Nash equilibrium. We first define a Nash equilibrium of the Lagrangian functions. Namely, we solve the system

$$\frac{\partial L^i}{\partial v^i}\big(x,v^i,\overline{v}^i,q^i\big)=0, \quad i=1,\dots,N. \tag{4.10}$$

This defines functions $\widehat{v}^i(x,q)$ where $q=(q^1,\dots,q^N)$. We define next the Hamiltonians

$$H^i(x,q)=L^i\big(x,\widehat{v}(x,q),q^i\big) \tag{4.11}$$

and

$$G^i(x,q) = g^i\big(x, \widehat{v}(x,q)\big). \tag{4.12}$$

We next introduce the system

$$\begin{aligned}
&-\Delta u^i + \alpha u^i = H^i(x, Du) + V^i_{[m]}(x), \quad x \in \mathcal{O},\\
&\left.\frac{\partial u^i}{\partial \nu}\right|_{\Gamma} = 0,\\
&-\Delta m^i + \operatorname{div}\big(G^i(x, Du)m^i\big) + \alpha m^i = \alpha m^i_0, \quad x \in \mathcal{O},\\
&\frac{\partial m^i}{\partial \nu} - G^i(x, Du)\cdot \nu(x) m^i = 0, \quad x \in \Gamma,
\end{aligned} \tag{4.13}$$

and define

$$\widehat{v}^i(x) = \widehat{v}^i\big(x, Du(x)\big), \qquad \widehat{p}^i(x) - m^i(x). \tag{4.14}$$

By construction, we have

$$m^i(x) = p^i_{\widehat{v}^i(\cdot), \overline{\widehat{v}}(\cdot)^i}(x), \tag{4.15}$$

$$u^i(x) = u^i_{\widehat{v}^i(\cdot), \overline{\widehat{v}}(\cdot)^i; \overline{m}^i(\cdot)}(x). \tag{4.16}$$

We can then state as follows.

Proposition 4.1 *A Nash equilibrium* $(\widehat{v}(\cdot), \widehat{p}(\cdot))$ *of functionals* (4.1) *in the sense of conditions* (4.4)–(4.5) *must satisfy the relations* (4.14).

Proof In view of (4.5) and the formula giving the Frechet differential (4.8), we must have

$$\frac{\partial L^i}{\partial v^i}\big(x, \widehat{v}^i(x), \overline{\widehat{v}}(x)^i, Du^i_{\widehat{v}^i(\cdot), \overline{\widehat{v}}(\cdot)^i; \overline{\widehat{p}}^i(\cdot)}(x)\big) = 0.$$

In view of (4.4), the functions $m^i(x)$ and $u^i(x)$ defined by (4.15) and (4.16) are solutions to (4.13), and conditions (4.14) are satisfied. This completes the proof. □

4.3 Examples

We give here an example of the functional $\Phi^i(m)$. We set

$$\Phi^i(m) = \frac{\gamma}{2}\left(\int_{\mathcal{O}} m^i(x)h^i(x)\mathrm{d}x - \frac{1}{N}\sum_{j=1}^{N}\int_{\mathcal{O}} m^j(x)h^j(x)\mathrm{d}x\right)^2. \tag{4.17}$$

When this functional is incorporated into the payoff (4.1), player i aims at equalizing a quantity of interest with all corresponding ones of other players. This functional has a Frechet differential in m^i given by

$$V^i_{[m]}(x) = \gamma\left(1 - \frac{1}{N}\right)\left(\int_{\mathcal{O}} m^i(\xi)h^i(\xi)\mathrm{d}\xi - \frac{1}{N}\sum_{j=1}^{N}\int_{\mathcal{O}} m^j(\xi)h^j(\xi)\mathrm{d}\xi\right)h^i(x). \tag{4.18}$$

4.4 Probabilistic Interpretation

We can give a probabilistic interpretation to the Nash game (4.1) in the sense of (4.4)–(4.5). We consider feedbacks $v^i(\cdot)$, and construct on a probability space $\Omega, \mathcal{A}, P$ trajectories $y^i(t) \in \overline{\mathcal{O}}$, which are independent and have probability densities $p^i(t)$ defined on $\mathcal{O}$. These densities as well as the feedbacks are decisions. Then we set

$$p^i = \alpha \int_0^{+\infty} \exp(-\alpha t p^i(t))\mathrm{d}t.$$

If we consider the functional (4.17), we have the interpretation

$$\Phi^i(p) = \frac{\gamma}{2}\left(\alpha E\int_0^{+\infty} \exp(-\alpha t h^i(y^i(t)))\mathrm{d}t - \frac{\alpha}{N}\sum_{j=1}^{N} E\int_0^{+\infty} \exp(-\alpha t h^j(y^j(t)))\mathrm{d}t\right)^2,$$

so the functional $J^i(v(\cdot); p(\cdot))$ defined by (4.1) has the following interpretation:

$$\begin{aligned} J^i(v(\cdot); p(\cdot)) &= \alpha E\int_0^{+\infty} \exp(-\alpha t f^i(y^i(t), v(y^i(t))))\mathrm{d}t \\ &\quad + \frac{\gamma}{2}\left(\alpha E\int_0^{+\infty} \exp(-\alpha t h^i(y^i(t)))\mathrm{d}t\right. \\ &\quad \left. - \frac{\alpha}{N}\sum_{j=1}^{N} E\int_0^{+\infty} \exp(-\alpha t h^j(y^j(t)))\mathrm{d}t\right)^2, \end{aligned} \tag{4.19}$$

in which

$$v(y^i(t)) = (v^1(y^i(t)), \dots, v^N(y^i(t))).$$

It is important to notice that, although the feedbacks relate to the different players, each player i considers that they operate on his trajectory $y^i(t)$. Moreover, condition

(4.4) means that player i sees his trajectory $y^i(t)$ as the solution to

$$\begin{aligned} \mathrm{d}y^i(t) &= g^i\big(y^i(t), v\big(y^i(t)\big)\big)\mathrm{d}t + \sqrt{2}\mathrm{d}w^i(t) - \nu\big(y^i(t)\big)\mathbb{1}_{y(t)\in\Gamma}\mathrm{d}\xi^i(t), \\ y^i(0) &= y_0^i, \end{aligned} \tag{4.20}$$

where the Wiener processes $w^i(\cdot)$ are independent standard, and y_0^i are independent random variables, also independent of the Wiener processes, with probability density m_0^i. Since $y^i(t)$ is a reflected process, the pair $y^i(t), \xi^i(t)$ has to be defined jointly, in a unique way.

Remark 4.1 In problem (4.19)–(4.20), it is important to emphasize that player i considers the trajectories of other players $y^j(t)$ as given. His own trajectory $y^i(t)$ is defined by (4.20), in which he takes into account all feedbacks. However, he does not take into account his own influence on the trajectories of other players. Taking into account this influence would be a much more complex problem.

5 Analytic Framework

We shall develop here a theory to solve systems of the type (4.13), and define the set of assumptions. This will extend the results given in [2]. However, many techniques are similar to those developed in this reference. For the convenience of the reader, we shall indicate the main steps without all the details. Since we shall treat boundary conditions with local charts, it will be helpful to replace the Laplacian operator by a general second order operator in the divergence form. So we consider functions $a_{kl}(x)$ $(k,l = 1,\dots,n)$ defined on $\mathbb{R}^n$, which satisfy

$$a_{kl}(\cdot) \text{ bounded}, \quad \sum_{k,l=1}^{n} a_{kl}(x)\xi_k\xi_l \geq \underline{a}|\xi|^2, \quad \forall \xi \in \mathbb{R}^n. \tag{5.1}$$

We shall consider the matrix $a(x)$, whose elements are the quantities $a_{kl}(x)$, and write

$$\underline{a}I \leq a(x) \leq \overline{a}I, \tag{5.2}$$

where I is the identity matrix. Note that $a(x)$ is not necessarily symmetric.

5.1 Assumptions

We denote by $\mathcal{O}$ a smooth bounded open domain of $\mathbb{R}^n$. We write $\Gamma = \partial\mathcal{O}$. We define the second order linear operator

$$A\varphi(x) = -\operatorname{div}\big(a(x)\operatorname{grad}\varphi(x)\big), \quad x \in \mathcal{O},$$

and the boundary operator

$$\frac{\partial \varphi}{\partial \nu_A}(x) = \nu(x) \cdot a(x) \operatorname{grad} \varphi(x), \quad x \in \Gamma,$$

where $\nu(x)$ is the unit pointed outward normal vector on a point $x \in \Gamma$. The adjoint operator is defined by

$$A^* \varphi(x) = -\operatorname{div}\big(a^*(x) \operatorname{grad} \varphi(x)\big), \quad x \in \mathcal{O},$$

where $a^*(x)$ is the transpose of the matrix $a(x)$. The corresponding boundary operator is

$$\frac{\partial \varphi}{\partial \nu_{A^*}}(x) = \nu(x) \cdot a^*(x) \operatorname{grad} \varphi(x), \quad x \in \Gamma.$$

For $i = 1, \dots, n$, we define the functions $H^i(x,q)$, $G^i(x,q)$, $q \in \mathbb{R}^{nN}$ with the following assumptions:

$$H^i(x,q) : \mathbb{R}^n \times \mathbb{R}^{nN} \to \mathbb{R}, \quad \text{measurable}, \tag{5.3}$$

$$|H^i(x,q)| \le K^i |q||q^i| + \sum_{j=1}^{i} K_j^i |q^j|^2 + k^i(x), \quad i = 1, \dots, N-1, \tag{5.4}$$

where q^i $(i = 1, \dots, N)$ are vectors of $\mathbb{R}^n$, representing the components of q. The functions $k^i(\cdot) \in L^p(\mathcal{O})$, $p > \frac{n}{2}$. We next assume

$$|H^N(x,q)| \le K^N |q|^2 + k^N(x), \quad k^N(\cdot) \in L^p(\mathcal{O}), \quad p > \frac{n}{2}. \tag{5.5}$$

We also assume that

$$|H^i(x,q)|_{q^i=0} \le C_0, \quad i = 1, \dots, N. \tag{5.6}$$

Concerning $G^i(x,q)$, we assume

$$G^i(x,q) : \mathbb{R}^n \times \mathbb{R}^{nN} \to \mathbb{R}^n, \quad \text{measurable}, \tag{5.7}$$

$$|G^i(x,q)| \le K|q| + K. \tag{5.8}$$

We next consider the functionals $V^i_{[m]}(\cdot) : L^1(\mathcal{O};\mathbb{R}^N) \to L^1(\mathcal{O})$, such that

$$\|V^i_{[m]}\|_{L^\infty(\mathcal{O})} \le l(\|m\|), \tag{5.9}$$

where

$$\|m\| = \|m\|_{L^1(\mathcal{O};\mathbb{R}^N)} = \sup_{i=1}^{N} \int_{\mathcal{O}} |m^i(x)| \mathrm{d}x.$$

We also assume the convergence property

$$\text{if } m_j \to m \text{ pointwise}, \ \|m_j\|_{L^\infty(\mathcal{O};\mathbb{R}^N)} \le C, \quad \text{then } V^i_{[m_j]} \to V^i_{[m]}; \text{ in } L^1(\mathcal{O}). \tag{5.10}$$

We finally consider

$$m_0^i \in L^p(\mathcal{O}), \quad p > \frac{n}{2}, \quad m_0^i \ge 0. \tag{5.11}$$

5.2 *Preliminaries*

We first state some technical results, the proof of which can be found in [2]. Without loss of generality, the assumptions (5.4)–(5.5) can be changed into

$$H^i(x,q) = Q^i(x,q)\cdot q^i + H_0^i(x,q), \quad i = 1,\dots,N \tag{5.12}$$

with

$$|Q^i(x,q)| \le K^i|q|, \tag{5.13}$$

$$Q^N(x,q) = Q^{N-1}(x,q), \tag{5.14}$$

$$|H_0^i(x,q)| \le \sum_{j=1}^{i} K_j^i |q^j|^2 + k^i(x), \quad i = 1,\dots,N, \tag{5.15}$$

in which all quantities have been defined in (5.4)–(5.5), except K_i^N $(i = 1,\dots, N-1)$ and K_N^N defined as follows:

$$K_i^N = K^N + \frac{K^{N-1}}{2}, \qquad K_N^N = K^N + K^{N-1}. \tag{5.16}$$

So, from now on, we assume that (5.12)–(5.15) hold.

We shall also use the following technical property. Define the function

$$\beta(x) = \exp x - x - 1.$$

Let $s \in \mathbb{R}^N$. The components are defined as s^i $(i = 1,\dots,N)$. Let

$$X^N(s) = \exp[\beta(\gamma^N s^N) + \beta(-\gamma^N s^N)],$$

where γ^N is a positive constant. We then define recursively

$$X^i(s) = \exp[X^{i+1}(s) + \beta(\gamma^i s^i) + \beta(-\gamma^i s^i)], \quad i = 1,\dots,N-1,$$

where γ^i are positive constants. We have the lemma below.

Lemma 5.1 *One has*

$$\frac{\partial X^i}{\partial s^j} = \begin{cases} 0, & \text{if } j < i, \\ X^i \cdots X^j \gamma^j (\exp(\gamma^j s^j) - \exp(-\gamma^j s^j)), & \text{if } j \geq i. \end{cases}$$

Hence

$$|X^i(s) - X^i(0)| \leq c(|s|)|s|^2, \tag{5.17}$$

$$X^i(s) \geq X^i(0) \geq 1, \tag{5.18}$$

where the constant c depends on the norm of the vector s and all constants $\gamma^1, \ldots, \gamma^N$. To avoid ambiguity later, we denote $X^i(0) = X_0^i$.

The proof is left to the reader.

5.3 Regularity Result

We are interested in the system

$$\begin{aligned} &Au^i + \alpha u^i = H^i(x, Du) + V^i_{[m]}(x), \quad x \in \mathcal{O}, \\ &\left.\frac{\partial u^i}{\partial \nu_A}\right|_\Gamma = 0, \\ &Am^i + \operatorname{div}\big(G^i(x, Du)m^i\big) + \alpha m^i = \alpha m_0^i, \quad x \in \mathcal{O}, \\ &\frac{\partial m^i}{\partial \nu_{A^*}} - G^i(x, Du) \cdot \nu(x) m^i = 0, \quad x \in \Gamma. \end{aligned} \tag{5.19}$$

We interpret (5.19) in the weak sense

$$\begin{aligned} &\int_{\mathcal{O}} a(x) Du^i(x) \cdot D\varphi^i(x) \mathrm{d}x + \alpha \int_{\mathcal{O}} u^i(x)\varphi^i(x) \mathrm{d}x \\ &\quad = \int_{\mathcal{O}} \big(H^i(x, Du) + V^i_{[m]}(x)\big)\varphi^i(x) \mathrm{d}x, \end{aligned} \tag{5.20}$$

$$\begin{aligned} &\int_{\mathcal{O}} a^*(x) Dm^i(x) \cdot D\psi^i(x) \mathrm{d}x - \int_{\mathcal{O}} m^i(x) G^i(x, Du) \cdot D\psi^i(x) \mathrm{d}x \\ &\quad + \alpha \int_{\mathcal{O}} m^i(x)\psi^i(x) \mathrm{d}x = \alpha \int_{\mathcal{O}} m_0^i(x)\psi^i(x) \mathrm{d}x \end{aligned} \tag{5.21}$$

for any pair $\varphi^i(\cdot) \in H^1 \cap L^\infty(\mathcal{O}),\ \psi^i \in W^{1,\infty}, i = 1, \ldots, N.$

We state the important regularity result concerning the u^i.

Theorem 5.1 *We assume that* (5.1) *and* (5.3)–(5.11) *hold. Suppose that there exists a solution* u, m *to the system* (5.20)–(5.21), *such that* $u, m \in H^1(\mathcal{O}; \mathbb{R}^N)$, $m \geq 0$. *Then one has*

$$u \in W^{1,r} \cap L^\infty(\mathcal{O}; \mathbb{R}^N), \quad 2 \leq r < r_0, \quad u \in C^{0,\delta}(\overline{\mathcal{O}}; \mathbb{R}^N), \quad 0 < \delta \leq \delta_0 < 1, \tag{5.22}$$

where the constants r_0, δ_0 *depend only on the constants in the assumptions and the data. They do not depend on the* H^1 *norm of* u, m. *The norm of* u *in the functional spaces* $W^{1,r} \cap L^\infty$ *and* $C^{0,\delta}$ *does not depend on the* H^1 *norm of* m.

Remark 5.1 This result extends the traditional additional results of regularity of H^1 solutions to (5.20). The functions m^i appear as an external factor. In view of the weak coupling, only the positivity of m^i is important.

6 A Priori Estimates

The proof of Theorem 5.1 will rely on a priori estimates. Although very close to the treatment in [2] which is done for Dirichlet problems, we develop the main steps of the proof. This will also be helpful at the existence phase. We will indeed consider an approximation procedure, and we shall have to check that the same estimates hold. That will be instrumental in passing to the limit.

6.1 Preliminary Steps

Taking $\psi^i = 1$ in (5.21), we obtain

$$\int_{\mathcal{O}} m^i(x)\mathrm{d}x = \int_{\mathcal{O}} m_0^i(x)\mathrm{d}x.$$

Since $m \geq 0$, we get immediately

$$m \in L^1(\mathcal{O}; \mathbb{R}^N), \quad \|m\|_{L^1(\mathcal{O};\mathbb{R}^N)} = \|m_0\|_{L^1(\mathcal{O};\mathbb{R}^N)}. \tag{6.1}$$

From the assumption (5.9), it follows that

$$\|V^i_{[m]}\|_{L^\infty(\mathcal{O})} \leq l(\|m_0\|). \tag{6.2}$$

Using the assumption (5.6) in the first equation of (5.19) and the standard maximum principle arguments for Neumann elliptic problems, we deduce easily

$$\|u^i\|_{L^\infty(\mathcal{O})} \leq \frac{C_0 + l(\|m_0\|)}{\alpha}. \tag{6.3}$$

Considering the vector $u(x)$ of components $u^i(x)$, we call

$$\|u\|_{L^\infty(\mathcal{O})} = \||u|\|_{L^\infty(\mathcal{O})},$$

where $|u|$ is the vector norm. Hence

$$\|u\|_{L^\infty(\mathcal{O})} \leq \sqrt{N}\frac{C_0 + l(\|m_0\|)}{\alpha} = \rho. \tag{6.4}$$

6.2 Basic Inequality

Thanks to (6.2), we can simply set $f^i(x) = V^i_{[m]}(x) - \alpha u^i(x)$, and consider that f^i is a given bounded function. We take advantage of the weak coupling of m and u in the first set of Eq. (5.20). We now consider a constant vector $c \in \mathbb{R}^N$. This constant vector will be chosen in specific applications of the basic inequality. The only thing that we require is $|c| \leq \rho$. Define next $\widetilde{u} = u - c$, and consider the functions $X^i(s)$ introduced in Sect. 5.2. We associate to these functions $X^i(x) = X^i(\widetilde{u}(x))$. This is a slight abuse of notation, to shorten the notation. The basic inequality is summarized in the following lemma.

Lemma 6.1 *Let $\Psi \in H^1(\mathcal{O}) \cup L^\infty(\mathcal{O})$, with $\Psi \geq 0$. We have the inequality*

$$\int_{\mathcal{O}} a(x)DX^1 \cdot D\Psi \mathrm{d}x + \underline{a}\int_{\mathcal{O}} \Psi |Du|^2 \mathrm{d}x \leq C(\rho)\int_{\mathcal{O}} \Psi \sum_{i=1}^{N}(|k^i| + |f^i|)\mathrm{d}x, \tag{6.5}$$

where $C(\rho)$ is a constant depending only on ρ and the various constants in (5.13)–(5.16). *This inequality is obtained for a specific choice of the constants γ^i in the definition of $X^i(s)$. This inequality is valid for any constant vector c with $|c| \leq \rho$.*

We note

$$|Du|^2 = \sum_{i=1}^{N} |Du^i|^2.$$

Proof The proof is rather technical. Details can be found in [2]. We only sketch here the main steps to facilitate the reading. Consider the functions $X^i(x)$. We have

$$DX^i = \sum_{j=i}^{N} \gamma^j X^i \cdots X^j(\exp(\gamma^j\widetilde{u}^j) - \exp(-\gamma^j\widetilde{u}^j))Du^j.$$

We take, in (5.20),

$$\varphi^i = \Psi\gamma^i(\exp(\gamma^i\widetilde{u}^i) - \exp(-\gamma^i\widetilde{u}^i))\prod_{j=1}^{i} X^j.$$

After tedious calculations, we obtain the expression

$$\sum_{i=1}^{N}\int_{\mathcal{O}} a(x)Du^i\cdot D\varphi^i\mathrm{d}x$$

$$=\int_{\mathcal{O}} a(x)DX^1\cdot D\Psi\mathrm{d}x+\sum_{i=1}^{N}\int_{\mathcal{O}}\Psi a(x)DF^i\cdot DF^i\prod_{j=1}^{i}X^j\mathrm{d}x$$

$$+\sum_{i=1}^{N}\int_{\mathcal{O}}\Psi a(x)Du^i\cdot Du^i\prod_{j=1}^{i}X^j(\gamma^i)^2\big(\exp(\gamma^i\widetilde{u}^i)+\exp(-\gamma^i\widetilde{u}^i)\big)\mathrm{d}x \quad (6.6)$$

with $F^i=\log X^i$. On the other hand, from (5.20), we have

$$\sum_{i=1}^{N}\int_{\mathcal{O}} a(x)Du^i\cdot D\varphi^i\mathrm{d}x=\sum_i\int_{\mathcal{O}}(H^i(x,Du)+f^i)\varphi^i\mathrm{d}x.$$

Using (5.12) and performing calculations, we can set

$$\sum_i\int_{\mathcal{O}}(H^i(x,Du)+f^i)\varphi^i\mathrm{d}x$$

$$=\int_{\mathcal{O}}\sum_{i=1}^{N-1}(Q^i-Q^{i-1})DF^i\prod_{j=1}^{i}X^j\mathrm{d}x$$

$$+\int_{\mathcal{O}}\Psi\sum_{i=1}^{N}(H_0^i(x,Du)+f^i)\gamma^i\big(\exp(\gamma^i\widetilde{u}^i)-\exp(-\gamma^i\widetilde{u}^i)\big)\prod_{j=1}^{i}X^j, \quad (6.7)$$

in which $Q^0=0$. Using the assumptions (5.13)–(5.16), we can check the inequality

$$\int_{\mathcal{O}} a(x)DX^1\cdot D\Psi\mathrm{d}x+\int_{\mathcal{O}}\Psi\sum_{j=1}^{N}|Du^j|^2B^j(x)\mathrm{d}x$$

$$\le\int_{\mathcal{O}}\Psi\sum_{i=1}^{N}(k^i+f^i)\gamma^i\big(\exp(\gamma^i\widetilde{u}^i)-\exp(-\gamma^i\widetilde{u}^i)\big)\prod_{j=1}^{i}X^j\mathrm{d}x,$$

where $B^j(x)$ is the long expression

$$B^j(x)=\underline{a}(\gamma^j)^2\big(\exp(\gamma^j\widetilde{u}^j)+\exp(-\gamma^j\widetilde{u}^j)\big)\prod_{h=1}^{j}X^h-\sum_{i=1}^{N-1}\frac{(K^i+K^{i-1})^2}{4\underline{a}^2}\prod_{h=1}^{i}X^h$$

$$-\sum_{i=j}^{N}K_j^i\gamma^i\big(\exp(\gamma^i\widetilde{u}^i)-\exp(-\gamma^i\widetilde{u}^i)\big)\prod_{h=1}^{i}X^h.$$

Rearranging the above expressions, we have

$$B^j(x) \geq \Bigg[\underline{a}(\gamma^j)^2 + \frac{\underline{a}(\gamma^j)^2 - 2\gamma^j K_j^j}{2}(\exp(\gamma^j \widetilde{u}^j) + \exp(-\gamma^j \widetilde{u}^j))$$

$$- \sum_{i=1}^{j} \frac{(K^i + K^{i-1})^2}{4\underline{a}^2} \sum_{i=j+1}^{N-1} \frac{(K^i + K^{i-1})^2}{4\underline{a}^2} \prod_{h=j+1}^{i} X^h$$

$$- \sum_{i=j+1}^{N} K_j^i \gamma^i (\exp(\gamma^i \widetilde{u}^i) - \exp(-\gamma^i \widetilde{u}^i)) \prod_{h=j+1}^{i} X^h \Bigg] \prod_{h=1}^{j} X^h.$$

We then chose recursively the constants γ^j, such that

$$\underline{a}\gamma^j - 2K_j^j > 0, \quad \gamma^j > 1,$$

$$\underline{a}(\gamma^j)^2 - 2\gamma^j K_j^j - \sum_{i=1}^{j} \frac{(K^i + K^{i-1})^2}{4\underline{a}^2}$$

$$> \sum_{i=j+1}^{N-1} \frac{(K^i + K^{i-1})^2}{4\underline{a}^2} \prod_{h=j+1}^{i} X^h - \sum_{i=j+1}^{N} K_j^i \gamma^i (\exp(\gamma^i \widetilde{u}^i)$$

$$- \exp(-\gamma^i \widetilde{u}^i)) \prod_{h=j+1}^{i} X^h.$$

This is possibly backward recursively, and the choice of these constants depends only on ρ and the various constants in the assumptions. With this choice of the constants, we get $B^j(x) \geq \underline{a}$ and the result follows easily. □

6.3 $W^{1,r}$ Estimates

We begin with the $W^{1,r}$ estimate, $2 \leq r < r_0$. Let $\tau(x)$ be a smooth function with $0 \leq \tau(x) \leq 1$ and

$$\tau(x) = 1, \quad \text{if } |x| \leq 1,$$
$$\tau(x) = 0, \quad \text{if } |x| \geq 2.$$

To any point x_0, we associate the ball of center x_0 and radius R, denoted by $B_R(x_0)$. We assume that $R \leq R_0$ but R can be arbitrarily small. We define the cut-off function

$$\tau_R(x) = \tau\left(\frac{x - x_0}{R}\right).$$

We then apply the basic inequality (6.5) with $\Psi = \tau_R^2$. We deduce easily

$$\underline{a}\int_{\mathcal{O}\cap B_R}|Du|^2\mathrm{d}x \leq \frac{\widehat{a}}{R}\int_{\mathcal{O}\cap B_{2R}}|DX^1|\mathrm{d}x + C(\rho)\int_{\mathcal{O}\cap B_{2R}}\sum_{i=1}^{N}\big(|k^i|+|f^i|\big)\mathrm{d}x. \tag{6.8}$$

We take $c = c_R$, to be defined below. We have

$$DX^1(x) = \sum_{j=1}^{N}\frac{\partial X^1}{\partial s^j}\big(u(x)-c_R\big)Du^j(x).$$

Using Lemma 5.1, we get

$$|DX^1(x)| \leq C^1(\rho)|Du(x)||u(x)-c_R|.$$

We have then

$$\int_{\mathcal{O}\cap B_{2R}}|DX^1|\mathrm{d}x \leq C^1(\rho)\Big(\int_{\mathcal{O}\cap B_{2R}}|Du|^{\mu}\mathrm{d}x\Big)^{\frac{1}{\mu}}\Big(\int_{\mathcal{O}\cap B_{2R}}|u-c_R|^{\lambda}\mathrm{d}x\Big)^{\frac{1}{\lambda}}$$

for any $\lambda > 1$ and $\frac{1}{\lambda}+\frac{1}{\mu} = 1$. We next define c_R. We consider points x_0, such that $|\mathcal{O}\cap B_R(x_0)| > 0$. Therefore, $|\mathcal{O}\cap B_{2R}(x_0)| > 0$. We consider two cases, that is, the case of $B_{2R}(x_0)\subset\mathcal{O}$, and the case of $B_{2R}(x_0)\cap(R^n-\mathcal{O})\neq\varnothing$. In the second case, by the smoothness of the domain, $\Gamma\cap B_{2R}(x_0)\neq\varnothing$. We pick a point $x_0'\in\Gamma\cap B_{2R}(x_0)$, and note that $B_{2R}(x_0)\subset B_{4R}(x_0')\subset B_{6R}(x_0)$. Again, from the smoothness of the domain, we have (the sphere condition)

$$\big|B_{4R}(x_0')\cap\mathcal{O}\big| \geq c_0R^n, \qquad \big|B_{4R}(x_0')\cap(R^n-\mathcal{O})\big| \geq c_0R^n,$$

where c_0 is a constant. We then define c_R by

$$c_R = \begin{cases} \frac{1}{|B_{2R}|}\int_{B_{2R}}u(x)\mathrm{d}x, & \text{if } B_{2R}(x_0)\subset\mathcal{O}, \\ \frac{1}{|B_{4R}(x_0')\cap\mathcal{O}|}\int_{B_{4R}(x_0')\cap\mathcal{O}}u(x)\mathrm{d}x, & \text{if } B_{2R}(x_0)\cap(R^n-\mathcal{O})\neq\varnothing. \end{cases}$$

We can state the Poincaré's inequality

$$\Big(\int_{\mathcal{O}B_{2R}}|u-c_R|^{\lambda}\mathrm{d}x\Big)^{\frac{1}{\lambda}} \leq c_1R^{n(\frac{1}{\lambda}-\frac{1}{\nu})+1}\Big(\int_{\mathcal{O}\cap B_{6R}}|Du|^{\nu}\mathrm{d}x\Big)^{\frac{1}{\nu}},$$

$$\forall\nu,\ \text{such that } 1\leq\nu\leq 2,\ n\Big(\frac{1}{\lambda}-\frac{1}{\nu}\Big)+1\geq 0.$$

We will apply this inequality with $n(\frac{1}{\lambda}-\frac{1}{\nu})+1 = 0$, i.e., $\nu = \frac{\lambda n}{n+\lambda}$. From the conditions $1\leq\nu\leq 2$, this is possible only when $n\geq 2$ and

$$\frac{n}{n-1}\leq\lambda\leq\frac{2n}{n-2}.$$

In that case, we have

$$\left(\int_{\mathcal{O}\cap B_{2R}} |u-c_R|^\lambda \mathrm{d}x\right)^{\frac{1}{\lambda}} \le c_1\left(\int_{\mathcal{O}\cap B_{6R}} |Du|^{\frac{\lambda n}{n+\lambda}} \mathrm{d}x\right)^{\frac{n+\lambda}{\lambda n}}.$$

We now chose λ, such that $\frac{\lambda n}{n+\lambda} = \mu = \frac{\lambda}{\lambda-1}$. This implies $\lambda = \frac{2n}{n-1}$, which is compatible with the restrictions on λ. Collecting the above results, we can assert that

$$\int_{\mathcal{O}\cap B_{2R}} |DX^1|\mathrm{d}x \le C^2(\rho)\left(\int_{\mathcal{O}\cap B_{6R}} |Du|^{\frac{2n}{n+1}} \mathrm{d}x\right)^{\frac{n+1}{n}},$$

where $C^2(\rho)$ is another constant, depending only on ρ. We next note that

$$\int_{\mathcal{O}\cap B_{2R}} \sum_{i=1}^{N} (|k^i| + |f^i|)\mathrm{d}x \le CR^{n(1-\frac{1}{p})}.$$

Therefore, collecting the above results, we can assert from (6.8) that

$$\int_{\mathcal{O}\cap B_R} |Du|^2\mathrm{d}x \le C^3(\rho)\left(R^{n(1-\frac{1}{p})} + \frac{1}{R}\left(\int_{\mathcal{O}\cap B_{6R}} |Du|^{\frac{2n}{n+1}} \mathrm{d}x\right)^{\frac{n+1}{n}}\right), \quad \forall R \le R_0. \tag{6.9}$$

Note that this inequality is trivial if $|\mathcal{O} \cap B_R(x_0)| = 0$. According to Gehring's result (see [2]), we assert that

$$\int_{\mathcal{O}} |Du|^r \mathrm{d}x \le C(r,\rho), \quad \forall 2 \le r < r_0, \tag{6.10}$$

where r_0 depends only on ρ and the data.

6.4 $C^{0,\delta}$ Estimates

We now turn to the Hölder regularity. To treat the Hölder regularity up to the boundary, we have to use local maps. The regularity is then reduced to interior regularity and regularity on balls centered on the boundary, which can be transformed into half-planes by a straightening operation. We shall again limit ourselves to the main ideas, leaving details to the reference [2]. We begin with the interior regularity.

Let $\widetilde{\mathcal{O}}$ be a smooth domain such that $\overline{\widetilde{\mathcal{O}}} \subset \mathcal{O}$. We shall prove the Hölder regularity on $\overline{\widetilde{\mathcal{O}}}$. Since $\widetilde{\mathcal{O}}$ is arbitrary, that will prove the Hölder regularity on $\mathcal{O}$. Let $x_0 \in \overline{\widetilde{\mathcal{O}}}$. We shall apply the Green function to the Dirichlet problem in $\mathcal{O}$. It is denoted by $G = G^{x_0}$ and defined by

$$\int_{\mathcal{O}} a(x)D\varphi \cdot DG\mathrm{d}x = \varphi(x_0), \quad \forall \varphi \in C_0^\infty(\mathcal{O}). \tag{6.11}$$

We shall use the following properties of Green functions (see [2] for details):

$$\begin{aligned} &G \in W_0^{1,\mu}(\mathcal{O}), \quad \forall \mu,\ 1 \le \mu < \frac{n}{n-1}, \\ &G \in L^{\nu}(\mathcal{O}), \quad \forall \nu,\ 1 \le \nu < \frac{n}{n-2}. \end{aligned} \tag{6.12}$$

Assume $n \ge 3$. Then

$$\begin{aligned} &c_0|x-x_0|^{2-n} \le G(x) \le c_1|x-x_0|^{2-n}, \\ &\forall x \in Q,\ \forall Q \text{ neighbourhood of } x_0 \text{ with } \overline{Q} \subset \mathcal{O}, \end{aligned} \tag{6.13}$$

where the constants c_0, c_1 depend only on $\underline{a}$ and $\overline{a}$.

We next consider the balls $B_R(x_0)$. We assume that $R \le R_0$, with $2R_0 < \text{dist}(\overline{\widetilde{\mathcal{O}}}, \mathbb{R}^n - \mathcal{O})$. This implies $\overline{B_{2R}(x_0)} \subset \mathcal{O}$. We consider the cut-off function $\tau_R(x)$ as that defined in Sect. 6.3.

In the basic inequality (6.5), we choose

$$c = c_R = \frac{1}{|B_{2R} - B_{\frac{R}{2}}|}\int_{B_{2R}-B_{\frac{R}{2}}} u\mathrm{d}x, \quad \Psi = G\tau_R^2,$$

so we get

$$\begin{aligned} &\int_{\mathcal{O}} a(x)DX^1 \cdot D(G\tau_R^2)\mathrm{d}x + \underline{a}\int_{\mathcal{O}} G\tau_R^2|Du|^2\mathrm{d}x \\ &\quad \le C(\rho)\int_{\mathcal{O}} G\tau_R^2\sum_{i=1}^N(|k^i|+|f^i|)\mathrm{d}x. \end{aligned} \tag{6.14}$$

Clearly $\int_{\mathcal{O}} G\tau_R^2|Du|^2\mathrm{d}x \ge \int_{B_{\frac{R}{2}}} G|Du|^2\mathrm{d}x$, and since $\frac{B_R(x_0)}{2} \subset \overline{B_{2R}(x_0)} \subset \mathcal{O}$, we can use the estimate (6.13) to assert

$$\underline{a}\int_{\mathcal{O}} G\tau_R^2|Du|^2\mathrm{d}x \ge C\int_{B_{\frac{R}{2}}}|Du|^2|x-x_0|^{2-n}\mathrm{d}x, \tag{6.15}$$

where C is a constant. Next

$$\int_{\mathcal{O}} G\tau_R^2\sum_{i=1}^N(|k^i|+|f^i|)\mathrm{d}x \le \sum_{i=1}^N\int_{B_{2R}} G|k^i|\mathrm{d}x + \sum_{i=1}^N\|f^i\|\int_{B_{2R}} G\mathrm{d}x,$$

and from the estimates (6.12) and Hölder's inequality,

$$\int_{B_{2R}} G\mathrm{d}x \le CR^{\frac{n}{\mu'}}, \quad \forall \mu' \text{ such that } \frac{n}{\mu'} < 2,$$

$$\int_{B_{2R}} G|k^i|\mathrm{d}x \leq CR^{n(\frac{1}{p'}-\frac{1}{\nu})}, \quad \forall \nu < \frac{n}{n-2}.$$

Collecting the above results, we can assert that

$$\int_{\mathcal{O}} G\tau_R^2 \sum_{i=1}^{N}(|k^i|+|f^i|)\mathrm{d}x \leq CR^{\beta}, \quad \beta < 2. \tag{6.16}$$

Next we have

$$\begin{aligned}\int_{\mathcal{O}} a(x)DX^1 \cdot D(G\tau_R^2)\mathrm{d}x &= \int_{\mathcal{O}} a(x)DX^1 \cdot DG\,\tau_R^2\mathrm{d}x \\ &\quad + 2\int_{\mathcal{O}} a(x)DX^1 \cdot D\tau_R\tau_R G\mathrm{d}x \\ &= \mathrm{Z} + \mathrm{I}.\end{aligned}$$

We have, as seen in the previous section,

$$|DX^1(x)| \leq C|Du(x)||u(x)-c_R|.$$

Therefore, using again the estimates on the Green function (6.13), we have

$$|\mathrm{I}| \leq C\int_{B_{2R}-B_{\frac{R}{2}}} |Du(x)|\frac{|u(x)-c_R|}{R}|x-x_0|^{2-n}\mathrm{d}x.$$

Note that

$$\begin{aligned}&\int_{B_{2R}-B_{\frac{R}{2}}} \frac{|u(x)-c_R|^2}{R^2}|x-x_0|^{2-n}\mathrm{d}x \\ &\leq CR^{-n}\int_{B_{2R}-B_{\frac{R}{2}}} |u(x)-c_R|^2\mathrm{d}x \\ &\leq CR^{2-n}\int_{B_{2R}-B_{\frac{R}{2}}} |Du|^2\mathrm{d}x \\ &\leq C\int_{B_{2R}-B_{\frac{R}{2}}} |Du|^2|x-x_0|^{2-n}\mathrm{d}x,\end{aligned}$$

by Poincaré's inequality. Therefore,

$$|\mathrm{I}| \leq C\int_{B_{2R}-B_{\frac{R}{2}}} |Du|^2|x-x_0|^{2-n}\mathrm{d}x. \tag{6.17}$$

We now turn to the term Z. Recalling the term $X_0^1 \geq 1$ (see (5.18)), we write

$$
\begin{aligned}
\mathrm{Z} &= \int_{\mathcal{O}} a(x) D\big(X^1 - X_0^1\big) \cdot DG\, \tau_R^2 \mathrm{d}x \\
&= \int_{\mathcal{O}} a(x) D\big(\tau_R^2 \big(X^1 - X_0^1\big)\big) \cdot DG \mathrm{d}x - 2\int_{\mathcal{O}} a(x) D\tau_R \cdot DG\big(X^1 - X_0^1\big)\tau_R \mathrm{d}x \\
&\geq -2\int_{\mathcal{O}} a(x) D\tau_R \cdot DG\big(X^1 - X_0^1\big)\tau_R \mathrm{d}x,
\end{aligned}
$$

where we have made use of the Green function's definition (6.11). Recalling (5.17), we have $|X^1(x) - X_0^1| \leq |u(x) - c_R|^2$. Therefore,

$$
\begin{aligned}
&\int_{\mathcal{O}} a(x) D\tau_R \cdot DG\big(X^1 - X_0^1\big)\tau_R \mathrm{d}x \\
&\quad \leq \frac{C}{R}\int_{B_{2R}-B_R} |u - c_R|^2 |DG| \tau_R \mathrm{d}x \\
&\quad \leq C\int_{B_{2R}-B_R} \frac{|u - c_R|^2}{R^2} G \mathrm{d}x + C\int_{B_{2R}-B_R} |u - c_R|^2 |DG|^2 G^{-1} \tau_R^2 \mathrm{d}x \\
&\quad \leq C\int_{B_{2R}-B_{\frac{R}{2}}} |Du|^2 |x - x_0|^{2-n} \mathrm{d}x + C\mathrm{Y}.
\end{aligned}
$$

Therefore, we have

$$
\mathrm{Z} \geq -C\int_{B_{2R}-B_{\frac{R}{2}}} |Du|^2 |x - x_0|^{2-n} \mathrm{d}x - C\mathrm{Y}. \tag{6.18}
$$

We now estimate

$$
\mathrm{Y} = \int_{B_{2R}-B_R} |u - c_R|^2 |DG|^2 G^{-1} \tau_R^2 \mathrm{d}x.
$$

We introduce a new cut-off function

$$
\xi(x) = \begin{cases} 0 & \text{for } |x| \leq \frac{1}{2}, \\ \tau(x) & \text{for } |x| \geq 1, \end{cases}
$$

and denote $\xi_R(x) = \xi(\frac{x - x_0}{R})$. Hence

$$
\begin{aligned}
&\xi_R(x) = \tau_R(x) \quad \text{on } B_{2R} - B_R, \\
&\xi_R(x) = 0 \quad \text{on } B_{\frac{R}{2}}.
\end{aligned}
$$

In the Green function equation (6.11), we take

$$
\varphi = G^{-\frac{1}{2}} |u - c_R|^2 \xi_R^2.
$$

Noting that $\varphi(x_0)=0$, it follows that

$$\int_{\mathcal{O}} G^{-\frac{1}{2}}a(x)D\big(|u-c_R|^2\xi_R^2\big)\cdot DG\mathrm{d}x=\frac{1}{2}\int_{\mathcal{O}} G^{-\frac{3}{2}}a(x)DG\cdot DG|u-c_R|^2\xi_R^2\mathrm{d}x. \tag{6.19}$$

Next, in (5.20), we take

$$\varphi^i=\big(u^i-c_R^i\big)G^{\frac{1}{2}}\xi_R^2.$$

We obtain, after rearrangements,

$$\begin{aligned}\sum_{i=1}^N\int_{\mathcal{O}} a(x)Du^i\cdot Du^iG^{\frac{1}{2}}\xi_R^2\mathrm{d}x&+\frac{1}{4}\int_{\mathcal{O}} a(x)D\big(|u-c_R|^2\xi_R^2\big)\cdot DGG^{-\frac{1}{2}}\mathrm{d}x\\ &-\frac{1}{2}\int_{\mathcal{O}} a(x)D\xi_R\cdot DG\xi_R|u-c_R|^2G^{-\frac{1}{2}}\mathrm{d}x\\ &+\int_{\mathcal{O}} a(x)D\big(|u-c_R|^2\big)\cdot D\xi_R\xi_RG^{\frac{1}{2}}\mathrm{d}x\\ &=\int_{\mathcal{O}}\sum_{i=1}^N\big(H^i+f^i\big)\big(u^i-c_R^i\big)G^{\frac{1}{2}}\xi_R^2\mathrm{d}x.\end{aligned}$$

Hence,

$$\begin{aligned}&\int_{\mathcal{O}} a(x)D\big(|u-c_R|^2\xi_R^2\big)\cdot DGG^{-\frac{1}{2}}\mathrm{d}x\\ &\quad\le 2\int_{\mathcal{O}} a(x)D\xi_R\cdot DG\xi_R|u-c_R|^2G^{-\frac{1}{2}}\mathrm{d}x\\ &\qquad+CR^{\frac{2-n}{2}}\int_{B_{2R}-B_{\frac{R}{2}}}|Du|^2\mathrm{d}x+CR^{1+\frac{n}{2}-\frac{n}{p}}.\end{aligned}$$

Therefore, from (6.19), we can write

$$\begin{aligned}&\frac{1}{2}\int_{\mathcal{O}} G^{-\frac{3}{2}}a(x)DG\cdot DG|u-c_R|^2\xi_R^2\mathrm{d}x\\ &\quad\le 2\int_{\mathcal{O}} a(x)D\xi_R\cdot DG\xi_R|u-c_R|^2G^{-\frac{1}{2}}\mathrm{d}x+CR^{\frac{2-n}{2}}\int_{B_{2R}-B_{\frac{R}{2}}}|Du|^2\mathrm{d}x\\ &\quad+CR^{1+\frac{n}{2}-\frac{n}{p}},\end{aligned}$$

from which one easily deduces

$$\begin{aligned}&\int_{\mathcal{O}} G^{-\frac{3}{2}}|DG|^2|u-c_R|^2\xi_R^2\mathrm{d}x\\ &\quad\le C\int_{B_{2R}-B_{\frac{R}{2}}}\frac{|u-c_R|^2}{R^2}G^{\frac{1}{2}}\mathrm{d}x+CR^{\frac{2-n}{2}}\int_{B_{2R}-B_{\frac{R}{2}}}|Du|^2\mathrm{d}x+CR^{1+\frac{n}{2}-\frac{n}{p}}.\end{aligned}$$

Hence

$$\int_{\mathcal{O}} G^{-\frac{3}{2}}|DG|^2|u-c_R|^2\xi_R^2 \mathrm{d}x \le CR^{\frac{2-n}{2}} \int_{B_{2R}-B_{\frac{R}{2}}} |Du|^2 \mathrm{d}x + CR^{1+\frac{n}{2}-\frac{n}{p}}.$$

Since $\tau_R = \xi_R$ on $B_{2R} - B_R$, we have

$$\begin{aligned}
\mathrm{Y} &= \int_{B_{2R}-B_R} |u-c_R|^2|DG|^2G^{-1}\xi_R^2 \mathrm{d}x \\
&\le \int_{B_{2R}-B_{\frac{R}{2}}} |u-c_R|^2|DG|^2G^{-1}\xi_R^2 \mathrm{d}x \\
&\le CR^{\frac{2-n}{2}} \int_{B_{2R}-B_{\frac{R}{2}}} |u-c_R|^2|DG|^2G^{-\frac{3}{2}}\xi_R^2 \mathrm{d}x \\
&\le CR^{2-n} \int_{B_{2R}-B_{\frac{R}{2}}} |Du|^2 \mathrm{d}x + CR^{2-\frac{n}{p}} \\
&\le C \int_{B_{2R}-B_{\frac{R}{2}}} |Du|^2|x-x_0|^{2-n} \mathrm{d}x + CR^{2-\frac{n}{p}}.
\end{aligned}$$

Therefore, from (6.18), we obtain

$$\mathrm{Z} \ge -C \int_{B_{2R}-B_{\frac{R}{2}}} |Du|^2|x-x_0|^{2-n} \mathrm{d}x - CR^{2-\frac{n}{p}}.$$

From (6.14)–(6.17), we obtain

$$\int_{B_{\frac{R}{2}}} |Du|^2|x-x_0|^{2-n} \mathrm{d}x \le C \int_{B_{2R}-B_{\frac{R}{2}}} |Du|^2|x-x_0|^{2-n} \mathrm{d}x - \mathrm{Z} + CR^{\beta}.$$

Noting that $0 < 2 - \frac{n}{p} < 2$, and changing the constant β to another possible constant strictly less than 2, we get

$$\int_{B_{\frac{R}{2}}} |Du|^2|x-x_0|^{2-n} \mathrm{d}x \le C \int_{B_{2R}-B_{\frac{R}{2}}} |Du|^2|x-x_0|^{2-n} \mathrm{d}x + CR^{\beta},$$

or

$$\int_{B_R} |Du|^2|x-x_0|^{2-n} \mathrm{d}x \le C \int_{B_{4R}-B_R} |Du|^2|x-x_0|^{2-n} \mathrm{d}x + CR^{\beta}, \quad \forall R \le \frac{R_0}{2}. \tag{6.20}$$

Now, going back to the basic inequality (6.5) and taking $\Psi = G$, we deduce

$$\int_{\mathcal{O}} a(x)DX^1 \cdot DG \mathrm{d}x + \underline{a} \int_{\mathcal{O}} G|Du|^2 \mathrm{d}x \le C(\rho) \int_{\mathcal{O}} G \sum_{i=1}^{N} (|k^i| + |f^i|) \mathrm{d}x.$$

From the definition of the Green function , the first integral is positive, and the third integral is bounded. Hence

$$\int_{\mathcal{O}} G|Du|^2 \mathrm{d}x \leq C,$$

and also $\int_{2R_0} G|Du|^2 \mathrm{d}x \leq C$. Hence

$$\int_{B_{\frac{R_0}{2}}} |Du|^2 |x - x_0|^{2-n} \mathrm{d}x \leq C. \tag{6.21}$$

From (6.20)–(6.21), using the hole filling technique (see [2]), we can find $\delta_0 \leq \frac{\beta}{2}$, depending only on the data and ρ, such that for $\delta < \delta_0$, one has

$$\mathbb{R}^{2-n-2\delta} \int_{B_R(x_0)} |Du|^2 \mathrm{d}x \leq C, \quad \forall R \leq \frac{R_0}{2},\ x_0 \in \overline{\tilde{\mathcal{O}}}.$$

From Hölder's inequality, it follows that

$$\int_{B_R(x_0)} |Du| \mathrm{d}x \leq C R^{n-1+\delta}, \quad \forall R \leq \frac{R_0}{2},\ x_0 \in \overline{\tilde{\mathcal{O}}}.$$

From Morrey's theorem, we obtain that $u^i \in C^{0,\delta}(\overline{\tilde{\mathcal{O}}})$.

So the interior Hölder regularity has been proven. To proceed on the closure, we consider a system of local maps, and prove the regularity on each of them. So we consider a ball B, centered on a point of the boundary, and we assume that there exists a diffeomorphism Ψ from B into $\mathbb{R}^n$, such that

$$\mathcal{O}^+ = \Psi(B \cap \mathcal{O}) \subset \{y \in \mathbb{R}^n \mid y_n > 0\},$$
$$\Gamma' = \Psi(B \cap \Gamma) \subset \{y \in \mathbb{R}^n \mid y_n = 0\}.$$

We also define the set obtained from $\mathcal{O}^+$ by reflection, namely,

$$\mathcal{O}^- = \{y \mid y_n < 0,\ (y_1, \ldots, y_{n-1}, -y_n) \in \mathcal{O}^+\},$$

and set

$$\mathcal{O}' = \mathcal{O}^+ \cup \mathcal{O}^- \cup \Gamma'.$$

Then $\mathcal{O}'$ is a bounded domain of $\mathbb{R}^n$. We consider, in (2.22), functions φ^i, which are in $H^1(\mathcal{O} \cap B)$, such that $\varphi^i|_{\mathcal{O} \cap \partial B} = 0$. These functions are extended by 0 on $\mathcal{O} - \mathcal{O} \cap B$. Therefore, (2.22) becomes

$$\int_{B \cap \mathcal{O}} a(x) Du^i \cdot D\varphi^i \mathrm{d}x = \int_{B \cap \mathcal{O}} (H^i(x, Du) + f^i)\varphi^i \mathrm{d}x. \tag{6.22}$$

We then make the change of coordinates $x = \Psi^{-1}(y)$. We call $v^i(y) = u^i(\Psi^{-1}(y))$. Consider the matrix

$$J_\Psi(x) = \text{matrix}\left(\frac{\partial \Psi_k}{\partial x_l}\right),$$

and set

$$\widetilde{a}(y) = \frac{J_\Psi(\Psi^{-1}(y))a(\Psi^{-1}(y))J_\Psi^*(\Psi^{-1}(y))}{|\det J_\Psi(\Psi^{-1}(y))|},$$

$$\widetilde{H^i}(y, Dv) = \frac{H^i(\Psi^{-1}(y), J_\Psi(\Psi^{-1}(y))Dv)}{|\det J_\Psi(\Psi^{-1}(y))|},$$

$$\widetilde{f}^i(y) = \frac{f^i(\Psi^{-1}(y))}{|\det J_\Psi(\Psi^{-1}(y))|}.$$

Naturally, the notation Dv refers to the gradient with respect to the variables v. Moreover,

$$J_\Psi\big(\Psi^{-1}(y)\big)Dv = \big(J_\Psi\big(\Psi^{-1}(y)\big)Dv^1, \dots, J_\Psi\big(\Psi^{-1}(y)\big)Dv^n\big),$$

so, in fact,

$$\widetilde{H^i}(y, q) = \frac{H^i(\Psi^{-1}(y), J_\Psi(\Psi^{-1}(y))q^1, \dots, J_\Psi(\Psi^{-1}(y))q^n)}{|\det J_\Psi(\Psi^{-1}(y))|}.$$

The system (2.19) can be written as

$$\int_{\mathcal{O}^+} \widetilde{a}(y)Dv^i \cdot D\widetilde{\varphi}^i \mathrm{d}y = \int_{\mathcal{O}^+} \big(\widetilde{H^i}(y, Dv) + \widetilde{f}^i\big)\widetilde{\varphi}^i \mathrm{d}y \tag{6.23}$$

for any $\widetilde{\varphi}^i(y) \in H^1 \cap L^\infty(\mathcal{O}^+)$, such that $\widetilde{\varphi}^i(y) = 0$ on $\Psi(\mathcal{O} \cap \partial B) = \partial\mathcal{O}^+ - \Gamma'$.

We then proceed with a reflexion procedure. Writing $y = (y', y_n)$, we define, for $y_n < 0$,

$$\begin{aligned}
\widetilde{a}_{kk}\big(y', y_n\big) &= \widetilde{a}_{kk}\big(y', -y_n\big), \quad \forall i,\\
\widetilde{a}_{kl}\big(y', y_n\big) &= \widetilde{a}_{kl}\big(y', -y_n\big), \quad \forall k, l \neq n,\\
\widetilde{a}_{kn}\big(y', y_n\big) &= \widetilde{a}_{kn}\big(y', -y_n\big), \quad \forall k \neq n,\\
\widetilde{H^i}\big(y', y_n; q^1, \dots, q^{n-1}, q^n\big) &= \widetilde{H^i}\big(y', -y_n; q^1, \dots, q^{n-1}, -q^n\big),\\
\widetilde{f}^i\big(y', y_n\big) &= \widetilde{f}^i\big(y', -y_n\big).
\end{aligned}$$

If we extend the solutions $v^i(y)$ to (6.23) for $y_n < 0$, by setting

$$v^i\big(y', y_n\big) = v^i\big(y', -y_n\big),$$

then it is easy to convince oneself that the functions $v^i(y)$ are in $H^1 \cap L^\infty(\mathcal{O}')$, and satisfy

$$\int_{\mathcal{O}'} \widetilde{a}(y) Dv^i \cdot D\widetilde{\varphi}^i \mathrm{d}y = \int_{\mathcal{O}'} (\widetilde{H}^i(y, Dv) + \widetilde{f}^i)\widetilde{\varphi}^i \mathrm{d}y, \quad \forall \widetilde{\varphi}^i \in H_0^1 \cap L^\infty(\mathcal{O}'). \tag{6.24}$$

Moreover, the functions $\widetilde{a}(y)$, $\widetilde{H}^i(y,q)$ and $\widetilde{f}^i(y)$ satisfy the same assumptions as $a(x)$, $H^i(x,q)$ and $f^i(x)$, respectively. Therefore, we can obtain the interior $C^{0,\delta}$ regularity of $v^i(y)$ on $\mathcal{O}'$. We thus obtain the $C^{0,\delta}$ regularity including points of the interior of Γ'. By taking a covering of the boundary Γ of $\mathcal{O}$ by a finite number of local maps, we complete the proof of the $C^{0,\delta}(\overline{\mathcal{O}})$ of the function u. This completes the proof of Theorem 5.1.

6.5 Alternative Assumptions

Assumptions (5.12)–(5.16) are not the only possible ones. Those were made in order to apply the method used in Lemma 6.1, which we call "exponential domination". They have been introduced in [6]. A certain form of exponential domination can be found already in [12]. An alternative condition replaces the growth condition for the Hamiltonians from below (resp. above) by the "sum coerciveness" of the Hamiltonians. This was first used in [3, 4], and thereafter in [6, 7]. In the case, the dimension $n = 2$ and the conditions, up to now, are better than those in the n-dimensional case. The first conditions are as usual. The conditions can be written as

$$|H^i(x,q)| \le K(|q|^2 + 1), \tag{6.25}$$

$$II^i(x,q) = H^{i0}(x,q) + q^i \cdot G(x,q), \tag{6.26}$$

$$|G(x,q)| \le K(|q| + 1). \tag{6.27}$$

In applications to the control theory, the term $q^i \cdot G(x,q)$ is derived from the dynamics, and the term $H^{i0}(x,q)$ is derived from the cost of the controls and the influence of nonmarket interaction.

In addition, from below (alternatively from above), the following "sum coerciveness" of the $H^{i0}(x,q)$ is assumed

$$\sum_{i=1}^{N} H^{i0}(x,q) \ge c_0 |Bq|^2 - K, \quad c_0 > 0, \tag{6.28}$$

and $B : \mathbb{R}^{nN} \to \mathbb{R}^m$ satisfies

$$|Bq| \le K(|q| + 1). \tag{6.29}$$

Of course, this applies to $B=$ identity, but in applications, B can be degenerate, i.e., $B^{-1}(0)$ may be nontrivial.

For G, we need the slightly stronger growth condition

$$|G(x,q)| \leq K(|Bq|+1). \tag{6.30}$$

In applications, B is the map, which assigns to the variables q the corresponding Nash equilibrium for controls in the Lagrangians (see (4.10)). In this context, to a certain extent, (6.30) and (6.28) are natural.

Finally, in [6, 7], for $n=2$, we assume that the above inequality holds. Then we have

$$H^{i0}(x,q) \leq K\big(|q^i||Bq|+1\big), \tag{6.31}$$

which also has a reasonable interpretation in control theory.

In this framework, in [6, 7], one obtains the $C^{0,\delta}$ regularity for Bellman systems.

$$Au^i + \alpha u^i = H^i(x, Du)$$

is recalled under the restriction $n=2$. The techniques can be used for the present mean field setting. Hence, the results of Theorem 5.1 will hold under the assumptions (6.25)–(6.31).

To get rid of the dimension condition, a partial progress was achieved in [8]. They use the same assumptions as (6.25)–(6.26), but they replace (6.31) with

$$H^i(x,q) \leq K\big(|q^i|^2 + q^i \cdot G_0(x,q)+1\big), \tag{6.32}$$

$$|G_0(x,q)| \leq K(|q|+1), \tag{6.33}$$

and $G_0(x,q)$ can be different from $G(x,q)$ above, which increases applicability. Then (6.25)–(6.30) and (6.32) can be used for our mean field setting for $n \geq 2$, in order to obtain Theorem 5.1.

Concerning weak solutions, [1] showed that the conditions (6.25) and (6.31) of the 2-dimensional case imply the existence of a weak solution $u \in L^\infty \cap H^1$ and the strong convergence of the approximations in H^1, also in dimension $n \geq 3$.

There are several slight generalizations. One may replace (6.26) and (6.28) by

$$\sum_{i=1}^N H^i(x,q) \geq c_0|Bq|^2 - K|Bq|\left|\sum_{i=1}^N q^i\right| - K\left|\sum_{i=1}^N q^i\right|^2 - K,$$

where the function $G(x,q)$ is not needed. Perturbations of type $|\sum_{i=1}^N q^i|^2$ are allowed in this setting.

6.6 *Full Regularity for u*

We can complete Theorem 5.1, and state the full regularity of u, provided that an additional assumption is made.

Theorem 6.1 *We make all the assumptions of Theorem* 5.1 *and*

$$a(x) \in W^{1,\infty}(\mathcal{O}), \quad k^i(x) \in L^\infty(\mathcal{O}), \quad i = 1, \dots, N. \tag{6.34}$$

Then $u \in W^{2,r}(\mathcal{O}; \mathbb{R}^N), \forall 1 \leq r < \infty$. *The norm of* u *in the functional space depends only on the data and the constants in the assumptions.*

Proof The proof is based on the linear theory of elliptic equations. It follows from a bootstrap argument based on the Miranda-Nirenberg interpolation result and the regularity theory of linear elliptic equations. The details can be found in [2]. □

7 Study of the Field Equations

By field equations, we consider Eq. (5.21).

7.1 Generic Equation

We shall make the assumptions of Theorem 6.1. We can then assume that the functions $G^i(x, Du)$ are bounded. From Theorem 6.1, the bound depends only on the data, not on the $H^1(\mathcal{O})$ norm of m.

We can see in Eq. (5.21) that there exists no coupling in the functions m^i. So it is sufficient to consider a generic problem

$$\int_{\mathcal{O}} a^*(x) Dm(x) \cdot D\psi(x) \mathrm{d}x - \int_{\mathcal{O}} m(x) G(x) \cdot D\psi(x) \mathrm{d}x + \alpha \int_{\mathcal{O}} m(x)\psi(x) \mathrm{d}x$$
$$= \alpha \int_{\mathcal{O}} m_0(x)\psi(x) \mathrm{d}x, \tag{7.1}$$

where $G(x)$ is bounded and $m^0 \geq 0$ is in $L^p(\mathcal{O})$, $p > \frac{n}{2}$. However, we assume that there exists a positive $H^1(\mathcal{O})$ solution to (7.1). The test function $\psi(x)$ in (7.1) can be taken in $H^1(\mathcal{O})$.

7.2 L^∞ Bound

An important step is as follows.

Proposition 7.1 *We make the assumptions of Theorem* 6.1. *A positive* $H^1(\mathcal{O})$ *solution to* (7.1) *is in* $L^\infty(\mathcal{O})$ *with a norm, which depends only on the data and the constants, and not on the* $H^1(\mathcal{O})$ *norm of* m.

Proof The proof relies on the properties of the Green function for the Neumann problem. For any $x_0 \in \mathcal{O}$, consider the solution $\Sigma = \Sigma^{x_0}$ to the equation

$$\int_{\mathcal{O}} a(x) D\Sigma \cdot D\psi \mathrm{d}x + \alpha \int_{\mathcal{O}} \Sigma \psi \mathrm{d}x = \alpha \psi(x_0), \quad \forall \psi \in H^1(\mathcal{O}) \cap C^0(\overline{\mathcal{O}}). \tag{7.2}$$

The function $\Sigma = \Sigma^{x_0}$ is the Green function associated to the point x_0. We have included the coefficient α for convenience. In (6.11), we had considered the Green function for the Dirichlet problem. We shall use properties, similar to (6.12),

$$\Sigma \in L^{\frac{n}{n-2}(1-s)}, \quad D\Sigma \in L^{\frac{n(1-s)}{n-1-s}}, \quad \forall 0 < s < 1. \tag{7.3}$$

We take $s < \frac{1}{n-1}$ having exponents strictly larger than 1. The second exponent is strictly less than 2, as soon as $n \geq 3$. We shall take $\psi = m$ in (7.2), and $\psi = \Sigma$ in (7.1). This is formal, since we do not have the smoothness required. The correct approach is to approximate Σ with smoother functions, in smoothing the Dirac measure which comes in (7.2). We skip this step, which is classical. Note that $\Sigma \geq 0$. Comparing the two relations, we obtain

$$\alpha m(x_0) = \alpha \int_{\mathcal{O}} m_0 \Sigma \mathrm{d}x + \int_{\mathcal{O}} mG \cdot D\Sigma \mathrm{d}x. \tag{7.4}$$

We stress that this writing is formal, since m is not continuous, and the third integral is not well defined. For the a priori estimates, it is sufficient.

We note first that

$$\int_{\mathcal{O}} m_0 \Sigma \mathrm{d}x \leq \|m_0\|_{L^p} \|\Sigma\|_{L^{\frac{p}{p-1}}}.$$

Using the first property (7.3), thanks to the assumption $p > \frac{n}{2}$, $\frac{p}{p-1} < \frac{n}{n-2}$ and the integral on the right-hand side is well defined.

Now, for any L,

$$\begin{aligned}\int_{\mathcal{O}} mG \cdot D\Sigma \mathrm{d}x &= \int_{\mathcal{O}\cap\{G\cdot D\Sigma \geq L\}} mG \cdot D\Sigma \mathrm{d}x + \int_{\mathcal{O}\cap\{G\cdot D\Sigma \leq L\}} mG \cdot D\Sigma \mathrm{d}x \\ &\leq L \int_{\mathcal{O}} m_0 \mathrm{d}x + \|m\|_\infty \int_{\mathcal{O}\cap\{G\cdot D\Sigma \geq L\}} G \cdot D\Sigma \mathrm{d}x.\end{aligned}$$

Set $z = (G \cdot D\Sigma)^+$. From the second property (7.3), we have

$$\int_{\mathcal{O}} z^{\frac{n(1-s)}{n-1-s}} \mathrm{d}x \leq C_s.$$

Therefore, we check easily that

$$\int_{\mathcal{O}\cap\{z \geq L\}} z \mathrm{d}x \leq C_s \frac{1}{L^{\frac{1-s(n-1)}{n-1-s}}}.$$

Collecting the above results, and choosing L sufficiently large, we deduce from (7.4) that $\|m\|_\infty \leq C$, where the constant depends only on the data, not on the $H^1(\mathcal{O})$ norm of m. This completes the proof. □

7.3 Regularity of m

We can write (7.1) as

$$\int_{\mathcal{O}} a^*(x)Dm(x) \cdot D\psi(x)\mathrm{d}x + \alpha \int_{\mathcal{O}} m(x)\psi(x)\mathrm{d}x$$
$$= \alpha \int_{\mathcal{O}} m_0\psi(x)\mathrm{d}x + \int_{\mathcal{O}} g(x) \cdot D\psi(x)\mathrm{d}x, \tag{7.5}$$

where $g(x)$ is a bounded function, with a bound depending only on the data. It follows immediately that the $H^1(\mathcal{O})$ norm of m depends only on the data and constants of the assumptions. We can then state it as follows.

Theorem 7.1 *We make the assumptions of Theorem* (6.1). *Then the solution m to* (7.5) *belongs to* $W^{2,p}(\mathcal{O}) \oplus W^{1,r}(\mathcal{O})$, $\forall r < \infty$.

The norm depends only on the data and the constants of the assumptions.

Proof This is an immediate consequence of the regularity of the solutions to the linear problems of type (7.5). □

8 Existence of Solutions

We can now address the issue of existence of solutions to the system (5.20)–(5.21), with smooth solutions and positive m^i. We shall assume, in addition to the assumptions of Theorem 6.1, that

$$H^i(x,q), \quad G^i(x,q) \text{ are continuous in } q \text{ (Caratheodory).} \tag{8.1}$$

8.1 Approximation Procedure

We begin by defining an approximation procedure. We introduce the following notations:

$$H^{i,\epsilon}(x,q) = \frac{H^i(x,q)}{1+\epsilon|H^i(x,q)|}, \qquad V^{i,\epsilon}_{[m]}(x) = V^i_{[\frac{m}{1+\epsilon|m|}]}(x), \tag{8.2}$$

and the function on R

$$h^{\epsilon}(\mu)=\frac{\mu^{+}}{1+\epsilon|\mu|}, \tag{8.3}$$

where $|m|$ is the norm of the vector m. Clearly

$$\left\|\frac{m}{1+\epsilon|m|}\right\|_{L^1(\mathcal{O};\mathbb{R}^N)}\leq\frac{|\mathcal{O}|}{\epsilon}.$$

Hence,

$$\|V^{i,\epsilon}_{[m]}\|_{L^\infty}\leq l\left(\frac{|\mathcal{O}|}{\epsilon}\right),\qquad |H^{i,\epsilon}(x,q)|\leq\frac{1}{\epsilon},\quad \forall m,x,q. \tag{8.4}$$

We then define a function T^ϵ from $H^1(\mathcal{O};\mathbb{R}^N)\times L^2(\mathcal{O};\mathbb{R}^N)$ into itself as follows. We write

$$(u,m)=T^\epsilon(v,\mu),$$

and u,m are the solutions to

$$\begin{aligned}&\int_{\mathcal{O}}a(x)Du^i(x)\cdot D\varphi^i(x)\mathrm{d}x+\alpha\int_{\mathcal{O}}u^i(x)\varphi^i(x)\mathrm{d}x\\&\quad=\int_{\mathcal{O}}\big(H^{i,\epsilon}(x,Dv)+V^{i,\epsilon}_{[\mu]}(x)\big)\varphi^i(x)\mathrm{d}x,\end{aligned} \tag{8.5}$$

$$\begin{aligned}&\int_{\mathcal{O}}a^*(x)Dm^i(x)\cdot D\psi^i(x)\mathrm{d}x+\alpha\int_{\mathcal{O}}m^i(x)\psi^i(x)\mathrm{d}x\\&\quad=\int_{\mathcal{O}}h^\epsilon\big(\mu^i(x)\big)G^i(x,Du)\cdot D\psi^i(x)\mathrm{d}x+\alpha\int_{\mathcal{O}}m^i_0(x)\psi^i(x)\mathrm{d}x.\end{aligned} \tag{8.6}$$

Note that the problems (8.5)–(8.6) are defined in sequence. In the right-hand side of (8.6), there is Du, not Dv. At any rate, u^i and m^i are solutions to linear problems. Using the linearity and the regularity theory of linear elliptic equations, we can assert that

$$u\in W^{2,r}\big(\mathcal{O};\mathbb{R}^N\big),\quad m\in W^{1,r}\big(\mathcal{O};\mathbb{R}^N\big)\oplus W^{2,p}\big(\mathcal{O};\mathbb{R}^N\big),\quad \forall r<\infty. \tag{8.7}$$

Moreover, the norm in these functional spaces is bounded by a fixed number, depending on ϵ, but not on the arguments v,μ. The map T^ϵ is continuous (thanks to (8.1)), and the image $T^\epsilon(v,\mu)$ remains in a fixed compact convex subset of $H^1(\mathcal{O};\mathbb{R}^N)\times L^2(\mathcal{O};\mathbb{R}^N)$. From Leray-Schauder theorem, the map T^ϵ has a fixed point. Therefore, we have obtained the following lemma.

Lemma 8.1 *Under the assumptions of Theorem* 6.1 *and* (8.1), *there exists a pair* u^ϵ, m^ϵ, *belonging to the functional spaces as in* (8.7) *and satisfying the system of* *equations*

$$\int_{\mathcal{O}} a(x)Du^{i,\epsilon}(x)\cdot D\varphi^i(x)\mathrm{d}x+\alpha\int_{\mathcal{O}} u^{i,\epsilon}(x)\varphi^i(x)\mathrm{d}x$$
$$=\int_{\mathcal{O}}\big(H^{i,\epsilon}\big(x,Du^{\epsilon}\big)+V^{i,\epsilon}_{[m^{\epsilon}]}(x)\big)\varphi^i(x)\mathrm{d}x, \tag{8.8}$$
$$\int_{\mathcal{O}} a^*(x)Dm^{i,\epsilon}(x)\cdot D\psi^i(x)\mathrm{d}x+\alpha\int_{\mathcal{O}} m^{i,\epsilon}(x)\psi^i(x)\mathrm{d}x$$
$$=\int_{\mathcal{O}} h^{\epsilon}\big(m^{i,\epsilon}(x)\big)G^i\big(x,Du^{\epsilon}\big)\cdot D\psi^i(x)\mathrm{d}x+\alpha\int_{\mathcal{O}} m_0^i(x)\psi^i(x)\mathrm{d}x. \tag{8.9}$$

8.2 Main Result

We can now state the main existence result.

Theorem 8.1 *Under the assumptions of Theorem* 6.1 *and* (8.1), *there exists a solution* (u,m) *to the system of Eqs.* (5.20)–(5.21), *such that*

$$u\in W^{2,r}\big(\mathcal{O};\mathbb{R}^N\big),\quad m\in W^{1,r}\big(\mathcal{O};\mathbb{R}^N\big)\oplus W^{2,p}\big(\mathcal{O};\mathbb{R}^N\big),\quad \forall r<\infty. \tag{8.10}$$

Proof The first thing to observe is that the fixed point $(u^{\epsilon},m^{\epsilon})$, i.e., the solution to (8.8)–(8.9), satisfies $m^{\epsilon}\geq 0$. This is easily seen by taking $\psi^i=(m^{i,\epsilon})^-$ in Eq. (8.9) and noting that

$$\int_{\mathcal{O}} h^{\epsilon}\big(m^{i,\epsilon}(x)\big)G^i\big(x,Du^{\epsilon}\big)\cdot D\big(m^{i,\epsilon}\big)^-(x)\mathrm{d}x=0.$$

Therefore, we can write (8.9) as follows:

$$\int_{\mathcal{O}} a^*(x)Dm^{i,\epsilon}(x)\cdot D\psi^i(x)\mathrm{d}x+\alpha\int_{\mathcal{O}} m^{i,\epsilon}(x)\psi^i(x)\mathrm{d}x$$
$$=\int_{\mathcal{O}}\frac{m^{i,\epsilon}(x)}{1+\epsilon|m^{i,\epsilon}(x)|}G^i\big(x,Du^{\epsilon}\big)\cdot D\psi^i(x)\mathrm{d}x+\alpha\int_{\mathcal{O}} m_0^i(x)\psi^i(x)\mathrm{d}x, \tag{8.11}$$

and $m^{i,\epsilon}\geq 0$. But then, by taking $\psi^i=1$, we get

$$\int_{\mathcal{O}} m^{i,\epsilon}(x)\mathrm{d}x=\int_{\mathcal{O}} m_0^i(x)\mathrm{d}x,$$

and we deduce $|V^{i,\epsilon}_{[m^{\epsilon}]}(x)|\leq l(\|m_0\|)$. Noting that $H^{i,\epsilon}(x,q)$ satisfies all the estimates of $H^i(x,q)$ and (5.4)–(5.6), we can apply all the techniques to (8.8) to derive the a priori estimates in Theorems 5.1 and 6.1. We can then assert that

$$\|u^{\epsilon}\|_{W^{2,r}(\mathcal{O},\mathbb{R}^N)}\leq C,\quad \forall r<\infty,$$

and the constant does not depend on ϵ. Similarly, the reasoning made in Proposition 7.1 and Theorem 7.1 carries over for m^ϵ in (8.11). We then obtain that

$$\|m^\epsilon\|_{W^{1,r_0}} \leq C, \quad r_0 > \frac{n}{2}.$$

We can then extract subsequences, still denoted u^ϵ, m^ϵ, such that, among other properties,

$$u^\epsilon \to u, \quad Du^\epsilon \to Du, \quad \text{pointwise } \|u^\epsilon\|_{L^\infty}, \|Du^\epsilon\|_{L^\infty} \leq C,$$

$$m^\epsilon \to m, \quad \text{pointwise } Dm^\epsilon \to Dm \text{ weakly in } L^2 \|m^\epsilon\| \leq C.$$

Using (5.10) and the continuity properties of $G^i(x,q)$, $H^i(x,q)$, it is easy to go to the limit as $\epsilon \to 0$ in Eqs. (8.8), (8.11), and obtain a solution to (5.20)–(5.21) with the regularity (8.10). This completes the proof of Theorem 8.1. □

References

1. Bensoussan, A., Bulicěk, M., Frehse, J.: Existence and Compactness for weak solutions to Bellman systems with critical growth. Discrete Contin. Dyn. Syst., Ser. B **17**(6), 1–21 (2012)
2. Bensoussan, A., Frehse, J.: Regularity Results for Nonlinear Elliptic Systems and Applications. Appl. Math. Sci., vol. 151. Springer, Berlin (2002)
3. Bensoussan, A., Frehse, J.: Ergodic Bellman systems for stochastic games. In: Elworty, K.D., Norrie Everitt, W., Bruce Lee, E., Dekker, M. (eds.) Markus Feestricht Volume Differential Equations, Dynamical Systems and Control Sciences. Lecture Notes in Pure and Appl. Math., vol. 152, pp. 411–421 (1993)
4. Bensoussan, A., Frehse, J.: Ergodic Bellman systems for stochastic games in arbitrary dimension. Proc. R. Soc. Lond. Ser. A, Math. Phys. Sci. **449**, 65–67 (1995)
5. Bensoussan, A., Frehse, J.: Smooth solutions of systems of quasilinear parabolic equations. ESAIM, COCV **8**, 169–193 (2002)
6. Bensoussan, A., Frehse, J., Vogelgesang, J.: Systems of Bellman equations to stochastic differential games with noncompact coupling. Discrete Contin. Dyn. Syst., Ser. A **274**, 1375–1390 (2010)
7. Bensoussan, A., Frehse, J., Vogelgesang, J.: Nash and Stackelberg differential games. Chin. Ann. Math. **33B**(3), 317–332 (2012)
8. Bulicěk, M., Frehse, J.: On nonlinear elliptic Bellman systems for a class of stochastic differential games in arbitrary dimension. Math. Models Methods Appl. Sci. **21**(1), 215–240 (2011)
9. Guéant, O., Lasry, J.M., Lions, P.L.: Mean field games and applications. In: Carmona, A.R., et al. (eds.) Paris-Princeton Lectures on Mathematical Sciences 2010, pp. 205–266 (2011)
10. Huang, M., Caines, P.E., Malhamé, R.P.: Large-population cost-coupled LQG problems with nonuniform agents: individual-mass behavior and decentralized ϵ-Nash equilibria. IEEE Trans. Autom. Control **52**(9), 1560–1571 (2007)
11. Huang, M., Caines, P.E., Malhamé, R.P.: An invariance principle in large population stochastic dynamic games. J. Syst. Sci. Complex. **20**(2), 162–172 (2007)
12. Ladyzhenskaya, O.A., Uraltseva, N.N.: Linear and Quasilinear Elliptic Equations. Academic Press, New York (1968)
13. Lasry, J.M., Lions, P.L.: Jeux champ moyen I. Le cas stationnaire. C. R. Acad. Sci., Ser. 1 Math. **343**, 619–625 (2006)
14. Lasry, J.M., Lions, P.L.: Jeux à champ moyen II. Horizn fini et contrôle optimal. C. R. Acad. Sci., Ser. 1 Math. **343**, 679–684 (2006)
15. Lasry, J.M., Lions, P.L.: Mean field games. Jpn. J. Math. **2**(1), 229–260 (2007)

The Rain on Underground Porous Media

Part I: Analysis of a Richards Model

Christine Bernardi, Adel Blouza, and Linda El Alaoui

Abstract The Richards equation models the water flow in a partially saturated underground porous medium under the surface. When it rains on the surface, boundary conditions of Signorini type must be considered on this part of the boundary. The authors first study this problem which results into a variational inequality and then propose a discretization by an implicit Euler's scheme in time and finite elements in space. The convergence of this discretization leads to the well-posedness of the problem.

Keywords Richards equation · Porous media · Euler's implicit scheme · Finite element discretization · Parabolic variational inequality

Mathematics Subject Classification 76S05 · 76M10 · 65M12

1 Introduction

The following equation:

$$\partial_t \widetilde{\Theta}(\psi) - \nabla \cdot K_w\big(\Theta(\psi)\big)\nabla(\psi + z) = 0 \tag{1.1}$$

models the flow of a wetting fluid, mainly water, in the underground surface, hence in an unsaturated medium (see [15] for the introduction of this type of models).

C. Bernardi (✉)
Laboratoire Jacques-Louis Lions, CNRS & Université Pierre et Marie Curie, BC 187, 4 Place Jussieu, 75252 Paris Cedex 05, France
e-mail: bernardi@ann.jussieu.fr

A. Blouza
Laboratoire de Mathématiques Raphaël Salem (UMR 6085 CNRS), Université de Rouen, Avenue de l'Université, BP 12, 76801 Saint-Étienne-du-Rouvray, France
e-mail: Adel.Blouza@univ-rouen.fr

L. El Alaoui
University Paris 13, Sorbonne Paris City, LAGA, CNRS (UMR 7539), 93430 Villetaneuse, France
e-mail: elalaoui@math.univ-paris13.fr

P.G. Ciarlet et al. (eds.), *Partial Differential Equations: Theory, Control and Approximation*, DOI 10.1007/978-3-642-41401-5_2,

In opposite to Darcy's or Brinkman's systems (see [14] for all these models), this equation, which is derived by combining Darcy's generalized equation with the mass conservation law, is highly nonlinear. This follows from the fact that, due to the presence of air above the surface, the porous medium is only partially saturated with water. The unknown ψ is the difference between the pressure of water and the atmospherical pressure.

This equation is usually provided with Dirichlet or Neumann type boundary conditions. Indeed, Neumann boundary conditions on the underground part of the boundary are linked to the draining of water outside of the domain, and Dirichlet boundary conditions on the surface are introduced to take into account the rain. However, when the porous media can no longer absorb the rainwater that falls, the upper surface of the domain allows to exfiltration and infiltration. In other words, the upper surface is divided into a saturated zone and an unsaturated zone. We assume that the re-infiltration process is negligible. This leads to variational inequalities of the following type:

$$-\psi \geq 0, \quad \boldsymbol{v}(\psi)\cdot\boldsymbol{n} \geq \boldsymbol{v}_r\cdot\boldsymbol{n}, \quad \psi\big(\boldsymbol{v}(\psi)\cdot\boldsymbol{n} - \boldsymbol{v}_r\cdot\boldsymbol{n}\big) = 0, \tag{1.2}$$

where $\boldsymbol{v}(\psi)$ is the flux

$$\boldsymbol{v}(\psi) = -K_w\big(\Theta(\psi)\big)\nabla(\psi + z), \tag{1.3}$$

and $\boldsymbol{n}$ stands for the unit outward normal vector to the surface, and $\boldsymbol{v}_r$ stands for a given rain fall rate. We refer to the thesis of Berninger [4] for the full derivation of this model from hydrology laws and more specifically to [4, Sect. 1.5] for the derivation of the boundary inequalities (1.2).

It is not so easy to give a mathematical sense to the system (1.1)–(1.2). As a standard, the key argument for the analysis of the problem (1.1) is to use Kirchhoff's change of unknowns. Indeed, after this transformation, the new equation fits the general framework proposed in [1] (see also [6] for the analysis of a different model). Thus, the existence and uniqueness of a solution to this equation with appropriate linear initial and boundary conditions can be derived from standard arguments. In order to handle the inequality in (1.2), we again use a variational formulation. We refer to [2] for the first analysis of very similar systems (see also [5]). We prove that the problem (1.1)–(1.2) is well-posed when the data are smooth enough but in the first step with a rather restrictive assumption on the coefficients.

The discretization of the problem (1.1) was proposed and/or studied in many papers with standard boundary conditions (see [3, 7, 13, 16, 18, 19] and [17] for a more general equation). However, it does not seem to be treated for the case of the boundary inequality (1.2). We propose here a discretization of system (1.1)–(1.2), in two steps as follows:

(i) We first use the Euler's implicit scheme to build a time semi-discrete problem, where one of the nonlinear terms is treated in an explicit way for simplicity.

(ii) We then construct a fully discrete problem that relies on the Galerkin method and finite elements in the spatial domain.

In both cases, we prove that the corresponding variational problem is well-posed.

To conclude, we prove that the solution to this discrete problem converges to a solution to the continuous one when the discretization parameters tend to zero. This ends the proof of our existence result, since no restrictive condition is needed here.

The outline of the paper is as follows.

In Sect. 2, we present the variational formulation of the full system, and investigate its well-posedness in appropriate Sobolev spaces.

Section 3 is devoted to the descriptions of the time semi-discrete problem and of the fully discrete problem. We check their well-posedness.

In Sect. 4, we investigate the convergence of the solution of the discrete problem to a solution of the continuous one.

2 The Continuous Problem and Its Well-Posedness

Let Ω be a bounded connected open set in $\mathbb{R}^d$ ($d = 2$ or 3), with a Lipschitz-continuous boundary $\partial\Omega$, and let $\boldsymbol{n}$ denote the unit outward normal vector to Ω on $\partial\Omega$. We assume that $\partial\Omega$ admits a partition without overlap into three parts Γ_B, Γ_F and Γ_G (these indices mean "bottom", "flux" and "ground", respectively), and that Γ_B has a positive measure. Let also T be a positive real number.

In order to perform the Kirchhoff's change of unknowns in the problem (1.1), we observe that, since the conductivity coefficient K_w is positive, the mapping

$$x \mapsto \mathcal{K}(x) = \int_0^x K_w\big(\Theta(\xi)\big)\mathrm{d}\xi$$

is one-to-one from $\mathbb{R}$ into itself. Thus, by setting

$$u = \mathcal{K}(\psi), \quad b(u) = \Theta \circ \mathcal{K}^{-1}(u), \quad k(\cdot) = K_w(\cdot),$$

and thanks to an appropriate choice of the function $\widetilde{\Theta}$, we derive the equation (more details are given in [3, Remark 2.1] for instance)

$$\alpha\partial_t u + \partial_t b(u) - \nabla \cdot \big(\nabla u + k \circ b(u)\boldsymbol{e}_z\big) = 0 \quad \text{in } \Omega \times [0, T],$$

where $-\boldsymbol{e}_z$ stands for the unit vector in the direction of gravity. Moreover, the Kirchhoff's change of unknowns has the further property of preserving the positivity: u is positive if and only if ψ is positive; u is negative if and only if ψ is negative. So, writing the inequality (1.2) in terms of the unknown u is easy.

As a consequence, from now on, we work with the following system:

$$\begin{cases} \alpha\partial_t u + \partial_t b(u) - \nabla\cdot(\nabla u + k\circ b(u)\boldsymbol{e}_z) = 0 & \text{in } \Omega\times[0,T],\\ u = u_B & \text{on } \Gamma_B\times[0,T],\\ -(\nabla u + k\circ b(u)\boldsymbol{e}_z)\cdot\boldsymbol{n} = f_F & \text{on } \Gamma_F\times[0,T],\\ u\le 0, \quad -(\nabla u + k\circ b(u)\boldsymbol{e}_z)\cdot\boldsymbol{n} \ge \boldsymbol{q}_r\cdot\boldsymbol{n}, & \\ u(\nabla u + k\circ b(u)\boldsymbol{e}_z + \boldsymbol{q}_r)\cdot\boldsymbol{n} = 0 & \text{on } \Gamma_G\times[0,T],\\ u|_{t=0} = u_0 & \text{in } \Omega. \end{cases} \tag{2.1}$$

The unknown is now the quantity u. The data are the Dirichlet boundary condition u_B on $\Gamma_B\times[0,T]$ and the initial condition u_0 on Ω, together with the boundary conditions f_F and $\boldsymbol{q}_r$ on the normal component of the flux, where f_F corresponds to the draining of water, and $\boldsymbol{q}_r$ corresponds to the rain. Finally, b and k are supposed to be known, while α is a positive constant. From now on, we assume that

(i) the function b is of class $\mathscr{C}^2$ on $\mathbb{R}$, with bounded and Lipschitz-continuous derivatives, and is nondecreasing,
(ii) the function $k\circ b$ is continuous, bounded, and uniformly Lipschitz-continuous on $\mathbb{R}$.

Remark 2.1 It must be noted that the parameter α has a physical meaning. Indeed, the function $\widetilde{\Theta}$ in (1.1) is usually the sum of Θ and a term linked to the saturation state. But it can also be considered as a regularization parameter, since it avoids the degeneracy of the equation, where the derivative of b vanishes. So, adding the term $\alpha\partial_t u$ is a standard technique in the analysis of such problems, which has been used with success for constructing effective numerical algorithms (see e.g., [12, 13]).

In what follows, we use the whole scale of Sobolev spaces $W^{m,p}(\Omega)$ with $m\ge 0$ and $1\le p\le +\infty$, equipped with the norm $\|\cdot\|_{W^{m,p}(\Omega)}$ and the seminorm $|\cdot|_{W^{m,p}(\Omega)}$, with the usual notation $H^m(\Omega)$ when $p=2$. As a standard, the range of $H^1(\Omega)$ by the trace operator on any part Γ of $\partial\Omega$ is denoted by $H^{\frac12}(\Gamma)$. For any separable Banach space E equipped with the norm $\|\cdot\|_E$, we denote by $\mathscr{C}^0(0,T;E)$ the space of continuous functions on $[0,T]$ with values in E. For each integer $m\ge 0$, we also introduce the space $H^m(0,T;E)$ as the space of measurable functions on $]0,T[$ with values in E, such that the mappings: $v\mapsto \|\partial_t^\ell v\|_E$, $0\le\ell\le m$, are square-integrable on $]0,T[$.

To write a variational formulation for the problem, we introduce the time-dependent subset

$$\mathbb{V}(t) = \big\{v\in H^1(\Omega);\ v|_{\Gamma_B} = u_B(\cdot,t) \text{ and } v|_{\Gamma_G}\le 0\big\}. \tag{2.2}$$

It is readily checked that each $\mathbb{V}(t)$ is closed and convex (see [4, Proposition 1.5.5]), when u_B belongs to $\mathscr{C}^0(0,T;H^{\frac12}(\Gamma_B))$. Thus, we are led to consider the following variational problem (with obvious notation for $L^2(0,T;\mathbb{V})$).

Find u in $L^2(0, T; \mathbb{V})$ with $\partial_t u$ in $L^2(0, T; L^2(\Omega))$, such that

$$u|_{t=0} = u_0, \tag{2.3}$$

and that, for a.e. t in $[0, T]$,

$$\begin{aligned}\forall v \in \mathbb{V}(t), \quad & \alpha \int_\Omega (\partial_t u)(\boldsymbol{x}, t)(v - u)(\boldsymbol{x}, t)\mathrm{d}\boldsymbol{x} + \int_\Omega \big(\partial_t b(u)\big)(\boldsymbol{x}, t)(v - u)(\boldsymbol{x}, t)\mathrm{d}\boldsymbol{x} \\ & + \int_\Omega \big(\nabla u + k \circ b(u)\boldsymbol{e}_z\big)(\boldsymbol{x}, t) \cdot \big(\nabla(v - u)\big)(\boldsymbol{x}, t)\mathrm{d}\boldsymbol{x} \\ & \geq - \int_{\Gamma_F} f_F(\boldsymbol{\tau}, t)(v - u)(\boldsymbol{\tau}, t)\mathrm{d}\boldsymbol{\tau} \\ & \quad - \int_{\Gamma_G} (\boldsymbol{q}_r \cdot \boldsymbol{n})(\boldsymbol{\tau}, t)(v - u)(\boldsymbol{\tau}, t)\mathrm{d}\boldsymbol{\tau},\end{aligned} \tag{2.4}$$

where $\boldsymbol{\tau}$ denotes the tangential coordinates on $\partial\Omega$. The reason for this follows.

Proposition 2.1 *The problems* (2.1) *and* (2.3)–(2.4) *are equivalent, and more precisely*:

(i) *Any solution to the problem* (2.1) *in* $L^2(0, T; H^1(\Omega)) \cap H^1(0, T; L^2(\Omega))$ *is a solution to* (2.3)–(2.4).
(ii) *Any solution to the problem* (2.3)–(2.4) *is a solution to the problem* (2.1) *in the distribution sense.*

Proof We check successively the two assertions of the proposition.

(1) Let u be any solution to (2.1) in $L^2(0, T; H^1(\Omega)) \cap H^1(0, T; L^2(\Omega))$. Obviously, it belongs to $L^2(0, T; \mathbb{V})$ and satisfies (2.3). Next, we observe that, for any v in $\mathbb{V}(t)$, the function $v - u$ vanishes on Γ_B. Multiplying the first line in (2.1) by this function and integrating it by parts on Ω, we have

$$\begin{aligned}& \alpha \int_\Omega (\partial_t u)(\boldsymbol{x}, t)(v - u)(\boldsymbol{x}, t)\mathrm{d}\boldsymbol{x} + \int_\Omega \big(\partial_t b(u)\big)(\boldsymbol{x}, t)(v - u)(\boldsymbol{x}, t)\mathrm{d}\boldsymbol{x} \\ & \quad + \int_\Omega \big(\nabla u + k \circ b(u)\boldsymbol{e}_z\big)(\boldsymbol{x}, t) \cdot \big(\nabla(v - u)\big)(\boldsymbol{x}, t)\mathrm{d}\boldsymbol{x} \\ & = \int_{\Gamma_F \cup \Gamma_G} \big(\nabla u + k \circ b(u)\boldsymbol{e}_z\big) \cdot \boldsymbol{n}(\boldsymbol{\tau})(v - u)(\boldsymbol{\tau}, t)\mathrm{d}\boldsymbol{\tau}.\end{aligned}$$

To conclude, we observe on Γ_G, either u is zero and $\nabla u + k \circ b(u)\boldsymbol{e}_z$ is smaller than $-q_\tau \cdot \boldsymbol{n}$, or u is not zero and $\nabla u + k \circ b(u)\boldsymbol{e}_z$ is equal to $-\boldsymbol{q}_r \cdot \boldsymbol{n}$. All these yield (2.4).

(2) Conversely, let u be any solution to (2.3)–(2.4).

(i) By noting that for any function w in $\mathscr{D}(\Omega)$, $(u + w)(\cdot, t)$ belongs to $\mathbb{V}(t)$. Taking v equal to $u \pm w$ in (2.4), we obtain the first line of (2.1) in the distribution sense.

(ii) The second line in (2.1) follows from the definition of $\mathbb{V}(t)$.

(iii) By taking v equal to $u \pm w$ for any w in $\mathscr{D}(\Omega \cup \Gamma_F)$, we also derive the third line in (2.1).

(iv) The fact that u is nonpositive on Γ_G, comes from the definition of $\mathbb{V}(t)$. On the other hand, the previous equations imply that for any v in $\mathbb{V}(t)$,

$$\int_{\Gamma_G} \big(\nabla u + k \circ b(u)\boldsymbol{e}_z\big) \cdot \boldsymbol{n}(\boldsymbol{\tau})(v-u)(\boldsymbol{\tau},t)\mathrm{d}\boldsymbol{\tau} \geq -\int_{\Gamma_G} (\boldsymbol{q}_r \cdot \boldsymbol{n})(\boldsymbol{\tau},t)(v-u)(\boldsymbol{\tau},t)\mathrm{d}\boldsymbol{\tau}.$$

Taking v equal to $u + w$, where w vanishes on Γ_B and is nonpositive on Γ_G, yields that $-(\nabla u + k \circ b(u)\boldsymbol{e}_z) \cdot \boldsymbol{n}$ is larger than $\boldsymbol{q}_r \cdot \boldsymbol{n}$. Finally, taking v equal to zero on Γ_G, leads to

$$\int_{\Gamma_G} \big(\nabla u + k \circ b(u)\boldsymbol{e}_z + \boldsymbol{q}_r\big) \cdot \boldsymbol{n}(\boldsymbol{\tau})u(\boldsymbol{\tau},t)\mathrm{d}\boldsymbol{\tau} \leq 0.$$

Since the two quantities u and $(\nabla u + k \circ b(u)\boldsymbol{e}_z + \boldsymbol{q}_r)$ are nonpositive on Γ_G, their product is zero.

(v) Finally the last line of (2.1) is written in (2.3).

Proving that the problem (2.3)–(2.4) is well-posed and is not at all obvious. We begin with the simpler result, i.e., the uniqueness of the solution. For brevity, we set

$$\mathbb{X} = L^2(0,T;\mathbb{V}) \cap H^1\big(0,T;L^2(\Omega)\big). \tag{2.5}$$

We also refer to [11, Chap. 1, Théorème 11.7] for the definition of the space $H_{00}^{\frac{1}{2}}(\Gamma_B)$. □

Proposition 2.2 *For any data u_B, f_F, $\boldsymbol{q}_r$ and u_0 satisfying*

$$\begin{aligned} &u_B \in H^1\big(0,T;H_{00}^{\frac{1}{2}}(\Gamma_B)\big), \quad f_F \in L^2\big(0,T;L^2(\Gamma_F)\big),\\ &\boldsymbol{q}_r \in L^2\big(0,T;L^2(\Gamma_G)^d\big), \quad u_0 \in H^1(\Omega), \end{aligned} \tag{2.6}$$

the problem (2.3)–(2.4) *has at most a solution in* $\mathbb{X}$.

Proof Let u_1 and u_2 be two solutions to the problem (2.3)–(2.4). Thus, the function $u = u_1 - u_2$ vanishes on Γ_B and at $t = 0$. Taking v equal to u_2 in the problem satisfied by u_1 and equal to u_1 in the problem satisfied by u_2, and subtracting the second problem from the first one, we obtain

$$\begin{aligned} \alpha \int_\Omega (\partial_t u)(\boldsymbol{x},t)u(\boldsymbol{x},t)\mathrm{d}\boldsymbol{x} &+ \int_\Omega \big(\partial_t b(u_1) - \partial_t b(u_2)\big)(\boldsymbol{x},t)u(\boldsymbol{x},t)\mathrm{d}\boldsymbol{x}\\ &+ \int_\Omega (\nabla u)^2(\boldsymbol{x},t)\mathrm{d}\boldsymbol{x} + \int_\Omega \big(k \circ b(u_1) - k \circ b(u_2)\big)(\boldsymbol{x},t)\boldsymbol{e}_z \cdot (\nabla u))(\boldsymbol{x},t)\mathrm{d}\boldsymbol{x} \leq 0. \end{aligned} \tag{2.7}$$

We integrate this inequality with respect to t and evaluate successively the four integrals.

(1) The first and third ones are obvious

$$\alpha \int_0^t \int_\Omega (\partial_t u)(\boldsymbol{x}, s) u(\boldsymbol{x}, s) \mathrm{d}\boldsymbol{x} \mathrm{d}s + \int_0^t \int_\Omega (\nabla u)^2(\boldsymbol{x}, s) \mathrm{d}\boldsymbol{x} \mathrm{d}s$$

$$= \frac{\alpha}{2} \|u(\cdot, t)\|^2_{L^2(\Omega)} + \int_0^t |u(\cdot, s)|^2_{H^1(\Omega)} \mathrm{d}s.$$

(2) To evaluate the second one, we use the decomposition

$$\int_\Omega \big(\partial_t b(u_1) - \partial_t b(u_2)\big)(\boldsymbol{x}, t) u(\boldsymbol{x}, t) \mathrm{d}\boldsymbol{x}$$

$$= \int_\Omega b'(u_1)(\boldsymbol{x}, t)(\partial_t u)(\boldsymbol{x}, t) u(\boldsymbol{x}, t) \mathrm{d}\boldsymbol{x}$$

$$+ \int_\Omega \big(b'(u_1) - b'(u_2)\big)(\boldsymbol{x}, t)(\partial_t u_2)(\boldsymbol{x}, t) u(\boldsymbol{x}, t) \mathrm{d}\boldsymbol{x},$$

and integrate the first term by parts with respect to t, which gives

$$\int_0^t \int_\Omega \big(\partial_t b(u_1) - \partial_t b(u_2)\big)(\boldsymbol{x}, s) u(\boldsymbol{x}, s) \mathrm{d}\boldsymbol{x} \mathrm{d}s$$

$$= \int_\Omega \frac{b'(u_1)(\boldsymbol{x}, t)}{2} u^2(\boldsymbol{x}, t) \mathrm{d}\boldsymbol{x}$$

$$- \frac{1}{2} \int_0^t \int_\Omega b''(u_1)(\boldsymbol{x}, s)(\partial_t u_1)(\boldsymbol{x}, s) u^2(\boldsymbol{x}, s) \mathrm{d}\boldsymbol{x} \mathrm{d}s$$

$$+ \int_0^t \int_\Omega \big(b'(u_1) - b'(u_2)\big)(\boldsymbol{x}, s)(\partial_t u_2)(\boldsymbol{x}, s) u(\boldsymbol{x}, s) \mathrm{d}\boldsymbol{x} \mathrm{d}s.$$

Next, the nonnegativity of b', the boundedness of b'' and the Lipschitz-continuity of b' yield

$$\int_0^t \int_\Omega \big(\partial_t b(u_1) - \partial_t b(u_2)\big)(\boldsymbol{x}, s) u(\boldsymbol{x}, s) \mathrm{d}\boldsymbol{x} \mathrm{d}s \geq -c(u_1, u_2) \int_0^t \|u(\cdot, s)\|^2_{L^4(\Omega)} \mathrm{d}s,$$

where $c(u_1, u_2) > 0$ depends on $\|\partial_t u_i\|_{L^2(0,T;L^2(\Omega))}$. Next, we use an interpolation inequality (see [11, Chap. 1, Proposition 2.3]) and the Poincaré-Friedrichs inequality

$$\|u\|_{L^4(\Omega)} \leq \|u\|^{1-\frac{d}{4}}_{L^2(\Omega)} (c|u|_{H^1(\Omega)})^{\frac{d}{4}} \leq c'\Big(1 - \frac{d}{4}\Big) \|u\|_{L^2(\Omega)} + \frac{d}{4} |u|_{H^1(\Omega)},$$

and conclude with a Young's inequality.

(3) Finally, to bound the last one, we combine the Lipschitz-continuity of $k \circ b$ together with a Young's inequality

$$\int_0^t \int_\Omega \big(k \circ b(u_1) - k \circ b(u_2)\big)(\boldsymbol{x}, s)\boldsymbol{e}_z \cdot (\nabla u)(\boldsymbol{x}, s)\mathrm{d}\boldsymbol{x}\mathrm{d}s$$

$$\leq \frac{1}{4}\Big(\int_0^t |u(\cdot, s)|^2_{H^1(\Omega)}\mathrm{d}s\Big) + c\Big(\int_0^t \|u(\cdot, s)\|^2_{L^2(\Omega)}\mathrm{d}s\Big).$$

All these give

$$\frac{\alpha}{2}\|u(\cdot, t)\|^2_{L^2(\Omega)} + \frac{1}{2}\int_0^t |u(\cdot, s)|^2_{H^1(\Omega)}\mathrm{d}s \leq c(u_1, u_2)\int_0^t \|u(\cdot, s)\|^2_{L^2(\Omega)}\mathrm{d}s.$$

Thus, applying Grönwall's lemma yields that u is zero, whence the uniqueness result follows.

Proving the existence is much more complex. We begin with a basic result. □

Lemma 2.1 *If the function u_B belongs to $\mathscr{C}^0(0, T; H^{\frac{1}{2}}_{00}(\Gamma_B))$, then for all t in $[0, T]$, the convex set $\mathbb{V}(t)$ is not empty.*

Proof Denoting by $\overline{u}_B(\cdot, t)$ the extension by zero of $u_B(\cdot, t)$ to $\partial\Omega$, we observe that any lifting of $\overline{u}_B(\cdot, t)$ in $H^1(\Omega)$ belongs to $\mathbb{V}(t)$, whence the desired result follows.

In the first step, we consider the linear problem, for any datum F in $L^2(0, T; L^2(\Omega))$.

Find u in $L^2(0, T; \mathbb{V})$ with $\partial_t u$ in $L^2(0, T; L^2(\Omega))$ satisfying (2.3) and such that, for a.e. t in $[0, T]$,

$$\forall v \in \mathbb{V}(t), \quad \alpha\int_\Omega (\partial_t u)(\boldsymbol{x}, t)(v - u)(\boldsymbol{x}, t)\mathrm{d}\boldsymbol{x} + \int_\Omega (\nabla u)(\boldsymbol{x}, t) \cdot \big(\nabla(v - u)\big)(\boldsymbol{x}, t)\mathrm{d}\boldsymbol{x}$$

$$\geq -\int_\Omega F(\boldsymbol{x}, t)(v - u)(\boldsymbol{x}, t)\mathrm{d}\boldsymbol{x} - \int_{\Gamma_F} f_F(\boldsymbol{\tau}, t)(v - u)(\boldsymbol{\tau}, t)\mathrm{d}\boldsymbol{\tau}$$

$$- \int_{\Gamma_G} (\boldsymbol{q}_r \cdot \boldsymbol{n})(\boldsymbol{\tau}, t)(v - u)(\boldsymbol{\tau}, t)\mathrm{d}\boldsymbol{\tau}. \tag{2.8}$$

However a weaker formulation of this problem can be derived by integrating with respect to t. It reads as follows.

Find u in $L^2(0, T; \mathbb{V})$ satisfying (2.3), such that

$$\forall v \in \mathbb{X}, \quad \alpha\int_0^T \int_\Omega (\partial_t u)(\boldsymbol{x}, t)(v - u)(\boldsymbol{x}, t)\mathrm{d}\boldsymbol{x}\mathrm{d}t$$

$$+ \int_0^T \int_\Omega (\nabla u)(\boldsymbol{x}, t) \cdot \big(\nabla(v - u)\big)(\boldsymbol{x}, t)\mathrm{d}\boldsymbol{x}\mathrm{d}t$$

$$\geq -\int_0^T \int_\Omega F(\boldsymbol{x},t)(v-u)(\boldsymbol{x},t)\mathrm{d}\boldsymbol{x}\mathrm{d}t$$
$$-\int_0^T \int_{\Gamma_F} f_F(\boldsymbol{\tau},t)(v-u)(\boldsymbol{\tau},t)\mathrm{d}\boldsymbol{\tau}\mathrm{d}t$$
$$-\int_0^T \int_{\Gamma_G} (\boldsymbol{q}_r\cdot\boldsymbol{n})(\boldsymbol{\tau},t)(v-u)(\boldsymbol{\tau},t)\mathrm{d}\boldsymbol{\tau}\mathrm{d}t. \qquad (2.9)$$

We recall in the next lemma the properties of this problem which are standard. □

Lemma 2.2 *Assume that the data u_B, f_F, $\boldsymbol{q}_r$ and u_0 satisfy* (2.6). *Then, for any F in $L^2(0,T;L^2(\Omega))$, the problem* (2.3)–(2.9) *has a unique solution u in $L^2(0,T;\mathbb{V})$.*

Proof It follows from Lemma 2.1 and the further assumption on u_B that $\mathbb{X}$ is a non-empty closed convex set. We also consider a lifting $\overline{u}_B$ of the extension by zero of u_B to $\partial\Omega$ in $H^1(0,T;H^1(\Omega))$. Then, it is readily checked that $u-\overline{u}_B$ is the solution to a problem, which satisfies all the assumptions in [10, Chap. 6, Theorem 2.2], whence the existence and uniqueness result follows.

Any solution to (2.3)–(2.8) is a solution to (2.3)–(2.9), but the converse property is not obvious in the general case (see [10, Chap. 6]). However, in our specific case, it is readily checked by a density argument that (2.9) is satisfied for any v in $L^2(0,T;\mathbb{V})$, so that problems (2.3)–(2.8) and (2.3)–(2.9) are fully equivalent.

To go further, we assume that the following compatibility condition holds:

$$u_0(\boldsymbol{x}) = u_B(\boldsymbol{x},0) \quad \text{for } \boldsymbol{x}\in\Gamma_B \quad \text{a.e.} \quad \text{and} \quad u_0(\boldsymbol{x})\leq 0 \quad \text{for } \boldsymbol{x}\in\Gamma_G \text{ a.e.} \qquad (2.10)$$

Moreover, we introduce a lifting u_B^* of an extension of u_B to $\partial\Omega$, which belongs to $H^1(0,T;\mathbb{V})$ and satisfies

$$u_B^*(\boldsymbol{x},0) = u_0(\boldsymbol{x}) \quad \text{for } x\in\Omega \quad \text{a.e.}, \qquad (2.11)$$

together with the stability property

$$\|u_B^*\|_{H^1(0,T;H^1(\Omega))} \leq c\|u_B\|_{H^1(0,T;H_{00}^{\frac{1}{2}}(\Gamma_B))}. \qquad (2.12)$$

Then, it is readily checked that u is a solution to the problem (2.3)–(2.4) if and only if the function $u^* = u - u_B^*$ is a solution to the following problem.

Find u^* in $L^2(0,T;\mathbb{V}_0)$ with $\partial_t u^*$ in $L^2(0,T;L^2(\Omega))$, such that

$$u^*|_{t=0} = 0, \qquad (2.13)$$

and that, for a.e. t in $[0,T]$,

$$\forall v\in\mathbb{V}_0, \quad \alpha\int_\Omega (\partial_t u^*)(\boldsymbol{x},t)\big(v-u^*\big)(\boldsymbol{x},t)\mathrm{d}\boldsymbol{x}$$
$$+\int_\Omega \big(\partial_t b_*(u^*)\big)(\boldsymbol{x},t)\big(v-u^*\big)(\boldsymbol{x},t)\mathrm{d}\boldsymbol{x}$$

$$+\int_{\Omega}\big(\nabla u^* + k\circ b_*(u^*)\boldsymbol{e}_z\big)(\boldsymbol{x},t)\cdot\big(\nabla(v-u^*)\big)(\boldsymbol{x},t)\mathrm{d}\boldsymbol{x}$$
$$\geq -\int_{\Omega} F_B(\boldsymbol{x},t)(v-u^*)(\boldsymbol{x},t)\mathrm{d}\boldsymbol{x} - \int_{\Gamma_F} f_F(\boldsymbol{\tau},t)(v-u^*)(\boldsymbol{\tau},t)\mathrm{d}\boldsymbol{\tau}$$
$$-\int_{\Gamma_G}(\boldsymbol{q}_r\cdot\boldsymbol{n})(\boldsymbol{\tau},t)(v-u^*)(\boldsymbol{\tau},t)\mathrm{d}\boldsymbol{\tau} \qquad (2.14)$$

with the definition of the subset $\mathbb{V}_0$,

$$\mathbb{V}_0 = \{v\in H^1(\Omega);\ v|_{\Gamma_B}=0 \text{ and } v|_{\Gamma_G}\leq 0\}, \qquad (2.15)$$

where the new application b_* is defined by $b_*(u^*) = b(u^* + u_B^*)$. The datum F_B is defined by, for a.e. t in $]0,T[$,

$$\int_{\Omega} F_B(\boldsymbol{x},t)v(\boldsymbol{x})\mathrm{d}\boldsymbol{x}$$
$$=\alpha\int_{\Omega}\big(\partial_t u_B^*\big)(\boldsymbol{x},t)v(\boldsymbol{x})\mathrm{d}\boldsymbol{x} + \int_{\Omega}\big(\nabla u_B^*\big)(\boldsymbol{x},t)\cdot(\nabla v)(\boldsymbol{x})\mathrm{d}\boldsymbol{x}, \qquad (2.16)$$

and clearly belongs to $L^2(0,T;\mathbb{W}')$, where $\mathbb{W}$ is the smallest linear space containing $\mathbb{V}_0$, namely

$$\mathbb{W} = \{v\in H^1(\Omega);\ v|_{\Gamma_B}=0\}. \qquad (2.17)$$

It can be noted that the existence result stated in Lemma 2.2 is still valid for any F in $L^2(0,T;\mathbb{W}')$.

We denote by $\mathcal{T}$ the operator, which associates with any pair (F,D), with F in $L^2(0,T;\mathbb{W}')$ and the datum $D=(0,f_F,\boldsymbol{q}_r,0)$ satisfying (2.6), the solution u to the problem (2.3)–(2.8). It follows from (2.13)–(2.14) that u^* satisfies

$$u^* - \mathcal{T}\big(F_B + F(u^*), D\big) = 0, \qquad (2.18)$$

where the quantity $F(u)$ is defined by duality, for a.e. t in $]0,T[$,

$$\langle F(u), v\rangle = \int_{\Omega}\big(\partial_t b_*(u)\big)(\boldsymbol{x},t)v(\boldsymbol{x})\mathrm{d}\boldsymbol{x} + \int_{\Omega} k\circ b_*(u)(\boldsymbol{x},t)\boldsymbol{e}_z\cdot(\nabla v)(\boldsymbol{x})\mathrm{d}\boldsymbol{x}. \qquad (2.19)$$

We first prove some further properties of the operator $\mathcal{T}$. □

Lemma 2.3 *The operator $\mathcal{T}$ is continuous from $L^2(0,T;\mathbb{W}')\times L^2(0,T;L^2(\Gamma_F))\times L^2(0,T;L^2(\Gamma_G)^d)$ into the space $L^2(0,T;\mathbb{V}_0)$. Moreover, the following estimate holds*:

$$\left(\int_0^T |\mathcal{T}(F,f_F,\boldsymbol{q}_r)(\cdot,t)|^2_{H^1(\Omega)}\mathrm{d}t\right)^{\frac{1}{2}}$$
$$\leq \|F\|_{L^2(0,T;\mathbb{W}')} + c\|f_F\|_{L^2(0,T:L^2(\Gamma_F))} + c\|\boldsymbol{q}_r\|_{L^2(0,T;L^2(\Gamma_G)^d)}. \qquad (2.20)$$

Proof We set $u = \mathcal{T}(F, f_F, \boldsymbol{q}_r)$ and only prove the estimate (indeed, it is readily checked that it implies the continuity property). We take v equal to $\frac{u}{2}$ in the problem (2.8). This obviously gives

$$\frac{\alpha}{2}\int_\Omega (\partial_t u^2)(\boldsymbol{x},t)\mathrm{d}\boldsymbol{x} + |u(\cdot,t)|^2_{H^1(\Omega)}$$
$$\le \big(\|F(\cdot,t)\|_{\mathbb{W}'} + c\|f_F(\cdot,t)\|_{L^2(\Gamma_F)} + c\|\boldsymbol{q}_r(\cdot,t)\|_{L^2(\Gamma_G)^d}\big)|u(\cdot,t)|_{H^1(\Omega)},$$

where c is the norm of the trace operator. Thus, integrating with respect to t gives the estimate (2.20). □

Lemma 2.4 *The operator $\mathcal{T}$ is continuous from $L^2(0,T;L^2(\Omega)) \times H^1(0,T;$ $L^2(\Gamma_F)) \times H^1(0,T;L^2(\Gamma_G)^d)$ into the space $H^1(0,T;L^2(\Omega))$. Moreover, the following estimate holds: for any positive ε,*

$$\alpha\|\partial_t\mathcal{T}(F, f_F, \boldsymbol{q}_r)\|_{L^2(0,T;L^2(\Omega))}$$
$$\le (1+\varepsilon)\|F\|_{L^2(0,T;L^2(\Omega))} + c\|f_F\|_{H^1(0,T:L^2(\Gamma_F))}$$
$$+ c\|\boldsymbol{q}_r\|_{H^1(0,T;L^2(\Gamma_G)^d)}. \tag{2.21}$$

Proof The continuity property of $\mathcal{T}$ is proved in [10, Chap. 6, Théorème 2.1]. Next, setting $u = \mathcal{T}(F, f_F, \boldsymbol{q}_r)$, we take v equal to $u - \eta\partial_t u$ in (2.8) for a positive η. Indeed, we have that:

(1) Since u vanishes on Γ_B, so does $\partial_t u$.

(2) Since u is nonpositive on Γ_G and $u(\boldsymbol{x}, t-\eta)$, which is close to $u(\boldsymbol{x},t) - \eta\partial_t u(\boldsymbol{x},t)$, is also nonpositive, there exists an $\eta > 0$, such that $u - \eta\partial_t u$ belongs to $\mathbb{V}_0$.

This yields

$$\alpha\|\partial_t u\|^2_{L^2(\Omega)} + \frac{1}{2}\partial_t|u|^2_{H^1(\Omega)}$$
$$\le \|F\|_{L^2(\Omega)}\|\partial_t u\|_{L^2(\Omega)} - \int_{\Gamma_F} f_F(\boldsymbol{\tau},t)\partial_t u(\boldsymbol{\tau},t)\mathrm{d}\boldsymbol{\tau}$$
$$- \int_{\Gamma_G}(\boldsymbol{q}_r\cdot\boldsymbol{n})(\boldsymbol{\tau},t)\partial_t u(\boldsymbol{\tau},t)\mathrm{d}\boldsymbol{\tau}.$$

To bound the first term, we use Young's inequality

$$\|F\|_{L^2(\Omega)}\|\partial_t u\|_{L^2(\Omega)} \le \frac{\alpha}{2}\|\partial_t u\|^2_{L^2(\Omega)} + \frac{1}{2\alpha}\|F\|^2_{L^2(\Omega)}.$$

To handle the last two integrals, we integrate them by parts with respect to t. For instance, we have, for any $\varepsilon > 0$,

$$\begin{aligned}
&\int_0^t \int_{\Gamma_F} f_F(\boldsymbol{\tau}, s)\partial_t u(\boldsymbol{\tau}, s) \mathrm{d}\boldsymbol{\tau} \mathrm{d}s \\
&= \int_{\Gamma_F} f_F(\boldsymbol{\tau}, t) u(\boldsymbol{\tau}, t) \mathrm{d}\boldsymbol{\tau} \mathrm{d}s - \int_0^t \int_{\Gamma_F} \partial_t f_F(\boldsymbol{\tau}, s) u(\boldsymbol{\tau}, s) \mathrm{d}\boldsymbol{\tau} \mathrm{d}s \\
&\le \frac{1}{4} |u(\cdot, t)|^2_{H^1(\Omega)} + c \| f_F(\cdot, t) \|^2_{L^2(\Gamma_F)} + c \| \partial_t f_F \|^2_{L^2(0,t;L^2(\Gamma_F))} \\
&\quad + \varepsilon \| u \|^2_{L^2(0,t;H^1(\Omega))}.
\end{aligned}$$

Thus, the desired estimate follows by combining all of those and using (2.20). □

We are thus in a position to prove the first existence result.

Theorem 2.1 *Assume that the coefficient* α *satisfies*

$$\frac{1}{\alpha} \| b' \|_{L^\infty(\mathbb{R})} < 1. \tag{2.22}$$

For any data u_B, f_F, $\boldsymbol{q}_r$ *and* u_0 *satisfying*

$$\begin{aligned}
&u_B \in H^1\big(0, T; H^{\frac{1}{2}}_{00}(\Gamma_B)\big), \quad f_F \in H^1\big(0, T; L^2(\Gamma_F)\big), \\
&\boldsymbol{q}_r \in H^1\big(0, T; L^2(\Gamma_G)^d\big), \quad u_0 \in H^1(\Omega)
\end{aligned} \tag{2.23}$$

and (2.10), *the problem* (2.3)–(2.4) *has at least a solution in* $\mathbb{X}$.

Proof We proceed in several steps.

(1) Let $\mathbb{X}_0$ be the space of functions of $\mathbb{X}$ vanishing at $t = 0$. We provide it with the norm

$$\| v \|_{\mathbb{X}_0} = \| \partial_t v \|_{L^2(0,T;L^2(\Omega))}.$$

It follows from the Lemma 2.4 that

$$\big\| \mathcal{T}\big(F_B + F(u^*), D\big) \big\|_{\mathbb{X}_0} \le \frac{1+\varepsilon}{\alpha} \big\| F(u^*) \big\|_{L^2(0,T;L^2(\Omega))} + c(D),$$

where the constant $c(D)$ only depends on the data u_B, f_F and $\boldsymbol{q}_r$. Due to the boundedness of b' and $k \circ b$ (see (2.19) for the definition of $F(u^*)$), we have

$$\big\| \mathcal{T}\big(F_B + F(u^*), D\big) \big\|_{\mathbb{X}_0} \le \frac{1+\varepsilon}{\alpha} \| b' \|_{L^\infty(\mathbb{R})} \| u^* \|_{\mathbb{X}_0} + c'(D).$$

Thus, due to (2.22), the application: $u^* \mapsto \mathcal{T}(F_B + F(u^*), D)$ maps the ball in $\mathbb{X}_0$ with radius R into itself for all R, such that, for an appropriate ε,

$$\left(1 - \frac{1+\varepsilon}{\alpha} \|b'\|_{L^\infty(\mathbb{R})}\right) R > c'(D). \tag{2.24}$$

(2) Since $\mathbb{X}_0$ is separable, there exists an increasing sequence of finite-dimensional spaces $\mathbb{X}_n$, which is dense in $\mathbb{X}_0$. If Π_n denotes the orthogonal projection operator (for the scalar product associated with the norm of $\mathbb{X}_0$) onto $\mathbb{X}_n$, the mapping: $u \mapsto \Pi_n \mathcal{T}(F_B + F(u), D)$ is continuous from $\mathbb{X}_n$ into itself. The same arguments as previously yield that it maps the ball of $\mathbb{X}_n$ with radius R into itself for all R satisfying (2.24). Thus, applying the Brouwer's fixed point theorem (see [9, Chap. IV, Theorem 1.1] for instance), implies that this mapping admits a fixed point in this same ball, namely, there exists a u_n in $\mathbb{X}_n$ satisfying the equation $u_n = \Pi_n \mathcal{T}(F_B + F(u_n), D)$. Moreover, it follows from Lemma 2.3 that this sequence is also bounded in $L^2(0, T; H^1(\Omega))$.

(3) The function u_n thus satisfies,

$$\begin{aligned}
\forall v \in \mathbb{X}_n, \quad & \alpha \int_\Omega (\partial_t u_n)(\boldsymbol{x}, t)(v - u_n)(\boldsymbol{x}, t) \mathrm{d}\boldsymbol{x} \\
& + \int_\Omega \big(\partial_t b_*(u_n)\big)(\boldsymbol{x}, t)(v - u_n)(\boldsymbol{x}, t) \mathrm{d}\boldsymbol{x} \\
& + \int_\Omega \big(\nabla u_n + k \circ b_*(u_n) \boldsymbol{e}_z\big)(\boldsymbol{x}, t) \cdot \big(\nabla(v - u_n)\big)(\boldsymbol{x}, t) \mathrm{d}\boldsymbol{x} \\
& \geq - \int_\Omega F_B(\boldsymbol{x}, t)(v - u_n)(\boldsymbol{x}, t) \mathrm{d}\boldsymbol{x} - \int_{\Gamma_F} f_F(\boldsymbol{\tau}, t)(v - u_n)(\boldsymbol{\tau}, t) \mathrm{d}\boldsymbol{\tau} \\
& - \int_{\Gamma_G} (\boldsymbol{q}_r \cdot \boldsymbol{n})(\boldsymbol{\tau}, t)(v - u_n)(\boldsymbol{\tau}, t) \mathrm{d}\boldsymbol{\tau}.
\end{aligned} \tag{2.25}$$

Moreover, due to the boundedness properties of the sequence $(u_n)_n$, there exists a subsequence still denoted by $(u_n)_n$ for simplicity, which converges to a function u^* of $\mathbb{X}_0$ weakly in $\mathbb{X}$ and strongly in $L^2(0, T; L^2(\Omega))$. Next, we observe that, for a fixed v in $\mathbb{X}_n$:

(i) The convergence of all terms in the right-hand side follows from the weak convergence in $L^2(0, T; \mathbb{W})$.

(ii) The convergence of the first term is derived by writing the expansion

$$\begin{aligned}
& \int_\Omega (\partial_t u_n)(\boldsymbol{x}, t)(v - u_n)(\boldsymbol{x}, t) \mathrm{d}\boldsymbol{x} \\
& \quad = \int_\Omega (\partial_t u^*)(\boldsymbol{x}, t)(v - u^*)(\boldsymbol{x}, t) \mathrm{d}\boldsymbol{x}
\end{aligned}$$

$$+\int_{\Omega} \partial_t \big(u_n - u^*\big)(\boldsymbol{x},t)\big(v - u^*\big)(\boldsymbol{x},t)\mathrm{d}\boldsymbol{x}$$

$$+\int_{\Omega} (\partial_t u_n)(\boldsymbol{x},t)\big(u^* - u_n\big)(\boldsymbol{x},t)\mathrm{d}\boldsymbol{x}$$

and by checking that the last two terms converge.

(iii) The convergence of the term $\int_{\Omega}(\nabla u_n)(\boldsymbol{x},t)\cdot(\nabla(v-u_n))(\boldsymbol{x},t)\mathrm{d}\boldsymbol{x}$ is obtained by using the weak lower semi-continuity of the norm $|u_n|_{H^1(\Omega)}$.

Moreover, the convergence of the nonlinear terms follows from the expansions

$$\begin{aligned}
&\int_{\Omega} \big(\partial_t b_*(u_n)\big)(\boldsymbol{x},t)(v - u_n)(\boldsymbol{x},t)\mathrm{d}\boldsymbol{x} \\
&\quad= \int_{\Omega} \big(\partial_t b_*\big(u^*\big)\big)(\boldsymbol{x},t)\big(v - u^*\big)(\boldsymbol{x},t)\mathrm{d}\boldsymbol{x} \\
&\qquad+ \int_{\Omega} \big(\partial_t b_*(u_n) - \partial_t b_*\big(u^*\big)\big)(\boldsymbol{x},t)\big(v - u^*\big)(\boldsymbol{x},t)\mathrm{d}\boldsymbol{x} \\
&\qquad+ \int_{\Omega} \big(\partial_t b_*(u_n)\big)(\boldsymbol{x},t)\big(u^* - u_n\big)(\boldsymbol{x},t)\mathrm{d}\boldsymbol{x}
\end{aligned}$$

and

$$\begin{aligned}
&\int_{\Omega} k \circ b_*(u_n)(\boldsymbol{x},t)\boldsymbol{e}_z \cdot \big(\nabla(v - u_n)\big)(\boldsymbol{x},t)\mathrm{d}\boldsymbol{x} \\
&\quad= \int_{\Omega} k \circ b_*\big(u^*\big)(\boldsymbol{x},t)\boldsymbol{e}_z \cdot \big(\nabla\big(v - u^*\big)\big)(\boldsymbol{x},t)\mathrm{d}\boldsymbol{x} \\
&\qquad+ \int_{\Omega} k \circ b_*\big(u^*\big)(\boldsymbol{x},t)\boldsymbol{e}_z \cdot \big(\nabla\big(u^* - u_n\big)\big)(\boldsymbol{x},t)\mathrm{d}\boldsymbol{x} \\
&\qquad+ \int_{\Omega} \big(k \circ b_*(u_n) - k \circ b_*\big(u^*\big)\big)(\boldsymbol{x},t)\boldsymbol{e}_z \cdot \big(\nabla(v - u_n)\big)(\boldsymbol{x},t)\mathrm{d}\boldsymbol{x},
\end{aligned}$$

combined with the Lipschitz-continuity of b' and $k \circ b$. Finally, using the density of the sequence $(\mathbb{X}_n)_n$ in $\mathbb{X}_0$, u^* is a solution to the problem (2.13)–(2.14). Thus, u is a solution to the problem (2.3)–(2.4).

Condition (2.22) is rather restrictive, since, in practical situations, α is small. However, this condition can be relaxed when b satisfies, for a positive constant b_0,

$$b'(\xi) \geq b_0 \quad \forall \xi \in \mathbb{R}. \tag{2.26}$$

Indeed, all the previous arguments are still valid when we replace α by $\alpha + b_0$ and replace the coefficient $b(\xi)$ by $b(\xi) - b_0\xi$. □

Corollary 2.1 *Assume that b satisfies* (2.26), *and that the coefficient α satisfies*

$$\frac{1}{\alpha + b_0} \|b' - b_0\|_{L^\infty(\mathbb{R})} < 1. \tag{2.27}$$

For any data u_B, f_F, $\boldsymbol{q}_r$ and u_0 satisfying (2.10) *and* (2.23), *the problem* (2.3)–(2.4) *has at least a solution in $\mathbb{X}$.*

Assume that b satisfies

$$\min_{\xi\in\mathbb{R}} b'(\xi) > 0, \qquad \max_{\xi\in\mathbb{R}} b'(\xi) < 2 \min_{\xi\in\mathbb{R}} b'(\xi). \tag{2.28}$$

Under this condition, the problem (2.3)–(2.4) has a solution even for $\alpha = 0$. We refer to [2] for another proof of this result of a similar problem.

3 The Discrete Problems

We present first the time semi-discrete problem constructed from the backward Euler's scheme. Next, we consider a finite element discretization of this problem relying on standard, conforming, finite element spaces.

3.1 A Time Semi-Discrete Problem

Since we intend to work with nonuniform time steps, we introduce a partition of the interval $[0, T]$ into subintervals $[t_{n-1}, t_n]$ $(1 \le n \le N)$, such that $0 = t_0 < t_1 < \cdots < t_N = T$. We denote by τ_n the time step $t_n - t_{n-1}$, by τ the N-tuple $(\tau_1, \ldots, \tau_N)$ and by $|\tau|$ the maximum of the τ_n $(1 \le n \le N)$.

As already hinted in Sect. 1, the time discretization mainly relies on a backward Euler's scheme, where the nonlinear term $k \circ b(u)$ is treated in an explicit way for simplicity. Thus, the semi-discrete problem reads as follows.

Find $(u^n)_{0\le n\le N}$ in $\prod_{n=0}^{N} \mathbb{V}(t_n)$, such that

$$u^0 = u_0 \quad \text{in } \Omega, \tag{3.1}$$

and for $1 \le n \le N$,

$$\forall v \in \mathbb{V}(t_n), \quad \alpha \int_\Omega \Big(\frac{u^n - u^{n-1}}{\tau_n}\Big)(\boldsymbol{x})\big(v - u^n\big)(\boldsymbol{x})\mathrm{d}\boldsymbol{x}$$

$$+ \int_\Omega \Big(\frac{b(u^n) - b(u^{n-1})}{\tau_n}\Big)(\boldsymbol{x})\big(v - u^n\big)(\boldsymbol{x})\mathrm{d}\boldsymbol{x}$$

$$+\int_{\Omega}\big(\nabla u^n + k \circ b\big(u^{n-1}\big)\big)(\boldsymbol{x})\boldsymbol{e}_z \cdot \nabla\big(v-u^n\big)\boldsymbol{x}\mathrm{d}\boldsymbol{x}$$

$$\geq -\int_{\Gamma_F} f_F(\boldsymbol{\tau}, t_n)\big(v-u^n\big)(\boldsymbol{\tau})\mathrm{d}\boldsymbol{\tau}$$

$$-\int_{\Gamma_G}(\boldsymbol{q}_r \cdot \boldsymbol{n})(\boldsymbol{\tau}, t_n)\big(v-u^n\big)(\boldsymbol{\tau})\mathrm{d}\boldsymbol{\tau}. \tag{3.2}$$

It can be noted that this problem makes sense when both f_F and $\boldsymbol{q}_r$ are continuous in time. Proving its well-posedness relies on rather different arguments as previously.

Theorem 3.1 *For any data u_B, f_F, $\boldsymbol{q}_r$ and u_0 satisfying*

$$\begin{aligned} &u_B \in H^1\big(0,T; H^{\frac{1}{2}}_{00}(\Gamma_B)\big), \quad f_F \in \mathscr{C}^0\big(0,T; L^2(\Gamma_F)\big),\\ &\boldsymbol{q}_r \in \mathscr{C}^0\big(0,T; L^2(\Gamma_G)^d\big), \quad u_0 \in H^1(\Omega), \end{aligned} \tag{3.3}$$

and (2.10), *for any nonnegative coefficient α, the problem* (3.1)–(3.2) *has a unique solution in $\prod_{n=0}^{N} \mathbb{V}(t_n)$.*

Proof We proceed by induction on n. Since u^0 is given by (3.1), we assume that u^{n-1} is known. We consider problem (3.2) for a fixed n, called $(3.2)_n$, that can equivalently be written as

$$\forall v \in \mathbb{V}(t_n), \quad \int_{\Omega}\big(\alpha u^n + b\big(u^n\big)\big)(\boldsymbol{x})\big(v-u^n\big)(\boldsymbol{x})\mathrm{d}\boldsymbol{x}$$

$$+\tau_n \int_{\Omega} \nabla u^n(\boldsymbol{x}) \cdot \nabla\big(v-u^n\big)(\boldsymbol{x})\mathrm{d}\boldsymbol{x}$$

$$\geq \int_{\Omega}\big(\alpha u^{n-1} + b\big(u^{n-1}\big)\big)(\boldsymbol{x})\big(v-u^n\big)(\boldsymbol{x})\mathrm{d}\boldsymbol{x}$$

$$-\tau_n \int_{\Omega} k \circ b\big(u^{n-1}\big)(\boldsymbol{x})\boldsymbol{e}_z \cdot \nabla\big(v-u^n\big)(\boldsymbol{x})\mathrm{d}\boldsymbol{x}$$

$$-\tau_n \int_{\Gamma_F} f_F(\boldsymbol{\tau}, t_n)\big(v-u^n\big)(\boldsymbol{\tau})\mathrm{d}\boldsymbol{\tau}$$

$$-\tau_n \int_{\Gamma_G}(\boldsymbol{q}_r \cdot \boldsymbol{n})(\boldsymbol{\tau}, t_n)\big(v-u^n\big)(\boldsymbol{\tau})\mathrm{d}\boldsymbol{\tau}.$$

Let us now set

$$\varphi(z) = \int_0^z \big(\alpha\zeta + b(\zeta)\big)\mathrm{d}\zeta, \qquad \Phi(v) = \int_{\Omega} \varphi\big(v(\boldsymbol{x})\big)\mathrm{d}\boldsymbol{x}.$$

It is readily checked that, since b' is nonnegative, both φ and Φ are convex, and moreover, that

$$D\Phi(u)\cdot\big(v-u^n\big)=\int_\Omega\big(\alpha u+b(u)\big)(\boldsymbol{x})\big(v-u^n\big)(\boldsymbol{x})\mathrm{d}\boldsymbol{x}.$$

Thus, taking

$$a(u,v)=\int_\Omega \nabla u(\boldsymbol{x})\cdot\nabla v(\boldsymbol{x})\mathrm{d}\boldsymbol{x},$$

$$\ell(v)=\int_\Omega\big(\alpha u^{n-1}+b\big(u^{n-1}\big)\big)(\boldsymbol{x})v(\boldsymbol{x})\mathrm{d}\boldsymbol{x}-\tau_n\int_\Omega k\circ b\big(u^{n-1}\big)(\boldsymbol{x})\boldsymbol{e}_z\cdot\nabla v(\boldsymbol{x})\mathrm{d}\boldsymbol{x}$$

$$-\tau_n\int_{\Gamma_F} f_F(\boldsymbol{\tau},t_n)v(\boldsymbol{\tau})\mathrm{d}\boldsymbol{\tau}-\tau_n\int_{\Gamma_G}(\boldsymbol{q}_r\cdot\boldsymbol{n})(\boldsymbol{\tau},t_n)v(\boldsymbol{\tau})\mathrm{d}\boldsymbol{\tau},$$

the problem $(3.2)_n$ can also be written as

$$D\Phi\big(u^n\big)\cdot\big(v-u^n\big)+a\big(u^n,v-u^n\big)-\ell\big(v-u^n\big)\geq 0,\quad \forall v\in\mathbb{V}(t_n).$$

We now set $\Psi(v)=\Phi(v)+J(v)$ with $J(v)=\frac{1}{2}a(v,v)-\ell(v)$. The problem $(3.2)_n$ can finally be written as

$$\forall v\in\mathbb{V}(t_n),\quad D\Psi\big(u^n\big)\cdot\big(v-u^n\big)\geq 0,$$

or

$$\forall v\in\mathbb{V}(t_n),\quad \Psi\big(u^n\big)\leq\Psi(v).$$

So it is equivalent to the minimization of a convex functional on the convex set $\mathbb{V}(t_n)$. Hence it admits a unique solution. This completes the proof. □

It can be noted that, in contrast with the continuous problem, the existence of a solution to the semi-discrete problem (3.1)–(3.2) does not require any limitation on α.

3.2 A Fully Discrete Problem

From now on, we assume that Ω is a polygon ($d=2$) or a polyhedron ($d=3$). Let $(\mathcal{T}_h)_h$ be a regular family of triangulations of Ω (by triangles or tetrahedra), in the sense that, for each h,

(i) $\overline{\Omega}$ is the union of all elements of $\mathcal{T}_h$.
(ii) The intersection of two different elements of $\mathcal{T}_h$, if not empty, is a vertex or a whole edge or a whole face of both of them.

(iii) The ratio of the diameter h_K of any element K of $\mathcal{T}_h$ to the diameter of its inscribed circle or sphere is smaller than a constant σ independent of h.

As usual, h stands for the maximum of the diameters h_K $(K \in \mathcal{T}_h)$. We make the further and nonrestrictive assumption that $\overline{\Gamma}_B$, $\overline{\Gamma}_F$ and $\overline{\Gamma}_G$ are the union of whole edges ($d = 2$) or whole faces ($d = 3$) of elements of $\mathcal{T}_h$. From now on, $c, c', \dots$ stand for generic constants that may vary from line to line and are always independent of τ and h.

We now introduce the finite element space

$$\overline{\mathbb{V}}_h = \{v_h \in H^1(\Omega);\ \forall K \in \mathcal{T}_h, v_h|_K \in \mathcal{P}_1(K)\}, \tag{3.4}$$

where $\mathcal{P}_1(K)$ is the space of restrictions to K of affine functions on $\mathbb{R}^d$. Let $\mathcal{I}_h$ denote the Lagrange interpolation operator at all the vertices of elements of $\mathcal{T}_h$ with values in $\overline{\mathbb{V}}_h$, and i_h^B denote the corresponding interpolation operator on Γ_B. Assuming that u_B is continuous where needed, we then define for each n $(0 \le n \le N)$, the subset of $\overline{\mathbb{V}}_h$,

$$\mathbb{V}_h(t_n) = \{v_h \in \overline{\mathbb{V}}_h;\ v_h|_{\Gamma_B} = i_h^B u_B(\cdot, t_n) \text{ and } v_h|_{\Gamma_G} \le 0\}. \tag{3.5}$$

We are thus in a position to write the discrete problem constructed from the problem (3.1)–(3.2) by the Galerkin method.

Find $(u_h^n)_{0 \le n \le N}$ in $\prod_{n=0}^N \mathbb{V}_h(t_n)$, such that

$$u_h^0 = \mathcal{I}_h u_0 \quad \text{in } \Omega, \tag{3.6}$$

and, for $1 \le n \le N$,

$$\begin{aligned}
\forall v_h \in \mathbb{V}_h(t_n),\quad & \alpha \int_\Omega \left(\frac{u_h^n - u_h^{n-1}}{\tau_n}\right)(\boldsymbol{x})\big(v_h - u_h^n\big)(\boldsymbol{x})\mathrm{d}\boldsymbol{x} \\
&+ \int_\Omega \left(\frac{b(u_h^n) - b(u_h^{n-1})}{\tau_n}\right)(\boldsymbol{x})\big(v_h - u_h^n\big)(\boldsymbol{x})\mathrm{d}\boldsymbol{x} \\
&+ \int_\Omega \big(\nabla u_h^n + k \circ b\big(u_h^{n-1}\big)\big)(\boldsymbol{x})\boldsymbol{e}_z \cdot \nabla\big(v_h - u_h^n\big)(\boldsymbol{x})\mathrm{d}\boldsymbol{x} \\
\ge & - \int_{\Gamma_F} f_F(\boldsymbol{\tau}, t_n)\big(v_h - u_h^n\big)(\boldsymbol{\tau})\mathrm{d}\boldsymbol{\tau} \\
& - \int_{\Gamma_G} (\boldsymbol{q}_r \cdot \boldsymbol{n})(\boldsymbol{\tau}, t_n)\big(v_h - u_h^n\big)(\boldsymbol{\tau})\mathrm{d}\boldsymbol{\tau}.
\end{aligned} \tag{3.7}$$

The proof of the next theorem is exactly the same as the proof of Theorem 3.1, so we omit it.

Theorem 3.2 *For any data u_B, f_F, $\boldsymbol{q}_r$ and u_0 satisfying* (2.10), (3.3) *and*

$$u_B \in \mathscr{C}^0(\overline{\Gamma}_B \times [0,T]), \quad u_0 \in \mathscr{C}^0(\overline{\Omega}) \tag{3.8}$$

for any nonnegative coefficient α, the problem (3.6)–(3.7) *has a unique solution.*

Here also the existence result is unconditional.

4 A Convergence Result

The aim of this section is to prove a convergence result for the solutions $(u_h^n)_{0\le n\le N}$ to the problem (3.6)–(3.7), when $|\tau|$ and h tend to zero. In order to do that, as in Sect. 2, we use the lifting u_B^* of u_B which satisfies (2.11)–(2.12), and assume moreover that it is continuous on $\overline{\Omega} \times [0,T]$. Indeed, if $(u_h^n)_{0\le n\le N}$ is a solution to (3.6)–(3.7), and the family $(u_h^{*n})_{0\le n\le N}$ with $u_h^{*n} = u_h^n - \mathcal{I}_h u_B^*(t_n)$ is a solution to the following problem:

Find $(u_h^{*n})_{0\le n\le N}$ in $\mathbb{V}_{h0}^{N+1}$, such that

$$u_h^{*0} = 0 \quad \text{in } \Omega, \tag{4.1}$$

and for $1 \le n \le N$,

$$\begin{aligned}
\forall v_h \in \mathbb{V}_{h0}, \quad & \alpha \int_\Omega \left(\frac{u_h^{*n} - u_h^{*n-1}}{\tau_n}\right)(\boldsymbol{x})\big(v_h - u_h^{*n}\big)(\boldsymbol{x})\mathrm{d}\boldsymbol{x} \\
& + \int_\Omega \left(\frac{b_{*n}(u_h^{*n}) - b_{*n-1}(u_h^{*n-1})}{\tau_n}\right)(\boldsymbol{x})\big(v_h - u_h^{*n}\big)(\boldsymbol{x})\mathrm{d}\boldsymbol{x} \\
& + \int_\Omega \big(\nabla u_h^{*n} + k \circ b_{*n-1}(u_h^{*n-1})\big)(\boldsymbol{x})\boldsymbol{e}_z \cdot \nabla\big(v_h - u_h^{*n}\big)(\boldsymbol{x})\mathrm{d}\boldsymbol{x} \\
& \ge -\int_\Omega F_{Bh}(\boldsymbol{x}, t_n)\big(v_h - u_h^{*n}\big)\mathrm{d}\boldsymbol{x} - \int_{\Gamma_F} f_F(\boldsymbol{\tau}, t_n)\big(v_h - u_h^{*n}\big)(\boldsymbol{\tau})\mathrm{d}\boldsymbol{\tau} \\
& - \int_{\Gamma_G} (\boldsymbol{q}_r \cdot \boldsymbol{n})(\boldsymbol{\tau}, t_n)\big(v_h - u_h^{*n}\big)(\boldsymbol{\tau})\mathrm{d}\boldsymbol{\tau},
\end{aligned} \tag{4.2}$$

where the convex set $\mathbb{V}_{h0}$ and the function F_{Bh} are defined, in analogy with (2.15)–(2.16), by

$$\mathbb{V}_{h0} = \overline{\mathbb{V}}_h \cap \mathbb{V}_0 \tag{4.3}$$

and

$$\begin{aligned}
& \int_\Omega F_{Bh}(\boldsymbol{x}, t)v(\boldsymbol{x})\mathrm{d}\boldsymbol{x} \\
& = \alpha \int_\Omega \big(\partial_t \mathcal{I}_h u_B^*\big)(\boldsymbol{x}, t)v(\boldsymbol{x})\mathrm{d}\boldsymbol{x} + \int_\Omega \big(\nabla \mathcal{I}_h u_B^*\big)(\boldsymbol{x}, t) \cdot (\nabla v)(\boldsymbol{x})\mathrm{d}\boldsymbol{x},
\end{aligned} \tag{4.4}$$

while each function b_{*n} is given by $b_{*n}(\xi) = b(\xi + \mathcal{I}_h u_B^*(\cdot, t_n))$. We now investigate the boundedness of the sequence $(u_h^{*n})_{0\le n\le N}$ in appropriate norms. We need a preliminary lemma for that.

Lemma 4.1 *For each part Γ of $\partial\Omega$, which is the union of whole edges ($d = 2$) or whole faces ($d = 3$) of elements of $\mathcal{T}_h$, the following inequality holds for all functions w_h in $\overline{\mathbb{V}}_h$:*

$$\|w_h\|_{H^{-\frac{1}{2}}(\Gamma)} \le c\|w_h\|_{L^2(\Omega)}. \tag{4.5}$$

Proof It relies on standard arguments. We have

$$\|w_h\|_{H^{-\frac{1}{2}}(\Gamma)} = \sup_{z\in H^{\frac{1}{2}}(\Gamma)} \frac{\int_\Gamma z(\boldsymbol{\tau})w_h(\boldsymbol{\tau})\mathrm{d}\boldsymbol{\tau}}{\|z\|_{H^{\frac{1}{2}}(\Gamma)}}.$$

Let e be any edge or face of an element K of $\mathcal{T}_h$ which is contained in Γ. Denoting by $\widehat{K}$ the reference triangle or tetrahedron, we have, with obvious notation for $\widehat{e}$, $\widehat{w}$, $\widehat{z}$,

$$\int_e z(\boldsymbol{\tau})w_h(\boldsymbol{\tau})\mathrm{d}\boldsymbol{\tau} \le c h_e^{d-1}\int_{\widehat{e}} \widehat{z}(\widehat{\boldsymbol{\tau}})\widehat{w}_h(\widehat{\boldsymbol{\tau}})\mathrm{d}\widehat{\boldsymbol{\tau}} \le c' h_K^{d-1}\|\widehat{z}\|_{L^2(\widehat{e})}\|\widehat{w}_h\|_{L^2(\widehat{e})}.$$

By using the equivalence of norms on $\mathcal{P}_1(\widehat{K})$ and an appropriate stable lifting operator $\widehat{\pi}$ which maps traces on $\widehat{e}$ into functions of K vanishing at the vertex of K which does not belong to $\overline{\Gamma}$, we derive

$$\begin{aligned}\int_e z(\boldsymbol{\tau})w_h(\boldsymbol{\tau})\mathrm{d}\boldsymbol{\tau} &\le c' h_K^{d-1}|\widehat{\pi}\widehat{z}|_{H^1(\widehat{K})}\|\widehat{w}_h\|_{L^2(\widehat{K})}\\ &\le c' h_K^{d-1} h_K^{1-\frac{d}{2}}|\pi z|_{H^1(K)} h_K^{-\frac{d}{2}}\|w_h\|_{L^2(K)},\end{aligned}$$

there also with an obvious definition of π. We conclude by summing this last inequality on e and by using a Cauchy-Schwarz inequality and the stability of $\widehat{\pi}$,

$$\int_\Gamma z(\boldsymbol{\tau})w_h(\boldsymbol{\tau})\mathrm{d}\boldsymbol{\tau} \le c\|z\|_{H^{\frac{1}{2}}(\Gamma)}\|w_h\|_{L^2(\Omega)},$$

whence the desired result follows. □

Lemma 4.2 *For any data u_B, f_F, $\boldsymbol{q}_r$ and u_0 satisfying*

$$\begin{aligned}&u_B \in H^1\big(0,T; H_{00}^{\frac{1}{2}}(\Gamma_B)\big), \quad f_F \in \mathscr{C}^0\big(0,T; H^{\frac{1}{2}}(\Gamma_F)\big),\\ &\boldsymbol{q}_r \in \mathscr{C}^0\big(0,T; H^{\frac{1}{2}}(\Gamma_G)^d\big), \quad u_0 \in H^1(\Omega)\end{aligned} \tag{4.6}$$

and (2.10), *the sequence* $(u_h^{*n})_{0\le n\le N}$ *satisfies the following inequality, for* $1 \le n \le N$,

$$\alpha \sum_{m=1}^{n} \tau_m \| \frac{u_h^{*m} - u_h^{*m-1}}{\tau_m} \|_{L^2(\Omega)}^2 + |u_h^{*n}|_{H^1(\Omega)}^2$$

$$\le c\big(1 + \|\mathcal{I}_h u_B^*\|_{H^1(0,T;H^1(\Omega))}^2 + \|f_F\|^2_{\mathscr{C}^0(0,T;H^{\frac{1}{2}}(\Gamma_F))} + \|\boldsymbol{q}_r\|^2_{\mathscr{C}^0(0,T;H^{\frac{1}{2}}(\Gamma_G)^d)}\big). \tag{4.7}$$

Proof Taking v equal to u_h^{*n-1} in (4.2), leads to

$$\alpha\tau_n \left\| \frac{u_h^{*n} - u_h^{*n-1}}{\tau_n} \right\|_{L^2(\Omega)}^2 + \int_\Omega \nabla u_h^{*n}(\boldsymbol{x}) \cdot \nabla\big(u_h^{*n} - u_h^{*n-1}\big)(\boldsymbol{x})\mathrm{d}\boldsymbol{x}$$

$$\le -\int_\Omega \left(\frac{b_{*n}(u_h^{*n}) - b_{*n-1}(u_h^{*n-1})}{\tau_n} \right)(\boldsymbol{x})\big(u_h^{*n} - u_h^{*n-1}\big)(\boldsymbol{x})\mathrm{d}\boldsymbol{x}$$

$$- \int_\Omega k \circ b_{*n-1}\big(u_h^{*n-1}\big)(\boldsymbol{x})\boldsymbol{e}_z \cdot \nabla\big(u_h^{*n} - u_h^{*n-1}\big)(\boldsymbol{x})\mathrm{d}\boldsymbol{x} + \big\langle \mathcal{G}, u_h^{*n} - u_h^{*n-1}\big\rangle,$$

where the data depending quantity $\mathcal{G}$ is defined by

$$\langle \mathcal{G}, v\rangle = -\int_\Omega F_{Bh}(\boldsymbol{x}, t_n)v(\boldsymbol{x})\mathrm{d}\boldsymbol{x} - \int_{\Gamma_F} f_F(\boldsymbol{\tau}, t_n)v(\boldsymbol{\tau})\mathrm{d}\boldsymbol{\tau} - \int_{\Gamma_G} (\boldsymbol{q}_r \cdot \boldsymbol{n})(\boldsymbol{\tau}, t_n)v(\boldsymbol{\tau})\mathrm{d}\boldsymbol{\tau}.$$

To handle the second term, we use the identity

$$\int_\Omega \nabla u_h^{*n} \cdot \nabla\big(u_h^{*n} - u_h^{*n-1}\big)(\boldsymbol{x})\mathrm{d}\boldsymbol{x}$$

$$= \frac{1}{2}\big(|u_h^{*n}|_{H^1(\Omega)}^2 + |u_h^{*n} - u_h^{*n-1}|_{H^1(\Omega)}^2 - |u_h^{*n-1}|_{H^1(\Omega)}^2\big).$$

To handle the third term, we write the expansion

$$\int_\Omega \left(\frac{b_{*n}(u_h^{*n}) - b_{*n-1}(u_h^{*n-1})}{\tau_n} \right)(\boldsymbol{x})\big(u_h^{*n} - u_h^{*n-1}\big)(\boldsymbol{x})\mathrm{d}\boldsymbol{x}$$

$$= \int_\Omega \left(\frac{b(u_h^{*n} + \mathcal{I}_h u_B^*(t_n)) - b(u_h^{*n-1} + \mathcal{I}_h u_B^*(t_n))}{\tau_n} \right)(\boldsymbol{x})\big(u_h^{*n} - u_h^{*n-1}\big)(\boldsymbol{x})\mathrm{d}\boldsymbol{x}$$

$$+ \int_\Omega \left(\frac{b(u_h^{*n-1} + \mathcal{I}_h u_B^*(t_n)) - b(u_h^{*n-1} + \mathcal{I}_h u_B^*(t_{n-1}))}{\tau_n} \right)(\boldsymbol{x})\big(u_h^{*n} - u_h^{*n-1}\big)$$

$$\times (\boldsymbol{x})\mathrm{d}\boldsymbol{x}.$$

By using the nonnegativity of b', together with the Lipschitz-continuity of b, we derive

$$\int_\Omega \Big(\frac{b_{*n}(u_h^{*n}) - b_{*n-1}(u_h^{*n-1})}{\tau_n}\Big)(\boldsymbol{x})\big(u_h^{*n} - u_h^{*n-1}\big)(\boldsymbol{x})\mathrm{d}\boldsymbol{x}$$

$$\leq \frac{\alpha}{4}\tau_n \Big\| \frac{u_h^{*n} - u_h^{*n-1}}{\tau_n} \Big\|^2_{L^2(\Omega)} + \frac{1}{\alpha}\tau_n \Big\| \frac{\mathcal{I}_h u_B^*(t_n) - \mathcal{I}_h u_B^*(t_{n-1})}{\tau_n} \Big\|^2_{L^2(\Omega)}.$$

Finally, evaluating the last term is an easy consequence of Lemma 4.1,

$$\big\langle \mathcal{G}, u_h^{*n} - u_h^{*n-1} \big\rangle \leq \frac{\alpha}{4}\tau_n \Big\| \frac{u_h^{*n} - u_h^{*n-1}}{\tau_n} \Big\|^2_{L^2(\Omega)}$$

$$+ c\tau_n \big(\|F_{Bh}(\cdot, t_n)\|^2_{L^2(\Omega)} + \|f_F(\cdot, t_n)\|^2_{H^{\frac{1}{2}}(\Gamma_F)}$$

$$+ \|\boldsymbol{q}_r(\cdot, t_n)\|^2_{H^{\frac{1}{2}}(\Gamma_G)^d} \big).$$

By combining, we obtain

$$\frac{\alpha}{2}\tau_n \Big\| \frac{u_h^{*n} - u_h^{*n-1}}{\tau_n} \Big\|^2_{L^2(\Omega)} + \frac{1}{2}|u_h^{*n}|^2_{H^1(\Omega)}$$

$$\leq \frac{1}{2}|u_h^{*n-1}|^2_{H^1(\Omega)}$$

$$+ c'\tau_n \big(\|F_{Bh}(\cdot, t_n)\|^2_{L^2(\Omega)} + \|f_F(\cdot, t_n)\|^2_{H^{\frac{1}{2}}(\Gamma_F)} + \|\boldsymbol{q}_r(\cdot, t_n)\|^2_{H^{\frac{1}{2}}(\Gamma_G)^d} \big)$$

$$- \int_\Omega k \circ b_{*n-1}\big(u_h^{*n-1}\big)(\boldsymbol{x})\boldsymbol{e}_z \cdot \nabla\big(u_h^{*n} - u_h^{*n-1}\big)(\boldsymbol{x})\mathrm{d}\boldsymbol{x}.$$

We sum up this inequality on n. To handle the last term, we observe that

$$- \sum_{m=1}^{n} \int_\Omega k \circ b_{*m-1}\big(u_h^{*m-1}\big)(\boldsymbol{x})\boldsymbol{e}_z \cdot \nabla\big(u_h^{*m} - u_h^{*m-1}\big)(\boldsymbol{x})\mathrm{d}\boldsymbol{x}$$

$$= - \int_\Omega k \circ b_{*n-1}\big(u_h^{*n-1}\big)(\boldsymbol{x})\boldsymbol{e}_z \cdot \nabla u_h^{*n}(\boldsymbol{x})\mathrm{d}\boldsymbol{x}$$

$$+ \sum_{m=1}^{n-1} \int_\Omega \big(k \circ b_{*m}\big(u_h^{*m}\big) - k \circ b_{*m-1}\big(u_h^{*m-1}\big)\big)(\boldsymbol{x})\boldsymbol{e}_z \cdot \nabla u_h^{*m}(\boldsymbol{x})\mathrm{d}\boldsymbol{x}.$$

Hence, thanks to the boundedness of k and the Lipschitz continuity of $k \circ b$, we derive

$$
\begin{aligned}
&-\sum_{m=1}^{n} \int_{\Omega} k \circ b_{*m-1}\big(u_h^{*m-1}\big)(\boldsymbol{x})\boldsymbol{e}_z \cdot \nabla\big(u_h^{*m} - u_h^{*m-1}\big)(\boldsymbol{x})\mathrm{d}\boldsymbol{x} \\
&\quad \leq c + \frac{1}{4}|u_h^{*n}|^2_{H^1(\Omega)} + \frac{\alpha}{4}\sum_{m=1}^{n-1} \tau_m \left\| \frac{u_h^{*m} - u_h^{*m-1}}{\tau_m} \right\|^2_{L^2(\Omega)} \\
&\quad + c' \sum_{m=1}^{n-1} \tau_m \left\| \frac{\mathcal{I}_h u_B^*(t_m) - \mathcal{I}_h u_B^*(t_m - 1)}{\tau_m} \right\|^2_{L^2(\Omega)} + c'' \sum_{m=1}^{n-1} \tau_m |u^{*m}|^2_{H^1(\Omega)}.
\end{aligned}
$$

We conclude by using the discrete Grönwall's lemma (see [8, Chap. V, Lemma 2.4]). □

Let us now introduce the function $u_{h\tau}^*$, which is affine on each interval $[t_{n-1}, t_n]$ $(1 \leq n \leq N)$, and equal to u_h^{*n} at time t_n $(0 \leq n \leq N)$. When the data u_B, f_F, $\boldsymbol{q}_r$ and u_0 satisfy

$$
\begin{aligned}
&u_B \in H^1\big(0, T; H^s(\Gamma_B)\big), \quad f_F \in \mathscr{C}^0\big(0, T; H^{\frac{1}{2}}(\Gamma_F)\big), \\
&\boldsymbol{q}_r \in \mathscr{C}^0\big(0, T; H^{\frac{1}{2}}(\Gamma_G)^d\big), \quad u_0 \in H^{s+\frac{1}{2}}(\Omega),
\end{aligned}
\tag{4.8}
$$

for some $s > \frac{d-1}{2}$ (in order to ensure the stability of the operator $\mathcal{I}_h$), it follows from Lemma 4.2 that this function belongs to the set $\mathbb{X}_0 = L^2(0, T; \mathbb{V}_0) \cap H^1(0, T; L^2(\Omega))$ (see (2.5) and (2.14)). More precisely, it satisfies

$$
\|u_{h\tau}^*\|_{L^2(0,T;H^1(\Omega))\cap H^1(0,T;L^2(\Omega))} \leq c(u_B, f_F, \boldsymbol{q}_r), \tag{4.9}
$$

where the constant $c(u_B, f_F, \boldsymbol{q}_r)$ only depends on the data. Thus, we are in a position to derive the next result.

Theorem 4.1 *For any data u_B, f_F, $\boldsymbol{q}_r$ and u_0 satisfying* (4.8) *and* (2.10), *and for any positive coefficient α, the problem* (2.3)–(2.4) *has at least a solution in $\mathbb{X}$.*

Proof Thanks to (4.9), the family of functions $u_{h\tau}^*$ is bounded in $\mathbb{X}_0$ independently of h and τ. Thus, there exist a sequence $(\mathcal{T}_{hk})_k$ of triangulations $\mathcal{T}_h$ and a sequence $(\tau_k)_k$ of parameters τ, such that the sequence $(u_k^*)_k$ converges to a function u^* of $\mathbb{X}_0$ weakly in $L^2(0, T; H^1(\Omega)) \cap H^1(0, T; L^2(\Omega))$ and strongly in $L^2(0, T; L^2(\Omega))$. We now intend to prove that u^* is a solution to the problem (2.13)–(2.14). Since it obviously satisfies (2.13), we now investigate the convergence of all terms in (4.2). For clarity, we keep the notation u_h^{*n} for $u_k^*(t_n)$.

(1) The convergence of the first term follows from the expansion

$$\begin{aligned}
\alpha \int_\Omega \Big(\frac{u_h^{*n} - u_h^{*n-1}}{\tau_n}\Big)(\boldsymbol{x})\big(v_h - u_h^{*n}\big)(\boldsymbol{x})\mathrm{d}\boldsymbol{x} \\
= \alpha \int_\Omega \big(\partial_t u^*\big)(\boldsymbol{x}, t_n)\big(v_h - u^*\big)(\boldsymbol{x}, t_n)\mathrm{d}\boldsymbol{x} \\
+ \alpha \int_\Omega \big(\partial_t (u_k^* - u^*)\big)(\boldsymbol{x}, t_n)\big(v_h - u^*\big)(\boldsymbol{x}, t_n)\mathrm{d}\boldsymbol{x} \\
+ \alpha \int_\Omega \big(\partial_t u_k^*\big)(\boldsymbol{x}, t_n)\big(u^* - u_h^{*n}\big)(\boldsymbol{x}, t_n)\mathrm{d}\boldsymbol{x}.
\end{aligned}$$

(2) To prove the convergence of the term

$$\int_\Omega \Big(\frac{b_{*n}(u_h^{*n}) - b_{*n-1}(u_h^{*n-1})}{\tau_n}\Big)(\boldsymbol{x})\big(v_h - u_h^{*n}\big)(\boldsymbol{x})\mathrm{d}\boldsymbol{x},$$

we use a rather complex expansion that we skip for brevity, combined with the dominated convergence theorem of Lebesgue. Indeed, since $(u_k^*)_k$ converges to a function u^* in $L^2(0, T; L^2(\Omega))$, it converges almost everywhere in $\Omega \times [0, T]$, so that $(b'(u_k^*))_k$ also converges a.e. to $b'(u^*)$. Thus, since b' is bounded, $(b'(u_k^*))_k$ also converges to $b'(u^*)$ in $L^2(0, T; L^2(\Omega))$.

(3) The convergence of the term $\int_\Omega \nabla u_h^{*n}(\boldsymbol{x}, t_n)\boldsymbol{e}_z \cdot \nabla(v_h - u_h^{*n})(\boldsymbol{x}, t_n)\mathrm{d}\boldsymbol{x}$ is a consequence of the weak lower semi-continuity of the norm.

(4) The convergence of the term $\int_\Omega k \circ b_{*n-1}(u_h^{*n-1})(\boldsymbol{x})\boldsymbol{e}_z \cdot \nabla(v_h - u_h^{*n})(\boldsymbol{x})\mathrm{d}\boldsymbol{x}$ is easily derived from the expansion

$$\begin{aligned}
\int_\Omega k \circ b_{*n-1}\big(u_h^{*n-1}\big)(\boldsymbol{x})\boldsymbol{e}_z \cdot \nabla\big(v_h - u_h^{*n}\big)(\boldsymbol{x})\mathrm{d}\boldsymbol{x} \\
= \int_\Omega k \circ b_*\big(u^*\big)(\boldsymbol{x}, t_n)\boldsymbol{e}_z \cdot \nabla\big(v_h - u^*\big)(\boldsymbol{x}, t_n)\mathrm{d}\boldsymbol{x} \\
+ \int_\Omega (k \circ b_{*n-1} - k \circ b_*)\big(u^*\big)(\boldsymbol{x}, t_n)\boldsymbol{e}_z \cdot \nabla\big(v_h - u^*\big)(\boldsymbol{x}, t_n)\mathrm{d}\boldsymbol{x} \\
+ \int_\Omega k \circ b_{*n-1}\big(u^*\big)(\boldsymbol{x}, t_n)\boldsymbol{e}_z \cdot \nabla\big(u^* - u_h^{*n}\big)(\boldsymbol{x})\mathrm{d}\boldsymbol{x} \\
+ \int_\Omega \big(k \circ b_{*n-1}\big(u_h^{*n-1}\big) - k \circ b_{*n-1}\big(u^*\big)\big)(\boldsymbol{x})\boldsymbol{e}_z \cdot \nabla\big(v_h - u_h^{*n}\big)(\boldsymbol{x}, t_n)\mathrm{d}\boldsymbol{x},
\end{aligned}$$

and from the dominated convergence theorem of Lebesgue.

(5) The convergence of all terms in the right-hand side of (4.2) is obviously derived from the weak convergence of the sequence $(u_k^*)_k$.

Finally, using the density of the union of the $\mathbb{V}_{h0}$ in $\mathbb{V}_0$, we derive that u^* is a solution to the problem (2.13)–(2.14). Thus, the function $u = u^* + u_B^*$ is a solution to the problem (2.3)–(2.4). □

Even if this requires a slightly different regularity of the data, Theorem 4.1 combined with Proposition 2.2 yields that, for any positive coefficient α, the problem (2.3)–(2.4) is well-posed in $\mathbb{X}$. Of course, this is a great improvement of the results in Sect. 2 and leads to considering that the discretization proposed in Sect. 3 is rather efficient. We shall check this in the second part of this work.

References

1. Alt, H.W., Luckhaus, S.: Quasilinear elliptic-parabolic differential equations. Math. Z. **183**, 311–341 (1983)
2. Alt, H.W., Luckhaus, S., Visintin, A.: On nonstationary flow through porous media. Ann. Mat. Pura Appl. **136**, 303–316 (1984)
3. Bernardi, C., El Alaoui, L., Mghazli, Z.: A posteriori analysis of a space and time discretization of a nonlinear model for the flow in variably saturated porous media. Submitted
4. Berninger, H.: Domain decomposition methods for elliptic problems with jumping nonlinearities and application to the Richards equation. Ph.D. Thesis, Freie Universität, Berlin, Germany (2007)
5. Brezzi, F., Hager, W.W., Raviart, P.A.: Error estimates for the finite element solution to variational inequalities. II. Mixed methods. Numer. Math. **31**, 1–16 (1978/1979)
6. Fabrié, P., Gallouët, T.: Modelling wells in porous media flows. Math. Models Methods Appl. Sci. **10**, 673–709 (2000)
7. Gabbouhy, M.: Analyse mathématique et simulation numérique des phénomènes d'écoulement et de transport en milieux poreux non saturés. Application à la région du Gharb. Ph.D. Thesis, University Ibn Tofail, Kénitra, Morocco (2000)
8. Girault, V., Raviart, P.A.: Finite Element Approximation of the Navier-Stokes Equations. Lecture Notes in Mathematics, vol. 749. Springer, Berlin (1979)
9. Girault, V., Raviart, P.A.: Finite Element Methods for Navier-Stokes Equations, Theory and Algorithms. Springer, Berlin (1986)
10. Glowinski, R., Lions, J.L., Trémolières, R.: Analyse numérique des inéquations variationnelles. 2. Applications aux phénomènes stationnaires et d'évolution, Collection. Méthodes Mathématiques de l'Informatique, vol. 5. Dunod, Paris (1976)
11. Lions, J.L., Magenes, E.: Problèmes aux limites non homogènes et applications, vol. I. Dunod, Paris (1968)
12. Nochetto, R.H., Verdi, C.: Approximation of degenerate parabolic problems using numerical integration. SIAM J. Numer. Anal. **25**, 784–814 (1988)
13. Radu, F., Pop, I.S., Knabner, P.: Order of convergence estimates for an Euler implicit, mixed finite element discretization of Richards' equation. SIAM J. Numer. Anal. **42**, 1452–1478 (2004)
14. Rajagopal, K.R.: On a hierarchy of approximate models for flows of incompressible fluids through porous solid. Math. Models Methods Appl. Sci. **17**, 215–252 (2007)
15. Richards, L.A.: Capillary conduction of liquids through porous mediums. Physics **1**, 318–333 (1931)
16. Schneid, E., Knabner, P., Radu, F.: A priori error estimates for a mixed finite element discretization of the Richards' equation. Numer. Math. **98**, 353–370 (2004)
17. Sochala, P., Ern, A.: Numerical methods for subsurface flows and coupling with runoff. (2013, to appear)

18. Sochala, P., Ern, A., Piperno, S.: Mass conservative BDF-discontinuous Galerkin/explicit finite volume schemes for coupling subsurface and overland flows. Comput. Methods Appl. Mech. Eng. **198**, 2122–2136 (2009)
19. Woodward, C.S., Dawson, C.N.: Analysis of expanded mixed finite element methods for a nonlinear parabolic equation modeling flow into variably saturated porous media. SIAM J. Numer. Anal. **37**, 701–724 (2000)

Finite Volume Multilevel Approximation of the Shallow Water Equations

Arthur Bousquet, Martine Marion, and Roger Temam

Abstract The authors consider a simple transport equation in one-dimensional space and the linearized shallow water equations in two-dimensional space, and describe and implement a multilevel finite-volume discretization in the context of the utilization of the incremental unknowns. The numerical stability of the method is proved in both cases.

Keywords Finite-volume methods · Multilevel methods · Shallow water equations · Stability analysis

Mathematics Subject Classification 65M60 · 65N21 · 65N99

1 Introduction

This article is closely related to and complements the article (see [1]), in which the authors implemented multilevel finite-volume discretizations of the shallow water equations in two-dimensional space, as a model for geophysical flows. The geophysical context is presented in [1] as well as practical issues concerning the implementation. In this article, we recall the motivation, present the algorithm, and

Project supported by the National Science Foundation (Nos. DMS 0906440, DMS 1206438) and the Research Fund of Indiana University.

A. Bousquet (✉) · M. Marion · R. Temam
The Institute for Scientific Computing and Applied Mathematics, Indiana University, Bloomington, IN 47405, USA
e-mail: arthbous@indiana.edu

M. Marion
e-mail: Martine.Marion@ec-lyon.fr

R. Temam
e-mail: temam@indiana.edu

M. Marion
Département Mathématique Informatique, Université de Lyon, Ecole Centrale de Lyon, CNRS UMR 5208, 36 avenue Guy de Collongue, 69134 Ecully Cedex, France

P.G. Ciarlet et al. (eds.), *Partial Differential Equations: Theory, Control and Approximation*, DOI 10.1007/978-3-642-41401-5_3,

discuss the numerical analysis of some variations of the algorithm, and in particular the stability in time.

The shallow water equations are a simplified model of the primitive equations (or PEs for short) of the atmosphere and the oceans. As shown in [20, 24], in rectangular geometry, the PEs can be expanded by using a certain vertical modal decomposition. With such a decomposition, we obtain an infinite system of coupled equations, which resemble the shallow water equations. See [6, 7] for the actual numerical resolution of these coupled systems. However, it appears in these articles that the problems to be solved are very difficult (demanding), and performable numerical methods are needed to tackle more and more realistic problems. We turned to multilevel finite-volume methods in [1], finite-volume methods are desirable for the treatment of complicated geometrical domains such as the oceans, and multilevel methods of the incremental unknown type are useful for the implementation of multilevel methods. Such methods have been introduced in the context of the nonlinear Galerkin method in [18] (see also [19]), finite differences in [23], and spectral methods and turbulence in [8]. As continuation of [1], this article explores the finite-volume implementation of the incremental unknowns.

Considering to simplify a rectangular geometry, we divide the domain into cells of size $\Delta x \times \Delta y$, which we regroup at the first level of increment, in cells of size $3\Delta x \times 3\Delta y$. The unknowns on the small cells being the original unknowns, we introduce for the coarse cells suitably averaged values of the unknowns. The dynamic strategy, which may take many different forms (see [1, 8]), consists in solving alternatively the system for a number of time steps on the fine mesh grid and then for a number of time steps, the system considered on the coarse mesh during which the increments as defined below, remain frozen. This coarsening can be repeated once more considering cells of size $9\Delta x \times 9\Delta y$, and possibly several times as the programming cost is repetitive and thus small, but we restrict ourselves in this article to one coarsening.

We have chosen to present the method for the shallow water (or SW for short) equations for the reasons mentioned above. We consider the SW equations without viscosity, linearized around a constant flow. The well-posedness of these linear hyperbolic equations has been established very recently (see [12]). We choose in this article one of many situations presented in [12], i.e., the fully supercritical case, since the boundary conditions depend on the nature of the flow (subcritical versus supercritical, subsonic versus supersonic). Other implementation of multilevel methods in geophysical fluid dynamics appear in [16]. See also [14, 15] for more developments on the primitive equations. Further developments along the lines of this work will appear in an article in [4].

Furthermore, some related results can be found in [2, 10, 11, 13, 17, 21, 25].

This article is organized as follows. We start in Sect. 2 with a simple model corresponding to a one-dimensional transport equation. We then proceed in Sect. 3 with the shallow water equation presenting first the equations (see Sect. 3.1), then the multilevel finite-volume discretization (see Sect. 3.2) and then the multilevel temporal discretization (see Sect. 3.3). In Sect. 4, we consider another related form of the algorithm. In Sects. 2 and 3, the algorithm on the coarse grid is the same as the algorithm on the fine grid (in space) with just a different spatial mesh. In this section, we consider another algorithm on which we started, where the spatial scheme on the coarse grid is obtained by averaging, in each coarse cell the equations for the corresponding fine cells. The study of the stability of the scheme in this case has not been completed yet. We present the analysis in one-dimensional space, for the simple transport equation (see Sect. 4.1) and for the one-dimensional linearized equation (see Sect. 4.2). The boundary condition is space periodicity and the stability analysis is conducted by the classical von Neumann method.

2 The One-Dimensional Case

We start with the one-dimensional space and consider the problem

$$\frac{\partial u}{\partial t}(x,t) + \frac{\partial u}{\partial x}(x,t) = f(x,t) \tag{2.1}$$

for $(x,t) \in (0,L) \times (0,T)$, with the boundary condition

$$u(0,t) = 0 \tag{2.2}$$

and the initial condition

$$u(x,0) = u^0(x). \tag{2.3}$$

We set $\mathcal{M} = (0,L)$ and $H = L^2(\mathcal{M})$, and also introduce the operator $Au = u_x$ with domain $D(A) = \{v \in H^1(\mathcal{M}), v(0) = 0\}$. Then for $f, f' \in L^1(0,T;H)$, $u^0 \in D(A)$, problem (2.1)–(2.3) possesses a unique solution u, such that

$$u \in C\big([0,T];H\big) \cap L^\infty\big(0,T;D(A)\big), \qquad \frac{\mathrm{d}u}{\mathrm{d}t} \in L^\infty\big(0,T;D(A)\big).$$

Our multilevel spatial discretization is presented in Sect. 2.1, while Sect. 2.2 deals with time and space discretization.

2.1 Multilevel Spatial Discretization

We consider, on the interval $(0,L)$, $3N$ cells $(k_i)_{1\le i\le 3N}$ of uniform length Δx with $3N\Delta x = L$. For $i = 0,\dots,3N$, we set

$$x_{i+\frac{1}{2}} = i\Delta x,$$

so that

$$k_i = (x_{i-\frac{1}{2}}, x_{i+\frac{1}{2}}).$$

We also introduce the center of each cell,

$$x_i = \frac{x_{i-\frac{1}{2}} + x_{i+\frac{1}{2}}}{2} = (i-1)\Delta x + \frac{\Delta x}{2}, \quad 1 \le i \le 3N.$$

The discrete unknowns are denoted by u_i ($1 \le i \le 3N$), and u_i is expected to be some approximation of the mean value of u over k_i. Equation (2.1) integrated over the cell k_i yields

$$\frac{\mathrm{d}}{\mathrm{d}t} \int_{k_i} u(x,t)\mathrm{d}x + u(x_{i+\frac{1}{2}},t) - u(x_{i-\frac{1}{2}},t) = \int_{k_i} f(x,t)\mathrm{d}x.$$

Here the term $u(x_{i+\frac{1}{2}},t)$ is approximated by $u_i(t)$ using an "upwind" scheme due to the direction of the characteristics for Eq. (2.1). Setting $f_i(t) = \frac{1}{\Delta x}\int_{k_i} f(x,t)\mathrm{d}x$, the upwind finite-volume discretization now reads

$$\frac{\mathrm{d}u_i}{\mathrm{d}t}(t) + \frac{u_i(t) - u_{i-1}(t)}{\Delta x} = f_i(t), \quad 1 \le i \le 3N, \tag{2.4}$$

where we have set

$$u_0(t) = 0. \tag{2.5}$$

These equations are supplemented with the initial condition

$$u_i(0) = \frac{1}{\Delta x}\int_{k_i} u^0(x)\mathrm{d}x, \quad 1 \le i \le 3N. \tag{2.6}$$

To rewrite the scheme in a more abstract form, we introduce the space V_h ($h = \Delta x$) of step functions u_h, which are constant on the intervals k_i, $i = 0, \ldots, 3N$ with $u_h|_{k_i} = u_i$ and $u_0 = 0$. Here to take into account the boundary condition, we have added the fictitious cell $k_0 = (-\Delta x, 0)$. The discrete space V_h is equipped with the norm induced by $L^2(\mathcal{M})$, that is,

$$|u_h|^2 = \Delta x \sum_{i=0}^{3N} |u_i|^2 = \Delta x \sum_{i=1}^{3N} |u_i|^2.$$

Next let us introduce the backward difference operator

$$\partial_h u_h = \frac{u_i - u_{i-1}}{\Delta x} \quad \text{on } k_i,\ 1 \le i \le 3N.$$

Then (2.4) can be rewritten as

$$\frac{\mathrm{d}u_h}{\mathrm{d}t} + \partial_h u_h = f_h$$

with $f_h|_{k_i} = f_i$.

We now introduce a coarser mesh consisting of the intervals K_l $(1 \leq l \leq N)$, with length $3\Delta x$ obtained as[1]

$$K_l = k_{3l-2} \cup k_{3l-1} \cup k_{3l} = (x_{3l-2-\frac{1}{2}}, x_{3l-\frac{1}{2}}). \tag{2.7}$$

Let $(u_i)_{1\leq i\leq 3N}$ still denote the approximation of u on the fine mesh $(k_i)_{1\leq i\leq 3N}$. Then an approximation of u on the coarse mesh is given by

$$U_l = \frac{1}{3}[u_{3l-2} + u_{3l-1} + u_{3l}], \quad 1 \leq l \leq N. \tag{2.8}$$

We introduce the incremental unknowns

$$Z_{3l-\alpha} = u_{3l-\alpha} - U_l \tag{2.9}$$

for $\alpha = 0, 1, 2,\ \ell = 1, \ldots, N$, so that

$$Z_{3\ell} + Z_{3\ell-1} + Z_{3\ell-2} = 0. \tag{2.10}$$

Remark 2.1 The definition of Z in (2.9) is at our disposal. In this case, Z are the order of Δx. For example, using Taylor's formula, we obtain

$$\begin{aligned}
Z_{3l-2} &= u_{3l-2} - \frac{1}{3}[u_{3l-2} + u_{3l-1} + u_{3l}] \\
&= \frac{1}{3}\big[2u_{3l-2} - \big(u_{3l-2} + \mathcal{O}(\Delta x)\big) - \big(u_{3l-2} + \mathcal{O}(\Delta x)\big)\big] \\
&= \mathcal{O}(\Delta x).
\end{aligned}$$

We will discuss elsewhere other definitions of the incremental unknown Z, and in particular those of order Δx^2 considered in [1].

The unknowns on the fine grid are thus written as the sum of the coarse grid unknowns $(U_l)_{1\leq l\leq N}$ and associated increments $(Z_i)_{1\leq i\leq 3N}$.

With this in mind, we consider a coarse grid discretization of the equation similar to (2.4), that is,

$$\frac{dU_\ell(t)}{dt} + \frac{1}{3\Delta x}\big(U_\ell(t) - U_{\ell-1}(t)\big) = F_\ell(t), \quad 1 \leq \ell \leq N \tag{2.11}$$

with

$$U_0(t) = 0, \tag{2.12}$$

$$F_\ell(t) = \frac{1}{3}\sum_{\alpha=0}^{2} f_{3\ell-\alpha}(t) \tag{2.13}$$

---

[1] Including, strictly speaking, the separation points.

and

$$U_\ell(0) = \frac{1}{3}\sum_{\alpha=0}^{2} u_{3\ell-\alpha}(0). \tag{2.14}$$

Independent of the equation under consideration and the numerical scheme, let us make the following algebraic observation: for $u_h \in V_h$, $u_h = (u_i)_{1\le i\le 3N}$, we have

$$\begin{aligned}
|u_h|^2 &= h\sum_{i=1}^{3N} u_i^2 = h\sum_{\alpha=0}^{2}\sum_{\ell=1}^{N} |u_{3\ell-\alpha}|^2\\
&= h\sum_{\alpha=0}^{2}\sum_{\ell=1}^{N} |U_\ell + Z_{3\ell-2}|^2\\
&= 3h\sum_{\ell=1}^{N} |U_\ell|^2 + h\sum_{i=1}^{3N}|Z_i|^2 \quad \big(\text{because of (2.10)}\big)\\
&= |U_h|^2 + |Z_h|^2.
\end{aligned} \tag{2.15}$$

In some sense, because of (2.10), the coarse component U and the increment Z are L^2-orthogonal.

2.2 Euler Implicit Time Discretization and Estimates

We define a time step Δt with $N_T\Delta t = T$, and set $t_n = n\Delta t$ for $0 \le n \le N_T$. We denote by $\{u_i^n, 1 \le i \le 3N, 0 \le n \le N_T\}$ the discrete unknowns. The value u_i^n is an expected approximation

$$u_i^n \simeq \frac{1}{\Delta x}\int_{k_i} u(x, t_n)\mathrm{d}x.$$

Our spatial discretization was presented in the previous section in (2.4)–(2.6), for the fine grid, and (2.11)–(2.14) for the coarse grid. We now discretize this equation in time by using the implicit Euler scheme with the time step $\frac{\Delta t}{p}$ on the fine mesh and time step Δt on the coarse mesh. More precisely, let $p > 1$ and $q > 1$ be two fixed integers. The multi-step discretization consists in alternating p steps on (2.4) with time step $\frac{\Delta t}{p}$, from t_n to t_{n+1} and then q steps on (2.11) with time step Δt, the incremental unknowns Z_i being frozen at t_{n+1} from t_{n+1} to t_{n+q+1}. Then, using equations (2.9), we can go back to the finer mesh for p steps from t_{n+q+1} to t_{n+q+2}. For simplicity, we suppose that N_T is a multiple of $q+1$, and set $N_q = \frac{N_T}{q+1}$.

Suppose that n is a multiple of $(q+1)$, and the $(u_i^n)_{1\le i\le 3N}$ are known. We introduce the discrete unknowns $u_i^{n+\frac{s}{p}}$ with $t_{n+\frac{s}{p}} = t_n + s\frac{\Delta t}{p}$ for $0 \le s \le p$ and

$1 \le i \le 3N$. We successively determine the $u_i^{n+\frac{s}{p}}$ $(1 \le i \le 3N,\ 1 \le s \le p)$ with p iterations of the following scheme:

$$\begin{cases} \frac{p}{\Delta t}(u_i^{n+\frac{s+1}{p}} - u_i^{n+\frac{s}{p}}) + \frac{1}{\Delta x}(u_i^{n+\frac{s+1}{p}} - u_{i-1}^{n+\frac{s+1}{p}}) = f_i^{n+\frac{s+1}{p}}, \\ u_0^{n+\frac{s+1}{p}} = 0 \end{cases} \tag{2.16}$$

for $1 \le i \le 3N, 0 \le s \le p-1$, where

$$f_i^{n+\frac{s+1}{p}} = \frac{1}{\frac{\Delta t}{p}} \frac{1}{\Delta x} \int_{(n+\frac{s}{p})\Delta t}^{(n+\frac{s+1}{p})\Delta t} \int_{k_i} f(x,t)\mathrm{d}x\mathrm{d}t. \tag{2.17}$$

It is convenient to introduce the step functions $u_h^{n+\frac{s}{p}}$, $f_h^{n+\frac{s}{p}}$ defined for $0 \le s \le p$ by

$$u_h^{n+\frac{s}{p}}(x) = u_i^{n+\frac{s}{p}}, \qquad f_h^{n+\frac{s}{p}}(x) = f_i^{n+\frac{s}{p}}, \quad x \in k_i,\ 1 \le i \le 3N.$$

We also introduce the backward difference operator ∂_h defined by

$$\partial_h g_i^n = \frac{g_i^n - g_{i-1}^n}{\Delta x} \quad \text{or} \quad \partial_h g(x) = \frac{g(x) - g(x-h)}{\Delta x},$$

so that (2.16) can now be rewritten as

$$\frac{p}{\Delta t}\left(u_h^{n+\frac{s+1}{p}} - u_h^{n+\frac{s}{p}}\right) + \partial_h u_h^{n+\frac{s+1}{p}} = f_h^{n+\frac{s+1}{p}}. \tag{2.18}$$

Our goal now is to estimate $|u_h^{n+1}|$ in terms of $|u_h^n|$. We take the scalar product in $L^2(\mathcal{M})$ of (2.18) with $2\frac{\Delta t}{p}u_h^{n+\frac{s+1}{p}}$. Denoting by $(\cdot,\cdot)$ the L^2 scalar product and using the well-known relation

$$2(a-b,a) = |a|^2 - |b|^2 + |a-b|^2,$$

we find

$$\begin{aligned} &|u_h^{n+\frac{s+1}{p}}|^2 - |u_h^{n+\frac{s}{p}}|^2 + |u_h^{n+\frac{s+1}{p}} - u_h^{n+\frac{s}{p}}|^2 + \frac{2\Delta t}{p}(\partial_h u_h^{n+\frac{s+1}{p}}, u_h^{n+\frac{s+1}{p}}) \\ &= \frac{2\Delta t}{p}(f_h^{n+\frac{s+1}{p}}, u_h^{n+\frac{s+1}{p}}). \end{aligned} \tag{2.19}$$

We have, for every $u_h \in V_h$,

$$2(\partial_h u_h, u_h) = |u_{3N}|^2 + \sum_{i=1}^{3N} |u_i - u_{i-1}|^2. \tag{2.20}$$

Indeed

$$2(\partial_h u_h, u_h) = 2\sum_{i=1}^{3N}(u_i - u_{i-1})u_i$$

$$= \sum_{i=1}^{3N}\left(|u_i|^2 - |u_{i-1}|^2 + |u_i - u_{i-1}|^2\right),$$

and (2.20) follows, since $u_0 = 0$.

Using (2.20) and Schwarz inequality, (2.19) yields

$$|u_h^{n+\frac{s+1}{p}}|^2 - |u_h^{n+\frac{s}{p}}|^2 + |u_h^{n+\frac{s+1}{p}} - u_h^{n+\frac{s}{p}}|^2$$

$$+ \frac{\Delta t}{p}\left[|u_{3N}^{n+\frac{s+1}{p}}|^2 + \sum_{i=1}^{3N}|u_i^{n+\frac{s+1}{p}} - u_{i-1}^{n+\frac{s+1}{p}}|^2\right]$$

$$\leq \frac{\Delta t}{p}|f_h^{n+\frac{s+1}{p}}|^2 + \frac{\Delta t}{p}|u_h^{n+\frac{s+1}{p}}|^2, \tag{2.21}$$

so that

$$\left(1 - \frac{\Delta t}{p}\right)|u_h^{n+\frac{s+1}{p}}|^2 \leq \frac{\Delta t}{p}|f_h^{n+\frac{s+1}{p}}|^2 + |u_h^{n+\frac{s}{p}}|^2. \tag{2.22}$$

This yields readily for $1 \leq s \leq p$,

$$|u_h^{n+\frac{s}{p}}|^2 \leq \frac{1}{(1-\frac{\Delta t}{p})^s}\left[|u_h^n|^2 + \frac{\Delta t}{p}\sum_{d=0}^{s-1}|f_h^{n+\frac{d+1}{p}}|^2\right]. \tag{2.23}$$

Here, in view of definition (2.17), we observe that

$$\frac{\Delta t}{p}|f_h^{n+\frac{d}{p}}|^2 = \frac{\Delta t}{p}\Delta x\sum_{i=1}^{3N}|f_i^{n+\frac{d}{p}}|^2 = \left(\int_{(n+\frac{d}{p})\Delta t}^{(n+\frac{d+1}{p})\Delta t}\int_0^L f(x,t)\mathrm{d}x\mathrm{d}t\right)^2$$

$$\leq \int_{(n+\frac{d}{p})\Delta t}^{(n+\frac{d+1}{p})\Delta t}\int_0^L |f(x,t)|^2\mathrm{d}x\mathrm{d}t.$$

By adding these inequalities for $d = 0, \ldots, p-1$, we obtain

$$\frac{\Delta t}{p}\sum_{d=0}^{p-1}|f_h^{n+\frac{d}{p}}|^2 \leq \int_{n\Delta t}^{(n+1)\Delta t}\int_0^L |f(x,t)|^2\mathrm{d}x\mathrm{d}t.$$

Combining this bound with (2.23) provides

$$|u_h^{n+\frac{s}{p}}|^2 \leq \frac{1}{(1-\frac{\Delta t}{p})^s}\left[|u_h^n|^2 + \int_{n\Delta t}^{(n+1)\Delta t}\int_0^L |f(x,t)|^2\mathrm{d}x\mathrm{d}t\right].$$

Since $1-x \geq 4^{-x}$ for $x \in [0, \frac{1}{2}]$, we see that, if $\frac{\Delta t}{p} \leq \frac{1}{2}$,

$$|u_h^{n+\frac{s}{p}}|^2 \leq 4^{\frac{s}{p}\Delta t}\Big[|u_h^n|^2 + \int_{n\Delta t}^{(n+1)\Delta t}\int_0^L |f(x,t)|^2 \mathrm{d}x\mathrm{d}t\Big]. \tag{2.24}$$

Here s varies between 1 and p, and therefore the bound for $s = p$ reads

$$|u_h^{n+1}|^2 \leq 4^{\Delta t}\Big[|u_h^n|^2 + \int_{n\Delta t}^{(n+1)\Delta t}\int_0^L |f(x,t)|^2 \mathrm{d}x\mathrm{d}t\Big]. \tag{2.25}$$

We now define the u_h^{n+s} for $2 \leq s \leq q+1$, by applying q-times the implicit Euler scheme to Eq. (2.11) with step Δt, that is,

$$\begin{cases} \dfrac{U_l^{n+s+1} - U_l^{n+s}}{\Delta t} + \dfrac{U_\ell^{n+s+1} - U_{\ell-1}^{n+s+1}}{3\Delta x} = F_l^{n+s+1}, \\ U_0^{n+s+1} = u_0^{n+s+1} = 0, \end{cases} \tag{2.26}$$

where

$$\begin{aligned} F_l^{n+s+1} &= \frac{1}{3}\big[f_{3l-2}^{n+s+1} + f_{3l-1}^{n+s+1} + f_{3l}^{n+s+1}\big] \\ &= \frac{1}{3\Delta t\Delta x}\int_{(n+s)\Delta t}^{(n+s+1)\Delta t}\int_{K_l} f(x,t)\mathrm{d}x\mathrm{d}t. \end{aligned} \tag{2.27}$$

As we said at the beginning of the section, the Z_i's are frozen between t_{n+1} and t_{n+q+1}, and therefore for $2 \leq s \leq q+1$, $1 \leq l \leq N$,

$$\begin{cases} U_l^{n+s} = \frac{1}{3}[u_{3l-2}^{n+s} + u_{3l-1}^{n+s} + u_{3l}^{n+s}], \\ Z_{3l-\alpha}^{n+s} = Z_{3l-\alpha}^{n+1} = u_{3l-\alpha}^{n+1} - U_l^{n+1}, \quad \alpha = 0, 1, 2. \end{cases} \tag{2.28}$$

We can invert this system (2.28) to obtain

$$u_{3l-\alpha}^{n+s} = U_l^{n+s} + Z_{3l-\alpha}^{n+1}, \quad \alpha = 0, 1, 2. \tag{2.29}$$

Classically these equations allow us to uniquely define the terms U_ℓ^{n+s+1}, when the terms U_ℓ^{n+1} are known. Then Eq. (2.29) allow us to compute the u_i^{n+s+1} ($i = 1, \ldots, 3N$, $s = 1, \ldots, q$).

To derive suitable a priori estimates, we multiply (2.26) by $6\Delta t\Delta x U_\ell^{n+s+1}$ and sum for $\ell = 1, \ldots, N$. Setting $\tau = n+s+1$, we find

$$\begin{aligned} &3\Delta x\sum_{\ell=1}^N(|U_\ell^\tau|^2 - |U_\ell^{\tau-1}|^2) + 3\Delta x\sum_{\ell=1}^N|U_l^\tau - U_l^{\tau-1}|^2 \\ &+ 2\Delta t|U_N^\tau|^2 + \Delta t\sum_{\ell=1}^N|U_\ell^\tau - U_{\ell-1}^\tau|^2 \end{aligned}$$

$$= 6\Delta t \Delta x \sum_{\ell=1}^{N} F_\ell^\tau U_\ell^\tau. \tag{2.30}$$

Hence, as for Eqs. (2.21)–(2.25),

$$|U_h^\tau|^2 \le 4^{\Delta t}\left[|U_h^{\tau-1}|^2 + \int_{(\tau-1)\Delta t}^{\tau\Delta t}\int_0^L |f(x,t)|^2 \mathrm{d}x\mathrm{d}t\right]. \tag{2.31}$$

We write Eq. (2.31) for $\tau = n+2,\ldots,n+q+1$, multiply the equation for $\tau = n+s$ by $4^{(q+1-s)\Delta t}$ and add for $s = 2,\ldots,q+1$. We obtain

$$|U_h^{n+q+1}|^2 \le 4^{q\Delta t}\left[|U_h^{n+1}|^2 + \int_{(n+1)\Delta t}^{(n+q+1)\Delta t}\int_0^L |f(x,t)|^2 \mathrm{d}x\mathrm{d}t\right]. \tag{2.32}$$

We add $|Z_h^{n+1}|^2$ to both sides and, in view of (2.15) and the second formula of (2.28), we find

$$|u_h^{n+q+1}|^2 \le 4^{q\Delta t}\left[|u_h^{n+1}|^2 + \int_{(n+1)\Delta t}^{(n+q+1)\Delta t}\int_0^L |f(x,t)|^2 \mathrm{d}x\mathrm{d}t\right]. \tag{2.33}$$

Taking into account (2.25), we find that

$$|u_h^{n+q+1}|^2 \le 4^{(q+1)\Delta t}\left[|u_h^{n}|^2 + \int_{n\Delta t}^{(n+q+1)\Delta t} |f(\cdot,t)|_2^2 \mathrm{d}t\right]. \tag{2.34}$$

More generally, we have the stability result

$$\begin{aligned} |u_h^m|^2 &\le 4^{m\Delta t}\left[|u_h^0|^2 + \int_0^{m\Delta t} |f(\cdot,t)|_2^2 \mathrm{d}t\right] \\ &\le 4^{T}\left[|u^0|^2 + \int_0^{T} |f(\cdot,t)|_{L^2}^2 \mathrm{d}t\right]. \end{aligned} \tag{2.35}$$

To summarize, we show the following result.

Theorem 2.1 *The multilevel scheme defined by Eqs.* (2.16) *and* (2.26) *is stable in* $L^\infty(0,T;L^2(\mathcal{M}))$ *in the sense of* (2.35).

3 The Linear Shallow Water Equations

We now want to extend the previous results to the more complex case of the shallow water equations linearized around a constant flow $(\tilde{u}_0, \tilde{v}_0, \tilde{\phi}_0)$ (see (3.2) below). As shown in [12] the boundary conditions, which can be associated with these equations, depend on the relative values of the velocities ($\tilde{u}_0^2, \tilde{v}_0^2 >$ (or $<$) $g\tilde{\phi}_0$), that is, whether these velocities are sub- or supercritical (sub- or supersonic). We consider here the case, where

$$\tilde{\phi}_0 > 0, \quad \tilde{u}_0 > \sqrt{g\tilde{\phi}_0}, \quad \tilde{v}_0 > \sqrt{g\tilde{\phi}_0}. \tag{3.1}$$

3.1 The Equations

We consider, in the domain $\mathcal{M} = (0, L_1) \times (0, L_2)$, the equations

$$\begin{cases} \frac{\partial u}{\partial t} + \tilde{u}_0 \frac{\partial u}{\partial x} + \tilde{v}_0 \frac{\partial u}{\partial y} + g \frac{\partial \phi}{\partial x} = f_u, \\ \frac{\partial v}{\partial t} + \tilde{u}_0 \frac{\partial v}{\partial x} + \tilde{v}_0 \frac{\partial v}{\partial y} + g \frac{\partial \phi}{\partial y} = f_v, \\ \frac{\partial \phi}{\partial t} + \tilde{u}_0 \frac{\partial \phi}{\partial x} + \tilde{v}_0 \frac{\partial \phi}{\partial y} + \tilde{\phi}_0 (\frac{\partial u}{\partial x} + \frac{\partial v}{\partial y}) = f_\phi. \end{cases} \tag{3.2}$$

Here (u, v) is the velocity, and ϕ is the potential height. The advecting velocities $\tilde{u}_0$, $\tilde{v}_0$ and the mean geopotential height $\tilde{\phi}_0$ are constants. $\mathbf{f} = (f_u, f_v, f_\phi)$ is the source term. For the subcritical flow under consideration, we supplement (3.2) with the boundary conditions,

$$\mathbf{u} = (u, v, \phi) = 0, \quad \text{at } \{x = 0\} \cup \{y = 0\}, \tag{3.3}$$

and the initial conditions

$$\mathbf{u} = (u, v, \phi) = \mathbf{u}^0 = \left(u^0, v^0, \phi^0\right), \quad \text{at } t = 0. \tag{3.4}$$

The system becomes

$$\frac{\mathrm{d}\mathbf{u}}{\mathrm{d}t} + \mathbf{A}\mathbf{u} = \mathbf{f},$$

where $\mathbf{A}\mathbf{u} = (A_1\mathbf{u}, A_2\mathbf{u}, A_3\mathbf{u})$ is given by

$$\begin{cases} A_1\mathbf{u} = \tilde{u}_0 \frac{\partial u}{\partial x} + \tilde{v}_0 \frac{\partial u}{\partial y} + g \frac{\partial \phi}{\partial x}, \\ A_2\mathbf{u} = \tilde{u}_0 \frac{\partial v}{\partial x} + \tilde{v}_0 \frac{\partial v}{\partial y} + g \frac{\partial \phi}{\partial y}, \\ A_3\mathbf{u} = \tilde{u}_0 \frac{\partial \phi}{\partial x} + \tilde{v}_0 \frac{\partial \phi}{\partial y} + \tilde{\phi}_0 (\frac{\partial u}{\partial x} + \frac{\partial v}{\partial y}). \end{cases} \tag{3.5}$$

It may also be convenient to decompose $\mathbf{A}$ with respect to its x and y derivatives, that is,

$$\mathbf{A} = \mathbf{A}^x + \mathbf{A}^y,$$
$$\mathbf{A}^x\mathbf{u} = \left(A_1^x\mathbf{u}, A_2^x\mathbf{u}, A_3^x\mathbf{u}\right), \quad \mathbf{A}^y\mathbf{u} = \left(A_1^y\mathbf{u}, A_2^y\mathbf{u}, A_3^y\mathbf{u}\right)$$

with

$$\mathbf{A}^x\mathbf{u} = \begin{cases} \tilde{u}_0 \frac{\partial u}{\partial x} + g \frac{\partial \phi}{\partial x}, \\ \tilde{u}_0 \frac{\partial v}{\partial x}, \\ \tilde{u}_0 \frac{\partial \phi}{\partial x} + \tilde{\phi}_0 \frac{\partial u}{\partial x}, \end{cases} \qquad \mathbf{A}^y\mathbf{u} = \begin{cases} \tilde{v}_0 \frac{\partial u}{\partial y}, \\ \tilde{v}_0 \frac{\partial v}{\partial y} + g \frac{\partial \phi}{\partial y}, \\ \tilde{v}_0 \frac{\partial \phi}{\partial y} + \tilde{\phi}_0 \frac{\partial v}{\partial y}. \end{cases}$$

We define the scalar product on $H = (L^2(\mathcal{M}))^3$ as follows: for $\mathbf{u} = (u, v, \phi)$, $\mathbf{u}' = (u', v', \phi')$, and we set

$$\langle \mathbf{u}, \mathbf{u}' \rangle = (u, u') + (v, v') + \frac{g}{\tilde{\phi}_0}(\phi, \phi'), \tag{3.6}$$

where $(\cdot, \cdot)$ denotes the standard scalar product on $L^2(\mathcal{M})$. Then the following positivity result for $\mathbf{A}$ holds.

Lemma 3.1 *Under the assumption* (3.1), *for all sufficiently smooth* $\mathbf{u}$ *satisfying* (3.3), *we have* $\langle \mathbf{A}\mathbf{u}, \mathbf{u} \rangle \geq 0$.

Proof We write

$$\langle \mathbf{A}\mathbf{u}, \mathbf{u} \rangle = \langle \mathbf{A}^x \mathbf{u}, \mathbf{u} \rangle + \langle \mathbf{A}^y \mathbf{u}, \mathbf{u} \rangle \tag{3.7}$$

with

$$\langle \mathbf{A}^x \mathbf{u}, \mathbf{u} \rangle = \iint_{\mathcal{M}} \left[\tilde{u}_0 u_x u + g\phi_x u + \tilde{u}_0 v_x v + \frac{g}{\tilde{\phi}_0} \tilde{u}_0 \phi_x \phi + g u_x \phi \right] \mathrm{d}x\mathrm{d}y,$$

$$\langle \mathbf{A}^y \mathbf{u}, \mathbf{u} \rangle = \iint_{\mathcal{M}} \left[\tilde{v}_0 v_y v + g\phi_y v + \tilde{v}_0 u_y u + \frac{g}{\tilde{\phi}_0} \tilde{v}_0 \phi_y \phi + g v_y \phi \right] \mathrm{d}x\mathrm{d}y.$$

Then

$$\begin{aligned} \langle \mathbf{A}^x \mathbf{u}, \mathbf{u} \rangle &= \frac{\tilde{u}_0}{2} \iint_{\mathcal{M}} \left[(u^2)_x + (v^2)_x + \frac{g}{\tilde{\phi}_0} (\phi^2)_x \right] \mathrm{d}x\mathrm{d}y + \iint_{\mathcal{M}} g(\phi u)_x \mathrm{d}x\mathrm{d}y \\ &= \frac{\tilde{u}_0}{2} \int_0^{L_2} \left[u^2 + v^2 + \frac{g}{\tilde{\phi}_0} \phi^2 \right]_{x=0}^{x=L_1} \mathrm{d}y + \int_0^{L_2} [g(\phi u)]_{x=0}^{x=L_1} \mathrm{d}y. \end{aligned} \tag{3.8}$$

Recall that $\mathbf{u} = \mathbf{0}$ at $x = 0$. Also the assumption (3.1) yields that

$$\frac{\tilde{u}_0}{2} u^2 + \frac{\tilde{u}_0}{2} g \frac{\phi^2}{\tilde{\phi}_0} + g\phi u$$

is pointwise positive. Therefore, we infer from (3.8) that $\langle \mathbf{A}^x \mathbf{u}, \mathbf{u} \rangle \geq 0$. A similar computation provides $\langle \mathbf{A}^y \mathbf{u}, \mathbf{u} \rangle \geq 0$ (since $\tilde{v}_0^2 > g\tilde{\phi}_0$). In view of (3.7), the proof of Lemma 3.1 is complete. □

Remark 3.1 The fact that the boundary and initial value problem (3.2)–(3.4) is well-posed is a recent result proved in [12]. The proof relies on the semigroup theory and necessitates in particular proving (by approximation) that $\langle \mathbf{A}\mathbf{u}, \mathbf{u} \rangle \geq 0$ for all $\mathbf{u} \in L^2(\mathcal{M})^3$, such that $\mathbf{A}\mathbf{u} \in L^2(\mathcal{M})^3$, and $\mathbf{u}$ satisfies (3.3). The fact that (3.3) makes sense for such $\mathbf{u}$'s results from a trace theorem also proved in [12].

3.2 Multilevel Finite-Volume Spatial Discretization

3.2.1 Finite-Volume Discretization

We decompose $\mathcal{M} = (0, L_1) \times (0, L_2)$ into $3N_1 \times 3N_2$ rectangles denoted by $(k_{i,j})_{1\le i\le 3N_1, 1\le j\le 3N_2}$ of size $\Delta x \times \Delta y$ with $3N_1\Delta x = L_1$ and $3N_2\Delta y = L_2$.

For $0 \le i \le 3N_1$ and for $0 \le j \le 3N_2$, let

$$x_{i+\frac{1}{2}} = i\Delta x \quad \text{and} \quad y_{j+\frac{1}{2}} = j\Delta y.$$

Then the rectangles $(k_{i,j})$ are, for $1 \le i \le 3N_1, 1 \le j \le 3N_2$,

$$k_{i,j} = (x_{i-\frac{1}{2}}, x_{i+\frac{1}{2}}) \times (y_{j-\frac{1}{2}}, y_{j+\frac{1}{2}}).$$

We also define the center (x_i, y_j) of each cell k_{ij},

$$\begin{cases} x_i = \frac{1}{2}(x_{i-\frac{1}{2}} + x_{i+\frac{1}{2}}) = (i-1)\Delta x + \frac{\Delta x}{2}, & 1 \le i \le 3N_1, \\ y_j = \frac{1}{2}(y_{j-\frac{1}{2}} + y_{j+\frac{1}{2}}) = (j-1)\Delta y + \frac{\Delta y}{2}, & 1 \le j \le 3N_2. \end{cases}$$

For the boundary conditions, we add fictitious cells on the west and south sides,

$$k_{0,j} = (-\Delta x, 0) \times (y_{j-\frac{1}{2}}, y_{j+\frac{1}{2}}), \quad \text{centered at} \left(x_0 = -\frac{\Delta x}{2}, y_j\right), \quad 1 \le j \le 3N_2$$

and

$$k_{i,0} = (x_{i-\frac{1}{2}}, x_{i+\frac{1}{2}}) \times (-\Delta y, 0), \quad \text{centered at} \left(x_i, y_0 = -\frac{\Delta y}{2}\right), \quad 1 \le i \le 3N_1.$$

The finite-volume scheme is found by integrating the equations (3.2) over each control volume $(k_{i,j})_{1\le i\le 3N_1, 1\le j\le 3N_2}$. The first equation yields for $1 \le i \le 3N_1, 1 \le j \le 3N_2$,

$$\begin{aligned} &\frac{\mathrm{d}}{\mathrm{d}t}\frac{1}{\Delta x\Delta y}\iint_{k_{i,j}} u(x,y,t)\mathrm{d}x\mathrm{d}y + \frac{\tilde{u}_0}{\Delta x\Delta y}\int_{y_{j-\frac{1}{2}}}^{y_{j+\frac{1}{2}}}\left[u(x_{i+\frac{1}{2}},y,t) - u(x_{i-\frac{1}{2}},y,t)\right]\mathrm{d}y \\ &+ \frac{\tilde{v}_0}{\Delta x\Delta y}\int_{x_{i-\frac{1}{2}}}^{x_{i+\frac{1}{2}}}\left[u(x,y_{j+\frac{1}{2}},t) - u(x,y_{j-\frac{1}{2}},t)\right]\mathrm{d}x \\ &+ \frac{g}{\Delta x\Delta y}\int_{y_{j-\frac{1}{2}}}^{y_{j+\frac{1}{2}}}\left[\phi(x_{i+\frac{1}{2}},y,t) - \phi(x_{i-\frac{1}{2}},y,t)\right]\mathrm{d}y = \int_{k_{i,j}} f_u(x,y,t)\mathrm{d}x\mathrm{d}y. \end{aligned}$$

Let us denote

V_h = {the space of step functions constant on $k_{i,j}$, $0 \le i \le 3N_1, 0 \le j \le 3N_2$ with $w_{|k_{i,j}} = w_{i,j}$ and $w_{0,j} = w_{i,0} = 0$}.

We approximate the unknown $\mathbf{u} = (u, v, \phi)$ with $\mathbf{u}_h \simeq \mathbf{u}_h(t) \in (V_h)^3 = \mathbf{V}_h$, and use an upwind scheme for the fluxes, since $\tilde{u}_0 > 0$ and $\tilde{v}_0 > 0$,

$$\mathbf{u}(x_{i+\frac{1}{2}}, y, t) \simeq \mathbf{u}_{i,j}(t), \quad y \in [y_{j-\frac{1}{2}}, y_{j+\frac{1}{2}}],$$
$$\mathbf{u}(x, y_{j+\frac{1}{2}}, t) \simeq \mathbf{u}_{i,j}(t), \quad x \in [x_{i-\frac{1}{2}}, x_{i+\frac{1}{2}}].$$

This gives the following semi-discrete equations for $1 \le i \le 3N_1$ and $1 \le j \le 3N_2$:

$$\begin{cases} \frac{\mathrm{d}}{\mathrm{d}t} u_{i,j} + \tilde{u}_0 \frac{u_{i,j}-u_{i-1,j}}{\Delta x} + \tilde{v}_0 \frac{u_{i,j}-u_{i,j-1}}{\Delta y} + g \frac{\phi_{i,j}-\phi_{i-1,j}}{\Delta x} = f_{u,i,j}, \\ \frac{\mathrm{d}}{\mathrm{d}t} v_{i,j} + \tilde{u}_0 \frac{v_{i,j}-v_{i-1,j}}{\Delta x} + \tilde{v}_0 \frac{v_{i,j}-v_{i,j-1}}{\Delta y} + g \frac{\phi_{i,j}-\phi_{i,j-1}}{\Delta y} = f_{v,i,j}, \\ \frac{\mathrm{d}}{\mathrm{d}t} \phi_{i,j} + \tilde{u}_0 \frac{\phi_{i,j}-\phi_{i-1,j}}{\Delta x} + \tilde{v}_0 \frac{\phi_{i,j}-\phi_{i,j-1}}{\Delta y} \\ \quad + \tilde{\phi}_0 (\frac{u_{i,j}-u_{i-1,j}}{\Delta x} + \frac{v_{i,j}-v_{i,j-1}}{\Delta y}) = f_{\phi,i,j}, \\ \mathbf{u}_{0,j} = \mathbf{u}_{i,0} = 0, \\ \mathbf{u}_{i,j}(0) = \mathbf{u}^0_{i,j}, \end{cases} \tag{3.9}$$

where $\mathbf{f} = (f_u, f_v, f_\phi)$, $\mathbf{u^0} = (u^0, v^0, \phi^0)$ and

$$\mathbf{f}_{i,j}(t) = \frac{1}{\Delta x \Delta y} \int_{k_{i,j}} \mathbf{f}(x, y, t) \mathrm{d}x\mathrm{d}y, \qquad \mathbf{u}^0_{i,j} = \frac{1}{\Delta x \Delta y} \int_{k_{i,j}} \mathbf{u}^0(x, y) \mathrm{d}x\mathrm{d}y. \tag{3.10}$$

Let us introduce the finite difference operators

$$\partial_{1h} g_h = \frac{1}{\Delta x}(g_{i,j} - g_{i-1,j}) \quad \text{on } k_{i,j},$$
$$\partial_{2h} g_h = \frac{1}{\Delta y}(g_{i,j} - g_{i,j-1}) \quad \text{on } k_{i,j}.$$

We can now define in an obvious way, based on (3.9) the finite difference operator $\mathbf{A}_h = (A_{1h}, A_{2h}, A_{3h})$, operating on $\mathbf{V}_h$

$$\begin{cases} A_{1h}\mathbf{u}_h = \tilde{u}_0 \partial_{1h} u_h + \tilde{v}_0 \partial_{2h} u_h + g \partial_{1h} \phi_h, \\ A_{2h}\mathbf{u}_h = \tilde{u}_0 \partial_{1h} v_h + \tilde{v}_0 \partial_{2h} v_h + g \partial_{2h} \phi_h, \\ A_{3h}\mathbf{u}_h = \tilde{u}_0 \partial_{1h} \phi_h + \tilde{v}_0 \partial_{2h} \phi_h + \tilde{\phi}_0 \partial_{1h} u_h + \tilde{\phi}_0 \partial_{2h} v_h \end{cases} \tag{3.11}$$

and its decomposition $\mathbf{A}_h = \mathbf{A}^x_h + \mathbf{A}^y_h$, to be used later on,

$$\begin{cases} A^x_h \mathbf{u}_h = (\tilde{u}_0 \partial_{1h} u_h + g \partial_{1h} \phi_h, \tilde{u}_0 \partial_{1h} v_h, \tilde{u}_0 \partial_{1h} \phi_h + \tilde{\phi}_0 \partial_{1h} u_h), \\ A^x_h \mathbf{u}_h = (\tilde{v}_0 \partial_{2h} u_h, \tilde{v}_0 \partial_{2h} v_h + g \partial_{2h} \phi_h, \tilde{v}_0 \partial_{2h} \phi_h + \tilde{\phi}_0 \partial_{2h} v_h). \end{cases} \tag{3.12}$$

Those are the discrete versions of $\mathbf{A}$, A_1, A_2, A_3, $\mathbf{A}^x$, $\mathbf{A}^y$.

We can now check that $\mathbf{A}_h$, the discrete version of $\mathbf{A}$, is positive like $\mathbf{A}$.

Lemma 3.2 *For all* $\mathbf{u}_h = (u_h, v_h, \phi_h) \in \mathbf{V}_h$, *we have*

$$\langle \mathbf{A}_h \mathbf{u}_h, \mathbf{u}_h \rangle \geq 0, \tag{3.13}$$

where $\langle \cdot, \cdot \rangle$ *is the scalar product on* $L^2(\mathcal{M})^3$, *given by* (3.6).

Proof We write

$$\langle \mathbf{A}_h \mathbf{u}_h, \mathbf{u}_h \rangle = \langle A_h^x \mathbf{u}_h, \mathbf{u}_h \rangle + \langle A_h^y \mathbf{u}_h, \mathbf{u}_h \rangle, \tag{3.14}$$

where

$$\begin{aligned} \left\langle A_h^x \mathbf{u}_h, \mathbf{u}_h \right\rangle &= (\tilde{u}_0 \partial_{1h} u_h, u_h) + (g \partial_{1h} \phi_h, u_h) + (\tilde{u}_0 \partial_{1h} v_h, v_h) \\ &\quad + \frac{g}{\tilde{\phi}_0} (\tilde{u}_0 \partial_{1h} \phi_h, \phi_h) + g(\partial_{1h} u_h, \phi_h), \\ \left\langle A_h^y \mathbf{u}_h, \mathbf{u}_h \right\rangle &= (\tilde{v}_0 \partial_{2h} u_h, u_h) + (g \partial_{2h} \phi_h, v_h) + (\tilde{v}_0 \partial_{2h} v_h, v_h) \\ &\quad + \frac{g}{\tilde{\phi}_0} (\tilde{v}_0 \partial_{2h} \phi_h, \phi_h) + g(\partial_{2h} v_h, \phi_h). \end{aligned}$$

We first remark that

$$\begin{aligned} (\tilde{u}_0 \partial_{1h} u_h, u_h) &= \frac{\tilde{u}_0}{2} \Delta y \sum_{i=1}^{3N_1} \sum_{j=1}^{3N_2} \left(|u_{i,j}|^2 - |u_{i-1,j}|^2 + |u_{i,j} - u_{i-1,j}|^2 \right) \\ &= \frac{\tilde{u}_0}{2} \Delta y \sum_{j=1}^{3N_2} \left(|u_{3N_1,j}|^2 + \sum_{i=1}^{3N_1} |u_{i,j} - u_{i-1,j}|^2 \right). \end{aligned} \tag{3.15}$$

Then we write

$$\begin{aligned} &(\phi_{i,j} - \phi_{i-1,j}) u_{i,j} + (u_{i,j} - u_{i-1,j}) \phi_{i,j} \\ &\quad = u_{i,j} \phi_{i,j} - u_{i-1,j} \phi_{i-1,j} + (u_{i,j} - u_{i-1,j})(\phi_{i,j} - \phi_{i-1,j}). \end{aligned}$$

Using these two formulas, we obtain

$$\begin{aligned} &(\tilde{u}_0 \partial_{1h} u_h, u_h) + \frac{g}{\tilde{\phi}_0} (\tilde{u}_0 \partial_{1h} \phi_h, \phi_h) + (g \partial_{1h} \phi_h, u_h) + g(\partial_{1h} u_h, \phi_h) \\ &= \frac{\tilde{u}_0}{2} \Delta y \sum_j \left(|u_{3N_1,j}|^2 + \frac{g}{\tilde{\phi}_0} |\phi_{3N_1,j}|^2 \right) \end{aligned}$$

$$+\Delta y \frac{\tilde{u}_0}{2} \sum_{i,j} |u_{i,j} - u_{i-1,j}|^2 + \Delta y \frac{g \tilde{u}_0}{2\tilde{\phi}_0} \sum_{i,j} |\phi_{i,j} - \phi_{i-1,j}|^2$$

$$+ g\Delta y \sum_{i,j} (u_{i,j} - u_{i-1,j})(\phi_{i,j} - \phi_{i-1,j}) + g\Delta y \sum_{j} u_{3N_1,j}\phi_{3N_1,j}.$$

Since $\tilde{u}_0 > 0$ and $\tilde{u}_0^2 > g\tilde{\phi}_0$, the expressions

$$\frac{\tilde{u}_0}{2}|u_{i,j} - u_{i-1,j}|^2 + \frac{g\tilde{u}_0}{2\tilde{\phi}_0}|\phi_{i,j} - \phi_{i-1,j}|^2 + g(u_{i,j} - u_{i-1,j})(\phi_{i,j} - u\phi_{i-1,j})$$

and

$$\frac{\tilde{u}_0}{2}|u_{3N_1,j}|^2 + \frac{g\tilde{u}_0}{2\tilde{\phi}_0}|\phi_{3N_1,j}|^2 + g u_{3N_1,j}\phi_{3N_1,j}$$

are positive and the corresponding sums are positive as well.

Finally, using also the analogue of (3.15) for v_h, we conclude that $\langle A_h^x \mathbf{u}_h, \mathbf{u}_h\rangle \geq 0$. Similarly, it can be checked that $\langle A_h^y \mathbf{u}_h, \mathbf{u}_h\rangle \geq 0$. Recalling (3.14), this completes the proof of Lemma 3.2. □

In fact, a perusal of the calculations above shows that we have proved the following useful lemma.

Lemma 3.3 *For every $\mathbf{u}_h \in \mathbf{V}_h$,*

$$\begin{cases} \langle \mathbf{A}_h^x \mathbf{u}_h, \mathbf{u}_h\rangle \geq \kappa_1 \Delta y \sum_{j=1}^{3N_2} \big[|\mathbf{u}_{3N1,j}|^2 + \sum_{i=1}^{3N_1} |\mathbf{u}_{i,j} - \mathbf{u}_{i-1,j}|^2\big], \\ \langle \mathbf{A}_h^y, \mathbf{u}_h, \mathbf{u}_h\rangle \geq \kappa_1 \Delta x \sum_{i=1}^{3N_1} \big[|\mathbf{u}_{i,3N_2}|^2 + \sum_{j=1}^{3N_1} |\mathbf{u}_{i,j} - \mathbf{u}_{i,j-1}|^2\big], \end{cases} \tag{3.16}$$

where the constant κ_1 depends on $\tilde{u}_0, \tilde{v}_0, \tilde{\phi}_0, g$ and in particular on the positive numbers $\tilde{u}_0^2 - g\tilde{\phi}_0, \tilde{v}_0^2 - g\tilde{\phi}_0$.

3.2.2 Multilevel Finite-Volume Discretization

We introduce the coarse mesh consisting of the rectangles K_{lm} $(1 \leq l \leq N_1, 1 \leq m \leq N_2)$,[2]

$$K_{lm} = \bigcup_{\alpha,\beta=0}^{2} k_{3l-\alpha,3m-\beta} = (x_{3l-2-\frac{1}{2}}, x_{3l-\frac{1}{2}}) \times (y_{3m-2-\frac{1}{2}}, y_{3m+\frac{1}{2}}).$$

We also define the fictitious rectangles $K_{0,m}, K_{l,0}$ $(l = 1, \ldots N_1,\ m = 1, \ldots, N_2)$, needed for the implementation of the boundary conditions, and they are defined as above with m or $l = 0$.

[2]Including, strictly speaking, the separation edges.

We introduce the space V_{3h} defined like V_h. If $u_h \in V_h$ and $u_h|_{k_{ij}} = u_{i,j}$, we define for $l = 1, \ldots, N_1,\ m = 1, \ldots, N_2$ the averages as

$$U_{l,m} = \frac{1}{9} \sum_{\alpha,\beta=0}^{2} u_{3l-\alpha,3m-\beta}, \tag{3.17}$$

and the incremental unknowns as

$$Z_{3l-\alpha,3m-\beta} = u_{3l-\alpha,3m-\beta} - U_{l,m}, \tag{3.18}$$

which satisfy of course

$$\sum_{\alpha,\beta=0}^{2} Z_{3l-\alpha,3m-\beta} = 0. \tag{3.19}$$

We note the following algebraic relations (using (3.19)):

$$\sum_{\alpha,\beta=0}^{2} |u_{3l-\alpha,3m-\beta}|^2 = 9|U_{l,m}|^2 + \sum_{\alpha,\beta=0}^{2} |Z_{3l-\alpha,3m-\beta}|^2. \tag{3.20}$$

Multiplying by $\Delta x \Delta y$ and adding for $l = 1, \ldots, N_1,\ m = 1, \ldots, N_2$, we find

$$|u_h|^2 = |U_h|^2 + |Z_h|^2, \tag{3.21}$$

where $|\cdot|$ is still the norm in $L^2(\mathcal{M})$, U_h is the step function equal to U_{lm} on $K_{l,m}$ and Z_h is the step function equal to $Z_{i,j}$ on $k_{i,j}$.

3.3 *Euler Implicit Time Discretization and Estimates*

We proceed to some extent as in the one-dimensional space. We define a time step Δt with $N_T \Delta t = T$, and set $t_n = n\Delta t$. We denote by

$$\mathbf{u}_h^n = \left\{\mathbf{u}_{i,j}^n,\ 1 \le i \le 3N_1,\ 1 \le j \le 3N_2\right\}$$

the discrete unknowns, where $\mathbf{u}_{i,j}^n$ is an expected approximation

$$\mathbf{u}_{i,j}^n \simeq \frac{1}{\Delta x \Delta y} \int_{k_{i,j}} \mathbf{u}(x, y, t_n) \mathrm{d}x \mathrm{d}y.$$

The spatial discretization has been presented in Sect. 3.2. We will now discretize the shallow water equations in time by using the implicit Euler scheme, and advance equation (3.9) for p steps in time on the fine mesh with a time step of $\frac{\Delta t}{p}$, where p (and q below) are two fixed integers larger than 1.

These steps will bring us, e.g., from t_n to t_{n+1}. We then perform q steps with a time step Δt bringing us from t_{n+1} to t_{n+q+1}. For simplicity, we suppose that N_T is a multiple of $q+1$, and we set $N_q = \frac{N_T}{q+1}$. The steps performed with the time step Δt will use the coarse mesh. We first consider in Sect. 3.3.1 the p steps performed with mesh $\frac{\Delta t}{p}$ on the fine grid. Then the q steps on the coarse grid are described in Sect. 3.3.2.

3.3.1 Scheme and Estimates on the Fine Grid

We start from Eqs. (3.9) and write thus for $s = 1, \ldots, p$,

$$\begin{cases} \frac{p}{\Delta t}\big(u_{i,j}^{n+\frac{s+1}{p}} - u_{i,j}^{n+\frac{s}{p}}\big) + \tilde{u}_0 \partial_{1h} u_{i,j}^{n+\frac{s+1}{p}} \\ \quad + \tilde{v}_0 \partial_{2h} u_{i,j}^{n+\frac{s+1}{p}} + g \partial_{1h} \phi_{i,j}^{n+\frac{s+1}{p}} = f_{u,i,j}^{n+\frac{s+1}{p}}, \\ \frac{p}{\Delta t}\big(v_{i,j}^{n+\frac{s+1}{p}} - v_{i,j}^{n+\frac{s}{p}}\big) + \tilde{u}_0 \partial_{1h} v_{i,j}^{n+\frac{s+1}{p}} \\ \quad + \tilde{v}_0 \partial_{2h} v_{i,j}^{n+(s+\frac{1}{p})} + g \partial_{2h} \phi_{i,j}^{n+\frac{s+1}{p}} = f_{v,i,j}^{n+\frac{s+1}{p}}, \\ \frac{p}{\Delta t}\big(\phi_{i,j}^{n+\frac{s+1}{p}} - \phi_{i,j}^{n+\frac{s}{p}}\big) + \tilde{u}_0 \partial_{1h} \phi_{i,j}^{n+\frac{s+1}{p}} \\ \quad + \tilde{v}_0 \partial_{2h} \phi_{i,j}^{n+\frac{s+1}{p}} + \tilde{\phi}_0\big(\partial_{1h} u_{i,j}^{n+\frac{s+1}{p}} + \partial_{2h} v_{i,j}^{n+\frac{s+1}{p}}\big) = f_{\phi,i,j}^{n+\frac{s+1}{p}}. \end{cases} \tag{3.22}$$

With the definition of $\mathbf{A}_h$ introduced in (3.11), Eq. (3.22) amount to

$$\frac{p}{\Delta t}\big(\mathbf{u}_h^\tau - \mathbf{u}_h^{\tau-\frac{1}{p}}\big) + A_h \mathbf{u}_h^\tau = \mathbf{f}_h^\tau. \tag{3.23}$$

Here we have set for simplicity $n + \frac{s+1}{p} = \tau$, $n + \frac{s}{p} = \tau - \frac{1}{p}$, $\mathbf{u}_h^\tau = (u_h^\tau, v_h^\tau, \phi_h^\tau)$, $\mathbf{f}_h^\tau = (f_{u,h}^\tau, f_{v,h}^\tau, f_{\phi,h}^\tau)$.

Taking the scalar product in $\mathbf{V}_h$ of each side of (3.23) with $2\frac{\Delta t}{p}\mathbf{u}^\tau$, we see that

$$\begin{aligned} &|\mathbf{u}_h^\tau|^2 - |\mathbf{u}_h^{\tau-\frac{1}{p}}|^2 + |\mathbf{u}_h^\tau - \mathbf{u}_h^{\tau-\frac{1}{p}}|^2 + 2\frac{\Delta t}{p}\langle \mathbf{A}_h \mathbf{u}_h^\tau, \mathbf{u}_h^\tau \rangle \\ &= \frac{2\Delta t}{p}\langle \mathbf{f}_h^\tau, \mathbf{u}_h^\tau \rangle \le \frac{\Delta t}{p}|\mathbf{f}_h^\tau|^2 + \frac{\Delta t}{p}|\mathbf{u}_h^\tau|^2. \end{aligned} \tag{3.24}$$

Hence thanks to Lemma 3.2 (comparing with (2.19)–(2.25)),

$$|\mathbf{u}_h^{n+\frac{s+1}{p}}|^2 \le \frac{1}{1-\frac{\Delta t}{p}}|\mathbf{u}_h^{n+\frac{s}{p}}|^2 + \frac{1}{1-\frac{\Delta t}{p}}\frac{\Delta t}{p}|\mathbf{f}_h^{n+\frac{s}{p}}|^2, \tag{3.25}$$

and for $\frac{\Delta t}{p} \leq \frac{1}{2}$ and $s = 1, \ldots, p$ (comparing with (2.25)),

$$\begin{aligned}
&|\mathbf{u}_h^{n+\frac{s}{p}}|^2 \leq 4^{\frac{s\Delta t}{p}} \kappa^n(\mathbf{u}^0, \mathbf{f}), \\
&\kappa^n(\mathbf{u}^0, \mathbf{f}) = |\mathbf{u}_h^0|^2 + \int_{n\Delta t}^{(n+1)\Delta t} \int_0^{L_2} \int_0^{L_1} |\mathbf{f}(x, y, t)|^2 \mathrm{d}x\mathrm{d}y\mathrm{d}t.
\end{aligned}$$

In particular, for $s = p$,

$$|\mathbf{u}_h^{n+1}|^2 \leq 4^{\Delta t} \kappa^n(\mathbf{u}^0, \mathbf{f}). \tag{3.26}$$

3.3.2 Scheme and Estimates on the Coarse Grid

We now consider the q a time-steps performed on the coarse grid with a time step Δt.

We discretize Eq. (3.9) in time, starting from time $t_{n+1} = (n + 1)\Delta t$ using the same scheme as for Eq. (3.22) but with a coarse mesh (comparing with (2.26)). We obtain

$$\frac{1}{\Delta t}(\mathbf{U}_h^\tau - \mathbf{U}_h^{\tau-1}) + \mathbf{A}_{3h}\mathbf{U}_h^\tau = \mathbf{F}_h^\tau, \tag{3.27}$$

where $\tau = n+s+1,\ s = 1, \ldots, q,\ \mathbf{U}_h^\tau = (U_{u,h}^\tau, U_{v,h}^\tau, U_{\phi,h}^\tau)$ and $\mathbf{U}_h \in \mathbf{V}_{3h}$ has components $\mathbf{U}_{i,j}$ on $K_{i,j}$ $(i = 0, \ldots, N_1,\ j = 0, \ldots, N_2)$. Finally, $\mathbf{F}_h^\tau$ has components $\mathbf{F}_{i,j}^\tau$ on $K_{i,j}$ with

$$\mathbf{F}_{i,j}^\tau = \frac{1}{\Delta t} \frac{1}{9\Delta x \Delta y} \int_{(\tau-1)\Delta t}^{\tau\Delta t} \int_{K_{i,j}} \mathbf{f}(x, y, t)\mathrm{d}x\mathrm{d}y\mathrm{d}t. \tag{3.28}$$

A priori estimates are obtained by taking the scalar product in $\mathbf{V}_{3h}$ of each side of (3.27) with $6\Delta t\mathbf{U}_h^\tau$. We find (comparing with (2.31))

$$|\mathbf{U}_h^\tau|^2 - |\mathbf{U}_h^{\tau-1}|^2 + |\mathbf{U}_h^\tau - \mathbf{U}_h^{\tau-1}|^2 + 2\Delta t(\mathbf{A}_{3h}\mathbf{U}_h^\tau, \mathbf{U}_h^\tau) = 2\Delta t(\mathbf{F}_h^\tau, \mathbf{U}_h^\tau),$$

and in view of Lemma 3.2 (for $\mathbf{A}_{3h}$),

$$\begin{aligned}
|\mathbf{U}_h^\tau|^2 &\leq |\mathbf{U}_h^{\tau-1}|^2 + 2\Delta t|\mathbf{F}_h^\tau||\mathbf{U}_h^\tau| \\
&\leq \Delta t|\mathbf{U}_h^\tau|^2 + |\mathbf{U}_h^{\tau-1}|^2 + \Delta t|\mathbf{F}_h^\tau|^2, \\
|\mathbf{U}_h^\tau|^2 &\leq \frac{1}{1-\Delta t}[|\mathbf{U}_h^{\tau-1}|^2 + |\mathbf{F}_h^\tau|^2] \\
&\leq \frac{1}{1-\Delta t}\left[|\mathbf{U}_h^{\tau-1}|^2 + \int_{(\tau-1)\Delta t}^{\tau\Delta t} |\mathbf{f}(\cdot, t)|_{L^2}^2 \mathrm{d}t\right].
\end{aligned}$$

Thus, for $\Delta t \leq \frac{1}{2}$,

$$\left|\mathbf{U}_h^{\tau}\right|^2 \leq 4^{\Delta t}\left[\left|\mathbf{U}_h^{\tau-1}\right|^2 + \int_{(\tau-1)\Delta t}^{\tau\Delta t} \left|\mathbf{f}(\cdot,t)\right|_{L^2}^2 \mathrm{d}t\right]. \tag{3.29}$$

We write Eq. (3.29) for $\tau = n+s+1,\ s = 1,\ldots,q$. We multiply the equation for $\tau = n+s+1$ by $4^{(q-s)\Delta t}$ and add these equations for $s = 1,\ldots,q$. We find

$$\left|\mathbf{U}_h^{n+q+1}\right|^2 \leq 4^{q\Delta t}\left[\left|\mathbf{U}_h^{n+1}\right|^2 + \int_{(n+1)\Delta t}^{(n+q+1)\Delta t} \left|\mathbf{f}(\cdot,t)\right|_{L^2}^2 \mathrm{d}t\right]. \tag{3.30}$$

During the steps from $(n+1)\Delta t$ to $(n+q+1)\Delta t$, the $\mathbf{Z}_h$ are frozen. Thus

$$\mathbf{Z}_h^{n+s+1} = \mathbf{Z}_h^{n+1}, \quad s = 1,\ldots,q, \tag{3.31}$$

and we recover the $\mathbf{u}_h^{n+s+1}$ in the form

$$\mathbf{u}_h^{n+s+1} = \mathbf{U}_h^{n+s+1} + \mathbf{Z}_h^{n+1}. \tag{3.32}$$

Then, because of (3.30) and (2.15),

$$\left|\mathbf{u}_h^{n+q+1}\right|^2 \leq 4^{q\Delta t}\left[\left|\mathbf{u}_h^{n+1}\right|^2 + \int_{(n+1)\Delta t}^{(n+q+1)\Delta t} \left|\mathbf{f}(\cdot,t)\right|_{L^2}^2 \mathrm{d}t\right]. \tag{3.33}$$

Combining (3.33) with (3.26), we find

$$\left|\mathbf{u}_h^{n+q+1}\right|^2 \leq 4^{(q+1)\Delta t}\left[\left|\mathbf{u}_h^{n}\right|^2 + \int_{n\Delta t}^{(n+q+1)\Delta t} \left|\mathbf{f}(\cdot,t)\right|_{L^2}^2 \mathrm{d}t\right]. \tag{3.34}$$

We can repeat the procedure for any interval of time $(n\Delta t, (n+q+1)\Delta t),\ n = 1,\ldots,N_q$, and arrive at the stability result

$$\begin{aligned}\left|\mathbf{u}_h^{m}\right|^2 &\leq 4^{m\Delta t}\left[\left|\mathbf{u}_h^{0}\right|^2 + \int_0^{m\Delta t} \left|\mathbf{f}(\cdot,t)\right|_{L^2}^2 \mathrm{d}t\right] \\ &\leq 4^{T}\left[\left|\mathbf{u}^{0}\right|^2 + \int_0^{T} \left|\mathbf{f}(\cdot,t)\right|_{L^2}^2 \mathrm{d}t\right]\end{aligned} \tag{3.35}$$

valid for $m = 1,\ldots,N_q$.

Theorem 3.1 *The multilevel scheme defined by Eqs.* (3.22) *and* (3.27) *is stable in* $L^\infty(0,T;L^2(\mathcal{M})^3)$ *in the sense of* (3.35).

4 Other Schemes and Other Methods

The coarse grid schemes that we have used in Sects. 2 and 3 amount to using the same schemes on the coarse grid as on the fine grid. Another possibility for the coarse grid is to average on each coarse grid the fine grid equations associated with the corresponding fine grids. These schemes are made explicit below. However, the study of the stability of these new schemes appears difficult, and we will only present the study of stability in the one-dimensional case for the simple transport equation (see Sect. 4.1), and for a one-dimensional shallow water equation (see Sect. 4.2). Furthermore, the boundary condition will be space periodicity, and the stability analysis is made by the von Neumann method (see [22]).

4.1 The One-Dimensional Case

We start with the one-dimensional space, and consider the same problem as (2.1), with $f = 0$,

$$\frac{\partial u}{\partial t}(x,t) + \frac{\partial u}{\partial x}(x,t) = 0 \tag{4.1}$$

for $(x,t) \in (0,L) \times (0,T)$, and with the space periodicity boundary condition, and the initial condition

$$u(x,0) = u^0(x). \tag{4.2}$$

On the fine grid, we will perform an approximation by the implicit Euler scheme in time and upwind finite-volume in space, so that the scheme will be very much like the one in (2.16) except that the second formula of (2.16) is replaced by the periodicity condition

$$u_0^{n+\frac{s+1}{p}} = u_{3N}^{n+\frac{s+1}{p}}. \tag{4.3}$$

We perform p steps with a time step $\frac{\Delta t}{p}$ and a space mesh $\Delta x = \frac{L}{3N}$. Then as explained below, we make q steps with a time step Δt and a mesh step $3\Delta x$. Thus we start again with the p steps.

4.1.1 The Fine Grid Scheme with a Small Time Step

The scheme reads

$$\frac{p}{\Delta t}\left(u_j^\tau - u_j^{\tau-\frac{1}{p}}\right) + \frac{1}{\Delta x}\left(u_j^\tau - u_{j-1}^\tau\right) = 0, \tag{4.4}$$

where $\tau = n + \frac{s}{p}$, $s = 1, \dots, p$, $j = 1, \dots, 3N$, u_j^τ is meant to be an approximation of $\frac{1}{\Delta x}\int_{k_j} u(x, \tau\Delta t)\mathrm{d}x$ with $k_j = ((j-1)h, jh)$ and $h = \Delta x$; $u_0^\tau = u_{3N}^\tau$ by periodicity.

We associate with a sequence v_j, and its Fourier transform (see [22, p. 38]) is as follows:

$$\hat{v}(\xi) = \frac{1}{2\pi} \sum_{j=-\infty}^{+\infty} \mathrm{e}^{-\mathrm{i}jh\xi} v_j h. \tag{4.5}$$

Below we will consider periodic sequences v_j, $j \in \mathbb{Z}$, $v_{j+3N} = v_j$, $h^* = \frac{2\pi}{3N}$ and define the discrete Fourier coefficients (see [5, 9, 22])

$$\hat{v}_m = \frac{1}{3N} \sum_{j=1}^{3N} \mathrm{e}^{-imjh^*} v_j, \quad m = 1, \ldots, 3N. \tag{4.6}$$

We then have the discrete Parseval formula

$$\sum_{m=1}^{3N} |\hat{v}_m|^2 = \frac{1}{3N} \sum_{j=1}^{3N} |v_j|^2 \tag{4.7}$$

(see the details in [3, 22]). Note that the sequence $\{\hat{v}_m\}$ is itself periodic with period $3N$, and if $(\sigma v)_j = v_{j-1}$, then

$$\hat{\sigma v}_m = \mathrm{e}^{-imh^*} \hat{v}_m. \tag{4.8}$$

Then (4.4) is rewritten as

$$\left(1 + \frac{\Delta t}{p\Delta x}\right) u_j^\tau - \frac{\Delta t}{p\Delta x} u_{j-1}^\tau = u_j^{\tau - \frac{1}{p}}, \tag{4.9}$$

that is, for the Fourier transforms defined as in (4.6), where $h^* = \frac{2\pi}{3N}$,

$$\left(1 + \frac{\Delta t}{p\Delta x}(1 - \mathrm{e}^{-imh^*})\right) \hat{u}_m^\tau = \hat{u}_m^{\tau - \frac{1}{p}}, \quad m = 1, \ldots, 3N. \tag{4.10}$$

Hence the amplification factor for the fine mesh is

$$g_{F,m} = \left[1 + \frac{\Delta t}{p\Delta x}(1 - \mathrm{e}^{-imh^*})\right]^{-1}, \quad m = 1, \ldots, 3N. \tag{4.11}$$

We observe that

$$g_{F,m}^{-1} = \left[1 + \frac{\Delta t}{p\Delta x}(1 - \cos(h^* m)) + \mathrm{i}\frac{\Delta t}{p\Delta x} \sin(h^* m)\right],$$

$$|g_{F,m}^{-1}|^2 = \left[1 + \frac{\Delta t}{p\Delta x}(1 - \cos(h^* m))\right]^2 + \left(\frac{\Delta t}{p\Delta x}\right)^2 \sin^2(h^* m),$$

$$= 1 + 2(1 - \cos(h^* m))\left(\left(\frac{\Delta t}{p\Delta x}\right)^2 + \frac{\Delta t}{p\Delta x}\right).$$

We conclude that

$$|g_{F,m}| \leq 1, \quad m = 1, \ldots, 3N. \tag{4.12}$$

Recall that $\tau = n + \frac{s}{p}$, $s = 1, \ldots, p$. Denoting by u_h^τ the piecewise constant function given by $u_h^\tau = u_j^\tau$ on k_j, (4.7) and (4.12) yield

$$\begin{aligned} |u_h^{n+\frac{s}{p}}|^2 &= \sum_{j=1}^{3N} \Delta x |u_j^{n+\frac{s}{p}}|^2 = 3N\Delta x \sum_{m=1}^{3N} |\hat{u}_m^{n+\frac{s}{p}}|^2 \\ &\leq 3N\Delta x \sum_{m=1}^{3N} |\hat{u}_m^n|^2 = |u_h^n|^2 \quad \text{for } s = 1, \ldots, q. \end{aligned}$$

In particular, for $s = p$,

$$|u_h^{n+1}|^2 \leq |u_h^n|^2, \tag{4.13}$$

and therefore these steps of the scheme (4.4) on the fine grid are stable for the L^2-norm.

4.1.2 The Coarse Grid Scheme with a "Large" Time Step

Considering first the analogue of (4.4) with a time step Δt and a space mesh Δx, we would write ($\tau = n + s + 1$ now, $s = 1, \ldots, q$)

$$\frac{1}{\Delta t}(u_j^\tau - u_j^{\tau-1}) + \frac{1}{\Delta x}(u_j^\tau - u_{j-1}^\tau) = 0. \tag{4.14}$$

To obtain the scheme with a time step Δt and a space mesh $3\Delta x$, we add (average) Eq. (4.14) corresponding to $j = 3l, 3l-1, 3l-2$.

Setting

$$U_l^\tau = \frac{1}{3}(u_{3l}^\tau + u_{3l-1}^\tau + u_{3l-2}^\tau), \tag{4.15}$$

we obtain

$$\frac{1}{\Delta t}(U_l^\tau - U_l^{\tau-1}) + \frac{1}{3\Delta x}(u_{3l}^\tau - u_{3l-3}^\tau) = 0 \tag{4.16}$$

for $l = 1, \ldots, N$.

We elaborate on the $u = U + Z$ decomposition (independent of the time step).

The $u = U + Z$ Decomposition Given the sequence u_j, $j = 1, \ldots, 3N$ ($u_0 = u_{3N}$), we define the sequence

$$U_\ell = \frac{1}{3}\sum_{\alpha=0}^{2} u_{3l-\alpha}, \quad l = 1, \ldots, N, \tag{4.17}$$

and the sequences

$$Z_{3l-\alpha} = u_{3l-\alpha} - U_l, \tag{4.18}$$

$\alpha = 0, 1, 2$, $\ell = 1, \ldots, N$. We observe that

$$\sum_{\alpha=0}^{2} Z_{3l-\alpha} = 0.$$

Now the multistep algorithm that we consider consists in freezing the Z during the step $n+2, \ldots, n+q+1$, that is,

$$Z_j^{n+s+1} = Z_j^{n+1}, \quad s = 1, \ldots, q, \quad j = 1, \ldots, 3N, \tag{4.19}$$

so that

$$Z_{3l-\alpha}^{\tau} = Z_{3l-\alpha}^{n+1} = u_{3l-\alpha}^{\tau} - U_l^{\tau}$$

for $\alpha = 0, 1, 2$, $\tau = n+s+1$, $s = 1, \ldots, q$. Hence $U_l^\tau - U_l^{\tau-1} = u_{3\ell-\alpha}^\tau - u_{3\ell-\alpha}^{\tau-1}$ for $\alpha = 0, 1, 2$, and for those values of τ. With $\alpha = 0$, (4.16) becomes

$$\frac{1}{\Delta t}\left(u_{3l}^\tau - u_{3l}^{\tau-1}\right) + \frac{1}{3\Delta x}\left(u_{3l}^\tau - u_{3l-3}^\tau\right) = 0. \tag{4.20}$$

That is, as in (4.9),

$$\left(1 + \frac{\Delta t}{3\Delta x}\right)u_{3l}^\tau - \frac{\Delta t}{3\Delta x}u_{3l-3}^\tau = u_{3l}^{\tau-1}. \tag{4.21}$$

Before we introduce the Fourier transform of (4.21) and the amplification function similar to the g_F, we have to elaborate a bit more on the $u = U + Z$ decomposition at the level of the Fourier transforms.

We write (independent of the time step τ), with $h^* = \frac{2\pi}{3N}$ for $m = 1, \ldots, 3N$,

$$\begin{aligned}\hat{u}_m &= \frac{1}{3N}\sum_{j=1}^{3N} u_j \mathrm{e}^{-ih^* jm} \\ &= \frac{1}{3N}\sum_{\ell=1}^{N}\left(u_{3l}\mathrm{e}^{-3ih^* lm} + u_{3l-1}\mathrm{e}^{-ih^*(3l-1)m} + u_{3l-2}\mathrm{e}^{-ih^*(3l-2)m}\right).\end{aligned}$$

We now introduce the partial Fourier sum of the type of (4.6),

$$\hat{u}_{(3l-\alpha),m} = \frac{1}{3N}\sum_{\ell=1}^{N} u_{3l-\alpha} e^{-ih^*3lm}. \tag{4.22}$$

We observe that this partial Fourier sum is periodic in m with period $3N$, and that Parseval relation similar to (4.7) holds,

$$\sum_{m=1}^{3N} |\hat{u}_{(3l-\alpha),m}|^2 = \frac{1}{3N}\sum_{\ell=1}^{N} |u_{3l-\alpha}|^2, \quad \alpha = 0, 1, 2. \tag{4.23}$$

We can hence write

$$\hat{u}_m = \hat{u}_{(3l),m} + e^{ih^*m}\hat{u}_{(3l-1),m} + e^{2ih^*m}\hat{u}_{(3l-2),m}. \tag{4.24}$$

Then

$$\hat{u}_{(3l-3),m} = \hat{u}_{(3l),m} e^{-3ih^*m}, \tag{4.25}$$

and now (4.21) yields by a partial Fourier transform

$$\left(1 + \frac{\Delta t}{3\Delta x}\right)\hat{u}^{\tau}_{(3l),m} - \frac{\Delta t}{3\Delta x} e^{-3ih^*m}\hat{u}^{\tau}_{(3l),m} = \hat{u}^{\tau-1}_{(3l),m}. \tag{4.26}$$

That is,

$$\hat{u}^{\tau}_{(3l),m} = g_{C,m}\hat{u}^{\tau-1}_{(3l),m}, \quad m = 1, \ldots, 3N, \tag{4.27}$$

corresponding to the amplification factor $g_{C,m}$ with

$$g^{-1}_{C,m} = 1 + \frac{\Delta t}{3\Delta x}\left(1 - e^{-3ih^*m}\right). \tag{4.28}$$

We can conclude as before that $|g^{-1}_{C,m}| \geq 1$,

$$|g_{C,m}| \leq 1, \quad m = 1, \ldots, 3N, \tag{4.29}$$

and thus the scheme (4.21), (4.26) is “stable”. Also

$$\hat{u}^{n+s+1}_{(3l),m} = g^{s}_{C,m}\hat{u}^{n+1}_{(3l),m}, \quad m = 1, \ldots, 3N, \ s = 1, \ldots, q. \tag{4.30}$$

The important point now is that we know nothing about the stability of the u^{τ}_{3l-1}, u^{τ}_{3l-2}, and we have to elaborate more to prove this stability.

In the similar way to (4.24), we write for $m = 1, \ldots, 3N$,

$$\hat{Z}_m = \hat{Z}_{(3l),m} + e^{ih^*m}\hat{Z}_{(3l-1),m} + e^{2ih^*m}\hat{Z}_{(3l-2),m}. \tag{4.31}$$

The relations

$$u_{3l-\alpha} = U_l + Z_{3l-\alpha}, \quad \alpha = 0, 1, 2$$

given by the partial discrete Fourier transform for $m = 1, \dots, 3N$,

$$\hat{u}_{(3l-\alpha),m} = \hat{U}_{(l),m} + \hat{Z}_{(3l-\alpha),m}. \tag{4.32}$$

Hence with (4.19) and (4.32),

$$\hat{u}^{n+s+1}_{(3l-\alpha),m} = \hat{U}^{n+s+1}_{(l),m} + \hat{Z}^{n+1}_{(3l-\alpha),m}. \tag{4.33}$$

Using (4.30), we obtain the expression of $\hat{U}^{n+s+1}_{(l),m}$ for $\alpha = 0$,

$$\hat{U}^{n+s+1}_{(l),m} = g^{s}_{C,m}\hat{u}^{n+1}_{(3l),m} - \hat{Z}^{n+1}_{(3l),m}, \quad m = 1, \dots, 3N. \tag{4.34}$$

There remains to express $\hat{Z}^{n+1}_{(3l)}$ in terms of the $\hat{u}^{n+1}_{(3l-\alpha),m}$, $\alpha = 0, 1, 2$.
We proceed in the physical space, independent of the time step τ, to have

$$U_l = \frac{1}{3}(u_{3l} + u_{3l-1} + u_{3l-2})$$

and

$$\begin{cases} Z_{3l} = u_{3l} - U_l = \frac{1}{3}(2u_{3l} - u_{3l-1} - u_{3l-2}), \\ Z_{3l-1} = u_{3l-1} - U_l = \frac{1}{3}(2u_{3l-1} - u_{3l} - u_{3l-2}), \\ Z_{3l-2} = u_{3l-2} - U_l = \frac{1}{3}(2u_{3l-2} - u_{3l} - u_{3l-1}). \end{cases} \tag{4.35}$$

Thus for the Fourier transforms, for $m = 1, \dots, 3N$,

$$\begin{cases} \hat{Z}_{(3l),m} = \frac{1}{3}(2\hat{u}_{(3l),m} - \hat{u}_{(3l-1),m} - \hat{u}_{(3l-2),m}), \\ \hat{Z}_{(3l-1),m} = \frac{1}{3}(2\hat{u}_{(3l-1),m} - \hat{u}_{(3l),m} - \hat{u}_{(3l-2),m}), \\ \hat{Z}_{(3l-2),m} = \frac{1}{3}(2\hat{u}_{(3l-2),m} - \hat{u}_{(3l),m} - \hat{u}_{(3l-1),m}). \end{cases} \tag{4.36}$$

This holds in particular at the time step $\tau = n + 1$.

Now we look for the expression of the $\hat{u}^{n+s+1}_{(3l-\alpha),m}$ $(\alpha = 0, 1, 2)$, in terms of the $\hat{u}^{n+1}_{(3l-\beta),m}$, that of $\hat{u}^{n+s+1}_{(3l),m}$ has been already found (see (4.30)).

By (4.32)–(4.34), (4.36) and (4.19),

$$\hat{u}^{n+s+1}_{(3l-1),m} = \big(g^{s}_{C,m} - 1\big)\hat{u}^{n+1}_{(3l),m} + \hat{u}^{n+1}_{(3l-1),m}, \tag{4.37}$$

$$\hat{u}^{n+s+1}_{(3l-2),m} = \big(g^{s}_{C,m} - 1\big)\hat{u}^{n+1}_{(3l),m} + \hat{u}^{n+1}_{(3l-2),m}. \tag{4.38}$$

We rewrite (4.30), (4.37)–(4.38) in matrical form,

$$\begin{pmatrix} \hat{u}^{n+s+1}_{(3l),m} \\ \hat{u}^{n+s+1}_{(3l-1),m} \\ \hat{u}^{n+s+1}_{(3l-2),m} \end{pmatrix} = G^{(s)}_{C,m} \begin{pmatrix} \hat{u}^{n+1}_{(3l),m} \\ \hat{u}^{n+1}_{(3l-1),m} \\ \hat{u}^{n+1}_{(3l-2),m} \end{pmatrix}, \quad m = 1, \ldots, 3N, \tag{4.39}$$

$$G^{(s)}_{C,m} = \begin{pmatrix} g^s_{C,m} & 0 & 0 \\ g^s_{C,m} - 1 & 1 & 0 \\ g^s_{C,m} - 1 & 0 & 1 \end{pmatrix}. \tag{4.40}$$

The passing from u^{n+1} to u^{n+s+1} is given in the matrical form by (4.39). The stability of the scheme for passing from u^{n+1} to u^{n+s+1} is equivalent to showing that the spectral radius of $G^{(s)}_{C,m}$ is not larger than 1 for $m = 1, \ldots, 3N$. The eigenvalues of $G^{(s)}_{C,m}$ are not larger than 1. These eigenvalues are 1, 1, $g^s_{C,m}$, and we have seen that $|g_{C,m}| \leq 1$.

More precisely, using that the spectral radius of $G^{(s)}_{C,m}$ is less than 1 and (4.23), we have

$$\begin{aligned} |u_h^{n+s+1}|^2 &= \sum_{\alpha=0}^{2} \sum_{\ell=1}^{N} \Delta x |u^{n+s+1}_{3\ell-\alpha}|^2 = 3N\Delta x \sum_{\alpha=0}^{2} \sum_{m=1}^{3N} |\hat{u}^{n+s+1}_{(3\ell-\alpha),m}|^2 \\ &\leq 3N\Delta x \sum_{\alpha=0}^{2} \sum_{m=1}^{3N} |\hat{u}^{n+1}_{(3\ell-\alpha),m}|^2 = |u_h^{n+1}|^2 \end{aligned} \tag{4.41}$$

and for $s = q$,

$$|u_h^{n+q+1}| \leq |u_h^{n+1}|. \tag{4.42}$$

Combining (4.13) and (4.42), we obtain the stability of the scheme.

Theorem 4.1 *The multilevel scheme defined by Eqs.* (4.4) *and* (4.16) *is stable in* $L^\infty(0, \infty; L^2(\mathcal{M}))$. *More precisely, for all* n,

$$|u_h^n| \leq |u^0|. \tag{4.43}$$

4.2 The Linearized 1D Shallow Water Equation

By restriction to 1 dimension, Eq. (3.2) with $f = 0$ become

$$\begin{cases} \frac{\partial u}{\partial t} + \tilde{u}_0 \frac{\partial u}{\partial x} + g \frac{\partial \phi}{\partial x} = 0, \\ \frac{\partial \phi}{\partial t} + \tilde{u}_0 \frac{\partial \phi}{\partial x} + \tilde{\phi}_0 \frac{\partial u}{\partial x} = 0. \end{cases} \tag{4.44}$$

We assume the background flow $(\tilde{u}_0, \tilde{\phi}_0)$ to be supersonic (supercritical), that is,

$$\tilde{u}_0 > \sqrt{g\tilde{\phi}_0}. \tag{4.45}$$

The boundary conditions are space periodicity, and the initial conditions are given such that they are similar as (3.4). The time and space meshes are the same as in Sects. 2.1 and 2.2.

4.2.1 The Fine Grid Scheme with a "Small" Time Step

The fine grid mesh scheme reads

$$\begin{cases} \frac{p}{\Delta t}(u_j^\tau - u_j^{\tau-\frac{1}{p}}) + \frac{\tilde{u}_0}{\Delta x}(u_j^\tau - u_{j-1}^\tau) + \frac{g}{\Delta x}(\phi_j^\tau - \phi_{j-1}^\tau) = 0, \\ \frac{p}{\Delta t}(\phi_j^\tau - \phi_j^{\tau-\frac{1}{p}}) + \frac{\tilde{u}_0}{\Delta x}(\phi_j^\tau - \phi_{j-1}^\tau) + \frac{\tilde{\phi}_0}{\Delta x}(u_j^\tau - u_{j-1}^\tau) = 0, \end{cases} \tag{4.46}$$

where $\tau = n + \frac{s}{p}$, $s = 1, \dots, p$, $j = 1, \dots, 3N$, $u_0^\tau = u_{3N}^\tau$, $\phi_0^\tau = \phi_{3N}^\tau$ by space periodicity.

We rewrite (4.46) in the form

$$\begin{cases} (1 + \frac{\tilde{u}_0}{p}\frac{\Delta t}{\Delta x})u_j^\tau - \frac{\tilde{u}_0}{p}\frac{\Delta t}{\Delta x}u_{j-1}^\tau + \frac{g}{p}\frac{\Delta t}{\Delta x}(\phi_j^\tau - \phi_{j-1}^\tau) = u_j^{\tau-\frac{1}{p}}, \\ (1 + \frac{\tilde{u}_0}{p}\frac{\Delta t}{\Delta x})\phi_j^\tau - \frac{\tilde{u}_0}{p}\frac{\Delta t}{\Delta x}\phi_{j-1}^\tau + \frac{\tilde{\phi}_0}{p}\frac{\Delta t}{\Delta x}(u_j^\tau - u_{j-1}^\tau) = \phi_j^{\tau-\frac{1}{p}}. \end{cases} \tag{4.47}$$

From this, we deduce for the Fourier transforms, for $m = 1, \dots, 3N$,

$$\begin{cases} (1 + \frac{\tilde{u}_0}{p}\frac{\Delta t}{\Delta x}(1 - \mathrm{e}^{-imh^*}))\hat{u}_m^\tau + \frac{g}{p}\frac{\Delta t}{\Delta x}\hat{\phi}_m^\tau(1 - \mathrm{e}^{-imh^*}) = \hat{u}_m^{\tau-\frac{1}{p}}, \\ (1 + \frac{\tilde{u}_0}{p}\frac{\Delta t}{\Delta x}(1 - \mathrm{e}^{-imh^*}))\hat{\phi}_m^\tau + \frac{\tilde{\phi}_0}{p}\frac{\Delta t}{\Delta x}\hat{u}_m^\tau(1 - \mathrm{e}^{-imh^*}) = \hat{\phi}_m^{\tau-\frac{1}{p}}, \end{cases} \tag{4.48}$$

that is,

$$\begin{pmatrix} \hat{u}_m^\tau \\ \hat{\phi}_m^\tau \end{pmatrix} = G_{F,m} \begin{pmatrix} \hat{u}_m^{\tau-\frac{1}{p}} \\ \hat{\phi}_m^{\tau-\frac{1}{p}} \end{pmatrix} \tag{4.49}$$

with

$$G_{F,m}^{-1} = \begin{pmatrix} 1 + \frac{\tilde{u}_0}{p}\frac{\Delta t}{\Delta x}(1 - \mathrm{e}^{-imh^*}) & \frac{g}{p}\frac{\Delta t}{\Delta x}(1 - \mathrm{e}^{-imh^*}) \\ \frac{\tilde{\phi}_0}{p}\frac{\Delta t}{\Delta x}(1 - \mathrm{e}^{-imh^*}) & 1 + \frac{\tilde{u}_0}{p}\frac{\Delta t}{\Delta x}(1 - \mathrm{e}^{-imh^*}) \end{pmatrix}.$$

The eigenvalues of $G_{F,m}^{-1}$ are easily computed

$$\rho_{\pm,m} = 1 + \Lambda_\pm(1 - \mathrm{e}^{-imh^*})$$

with

$$\Lambda_{\pm} = \frac{1}{p}(\tilde{u}_0 \pm \sqrt{g\tilde{\phi}_0})\frac{\Delta t}{\Delta x}.$$

We have

$$|\rho_{\pm,m}|^2 = 1 + 2\big(1 - \cos(h^* m)\big)\big(\Lambda_{\pm}^2 + \Lambda_{\pm}\big).$$

The condition $\tilde{u}_0 > \sqrt{g\tilde{\phi}_0}$ implies $\Lambda_{\pm} > 0$, and thus

$$|\rho_{\pm,m}| \geq 1, \quad m = 1, \ldots, 3N.$$

Hence, setting $\mathbf{u} = (u, \phi)$ (comparing with (4.13)), we have

$$\left|\mathbf{u}_h^{n+1}\right|^2 \leq \left|\mathbf{u}_h^n\right|^2, \tag{4.50}$$

so that these steps of the small step scheme (4.46) are stable.

4.2.2 The Coarse Grid Scheme with a "Large" Time Step

We define the cell averages

$$U_l = \frac{1}{3}(u_{3l} + u_{3l-1} + u_{3l-2}),$$

$$\Phi_l = \frac{1}{3}(\phi_{3l} + \phi_{3l-1} + \phi_{3l-2})$$

and the incremental unknowns

$$Z_{3l-\alpha}^u = u_{3l-\alpha} - U_l,$$

$$Z_{3l-\alpha}^\phi = \phi_{3l-\alpha} - \Phi_l.$$

The analogue of scheme (4.16) reads

$$\begin{cases} \frac{1}{\Delta t}(U_l^\tau - U_l^{\tau-1}) + \frac{\tilde{u}_0}{3\Delta x}(u_{3l}^\tau - u_{3l-3}^\tau) + \frac{g}{3\Delta x}(\phi_{3l}^\tau - \phi_{3l-3}^\tau) = 0, \\ \frac{1}{\Delta t}(\Phi_l^\tau - \Phi_l^{\tau-1}) + \frac{\tilde{u}_0}{3\Delta x}(\phi_{3l}^\tau - \phi_{3l-3}^\tau) + \frac{\tilde{\phi}_0}{3\Delta x}(u_{3l}^\tau - u_{3l-3}^\tau) = 0 \end{cases} \tag{4.51}$$

for $\tau = n + s + 1$, $s = 1, \ldots, q$ and $l = 1, \ldots, N$.

Observing as in (4.19) that

$$Z_j^{u,n+s+1} = Z_j^{u,n+1}, \qquad Z_j^{\phi,n+s+1} = Z_j^{\phi,n+1} \tag{4.52}$$

for $s = 1, \ldots, q$, $j = 1, \ldots, 3N$ and thus that

$$U_l^\tau - U_l^{\tau-1} = u_{3l}^\tau - u_{3l}^{\tau-1},$$

$$\Phi_l^\tau - \Phi_l^{\tau-1} = \phi_{3l}^\tau - \phi_{3l}^{\tau-1},$$

(4.51) yields

$$\begin{cases} \frac{1}{\Delta t}(u_{3l}^\tau - u_{3l}^{\tau-1}) + \frac{\tilde{u}_0}{3\Delta x}(u_{3l}^\tau - u_{3l-3}^\tau) + \frac{g}{3\Delta x}(\phi_{3l}^\tau - \phi_{3l-3}^\tau) = 0, \\ \frac{1}{\Delta t}(\phi_{3l}^\tau - \phi_{3l}^{\tau-1}) + \frac{\tilde{u}_0}{3\Delta x}(\phi_{3l}^\tau - \phi_{3l-3}^\tau) + \frac{\tilde{\phi}_0}{3\Delta x}(u_{3l}^\tau - u_{3l-3}^\tau) = 0. \end{cases} \tag{4.53}$$

Hence, for the partial Fourier transforms for $m = 1, \dots, 3N$ (comparing with (4.27)),

$$\begin{pmatrix} \hat{u}_{(3l),m}^\tau \\ \hat{\phi}_{(3l),m}^\tau \end{pmatrix} = G_{C,m} \begin{pmatrix} \hat{u}_{(3l),m}^{\tau-1} \\ \hat{\phi}_{(3l),m}^{\tau-1} \end{pmatrix} \tag{4.54}$$

with

$$G_{C,m}^{-1} = \begin{pmatrix} 1 + \frac{\tilde{u}_0}{3}\frac{\Delta t}{\Delta x}(1 - \mathrm{e}^{-3ih^*m}) & \frac{g}{3}\frac{\Delta t}{\Delta x}(1 - \mathrm{e}^{-3ih^*m}) \\ \frac{\tilde{\phi}_0}{3}\frac{\Delta t}{\Delta x}(1 - \mathrm{e}^{-3ih^*m}) & 1 + \frac{\tilde{u}_0}{3}\frac{\Delta t}{\Delta x}(1 - \mathrm{e}^{-3ih^*m}) \end{pmatrix},$$

where $G_{C,m}^{-1}$ is very similar to $G_{F,m}^{-1}$, and we prove in the same way that its eigenvalues are larger than or equal to 1 in magnitude.

For the moment, we infer from (4.54) that

$$\begin{pmatrix} \hat{u}_{(3l),m}^{n+s+1} \\ \hat{\phi}_{(3l),m}^{n+s+1} \end{pmatrix} = G_{C,m}^s \begin{pmatrix} \hat{u}_{(3l),m}^{n+1} \\ \hat{\phi}_{(3l),m}^{n+1} \end{pmatrix}. \tag{4.55}$$

Then by (4.55),

$$\begin{aligned} \begin{pmatrix} \hat{U}_{(l),m}^{n+s+1} \\ \hat{\Phi}_{(l),m}^{n+s+1} \end{pmatrix} &= \begin{pmatrix} \hat{u}_{(3l),m}^{n+s+1} \\ \hat{\phi}_{(3l),m}^{n+s+1} \end{pmatrix} - \begin{pmatrix} \hat{Z}_{(3l),m}^{u,n+s+1} \\ \hat{Z}_{(3l),m}^{\phi,n+s+1} \end{pmatrix} \\ &= G_{C,m}^s \begin{pmatrix} \hat{u}_{(3l),m}^{n+1} \\ \hat{\phi}_{(3l),m}^{n+1} \end{pmatrix} - \begin{pmatrix} \hat{Z}_{(3l),m}^{u,n+1} \\ \hat{Z}_{(3l),m}^{\phi,n+1} \end{pmatrix}. \end{aligned} \tag{4.56}$$

We then need to express $\hat{u}_{(3l-\alpha),m}^{n+s+1}$, $\hat{\phi}_{(3l-\alpha),m}^{n+s+1}$ in terms of $\hat{u}_{(3l-\beta),m}^{n+1}$, $\hat{\phi}_{(3l-\beta),m}^{n+1}$, $\alpha = 1, 2$, $\beta = 0, 1, 2$. We write as in Eqs. (4.37)–(4.38),

$$\begin{pmatrix} \hat{u}_{(3l-1),m}^{n+s+1} \\ \hat{\phi}_{(3l-1),m}^{n+s+1} \end{pmatrix} = (G_{C,m}^s - I) \begin{pmatrix} \hat{u}_{(3l),m}^{n+1} \\ \hat{\phi}_{(3l)}^{n+1} \end{pmatrix} + \begin{pmatrix} \hat{u}_{(3l-1),m}^{n+1} \\ \hat{\phi}_{(3l-1),m}^{n+1} \end{pmatrix}, \tag{4.57}$$

$$\begin{pmatrix} \hat{u}_{(3l-2),m}^{n+s+1} \\ \hat{\phi}_{(3l-2),m}^{n+s+1} \end{pmatrix} = (G_{C,m}^s - I) \begin{pmatrix} \hat{u}_{(3l),m}^{n+1} \\ \hat{\phi}_{(3l),m}^{n+1} \end{pmatrix} + \begin{pmatrix} \hat{u}_{(3l-2),m}^{n+1} \\ \hat{\phi}_{(3l-2),m}^{n+1} \end{pmatrix}. \tag{4.58}$$

In the end,

$$\begin{pmatrix} \hat{u}^{n+s+1}_{(3l),m} \\ \hat{\phi}^{n+s+1}_{(3l),m} \\ \hat{u}^{n+s+1}_{(3l-1),m} \\ \hat{\phi}^{n+s+1}_{(3l-1),m} \\ \hat{u}^{n+s+1}_{(3l-2),m} \\ \hat{\phi}^{n+s+1}_{(3l-2),m} \end{pmatrix} = \mathcal{G}^{(s)}_{C,m} \begin{pmatrix} \hat{u}^{n+1}_{(3l),m} \\ \hat{\phi}^{n+1}_{(3l),m} \\ \hat{u}^{n+1}_{(3l-1),m} \\ \hat{\phi}^{n+1}_{(3l-1),m} \\ \hat{u}^{n+1}_{(3l-2),m} \\ \hat{\phi}^{n+1}_{(3l-2),m} \end{pmatrix}, \quad m = 1, \ldots, 3N \tag{4.59}$$

with

$$\mathcal{G}^{(s)}_{C,m} = \begin{pmatrix} G^{s}_{C,m} & 0 & 0 \\ G^{s}_{C,m} - I & I & 0 \\ G^{s}_{C,m} - I & 0 & I \end{pmatrix}.$$

All the eigenvalues of $\mathcal{G}^{(s)}_{C,m}$ are less than or equal to 1, which ensures the stability of the scheme (4.51) going from $t = (n+1)\Delta t$ to $t = (n+s+1)\Delta t$.

Then we have

$$\left|\mathbf{u}_h^{n+s+1}\right| \le \left|\mathbf{u}_h^{n+1}\right|, \quad \text{for } s = 1, \ldots, q. \tag{4.60}$$

Theorem 4.2 *The multilevel scheme defined by Eqs.* (4.46) *and* (4.51) *is stable in* $L^\infty(0,\infty; L^2(\mathcal{M})^2)$. *More precisely, for all n,*

$$\left|\mathbf{u}_h^{n}\right| \le \left|\mathbf{u}^{0}\right|. \tag{4.61}$$

References

1. Adamy, K., Bousquet, A., Faure, S., et al.: A multilevel method for finite-volume discretization of the two-dimensional nonlinear shallow-water equations. Ocean Model. **33**, 235–256 (2010). doi:10.1016/j.ocemod.2010.02.006
2. Adamy, K., Pham, D.: A finite-volume implicit Euler scheme for the linearized shallow water equations: stability and convergence. Numer. Funct. Anal. Optim. **27**(7–8), 757–783 (2006)
3. Bellanger, M.: Traitement du Signal. Dunod, Paris (2006)
4. Bousquet, A., Marion, M., Temam, R.: Finite volume multilevel approximation of the shallow water equations II. (2013, in preparation)
5. Canuto, C., Hussaini, M.Y., Quarteroni, A., Zhang, T.A.: Spectral Methods, Evolution to Complex Geometries and Applications to Fluid Dynamics, Scientific Computation. Springer, Berlin (2007)
6. Chen, Q., Shiue, M.C., Temam, R.: The barotropic mode for the primitive equations. J. Sci. Comput. **45**, 167–199 (2010). doi:10.1007/s10915-009-9343-8. Special issue in memory of David Gottlieb
7. Chen, Q., Shiue, M.C., Temam, R., Tribbia, J.: Numerical approximation of the inviscid 3D Primitive equations in a limited domain. Modél. Math. Anal. Numér. **45**, 619–646 (2012). doi:10.105/m2an/2011058

8. Dubois, T., Jauberteau, F., Temam, R.: Dynamic, Multilevel Methods and the Numerical Simulation of Turbulence. Cambridge University Press, Cambridge (1999)
9. Dautray, R., Lions, J.L.: Mathematical Analysis and Numerical Methods for Science and Technology. Springer, Berlin (1990–1992)
10. Eymard, R., Gallouet, T., Herbin, R.: Finite volume methods. In: Ciarlet, P.G., Lions, J.L. (eds.) Handbook of Numerical Analysis, vol. VII, pp. 713–1020. North-Holland, Amsterdam (2002)
11. Gie, G.M., Temam, R.: Cell centered finite-volume methods using Taylor series expansion scheme without fictitious domains. Int. J. Numer. Anal. Model. **7**(1), 1–29 (2010)
12. Huang, A., Temam, R.: The linearized 2D inviscid shallow water equations in a rectangle: boundary conditions and well-posedness. (2013, to appear)
13. Leveque, R.J.: Finite Volume Methods for Hyperbolic Problems. Cambridge Texts in Applied Mathematics. Cambridge University Press, Cambridge (2002)
14. Lions, J.L., Temam, R., Wang, S.: Models of the coupled atmosphere and ocean (CAO I). Comput. Mech. Adv. **1**, 5–54 (1993)
15. Lions, J.L., Temam, R., Wang, S.: Numerical analysis of the coupled models of atmosphere and ocean (CAO II). Comput. Mech. Adv. **1**, 55–119 (1993)
16. Lions, J.L., Temam, R., Wang, S.: Splitting up methods and numerical analysis of some multiscale problems. Comput. Fluid Dyn. J. **5**(2), 157–202 (1996). Special issue dedicated to A. Jameson
17. Marchuk, G.I.: Methods of numerical mathematics, 2nd edn. Applications of Mathematics, vol. 2. Springer, New York (1982). Translated from the Russian by Arthur A. Brown
18. Marion, M., Temam, R.: Nonlinear Galerkin methods. SIAM J. Numer. Anal. **26**, 1139–1157 (1989)
19. Marion, M., Temam, R.: Navier-Stokes equations, theory and approximation. In: Ciarlet, P.G., Lions, J.L. (eds.) Handbook of Numerical Analysis, vol. VI, pp. 503–689. North-Holland, Amsterdam (1998)
20. Rousseau, A., Temam, R., Tribbia, J.: The 3D primitive equations in the absence of viscosity: boundary conditions and well-posedness in the linearized case. J. Math. Pures Appl. **89**(3), 297–319 (2008). doi:10.1016/j.matpur.2007.12.001
21. Rousseau, A., Temam, R., Tribbia, J.: Boundary value problems for the inviscid primitive equations in limited domains. In: Temam, R.M., Tribbia, J.J., Ciarlet, P.G. (eds.) Computational Methods for the Atmosphere and the Oceans, Handbook of Numerical Analysis, vol. XIV. Elsevier, Amsterdam (2008)
22. Strikwerda, J.C.: Finite Difference Schemes and Partial Differential Equations, 2nd edn. SIAM, Philadelphia (2004)
23. Temam, R.: Inertial manifolds and multigrid methods. SIAM J. Math. Anal. **21**, 154–178 (1990)
24. Temam, R., Tribbia, J.: Open boundary conditions for the primitive and Boussinesq equations. J. Atmos. Sci. **60**, 2647–2660 (2003)
25. Yanenko, N.N.: The Method of Fractional Steps, The Solution of Problems of Mathematical Physics in Several Variables. Springer, New York (1971). Translated from the Russian by T. Cheron. English translation edited by M. Holt

Non-Gaussian Test Models for Prediction and State Estimation with Model Errors

Michal Branicki, Nan Chen, and Andrew J. Majda

Abstract Turbulent dynamical systems involve dynamics with both a large dimensional phase space and a large number of positive Lyapunov exponents. Such systems are ubiquitous in applications in contemporary science and engineering where the statistical ensemble prediction and the real time filtering/state estimation are needed despite the underlying complexity of the system. Statistically exactly solvable test models have a crucial role to provide firm mathematical underpinning or new algorithms for vastly more complex scientific phenomena. Here, a class of statistically exactly solvable non-Gaussian test models is introduced, where a generalized Feynman-Kac formulation reduces the exact behavior of conditional statistical moments to the solution to inhomogeneous Fokker-Planck equations modified by linear lower order coupling and source terms. This procedure is applied to a test model with hidden instabilities and is combined with information theory to address two important issues in the contemporary statistical prediction of turbulent dynamical systems: the coarse-grained ensemble prediction in a perfect model and the improving long range forecasting in imperfect models. The models discussed here should be useful for many other applications and algorithms for the real time prediction and the state estimation.

Keywords Prediction · Model error · Information theory · Feynman-Kac framework · Fokker-Planck · Turbulent dynamical systems

Mathematics Subject Classification 60G25 · 60H10 · 60H30 · 82C31 · 94A15

Project supported by the Office of Naval Research (ONR) Grants (No. ONR DRI N00014-10-1-0554) and the DOD-MURI award "Physics Constrained Stochastic-Statistical Models for Extended Range Environmental Prediction".

M. Branicki (✉) · N. Chen · A.J. Majda
Department of Mathematics and Center for Atmosphere Ocean Science, Courant Institute of Mathematical Sciences, New York University, New York, NY, USA
e-mail: branicki@cims.nyu.edu

N. Chen
e-mail: chennan@cims.nyu.edu

A.J. Majda
e-mail: jonjon@cims.nyu.edu

P.G. Ciarlet et al. (eds.), *Partial Differential Equations: Theory, Control and Approximation*, DOI 10.1007/978-3-642-41401-5_4,

1 Introduction

Turbulent dynamical systems involve dynamics with both a large dimensional phase space and a large number of positive Lyapunov exponents. Such extremely complex systems are ubiquitous in many disciplines of contemporary science and engineering such as climate-atmosphere-ocean science, neural science, material science, and engineering turbulence. Wide contemporary interest topics involve the statistical ensemble prediction (see [31]) and the real time state estimation/filtering (see [34]) for the extremely complex systems while coping with the fundamental limitations of model error and the curse of small ensemble size (see [22]).

An important role of mathematics in applied sciences is to develop simple and accurate (or easily solvable) test models with unambiguous mathematical features which nevertheless capture crucial features of vastly more complex systems in science and engineering. Such models provide the firm underpinning for both the advancing scientific understanding and the developing new numerical or statistical understanding. One of the authors developed this approach with various collaborators over the past few years for paradigm problems for turbulent dynamical systems. For example, simple statistically exactly solvable test models were developed for slow-fast systems (see [12, 13]), turbulent tracers (see [2, 14, 21, 33]) and as stochastic parameterization algorithms for the real time filtering of turbulent dynamical systems with the judicious model error (see [8, 9, 15, 34, 35]). Such models were utilized as unambiguous test models for improving prediction with imperfect models in climate science through the empirical information theory (see [5, 10, 28–30]) and for testing algorithms for uncertainty quantification (see [4, 5, 25]).

Here, we study non-Gaussian statistics in a class of test models which are statistically exactly solvable through a generalized Feynman-Kac formula (see [16, 21]) which reduces the exact behavior of conditional statistical moments to the solution to inhomogeneous Fokker-Planck equations modified by linear lower-order coupling terms and source terms. This exact procedure is developed in Sect. 2 below and involves only the marginal averaging and the integration by parts. In Sect. 3, elementary test models are introduced where the general procedure from Sect. 2 can be evaluated through elementary numerical solutions to the coupled generalized Fokker-Planck equations (CGFPE). Section 4 contains a brief introduction to the use of information theory to the quantify model error in a framework adapted to the present context. Section 5 contains two applications of the material in Sects. 3–4 to the statistical ensemble forecasting: the first application involves the coarse-grained ensemble prediction in a perfect model with hidden instabilities; the second application involves the use of imperfect models for the long range forecasting.

2 Test Models with Exactly Solvable Conditional Moments

We consider a special class of test models and illustrate that the evolution of the exact conditional statistical moments can be calculated through the solution to coupled generalized Fokker-Planck equations (CGFPE). Our elementary derivation follows the philosophy of the generalized Feynman-Kac framework (see [16, 21]) although we do not know any specific reference for the general principle developed below.

Consider a vector $\boldsymbol{u} \in \mathbb{R}^M$ partitioned into components $\boldsymbol{u} = (\boldsymbol{u}_{\mathrm{I}}, \boldsymbol{u}_{\mathrm{II}})$ with $\boldsymbol{u}_{\mathrm{I}} \in \mathbb{R}^{M_{\mathrm{I}}}$, $\boldsymbol{u}_{\mathrm{II}} \in \mathbb{R}^{M_{\mathrm{II}}}$, and $M = M_{\mathrm{I}} + M_{\mathrm{II}}$. We focus on the special class of test models given by the system of (Itô) SDE's,

$$\begin{cases} \mathrm{d}\boldsymbol{u}_{\mathrm{I}} = F_{\mathrm{I}}(\boldsymbol{u}_{\mathrm{I}}, t)\mathrm{d}t + \sigma_{\mathrm{I}}(\boldsymbol{u}_{\mathrm{I}}, t)\mathrm{d}W_{\mathrm{I}}(t), \\ \mathrm{d}\boldsymbol{u}_{\mathrm{II}} = (F_{\mathrm{II}}(\boldsymbol{u}_{\mathrm{I}}, t) + \Gamma(\boldsymbol{u}_{\mathrm{I}}, t)\boldsymbol{u}_{\mathrm{II}})\mathrm{d}t + \sigma_{\mathrm{II}}(\boldsymbol{u}_{\mathrm{I}}, t)\mathrm{d}W_{\mathrm{I}}(t) \\ \qquad\quad + \sigma_{\mathrm{II,A}}(\boldsymbol{u}_{\mathrm{I}}, t)\mathrm{d}W_{\mathrm{II,A}}(t) + (\sigma_{\mathrm{II,0}} + \sigma_{\mathrm{II,M}}(\boldsymbol{u}_{\mathrm{I}}, t)\boldsymbol{u}_{\mathrm{II}})\mathrm{d}W_{\mathrm{II,M}}, \end{cases} \tag{2.1}$$

where W_{I} is an M_{I}-dimensional Wiener process, and $W_{\mathrm{II,A}}$, $W_{\mathrm{II,0}}$, $W_{\mathrm{II,M}}$ are independent M_{II}-dimensional Wiener processes. Note that the dynamics of $\boldsymbol{u}_{\mathrm{I}}$ is arbitrary while the dynamics of $\boldsymbol{u}_{\mathrm{II}}$ is quasilinear, i.e., it is linear in $\boldsymbol{u}_{\mathrm{II}}$ in both the drift and the noise with general nonlinear coefficients depending on $\boldsymbol{u}_{\mathrm{I}}$. Also, note that the noise for $\boldsymbol{u}_{\mathrm{I}}$ and $\boldsymbol{u}_{\mathrm{II}}$ can be correlated through W_{I} appearing in the equations for both $\boldsymbol{u}_{\mathrm{I}}$ and $\boldsymbol{u}_{\mathrm{II}}$. All of the nonlinear test models for slow-fast systems (see [12, 13]), turbulent tracers (see [2, 14, 21, 33]) and exactly solvable stochastic parameterized filters (see [8, 9, 15, 34, 35]) have the structural form as in (2.1). Such systems are known to have exactly solvable non-Gaussian statistics for filters, where $\boldsymbol{u}_{\mathrm{I}}$ is observed conditionally over a time interval [1, 20]. Below, we derive explicit closed equations for the evolution of conditional moments of $\boldsymbol{u}_2$ through CGFPE.

The Fokker-Planck equation for the probability density $p(\boldsymbol{u}_{\mathrm{I}}, \boldsymbol{u}_{\mathrm{II}}, t)$ associated with (2.1) is given by [7, 36]

$$\begin{aligned} p_t = &-\nabla_{\mathrm{I}} \cdot (F_{\mathrm{I}} p) - \nabla_{\mathrm{II}} \cdot \big((F_{\mathrm{II}} + \Gamma \boldsymbol{u}_{\mathrm{II}})p\big) + \frac{1}{2}\nabla \cdot \nabla(Qp) \\ &+ \frac{1}{2}\nabla_{\mathrm{II}} \cdot \nabla_{\mathrm{II}}(Q_{\mathrm{A}} p) + \frac{1}{2}\nabla_{\mathrm{II}} \cdot \nabla_{\mathrm{II}}(Q_{\mathrm{M}} p), \end{aligned} \tag{2.2}$$

where $\nabla = (\nabla_{\mathrm{I}}, \nabla_{\mathrm{II}})$, and

$$\begin{aligned} &Q = (\sigma_{\mathrm{I}}, \sigma_{\mathrm{II}}) \otimes \big(\sigma_{\mathrm{I}}^{\mathrm{T}}, \sigma_{\mathrm{II}}^{\mathrm{T}}\big), \qquad Q_{\mathrm{A}} = \sigma_{\mathrm{II,A}} \otimes \sigma_{\mathrm{II,A}}^{\mathrm{T}}, \\ &Q_{\mathrm{M}} = (\sigma_{\mathrm{II,0}} + \sigma_{\mathrm{II,M}}\boldsymbol{u}_{\mathrm{II}}) \otimes \big(\sigma_{\mathrm{II,0}}^{\mathrm{T}} + \boldsymbol{u}_{\mathrm{II}}^{\mathrm{T}}\sigma_{\mathrm{II,M}}^{\mathrm{T}}\big). \end{aligned} \tag{2.3}$$

We are interested in developing exact statistical approximations for $p(\boldsymbol{u}_{\mathrm{I}}, \boldsymbol{u}_{\mathrm{II}}, t)$ which, by Bayes theorem, can be written as

$$p(\boldsymbol{u}_{\mathrm{I}}, \boldsymbol{u}_{\mathrm{II}}, t) = p(\boldsymbol{u}_{\mathrm{II}} \mid \boldsymbol{u}_{\mathrm{I}}, t)\pi(\boldsymbol{u}_{\mathrm{I}}, t), \tag{2.4}$$

where $\pi(\boldsymbol{u}_{\mathrm{I}}, t)$ is the marginal distribution

$$\pi(\boldsymbol{u}_{\mathrm{I}}, t) \equiv \int p(\boldsymbol{u}_{\mathrm{I}}, \boldsymbol{u}_{\mathrm{II}}, t)\mathrm{d}\boldsymbol{u}_{\mathrm{II}}. \tag{2.5}$$

We first integrate (2.2) with respect to $\boldsymbol{u}_{\mathrm{II}}$ and use the divergence theorem to verify that the marginal density $\pi(\boldsymbol{u}_{\mathrm{I}}, t)$ satisfies the Fokker-Planck equation

$$\pi_t = \mathcal{L}_{\mathrm{FP,I}}\pi \tag{2.6}$$

with

$$\mathcal{L}_{\mathrm{FP,I}}\pi = -\nabla_{\mathrm{I}} \cdot (F_{\mathrm{I}}\pi) + \frac{1}{2}\nabla_{\mathrm{I}} \cdot \nabla_{\mathrm{I}}(Q_{\mathrm{I}}\pi), \quad Q_{\mathrm{I}} = \sigma_{\mathrm{I}} \otimes \sigma_{\mathrm{I}}^{\mathrm{T}}. \tag{2.7}$$

Next, we derive the closed system of coupled generalized Fokker-Planck equations (CGFPE) for the conditional moments

$$\mathcal{M}_{\boldsymbol{\alpha}}(\boldsymbol{u}_{\mathrm{I}}, t) \equiv \int \boldsymbol{u}_{\mathrm{II}}^{\boldsymbol{\alpha}} p(\boldsymbol{u}_{\mathrm{I}}, \boldsymbol{u}_{\mathrm{II}}, t)\mathrm{d}\boldsymbol{u}_{\mathrm{II}} = \pi(\boldsymbol{u}_{\mathrm{I}}, t) \int \boldsymbol{u}_{\mathrm{II}}^{\boldsymbol{\alpha}} p(\boldsymbol{u}_{\mathrm{II}} \mid \boldsymbol{u}_{\mathrm{I}}, t)\mathrm{d}\boldsymbol{u}_{\mathrm{II}}. \tag{2.8}$$

Note that $\mathcal{M}_0(\boldsymbol{u}_{\mathrm{I}}, t) = \pi(\boldsymbol{u}_{\mathrm{I}}, t)$ is just the marginal density of (2.1) in $\boldsymbol{u}_{\mathrm{I}}$. Here and hereafter, we use the standard multi-index notation $\boldsymbol{\alpha} = (\alpha_1, \alpha_2, \ldots, \alpha_{M_{\mathrm{II}}}) \in \mathbb{R}^{M_{\mathrm{II}}}$ with

$$\boldsymbol{u}_{\mathrm{II}}^{\boldsymbol{\alpha}} \equiv (\boldsymbol{u}_{\mathrm{II}})_1^{\alpha_1} (\boldsymbol{u}_{\mathrm{II}})_2^{\alpha_2} \cdots (\boldsymbol{u}_{\mathrm{II}})_{M_{\mathrm{II}}}^{\alpha_{M_{\mathrm{II}}}}. \tag{2.9}$$

We have the following general principles for computing the vector $\boldsymbol{\mathcal{M}}_{\boldsymbol{\alpha}}(\boldsymbol{u}_{\mathrm{I}}, t) \equiv (\mathcal{M}_{\boldsymbol{\alpha}}(\boldsymbol{u}_{\mathrm{I}}, t))$ $(|\boldsymbol{\alpha}| = N)$ of conditional moments of order N.

Proposition 2.1 (Generalized Feynman-Kac Framework) *The vector $\boldsymbol{\mathcal{M}}_N(\boldsymbol{u}_{\mathrm{I}}, t)$ of conditional moments of order N associated with the probability density of* (2.1) *satisfies the system of coupled generalized Fokker-Planck equations* (*CGFPE*)

$$\begin{aligned}\frac{\partial \boldsymbol{\mathcal{M}}_N(\boldsymbol{u}_{\mathrm{I}}, t)}{\partial t} &= \mathcal{L}_{\mathrm{FP}}\boldsymbol{\mathcal{M}}_N(\boldsymbol{u}_{\mathrm{I}}, t) + \mathscr{L}_N(\boldsymbol{u}_{\mathrm{I}}, t)\boldsymbol{\mathcal{M}}_N(\boldsymbol{u}_{\mathrm{I}}, t) \\ &\quad + \mathscr{F}_N\big(\boldsymbol{u}_{\mathrm{I}}, \boldsymbol{\mathcal{M}}_{N-1}(\boldsymbol{u}_{\mathrm{I}}, t), \nabla_{\mathrm{I}}\boldsymbol{\mathcal{M}}_{N-1}(\boldsymbol{u}_{\mathrm{I}}, t), \boldsymbol{\mathcal{M}}_{N-2}(\boldsymbol{u}_{\mathrm{I}}, t)\big)\end{aligned} \tag{2.10}$$

with the convention $\boldsymbol{\mathcal{M}}_{-2} = \boldsymbol{\mathcal{M}}_{-1} = 0$, where $\mathscr{F}_N$ is an explicit linear function with coefficients depending on $\boldsymbol{u}_{\mathrm{I}}$ of the lower order moments, $\mathscr{L}_N$ is an $N \times N$ Feynman-Kac matrix potential which is an explicit linear function with coefficients depending on $\boldsymbol{u}_{\mathrm{I}}$ of the quantities

$$\Gamma(\boldsymbol{u}_{\mathrm{I}}, t), \quad Q_{\mathrm{II,M}} = \sigma_{\mathrm{II,M}} \otimes \sigma_{\mathrm{II,M}}^{\mathrm{T}}, \tag{2.11}$$

which vanishes when $\Gamma = 0$ and $Q_{\mathrm{II,M}} = 0$.

The proof below immediately yields explicit formulas for $\mathscr{L}_N$ and $\mathscr{F}_N$ in any concrete application (see Sect. 3). But a general notation for these coefficients will be tedious and unnecessary to develop here. The advantage of CGFPE in (2.10) is that high resolution numerical integrators can be developed for (2.10) to find these statistics provided that M_{I} is low-dimensional or has the special algebraic structure (see Sect. 3).

The sketch of the proof below emphasizes the main contributions to the operator $\mathscr{L}_N$ in (2.10). As in the derivation of (2.6), we first multiply the Fokker-Planck equation (2.2) by $\boldsymbol{u}_{\rm II}^{\boldsymbol{\alpha}}$ and integrate with respect to $\boldsymbol{u}_{\rm II}$ to obtain

$$\begin{aligned}\frac{\partial \mathcal{M}_N(\boldsymbol{u}_{\rm I},t)}{\partial t} &= \mathcal{L}_{\rm FP}\mathcal{M}_N(\boldsymbol{u}_{\rm I},t) - \int \boldsymbol{u}_{\rm II}^{\boldsymbol{\alpha}}\cdot\nabla_{\rm II}\big(\Gamma(\boldsymbol{u}_{\rm I},t)\boldsymbol{u}_{\rm II}p\big)d\boldsymbol{u}_{\rm II}\\ &\quad+\frac{1}{2}\int \boldsymbol{u}_{\rm II}^{\boldsymbol{\alpha}}\cdot\nabla_{\rm II}\cdot\nabla_{\rm II}\big(\sigma_{\rm II,M}\boldsymbol{u}_{\rm II}\otimes\boldsymbol{u}_{\rm II}^{\rm T}\sigma_{\rm II,M}^{\rm T}p\big)d\boldsymbol{u}_{\rm II}+\cdots,\end{aligned}\tag{2.12}$$

where "$+\cdots$" denotes all the remaining terms which define the recursive source term $\mathscr{F}_N$. We simplify (2.12) by using the integration by parts of the last two terms on the right hand side, namely,

$$\begin{aligned}-\int \boldsymbol{u}_{\rm II}^{\boldsymbol{\alpha}}\cdot\nabla_{\rm II}\big(\Gamma(\boldsymbol{u}_{\rm I},t)\boldsymbol{u}_{\rm II}p\big)d\boldsymbol{u}_{\rm II} &= \int \nabla_{\rm II}\boldsymbol{u}_{\rm II}^{\boldsymbol{\alpha}}\cdot\big(\Gamma(\boldsymbol{u}_{\rm I},t)\boldsymbol{u}_{\rm II}\big)p\,d\boldsymbol{u}_{\rm II}\\ &= \mathscr{L}_N^{(1,2)}\mathcal{M}_N(\boldsymbol{u}_{\rm I},t)\end{aligned}\tag{2.13}$$

and

$$\begin{aligned}&\frac{1}{2}\int \boldsymbol{u}_{\rm II}^{\boldsymbol{\alpha}}\cdot\nabla_{\rm II}\cdot\nabla_{\rm II}\big(\sigma_{\rm II,M}\boldsymbol{u}_{\rm II}\otimes\boldsymbol{u}_{\rm II}^{\rm T}\sigma_{\rm II,M}^{\rm T}p\big)d\boldsymbol{u}_{\rm II}\\ &\quad=\frac{1}{2}\int \nabla_{\rm II}\cdot\nabla_{\rm II}\boldsymbol{u}_{\rm II}^{\boldsymbol{\alpha}}\cdot\big(\sigma_{\rm II,M}\boldsymbol{u}_{\rm II}\otimes\boldsymbol{u}_{\rm II}^{\rm T}\sigma_{\rm II,M}^{\rm T}\big)p\,d\boldsymbol{u}_{\rm II}\\ &\quad=\mathscr{L}_N^{(2,2)}\mathcal{M}_N(\boldsymbol{u}_{\rm I},t),\end{aligned}\tag{2.14}$$

so that $\mathscr{L}_N=\mathscr{L}_N^{(1,2)}+\mathscr{L}_N^{(2,2)}$ in (2.10). The remaining terms in "$+\cdots$" are explicitly computed by a similar integration by parts to define $\mathscr{F}_N$. The correlated noise terms in (2.1) involving $W_{\rm I}$ which defines the noise Q in (2.2) determine the dependence on $\nabla\mathcal{M}_{N-1}(\boldsymbol{u}_{\rm I},t)$ in $\mathscr{F}_N$ since they have the typical form

$$\begin{aligned}-\int \boldsymbol{u}_{\rm II}^{\boldsymbol{\alpha}}\nabla_{\rm I}\cdot\nabla_{\rm II}\big(\sigma_{\rm I}\sigma_{\rm II}^{\rm T}p\big)d\boldsymbol{u}_{\rm II} &= \nabla_{\rm I}\cdot\int \nabla_{\rm II}\boldsymbol{u}_{\rm II}^{\boldsymbol{\alpha}}\big(\sigma_{\rm I}\sigma_{\rm II}^{\rm T}p\big)d\boldsymbol{u}_{\rm II}\\ &= \mathscr{F}_N\big(\boldsymbol{u}_{\rm I},\nabla_{\rm I}\mathcal{M}_{N-1}(\boldsymbol{u}_{\rm I},t)\big).\end{aligned}\tag{2.15}$$

It is worth pointing out that $\mathscr{F}_N$ depends only on the point-wise values of $\mathcal{M}_{N-1}(\boldsymbol{u}_{\rm I},t)$ and $\mathcal{M}_{N-2}(\boldsymbol{u}_{\rm I},t)$ if there are non-correlated noise interactions and $\sigma_{\rm II}=0$.

3 Application of the Conditional Moment PDE's to a Non-Gaussian Test Model

We develop the simplest non-Gaussian test model, where we can explicitly evaluate non-trivial statistical features utilizing the coupled system of PDE's in (2.10) from Sect. 2 for the conditional moments $\mathcal{M}_{\boldsymbol{\alpha}}(\boldsymbol{u}_{\rm I},t)$. We then derive and validate a numerical procedure for the accurate numerical solution to the closed system of equations in (2.10) for the conditional moments in several stringent test problems. This explicit solution procedure is applied in Sect. 5 to understand the role of coarse-graining and non-Gaussian statistics with the model error in ensemble predictions.

Clearly, the simplest models considered with the structure as in (2.1) have $M_{\mathrm{I}} = M_{\mathrm{II}} = 1$ so that the recursion formulas in (2.10) involve scalar fields and the CGFPE are integrated in a single spatial dimension. For $\boldsymbol{u}_{\mathrm{I}}$, we choose the general nonlinear scalar Itô SDE

$$\mathrm{d}u_{\mathrm{I}} = F_{\mathrm{I}}(u_{\mathrm{I}}, t)\mathrm{d}t + \sigma_{\mathrm{I}}(u_{\mathrm{I}}, t)\mathrm{d}W_{\mathrm{I}}, \tag{3.1}$$

while for u_{II}, we utilize the quasi-linear equation

$$\mathrm{d}u_{\mathrm{II}} = \big(-u_{\mathrm{I}}u_{\mathrm{II}} + f(t)\big)\mathrm{d}t + \sigma_{\mathrm{II}}\mathrm{d}W_{\mathrm{II}}, \tag{3.2}$$

where $f(t)$ does not depend on u_{I}, and the noise σ_{II} is constant. Note that u_{I} enters in (3.2) as a multiplicative coefficient and fluctuations in u_{I} can show the growth and intermittent instabilities with highly non-Gaussian behavior even when u_{I} in (3.1) has a positive mean (see [3, 5, 25]). The stochastic models for u_{I} in (3.1) will vary from linear stochastic models (a special case of the SPEKF models for filtering (see [9, 25, 34, 35])) to cubic nonlinear models with additive and multiplicative noise (see [25]). For the systems with dynamics as in (3.1)–(3.2), the closed equations for the conditional moments $\mathcal{M}_{\boldsymbol{\alpha}}$ in (2.8) become

$$\begin{aligned}\frac{\partial}{\partial t}\mathcal{M}_N(u_{\mathrm{I}}, t) = {}& \mathcal{L}_{\mathrm{FP}}\mathcal{M}_N(u_{\mathrm{I}}, t) - Nu_{\mathrm{I}}\mathcal{M}_N(u_{\mathrm{I}}, t) + Nf(t)\mathcal{M}_{N-1}(u_{\mathrm{I}}, t)\\ &+ \frac{1}{2}N(N-1)\sigma_{\mathrm{II}}^2\mathcal{M}_{N-2}(u_{\mathrm{I}}, t),\end{aligned} \tag{3.3}$$

where $N = 0, 1, \ldots, N_{\max}$ and $\mathcal{M}_{-2} = \mathcal{M}_{-1} = 0$. Such models illustrate a wide range of intermittent non-Gaussian behavior mimicking one in vastly more complex systems (see [22]). These simple revealing models will be used in Sect. 5 to study various new aspects of model error in ensemble predictions for non-Gaussian turbulent systems.

3.1 *Validation of a Numerical Method for Solving the CGFPE*

Determination of the time evolution of the conditional moments $\mathcal{M}_{\boldsymbol{\alpha}}$ in (2.8) requires an accurate numerical procedure for solving the inhomogeneous system of coupled Fokker-Planck equations (CGFPE) in (2.10). The algorithms discussed below is applied to the case $u_{\mathrm{I}} \in \mathbb{R}$ (i.e., $M_{\mathrm{I}} = 1$) which is sufficient for our purposes and leads to many new insights on the model error in imperfect ensemble predictions of turbulent systems with positive Lyapunov exponents, as discussed in Sect. 5. Similar to the case of the homogeneous Fokker-Planck equation, solving the inhomogeneous CGFPE system (2.10) for $M_{\mathrm{I}} \geqslant 3$ poses a formidable challenge which, for convenience, is unnecessary here.

Here, the coupled system in (2.10) is solved by the third-order temporal discretization through the backward differentiation formulas (see [17]) and the second-order spatial discretization via the finite volume method (see [19] and see Appendix A for details). The performance of the numerical procedure for solving CGFPE in one spatial dimension (i.e., $u_{\mathrm{I}} \in \mathbb{R}$ in (2.10)) is tested in the following widely varying dynamical configurations:

(i) Dynamics with time-invariant statistics on the attractor/equilibrium with

- (a) nearly Gaussian marginal equilibrium PDFs in u_{II} and linear Gaussian dynamics for u_{I} in (3.1),
- (b) fat-tailed marginal equilibrium PDFs in u_{II} and linear Gaussian dynamics for u_{I} in (3.1),
- (c) highly non-Gaussian marginal equilibrium PDFs in u_{II} and cubic dynamics for u_{I} in (3.1) with highly skewed equilibrium PDFs.

(ii) Dynamics with the time-periodic statistics on the attractor with the time-periodic regime switching between nearly Gaussian and highly skewed regimes with cubic dynamics for u_{I} in (3.1) and highly non-Gaussian dynamics of u_{II} in (3.2).

Below, we introduce the relevant test models in Sect. 3.1.1 and provide the evidence for the good accuracy of the developed technique in Sect. 3.1.2, as well as its advantages over the direct Monte Carlo sampling.

3.1.1 Non-Gaussian Test Models for Validating CGFPE

We consider two non-Gaussian models with intermittent instabilities and with the structure as in (3.1)–(3.2), where we adopt the following notations:

$$u_{\mathrm{I}} = \gamma, \qquad u_{\mathrm{II}} = u.$$

The first model is a simplified version of the SPEKF model developed originally for filtering turbulent systems with stochastically parameterized unresolved variables (see [8, 9, 15, 34, 35]) and is given by

$$\begin{aligned} &\text{(a)} \quad \mathrm{d}\gamma = \big(-d_\gamma(\gamma - \widehat{\gamma}) + f_\gamma(t)\big)\mathrm{d}t + \sigma_\gamma \mathrm{d}W_\gamma, \\ &\text{(b)} \quad \mathrm{d}u = \big(-\gamma u + f_u(t)\big)\mathrm{d}t + \sigma_u \mathrm{d}W_u. \end{aligned} \tag{3.4}$$

Note that despite the Gaussian dynamics of the damping fluctuation γ, the dynamics of u in (3.4) can be highly non-Gaussian with intermittently positive Lyapunov exponents even when the equilibrium mean $\widehat{\gamma}$ is positive (see [3, 4, 25]). The system (3.4) possesses a wide range of turbulent dynamical regimes ranging from highly non-Gaussian dynamics with intermittency and fat-tailed marginal PDFs for u to laminar regimes with nearly Gaussian statistics. A detailed discussion of properties of this system can be found in [3, 5]. In the numerical tests discussed in the next section, we examine the accuracy of the numerical algorithm for solving CGFPE in

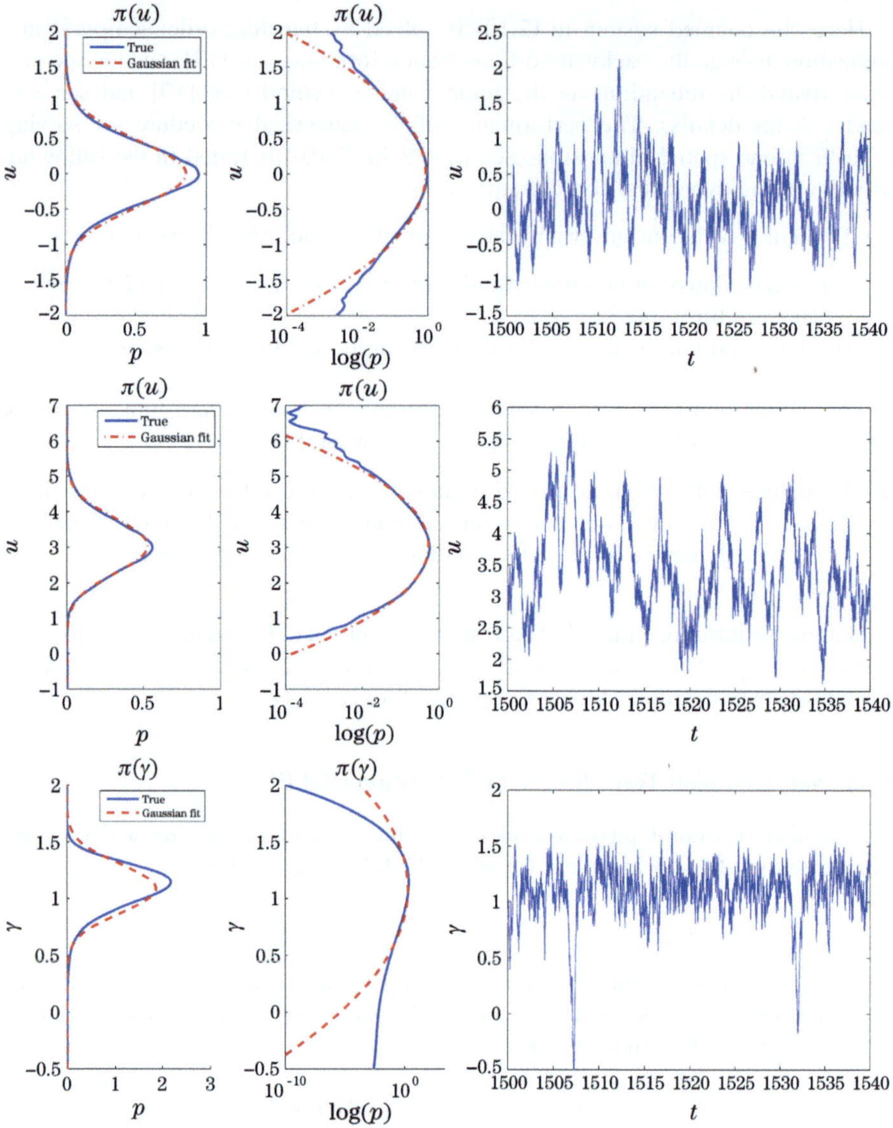

Fig. 1 (*Top*) Marginal statistics $p_{\text{eq}}(u)$ and a path-wise solution $u(t)$ on the attractor of the system (3.4) in the non-Gaussian regime with invariant measure characterized by intermittent transient instabilities and fat-tailed marginal PDFs (dynamics of γ is Gaussian in this model). (*Bottom*) Marginal statistics $p_{\text{eq}}(\gamma)$ and $p_{\text{eq}}(u)$, and path-wise solutions $\gamma(t)$ and $u(t)$ on the attractor of the system (3.5) in the non-Gaussian regime with the regime switching in the path-wise dynamics despite a unimodal, skewed marginal PDF in γ

the dynamical regime characterized by a highly intermittent marginal dynamics in u associated with fat-tailed marginal equilibrium PDFs for u (see Fig. 1 for examples of such dynamics).

The second model which we examine, has a cubic nonlinearity in the dynamics of the damping fluctuation γ, and is given by

$$\begin{aligned} &\text{(a)} \quad \mathrm{d}\gamma = \left[-a\gamma + b\gamma^2 - c\gamma^3 + f_\gamma(t)\right]\mathrm{d}t + (A - B\gamma)\mathrm{d}W_C + \sigma_\gamma \mathrm{d}W_\gamma, \\ &\text{(b)} \quad \mathrm{d}u = \left(-\gamma u + f_u(t)\right)\mathrm{d}t + \sigma_u \mathrm{d}W_u. \end{aligned} \tag{3.5}$$

The above nonlinear model for γ with correlated additive and multiplicative noise W_C and exactly solvable equilibrium statistics was first derived in [26] as a normal form for a single low-frequency variable in climate models, where the noise correlations arise through advection of the large scales by the small scales and simultaneously strong cubic damping. The nonlinear dynamics of γ has many interesting features which were studied in detail elsewhere (see [25]). Here, we consider a more complex problem with the dynamics of u in (a) of (3.5) coupled with γ through the quadratic nonlinearity. In the numerical tests below, we focus on the particularly interesting regime, where the damping fluctuation γ exhibits the regime switching despite the unimodality of the associated equilibrium statistics (see Fig. 1). This configuration represents the simplest possible test model for the analogous behavior occurring in comprehensive climate models (see [23, 27]). Another important configuration of (3.5) tested below with relevance to atmospheric/climate dynamics corresponds to time-periodic transitions in γ between a highly skewed and a nearly Gaussian phase in γ with the dynamics in u remaining highly non-Gaussian throughout the evolution (see Fig. 2 for an illustration of such dynamics).

The above two non-Gaussian models are utilized below to validate the accuracy of our numerical method for solving the CGFPE system (2.10). This framework is then used to analyze the model error in imperfect predictions of turbulent non-Gaussian systems in Sect. 5.

3.1.2 Numerical Tests

We use the test models introduced in the previous section to analyze the performance of the numerical scheme for solving the CGFPE system (2.10) in one-spatial dimension. In order to assess the accuracy of the algorithm, we consider the following two types of the relative error in the conditional moments: the point-wise relative error in the N-th conditional moment

$$\epsilon_N(\gamma, t) = \left| \frac{\mathcal{M}_N^{\text{CGFPE}}(\gamma, t) - \mathcal{M}_N^{\text{ref}}(\gamma, t)}{\mathcal{M}_N^{\text{ref}}(\gamma, t)} \right| \tag{3.6}$$

and the L^2 relative error for each fixed time

$$\epsilon_N(t) = \frac{\|\mathcal{M}_N^{\text{CGFPE}}(\gamma, t) - \mathcal{M}_N^{\text{ref}}(\gamma, t)\|_{L^2}}{\|\mathcal{M}_N^{\text{ref}}(\gamma, t)\|_{L^2}}. \tag{3.7}$$

The reference values for the conditional moments, $\mathcal{M}_N^{\text{ref}}$ in the above formulas, are obtained from either the analytical solutions (in the case of system (3.4)) through

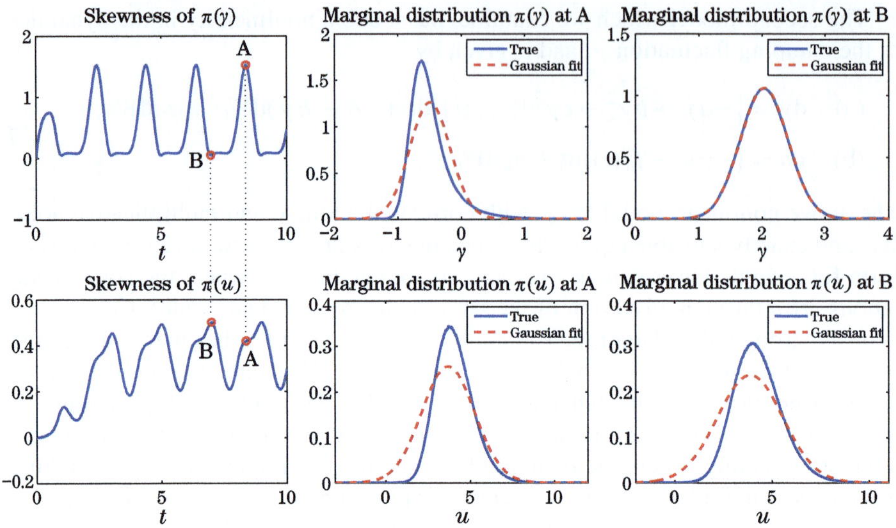

Fig. 2 (*Top*) Time-periodic evolution of the skewness of the marginal dynamics of γ in the non–Gaussian system (3.5) with cubic nonlinearity in γ in the configuration, where γ cycles between a highly skewed (*top middle*) and a nearly Gaussian (*top right*) phase. The phases of high/low skewness in the marginal statistics of γ are correlated with those in the marginal statistics of u. However, note that the dynamics of u remains highly non-Gaussian throughout the evolution. The snapshots of the marginal PDFs in u on *the bottom* are shown for the times indicated on *the top panel*

the formulas derived in [9]), or via the Monte Carlo estimates. The conditional moments are normalized in the standard fashion, with the conditional mean, variance, skewness and kurtosis given by

$$\widetilde{\mathcal{M}}_0(\gamma,t)=\mathcal{M}_0(\gamma,t), \qquad \widetilde{\mathcal{M}}_1(\gamma,t)=\mathcal{M}_1(\gamma,t), \tag{3.8}$$

$$\widetilde{\mathcal{M}}_2(\gamma,t)=\int \big(u(t)-\mathcal{M}_1(\gamma,t)\big)^2 p(u,\gamma,t)\mathrm{d}u=\mathcal{M}_2(\gamma,t)-\mathcal{M}_1^2(\gamma,t), \tag{3.9}$$

$$\begin{aligned}\widetilde{\mathcal{M}}_3(\gamma,t)&=\frac{1}{\widetilde{\mathcal{M}}_2^{\frac{3}{2}}(\gamma,t)}\int \big(u(t)-\mathcal{M}_1(\gamma,t)\big)^3 p(u,\gamma,t)\mathrm{d}u\\ &=\frac{\mathcal{M}_3(\gamma,t)-3\mathcal{M}_1(\gamma,t)\mathcal{M}_2(\gamma,t)+2\mathcal{M}_1^3(\gamma,t)}{\widetilde{\mathcal{M}}_2^{\frac{3}{2}}(\gamma,t)},\end{aligned} \tag{3.10}$$

$$\begin{aligned}\widetilde{\mathcal{M}}_4(\gamma,t)&=\frac{1}{\widetilde{\mathcal{M}}_2^{2}(\gamma,t)}\int \big(u(t)-\mathcal{M}_1(\gamma,t)\big)^4 p(u,\gamma,t)\mathrm{d}u\\ &=\frac{\mathcal{M}_4(\gamma,t)-4\mathcal{M}_1(\gamma,t)\mathcal{M}_3(\gamma,t)+6\mathcal{M}_1^2(\gamma,t)\mathcal{M}_2(\gamma,t)-3\mathcal{M}_1^4(\gamma,t)}{\widetilde{\mathcal{M}}_2^{2}(\gamma,t)},\end{aligned} \tag{3.11}$$

respectively.

Table 1 Relative errors ϵ_N in (3.7), in the conditional moments $\mathcal{M}_0$–$\mathcal{M}_4$ in (3.8)–(3.11) at equilibrium for the two test models (3.4)–(3.5) with the reference input obtained from Monte Carlo estimates from 10^7 runs

	$\mathcal{M}_0$	$\mathcal{M}_1$	$\mathcal{M}_2$	$\mathcal{M}_3$	$\mathcal{M}_4$
System (3.4): Nearly Gaussian reg.	0.0031		0.0241		0.0494
System (3.4): Fat algebraic tail reg.	0.0225		0.0202		0.0593
System (3.5): High skewness reg.	0.0179	0.0181	0.0183	0.0196	0.0236

Table 2 Relative errors (3.7) in the conditional moments $\mathcal{M}_0$, $\mathcal{M}_1$ and $\mathcal{M}_3$ at equilibrium for the two test models (3.4)–(3.5) with the reference input obtained from analytical solutions

	$\mathcal{M}_0$	$\mathcal{M}_1$	$\mathcal{M}_2$	$\mathcal{M}_3$	$\mathcal{M}_4$
System (3.4): Nearly Gaussian reg.	2.1520×10^{-5}	0		0	
System (3.4): Fat algebraic tail reg.	3.6825×10^{-6}	0		0	
System (3.5): High skewness reg.	0.0018				

Table 3 Relative errors in time-periodic conditional moments $\mathcal{M}_0$–$\mathcal{M}_4$ in (3.8)–(3.11) for the test model (3.5) in the regime with transitions (see Fig. 1) between highly skewed and nearly Gaussian marginal densities $\pi_{att}(\gamma)$; the reference input obtained from Monte Carlo estimates from 10^7 runs

	$\mathcal{M}_0$	$\mathcal{M}_1$	$\mathcal{M}_2$	$\mathcal{M}_3$	$\mathcal{M}_4$
$t_* = 7.00$	0.0185	0.0199	0.0218	0.0242	0.0271
$t_* = 8.40$	0.0299	0.0337	0.0400	0.0447	0.0561
$t_* = 7.70$	0.0309	0.0316	0.0321	0.0327	0.0332
$t_* = 9.00$	0.0182	0.0196	0.0229	0.0275	0.0330

The L^2 errors for the two test models discussed in the previous section and parameters as specified below are listed in Tables 1, 2, 3. Note that the errors in the conditional moments do not exceed 6% for the wide range of dynamical regimes considered. Moreover, the comparison of the results in Tables 1 and 2 shows that the numerical algorithm developed here is more efficient and accurate than the Monte Carlo estimates, even when a relatively large sample size ($\sim 10^7$) is used in the MC simulations.

In Fig. 3, we illustrate the performance of the algorithm for computing the conditional moments in (2.10) associated with the conditional equilibrium density $p_{\text{eq}}(u \mid \gamma)$ for the system (3.4). The system parameters $(d_\gamma, \sigma_\gamma, \widehat{\gamma}, \sigma_u)$ in (3.4) are chosen to represent the non-Gaussian dynamics in the regime with intermittent instabilities and a fat-tailed marginal equilibrium PDF in u. In particular, we choose

$$\sigma_\gamma = 10, \quad d_\gamma = 10, \quad \widehat{\gamma} = 3, \quad f_u = f_\gamma = 0$$

(see Fig. 1 for an example of the corresponding dynamics).

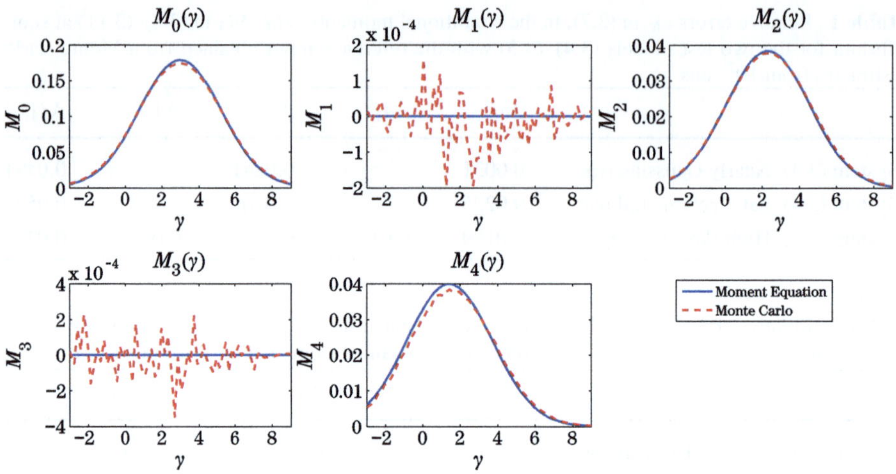

Fig. 3 Equilibrium conditional statistics of the system (3.4) with Gaussian damping fluctuations, intermittent instabilities and fat-tailed marginal PDFs in u. Unnormalized conditional moments, $\mathcal{M}_0(\gamma) - \mathcal{M}_4(\gamma)$, (2.8) of u at the equilibrium of the two-dimensional non-Gaussian turbulent system (3.4) with intermittent instabilities due to Gaussian damping fluctuations; the results of CGFPE (2.10) and Monte Carlo estimates from 10^7 runs are compared. In the dynamical regime shown in the marginal equilibrium PDF, $p_{\mathrm{eq}}(u)$, is symmetric and fat-tailed due to these intermittent instabilities (see Fig. 1). Note the errors in the Monte Carlo estimates in the odd moments

In Figs. 5 and 6, we illustrate the performance of our algorithm for computing the conditional moments $\mathcal{M}_\alpha(\gamma, t)$ of u in the system (3.5) with the cubic nonlinearity in γ which is coupled multiplicatively to the dynamics in u. Here, we consider two distinct configurations. For the constant forcing, we choose the parameters in (3.5) in such a way that γ displays regime switching with the unimodal, highly skewed marginal equilibrium PDF for γ, while the marginal dynamics of u is highly non-Gaussian and second-order stable. This dynamical configuration can be achieved by setting, for example,

$$
\begin{aligned}
&a = 1, \quad b = 1, \quad c = 1, \\
&A = 0.5, \quad B = -2, \\
&\sigma = 1, \quad f_u = 1, \quad f_\gamma = 3
\end{aligned}
$$

(see Fig. 1 for an illustration of such dynamics). For the time-periodic forcing, when the dynamics in γ cycles between a highly skewed and a nearly Gaussian phase while u remains highly non-Gaussian, we set

$$
\begin{aligned}
&a = 1, \quad b = 1, \quad c = 1, \\
&A = 0.5, \quad B = -0.5, \\
&\sigma = 0.5, \quad f_u = -0.5
\end{aligned}
$$

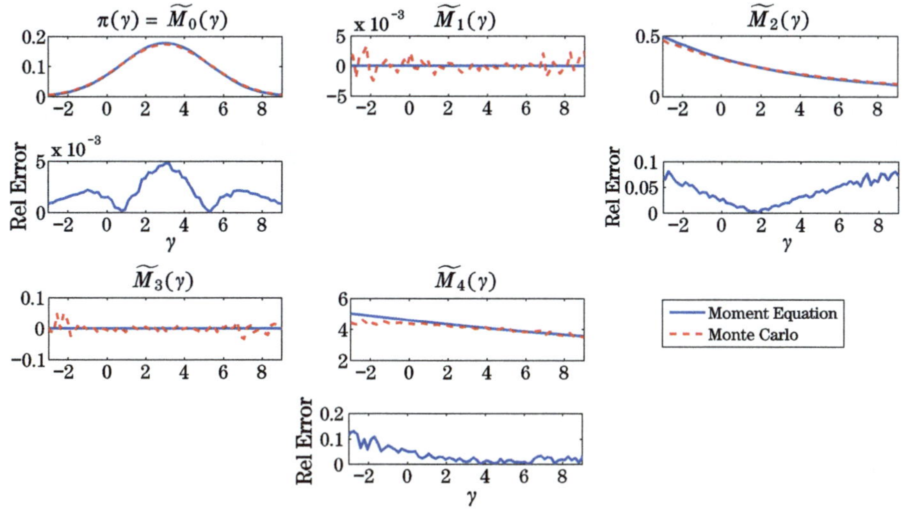

Fig. 4 The same as that in Fig. 3 but for centered, normalized conditional moments $\widetilde{\mathcal{M}}_0(\gamma)$ – $\widetilde{\mathcal{M}}_4(\gamma)$, which correspond to marginal density $\pi(\gamma)$, the conditional mean, variance, skewness and kurtosis given by (3.8)–(3.11), respectively

with the time-periodic forcing in γ given by

$$f_\gamma(t) = 6.5\sin\left(\pi t - \frac{\pi}{2}\right) + 2.5.$$

Based on the results summarized in Figs. 3, 4, 5, 6 and Tables 1, 2, 3, we make the following points:

(1) The numerical algorithm for solving the coupled system (2.10) in the CGFPE framework with $u_{\mathrm{I}} \in \mathbb{R}$ provides robust and accurate estimates for the conditional moments (2.8).

(2) The discrepancies between the estimates obtained from (2.10) and the direct Monte Carlo estimates with a large sample size ($\sim 10^7$) are below 6% for both time-periodic and time-invariant attractor statistics.

(3) The largest discrepancies in the normalized conditional moments obtained from CGFPE and Monte Carlo estimates in the normalized moments occur in tail regions, where the corresponding probability densities are very small.

(4) The developed algorithm for solving the CGFPE system (2.10) is more efficient and more accurate than the Monte Carlo estimates with relatively large sample sizes ($\sim 10^7$).

4 Quantifying Model Error Through Empirical Information Theory

As discussed extensively recently (see [5, 11, 25, 28–30]), a very natural way to quantify the model error in statistical solutions to complex systems is through the

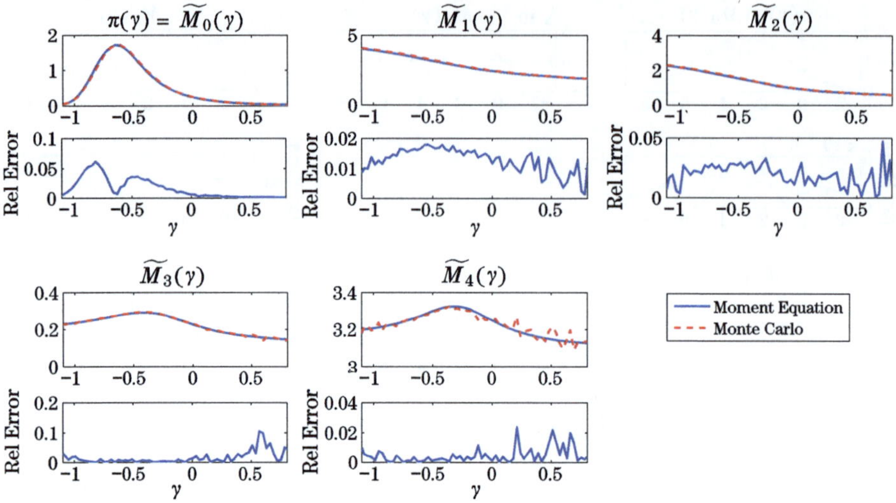

Fig. 5 Snapshots of time-periodic conditional statistics on the attractor of the system (3.5); the cubic nonlinearity in damping fluctuations, the highly skewed PDF phase. Normalized conditional moments $\widetilde{\mathcal{M}}_0(\gamma,t_*) - \widetilde{\mathcal{M}}_4(\gamma,t_*)$ in (3.8)–(3.11) of u on the time-periodic attractor of the two-dimensional non-Gaussian turbulent system (3.5) with cubic dynamics of damping fluctuations γ; the results obtained via CGFPE (2.10) and Monte Carlo simulations with 10^7 sample runs are compared at time $t_* = 8.4$ which corresponds to the highly non-Gaussian phase with highly skewed marginal PDFs, $\pi_{att}(u,t_*)$, $\pi_{att}(\gamma,t_*)$ (see Fig. 2). The normalized conditional moments are the conditional mean, variance skewness and kurtosis

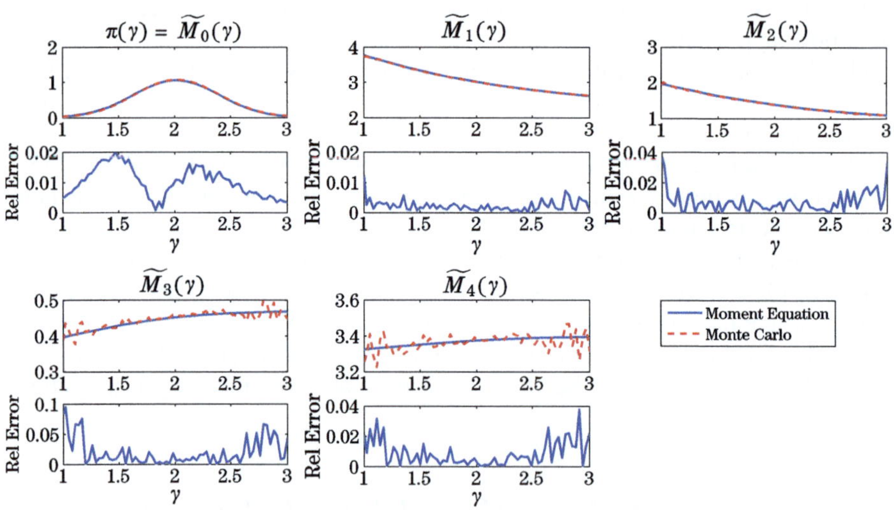

Fig. 6 The same as that in Fig. 5, but showing the normalized conditional moments $\widetilde{\mathcal{M}}_1(\gamma,t_*) - \widetilde{\mathcal{M}}_4(\gamma,t_*)$ (i.e., the conditional mean, variance, skewness and kurtosis), at $t_* = 7$ which corresponds to the nearly Gaussian phase in γ, but has a highly skewed marginal $\pi_{att}(u,t_*)$ (see also Fig. 2)

relative entropy $\mathcal{P}(p,q) \geqslant 0$ for two probability measures p and q given by

$$\mathcal{P}(p,q) = \int p \ln \frac{p}{q} = -\mathscr{S}(p) - \int p \ln q, \tag{4.1}$$

where

$$\mathscr{S}(p) = -\int p \ln p \tag{4.2}$$

is the Shannon entropy of the probability measure p. The relative entropy $\mathcal{P}(p,q)$ measures the lack of information in q about the probability measure p. If p is the perfect density and p_{M}, $\mathrm{M} \in \mathscr{M}$ is a class of probability densities, then M_1 is a better model than M_2 provided that

$$\mathcal{P}(p, p_{\mathrm{M}_1}) < \mathcal{P}(p, p_{\mathrm{M}_2}), \tag{4.3}$$

and the best model $\mathrm{M}* \in \mathscr{M}$ satisfies

$$\mathcal{P}(p, p_{\mathrm{M}*}) = \min_{\mathrm{M} \in \mathscr{M}} \mathcal{P}(p, p_{\mathrm{M}}). \tag{4.4}$$

There are extensive applications of information theory to improve imperfect models in climate science developed recently (see [5, 11, 25, 28–30]); the interested reader can refer to these references. The goal here is to develop and illustrate this perspective of the information theory on the model error for direct application to the estimate of model error for the setup developed above in Sects. 2–3. These formulas are utilized in Sect. 5 below.

We consider a probability density for the perfect model $p(\boldsymbol{u}_{\mathrm{I}}, \boldsymbol{u}_{\mathrm{II}})$ which can be written by Bayes theorem as

$$p(\boldsymbol{u}_{\mathrm{I}}, \boldsymbol{u}_{\mathrm{II}}) = p(\boldsymbol{u}_{\mathrm{II}} \mid \boldsymbol{u}_{\mathrm{I}}) \pi(\boldsymbol{u}_{\mathrm{I}}), \tag{4.5}$$

here and hereafter, $\pi(\boldsymbol{u}_{\mathrm{I}})$ is the marginal

$$\pi(\boldsymbol{u}_{\mathrm{I}}) = \int p(\boldsymbol{u}_{\mathrm{I}}, \boldsymbol{u}_{\mathrm{II}}) \mathrm{d}\boldsymbol{u}_{\mathrm{II}}. \tag{4.6}$$

From the CGFPE procedure developed in Sects. 2–3, we have exact expressions for the conditional moments up to some order L for $p(\boldsymbol{u}_{\mathrm{II}} \mid \boldsymbol{u}_{\mathrm{I}})$ evolving in time already, this is a source of information loss through the coarse graining of $p(\boldsymbol{u}_{\mathrm{I}}, \boldsymbol{u}_{\mathrm{II}})$. To quantify this information loss by measuring only the conditional moments up to order L, let

$$p_L(\boldsymbol{u}_{\mathrm{I}}, \boldsymbol{u}_{\mathrm{II}}) = p_L(\boldsymbol{u}_{\mathrm{II}} \mid \boldsymbol{u}_{\mathrm{I}}) \pi(\boldsymbol{u}_{\mathrm{I}}), \tag{4.7}$$

where for each value $\boldsymbol{u}_{\mathrm{I}}$, the conditional density $p_L(\boldsymbol{u}_{\mathrm{II}} \mid \boldsymbol{u}_{\mathrm{I}})$ satisfies the maximum entropy (least biased) criterion (see [24, 31, 32])

$$\mathcal{S}\big(p_L(\boldsymbol{u}_{\mathrm{II}} \mid \boldsymbol{u}_{\mathrm{I}})\big) = \max_{\pi_L \in \mathfrak{L}} \mathcal{S}\big(\pi_L(\boldsymbol{u}_{\mathrm{II}})\big), \tag{4.8}$$

where $\mathfrak{L}$ is a class of marginal densities π_L with identical moments up to order L, i.e.,

$$\int \boldsymbol{u}_{\mathrm{II}}^{\boldsymbol{\alpha}} \pi_L(\boldsymbol{u}_{\mathrm{II}}) \mathrm{d}\boldsymbol{u}_{\mathrm{II}} = \int \boldsymbol{u}_{\mathrm{II}}^{\boldsymbol{\alpha}} p_L(\boldsymbol{u}_{\mathrm{I}}, \boldsymbol{u}_{\mathrm{II}}) \mathrm{d}\boldsymbol{u}_{\mathrm{II}} = \int \boldsymbol{u}_{\mathrm{II}}^{\boldsymbol{\alpha}} p(\boldsymbol{u}_{\mathrm{I}}, \boldsymbol{u}_{\mathrm{II}}) \mathrm{d}\boldsymbol{u}_{\mathrm{II}}, \quad |\boldsymbol{\alpha}| \leqslant L. \tag{4.9}$$

Below and in Sect. 5, we will always apply the variational problem in (4.8) for $L = 2$ which guarantees that $p_L(\boldsymbol{u}_{\mathrm{II}} \mid \boldsymbol{u}_{\mathrm{I}})$ is a Gaussian density with the specified conditional mean and variance. In general, for L even and $L > 2$, it is a subtle issue as to whether the solution to the variational problem (4.8) exists (see [34]), but here we tacitly assume this. We remark here that highly non-Gaussian densities can have Gaussian conditional densities like $p_L(\boldsymbol{u}_{\mathrm{II}} \mid \boldsymbol{u}_{\mathrm{I}})$ as discussed in Sect. 5.

Natural imperfect densities with the model error have the form

$$p_L^{\mathrm{M}}(\boldsymbol{u}_{\mathrm{I}}, \boldsymbol{u}_{\mathrm{II}}) = p_L^{\mathrm{M}}(\boldsymbol{u}_{\mathrm{II}} \mid \boldsymbol{u}_{\mathrm{I}}) \pi^{\mathrm{M}}(\boldsymbol{u}_{\mathrm{I}}). \tag{4.10}$$

The simplest model with the model error is a Gaussian density $p_G(\boldsymbol{u}_{\mathrm{I}}, \boldsymbol{u}_{\mathrm{II}})$ which is defined by its mean and variance; the standard regression formula for Gaussian densities (see [7]) automatically guarantees that the form in (4.10) is applied with $L = 2$ in this important case.

Another important way of generating an imperfect model with the form (4.10) is to have a different model (see [22, 25]) for the stochastic dynamics of $\boldsymbol{u}_{\mathrm{I}}$ rather than that in (2.1) and to compute the conditional moments up to order L in the approximate model through CGFPE so that the model approximations automatically have the form (4.10) (see Sect. 5 below).

Here we have a precise way to quantify the model error in an imperfect model in the present setup.

Proposition 4.1 *Given the perfect model distribution $p(\boldsymbol{u}_{\mathrm{I}}, \boldsymbol{u}_{\mathrm{II}})$ with its conditional approximation $p_L(\boldsymbol{u}_{\mathrm{I}}, \boldsymbol{u}_{\mathrm{II}})$ in* (4.7) *and the imperfect model density $p_L^{\mathrm{M}}(\boldsymbol{u}_{\mathrm{I}}, \boldsymbol{u}_{\mathrm{II}})$ defined in* (4.10), *we have*

$$\begin{aligned} \mathcal{P}\big(p(\boldsymbol{u}_{\mathrm{I}}, \boldsymbol{u}_{\mathrm{II}}), p_L^{\mathrm{M}}(\boldsymbol{u}_{\mathrm{I}}, \boldsymbol{u}_{\mathrm{II}})\big) &= \mathcal{P}\big(p(\boldsymbol{u}_{\mathrm{I}}, \boldsymbol{u}_{\mathrm{II}}), p_L(\boldsymbol{u}_{\mathrm{I}}, \boldsymbol{u}_{\mathrm{II}})\big) \\ &\quad + \mathcal{P}\big(p_L(\boldsymbol{u}_{\mathrm{I}}, \boldsymbol{u}_{\mathrm{II}}), p_L^{\mathrm{M}}(\boldsymbol{u}_{\mathrm{I}}, \boldsymbol{u}_{\mathrm{II}})\big), \end{aligned} \tag{4.11}$$

where

$$\begin{aligned} 0 &\leqslant \mathcal{P}\big(p(\boldsymbol{u}_{\mathrm{I}}, \boldsymbol{u}_{\mathrm{II}}), p_L(\boldsymbol{u}_{\mathrm{I}}, \boldsymbol{u}_{\mathrm{II}})\big) \\ &= \int \pi(\boldsymbol{u}_{\mathrm{I}}) \big[\mathcal{S}\big(p_L(\boldsymbol{u}_{\mathrm{II}} \mid \boldsymbol{u}_{\mathrm{I}})\big) - \mathcal{S}\big(p(\boldsymbol{u}_{\mathrm{II}} \mid \boldsymbol{u}_{\mathrm{I}})\big)\big] \mathrm{d}\boldsymbol{u}_{\mathrm{I}} \\ &= \int \pi(\boldsymbol{u}_{\mathrm{I}}) \mathcal{P}\big(p(\boldsymbol{u}_{\mathrm{II}} \mid \boldsymbol{u}_{\mathrm{I}}), p_L(\boldsymbol{u}_{\mathrm{II}} \mid \boldsymbol{u}_{\mathrm{I}})\big) \mathrm{d}\boldsymbol{u}_{\mathrm{I}} \end{aligned} \tag{4.12}$$

and

$$
\begin{aligned}
0 &\leqslant \mathcal{P}\big(p_L(\boldsymbol{u}_{\mathrm{I}},\boldsymbol{u}_{\mathrm{II}}),\, p_L^{\mathrm{M}}(\boldsymbol{u}_{\mathrm{I}},\boldsymbol{u}_{\mathrm{II}})\big) \\
&= \mathcal{P}\big(\pi(\boldsymbol{u}_{\mathrm{I}}),\pi^{\mathrm{M}}(\boldsymbol{u}_{\mathrm{I}})\big) + \int \pi(\boldsymbol{u}_{\mathrm{I}})\big[\mathcal{S}\big(p_L^{\mathrm{M}}(\boldsymbol{u}_{\mathrm{II}}\,|\,\boldsymbol{u}_{\mathrm{I}})\big) - \mathcal{S}\big(p_L^{\mathrm{M}}(\boldsymbol{u}_{\mathrm{II}}\,|\,\boldsymbol{u}_{\mathrm{I}})\big)\big]\mathrm{d}\boldsymbol{u}_{\mathrm{I}} \\
&= \mathcal{P}\big(\pi(\boldsymbol{u}_{\mathrm{I}}),\pi^{\mathrm{M}}(\boldsymbol{u}_{\mathrm{I}})\big) + \int \pi(\boldsymbol{u}_{\mathrm{I}})\mathcal{P}\big(p(\boldsymbol{u}_{\mathrm{II}}\,|\,\boldsymbol{u}_{\mathrm{I}}),\, p_L^{\mathrm{M}}(\boldsymbol{u}_{\mathrm{II}}\,|\,\boldsymbol{u}_{\mathrm{I}})\big)\mathrm{d}\boldsymbol{u}_{\mathrm{I}}.
\end{aligned}
\tag{4.13}
$$

In particular, $\mathcal{P}(p(\boldsymbol{u}_{\mathrm{I}},\boldsymbol{u}_{\mathrm{II}}),\, p_L(\boldsymbol{u}_{\mathrm{I}},\boldsymbol{u}_{\mathrm{II}}))$ *quantifies an intrinsic information barrier* (*see* [5, 11, 25, 29, 30]) *for all imperfect model densities with the form as in* (4.10).

To prove Proposition 4.1, first, utilize the general identity (see [6]) to calculate

$$
\begin{aligned}
&\mathcal{P}\big(p_L(\boldsymbol{u}_{\mathrm{I}},\boldsymbol{u}_{\mathrm{II}}),\, p_L^{\mathrm{M}}(\boldsymbol{u}_{\mathrm{I}},\boldsymbol{u}_{\mathrm{II}})\big) \\
&\quad = \mathcal{P}\big(\pi(\boldsymbol{u}_{\mathrm{I}}),\pi^{\mathrm{M}}(\boldsymbol{u}_{\mathrm{I}})\big) + \int \pi(\boldsymbol{u}_{\mathrm{I}})\mathcal{P}\big(p(\boldsymbol{u}_{\mathrm{II}}\,|\,\boldsymbol{u}_{\mathrm{I}}),\, p_L^{\mathrm{M}}(\boldsymbol{u}_{\mathrm{II}}\,|\,\boldsymbol{u}_{\mathrm{I}})\big)\mathrm{d}\boldsymbol{u}_{\mathrm{I}},
\end{aligned}
\tag{4.14}
$$

which is easily verified by the reader. Next, for each $\boldsymbol{u}_{\mathrm{I}}$, use the general identity for least biased densities, which follows from the max-entropy principle in (4.8) (see [24, Chap. 2])

$$
\mathcal{P}\big(p(\boldsymbol{u}_{\mathrm{I}},\boldsymbol{u}_{\mathrm{II}}),\, p_L^{\mathrm{M}}(\boldsymbol{u}_{\mathrm{I}},\boldsymbol{u}_{\mathrm{II}})\big) = \mathcal{P}\big(p(\boldsymbol{u}_{\mathrm{I}},\boldsymbol{u}_{\mathrm{II}}),\, p_L^{\mathrm{M}}(\boldsymbol{u}_{\mathrm{I}},\boldsymbol{u}_{\mathrm{II}})\big) + \mathcal{P}\big(p(\boldsymbol{u}_{\mathrm{I}},\boldsymbol{u}_{\mathrm{II}}),\, p_L^{\mathrm{M}}(\boldsymbol{u}_{\mathrm{I}},\boldsymbol{u}_{\mathrm{II}})\big),
\tag{4.15}
$$

and insert this in (4.14). Finally, computing $\mathcal{P}(p(\boldsymbol{u}_{\mathrm{I}},\boldsymbol{u}_{\mathrm{II}}),\, p_L(\boldsymbol{u}_{\mathrm{I}},\boldsymbol{u}_{\mathrm{II}}))$ and $\mathcal{P}(p_L(\boldsymbol{u}_{\mathrm{I}},\boldsymbol{u}_{\mathrm{II}}),\, p_L^{\mathrm{M}}(\boldsymbol{u}_{\mathrm{I}},\boldsymbol{u}_{\mathrm{II}}))$ by the formula in (4.14) once again, with simple algebra, we arrive at the required formulas in (4.11)–(4.13).

5 Non-Gaussian Test Models for Statistical Prediction with Model Errors

We apply the material developed in Sects. 3–4 with $L = 2$ to gain new insight into statistical predictions with the effects of coarse-graining and model errors in the non-Gaussian setting. In the first part of this section, we consider the effect of model errors through coarse-graining the statistics in a perfect model setting (see [18]) for short, medium, and long range forecasting. In the second part of this section, we consider the effect of model errors in the dynamics of $\boldsymbol{u}_{\mathrm{I}}$ (see [25]) on the long range forecasting skill. The errors in both the full probability density and the marginal densities in $\boldsymbol{u}_{\mathrm{II}}$ are considered.

5.1 Choice of Initial Statistical Conditions

As already mentioned in Sect. 4, we are particularly interested in assessing the model error due to various coarse-grainings of the perfect statistics. These model errors arise naturally either when deriving the approximate least-biased conditional densities through estimating the conditional moments in the CGFPE framework of Sect. 2, or when deriving the Gaussian estimators of non-Gaussian densities. The effects of initial conditions are clearly important in the short and medium range predictions, for both the perfect and the coarse-grained statistics, and the choice of a representative set of statistical initial conditions requires some care.

In the following sections, we consider the least-biased conditionally Gaussian estimators (i.e., $L=2$ in Sect. 4) of the true statistics $p(u,\gamma,t)$, leading to the non-Gaussian densities $p_2(u,\gamma,t)$, as well as fully Gaussian approximations $p_G(u,\gamma,t)$ of the true non-Gaussian statistics $p(u,\gamma,t)$. Therefore, in order to compare the effects of coarse-graining the structure of the PDFs in a standardized setting, we consider the initial joint densities with identical second-order moments, i.e., any two initial densities, $\widetilde{p}_i$ and $\widetilde{p}_j$, satisfy

$$\int u^\alpha \gamma^\beta \widetilde{p}_i(u,\gamma)\mathrm{d}u\mathrm{d}\gamma = \int u^\alpha \gamma^\beta \widetilde{p}_j(u,\gamma)\mathrm{d}u\mathrm{d}\gamma, \quad 0 \leqslant \alpha+\beta \leqslant 2. \tag{5.1}$$

For simplicity, we choose the initial densities with uncorrelated variables,

$$\widetilde{p}_i(u,\gamma) = \widetilde{\pi}_i(u)\widetilde{\pi}_i(\gamma),$$

where the marginal densities $\widetilde{\pi}_i(u)$ and $\widetilde{\pi}_i(\gamma)$ are given by the mixtures of simple densities (see Appendix B for more details). This procedure is sufficient for the present purposes and reduces the complexity of exposition. An analogous procedure can be used to generate PDFs with correlated variables by, for example, changing the coordinate frame. Such a step might be necessary when studying the model error in filtering problems.

The following sets of non-Gaussian initial conditions, shown in Fig. 7 and constructed in the way described above, are used in the suite of tests discussed next (see also Appendix B):

(1) $\widetilde{p}_1(u,\gamma)$: Nearly Gaussian PDF with the Gaussian marginal in u and a weakly sub-Gaussian marginal in γ.
(2) $\widetilde{p}_2(u,\gamma)$: PDF with a bimodal marginal in u and a weakly skewed marginal in γ.
(3) $\widetilde{p}_3(u,\gamma)$: Multimodal PDF with a bimodal marginal in u and a tri-modal marginal in γ.
(4) $\widetilde{p}_4(u,\gamma)$: PDF with a highly skewed marginal in u and a bimodal marginal in γ.
(5) $\widetilde{p}_5(u,\gamma)$: PDF with a weakly skewed marginal in u and a highly skewed marginal in γ.
(6) $\widetilde{p}_6(u,\gamma)$: Multimodal PDF with a Gaussian marginal in u and a tri-modal marginal in γ.
(7) $\widetilde{p}_7(u,\gamma)$: Multimodal PDF with a bimodal marginal in u and a Gaussian marginal γ.

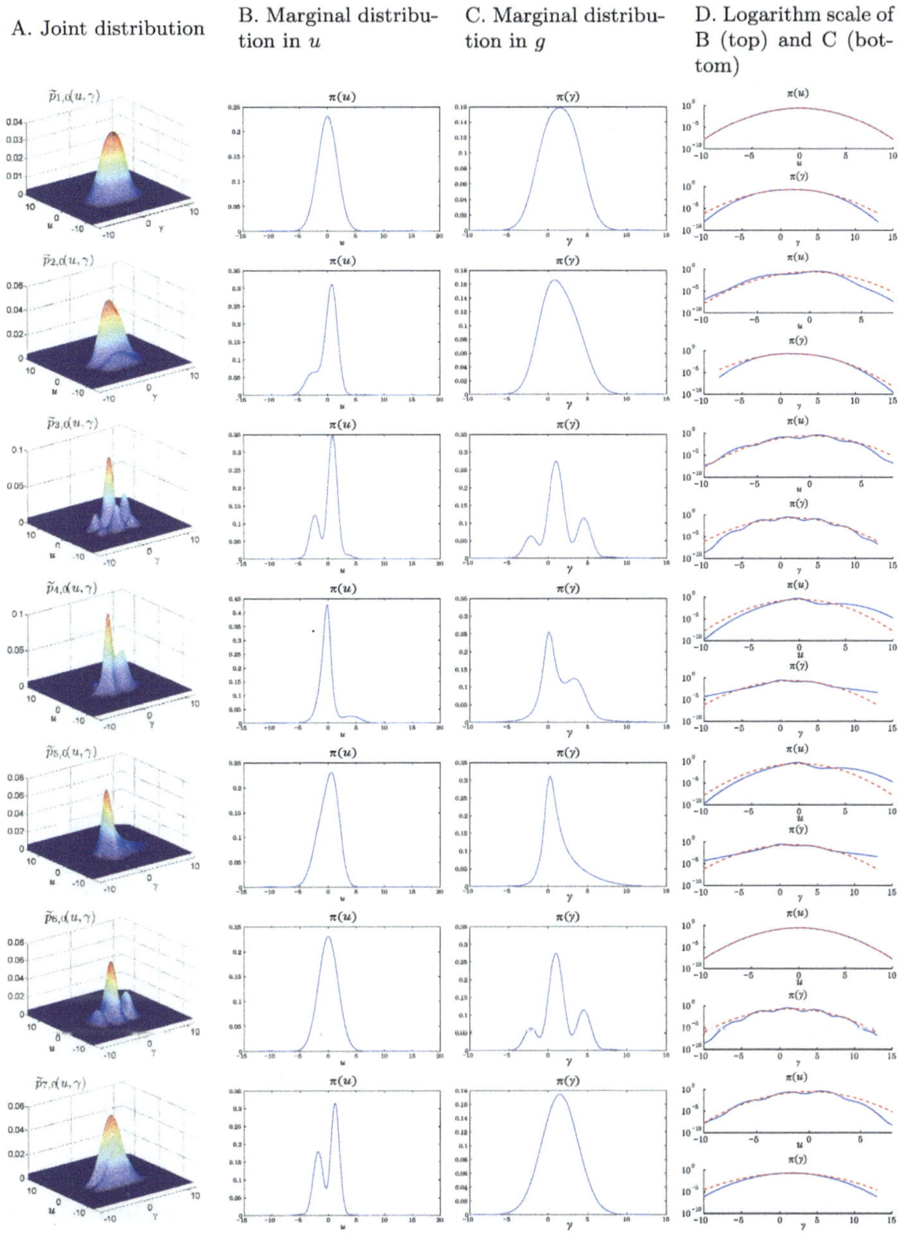

Fig. 7 The set of seven non-Gaussian initial conditions with identical second-order statistics used in the tests in Figs. 9–20 (see Sect. 5.1 and Appendix B for more details)

5.2 Ensemble Prediction with Model Error Due to Coarse-Graining the Perfect Dynamics

We consider the dynamics of the same non-Gaussian system (3.4) with intermittent instabilities as in Sect. 3.1.1 which has the general structure as in (3.1)–(3.2). The wide range of interesting turbulent dynamical regimes (see [3–5, 25]) makes this statistically exactly solvable system an unambiguous tested for studying the effects of model errors introduced through various coarse-grainings of the perfect density $p(u,\gamma,t)$ as discussed in Sect. 4. In this section, following the methodology introduced in Sect. 4, we focus on the model error arising from two particular coarse-grainings of the perfect model density $p(u,\gamma,t)$:

(1) $p_2(u,\gamma,t)$: Non-Gaussian density obtained through the least-biased conditionally Gaussian approximation of the true conditional densities such that the true density $p(u,\gamma,t)$ and the coarse-grained density $p_2(u,\gamma,t)$ have the same first two conditional moments, i.e., for each fixed γ and t, we set

$$\mathcal{S}\big(p_2(u\mid\gamma,t)\big)=\max_{\mathcal{M}_{N,2}=\mathcal{M}_N}\mathcal{S}\big(q(u)\big),$$

where

$$\mathcal{M}_N=\int u^N p(u\mid\gamma,t)\mathrm{d}u,\qquad \mathcal{M}_{N,2}=\int u^N q(u)\mathrm{d}u,\quad 0\leqslant n\leqslant 2.$$

Note that, despite the Gaussian approximations for the conditional densities $p_2(u|\gamma,t)$, the coarse-grained joint and marginal densities

$$p_2(u,\gamma,t)=p_2(u\mid\gamma,t)\pi(\gamma,t),\qquad \pi_2(u,t)=\int p_2(u,\gamma,t)\mathrm{d}\gamma$$

can be highly non-Gaussian.

(2) $p_G(u,\gamma,t)$: Gaussian approximation of the joint density $p(u,\gamma,t)$. The error in the Gaussian estimators $p_G(u,\gamma,t)$ and $\pi_G(u,t)=\int p_G(u,\gamma,t)\mathrm{d}\gamma$, arises from the least-biased approximation of the true non-Gaussian density $p(u,\gamma,t)$, which for each fixed t, maximizes the entropy

$$\mathcal{S}\big(p_G(u,\gamma,t)\big)=\max_{\mathcal{M}_{ij,G}=\mathcal{M}_{ij}}\mathcal{S}\big(q(u,\gamma)\big),$$

subject to the following moment constraints:

$$\mathcal{M}_{i,j}=\int u^i\gamma^j p(u,\gamma,t)\mathrm{d}u\mathrm{d}\gamma,\qquad \mathcal{M}_{ij,G}=\int u^i\gamma^j q(u,\gamma)\mathrm{d}u\mathrm{d}\gamma,$$
$$0\leqslant i+j\leqslant 2.$$

In the above set-up, the conditional approximations p_2 and π_2 represent the best possible (least-biased) estimates for the true joint and marginal densities, given the first two conditional moments. Thus, the errors $\mathcal{P}(p,p_2)$ and $\mathcal{P}(\pi,\pi_2)$ represent the intrinsic information barriers which can not be overcome by models based on utilizing two-moment approximations of the true densities (see Proposition 4.1 in Sect. 4).

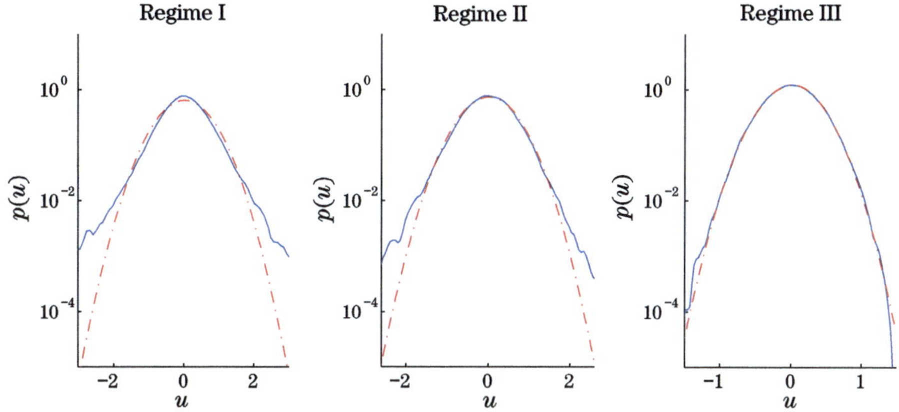

Fig. 8 Three dynamical regimes of the non-Gaussian system (3.4) characterized by different equilibrium marginal densities $\pi_{\text{eq}}(u)$ used for studying the model error in coarse-grained densities in Sect. 5.2 (see Figs. 9–11). Regimes I–II of (3.4) are characterized by intermittent dynamics of u due to transient instabilities induced by the damping fluctuation γ

In Figs. 9, 10, 11, we show the evolution of the model error (4.11) due to different coarse-grainings in p_2 and p_G in the following three dynamical regimes of the system (3.4) with Gaussian damping fluctuations (see also Fig. 8):

Regime I (see Fig. 11) A regime with plentiful, short-lasting transient instabilities in the resolved component $u(t)$ with fat-tailed marginal equilibrium densities $\pi(u)$, where, the parameters used in (3.4) are

$$\hat{\gamma}=2, \quad \sigma_\gamma=d_\gamma=10, \quad \sigma_u=1, \quad f_u=0.$$

Regime II (see Fig. 10) A regime with intermittent large-amplitude bursts of the instability in $u(t)$ with fat-tailed marginal equilibrium densities $\pi(u)$, where, the parameters used in (3.4) are

$$\hat{\gamma}=2, \quad \sigma_\gamma=d_\gamma=2, \quad \sigma_u=1, \quad f_u=0.$$

Regime III (see Fig. 9) A regime with nearly Gaussian marginal equilibrium density $\pi(u)$, where, the parameters used in (3.4) are

$$\hat{\gamma}=7, \quad \sigma_\gamma=d_\gamma=1, \quad \sigma_u=1, \quad f_u=0.$$

In each regime, the model error in the ensemble predictions is examined for the set of seven different initial densities introduced in Sect. 5.1 and Fig. 7 with identical second-order statistics. The evolution of the true density $p(u,\gamma,t)$ is estimated via Monte Carlo simulations with 10^7 samples, while the coarse-grained joint densities p_2, p_G, and their marginals π_2, π_G are computed according to the moment-constrained maximum entropy principle in (4.8) using the conditional moments computed from the CGFPE procedure (2.10).

The top row in Figs. 9–11 shows the evolution of the model error in the Gaussian estimators $p_G(u,\gamma,t)$ and $\pi_G(u,t)$ of the true density. The intrinsic information

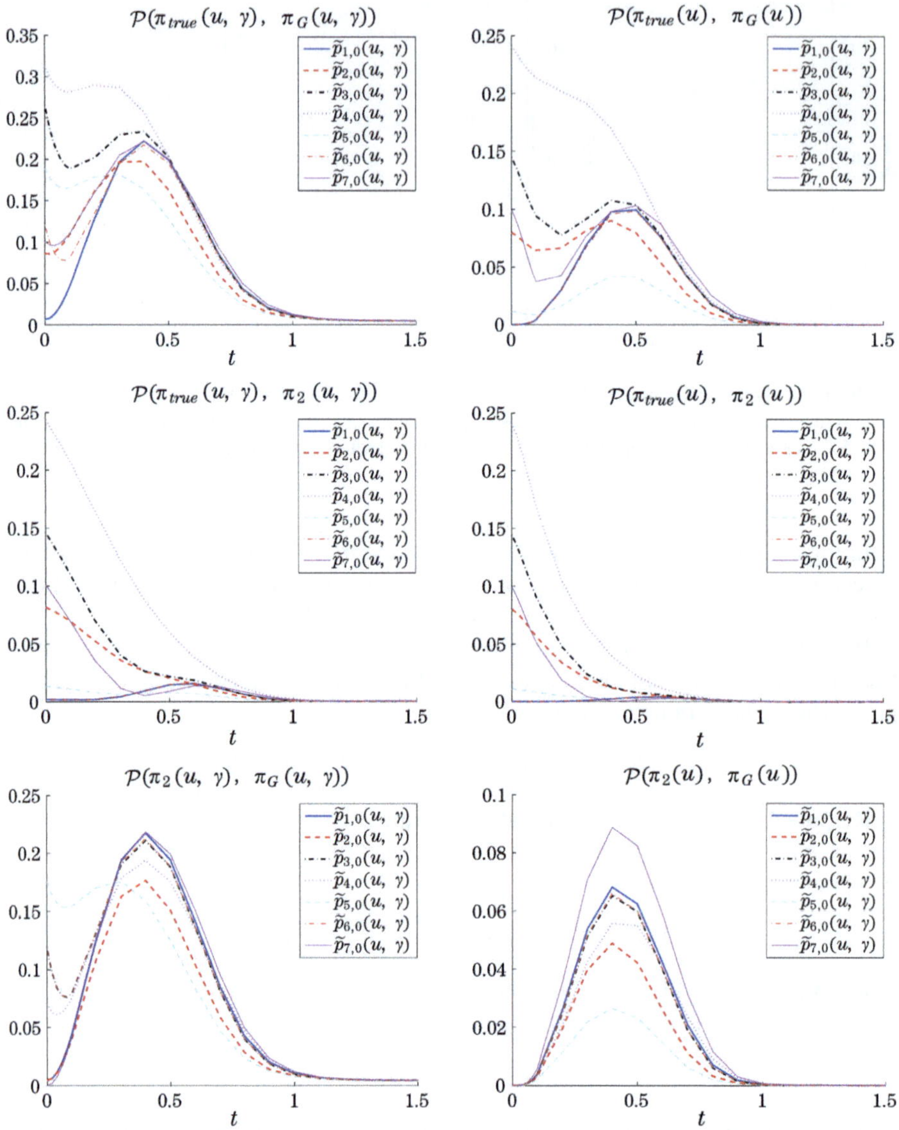

Fig. 9 Model errors due to coarse-graining the perfect dynamics of the system (3.4) in the nearly Gaussian regime (Regime III in Fig. 8). (*Top two rows*) Evolution of the model errors (4.11) due to different coarse-grainings of the perfect dynamics in the system (3.4) with Gaussian damping fluctuations. The non-Gaussian joint and marginal densities, p_2 and π_2, are obtained through the Gaussian coarse-graining of the conditional statistics $p(u \mid \gamma)$ (see Sect. 3–4), while p_G and π_G are the joint and the marginal densities of the Gaussian estimators (see Sect. 4). The information barrier (*bottom row*) equals $\mathcal{P}(p, p_G) - \mathcal{P}(p_2, p_G)$ (see (4.11)). The respective statistical initial conditions, all with the same second-order moments, are described in Sect. 5.1 and shown in Fig. 7

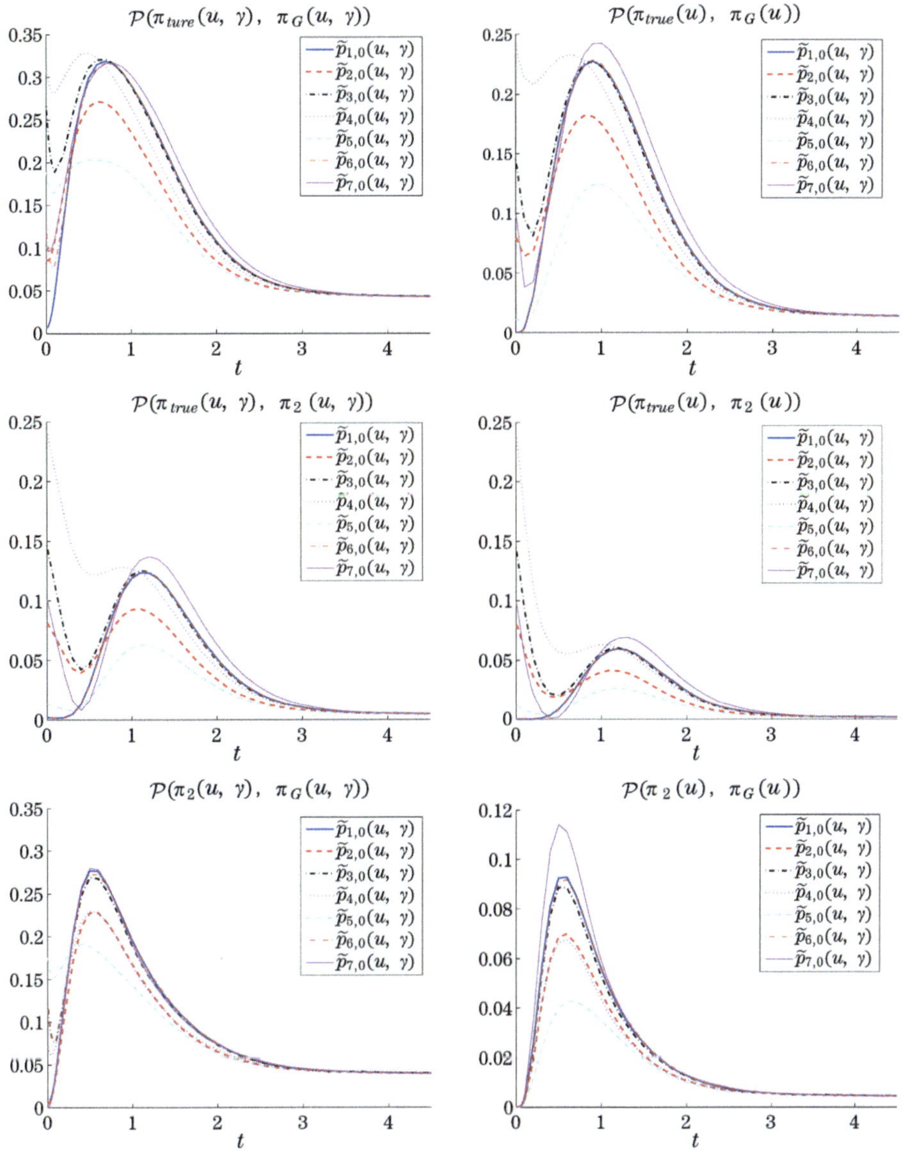

Fig. 10 Model errors due to coarse-graining perfect dynamics; the system (3.4) in the regime with intermittent large amplitude instabilities. (*Top two rows*) Evolution of the model errors (4.11) due to different coarse-grainings of the perfect dynamics in the system (3.4) with Gaussian damping fluctuations. The non-Gaussian joint and marginal densities p_2 and π_2, are obtained through the Gaussian coarse-graining of the conditional statistics $p(u \mid \gamma)$ (see Sect. 3–4), while p_G and π_G are the joint and the marginal densities of the Gaussian estimators (see Sect. 4). The information barrier (*bottom row*) equals $\mathcal{P}(p, p_G) - \mathcal{P}(p_2, p_G)$ (see (4.11)). The respective statistical initial conditions, all with the same second-order moments, are described in Sect. 5.1 and shown in Fig. 7

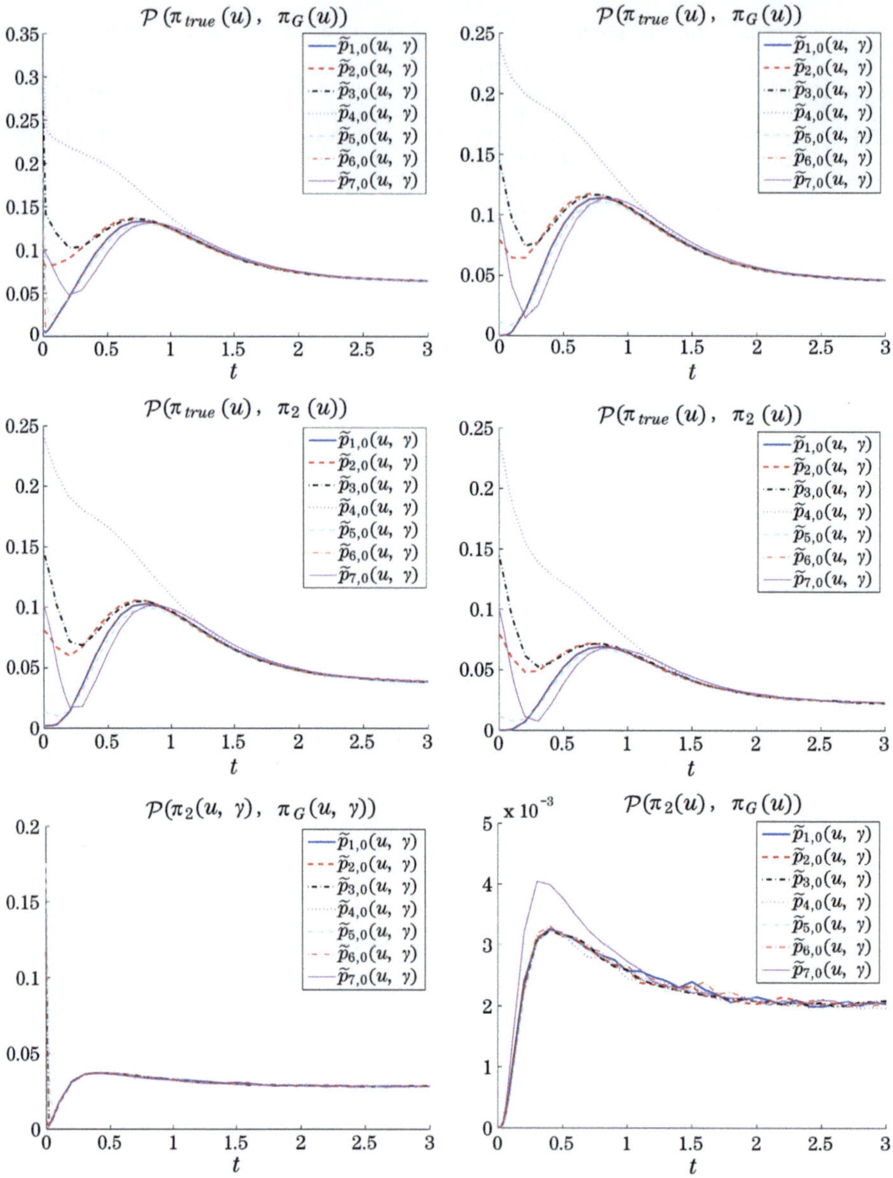

Fig. 11 Model errors due to coarse-graining perfect dynamics; the system (3.4) in the regime with abundant transient instabilities. (*Top two rows*) Evolution of the model errors (4.11) due to different coarse-grainings of the perfect dynamics in the system (3.4) with Gaussian damping fluctuations. The non-Gaussian joint and marginal densities p_2 and π_2, are obtained through the Gaussian coarse-graining of the conditional statistics $p(u \mid \gamma)$ (see Sect. 3–4), while p_G and π_G are the joint and the marginal densities of the Gaussian estimators (see Sect. 4). The information barrier (*bottom row*) equals $\mathcal{P}(p, p_G) - \mathcal{P}(p_2, p_G)$ (see (4.11)). The respective statistical initial conditions, all with the same second-order moments, are described in Sect. 5.1 and shown in Fig. 7

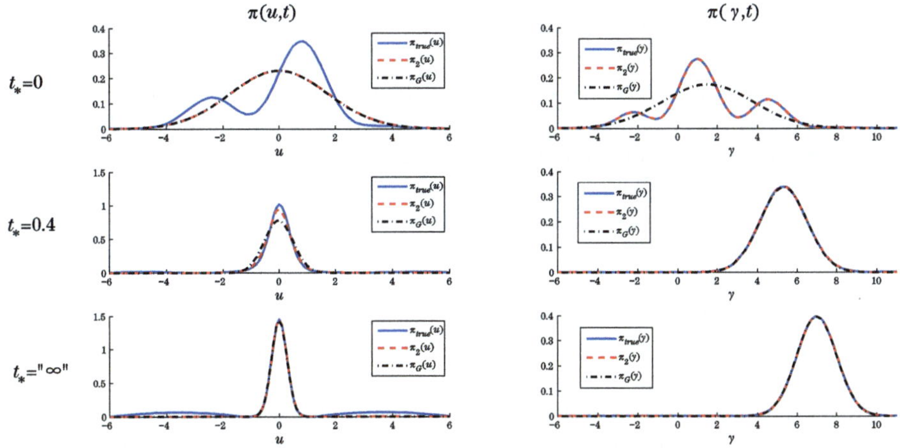

Fig. 12 Three distinct stages in the statistical evolution of the system (3.4) illustrated for the regime with nearly Gaussian dynamics and highly non-Gaussian multimodal initial statistical conditions $\widetilde{p}_3(u,\gamma)$ (see Fig. 7). These three stages exist regardless of the dynamical regime of (3.4) and the form of the initial conditions (not shown). (*Top*) The initial configuration projected on the marginal densities at $t_* = 0$. (*Middle*) The fat-tailed phase in the marginal $\pi(u,t)$ corresponding to the large error phase in the coarse grained models (see Figs. 9–11). (*Bottom*) Equilibrium marginal statistics on the attractor in the regime with nearly Gaussian statistics (see Regime III in Fig. 8)

barrier in the Gaussian approximation (see Proposition 4.1), represented by the lack of information in the least-biased density p_2, based on two conditional moment constraints, is shown for each regime in the middle row. It can be seen in Figs. 9–11 that the common feature of the model error evolution in all the examined regimes of (3.4) is the presence of a large error at the intermediate lead times. The source of this phenomenon is illustrated in Fig. 12 in Regime III of (3.4) with nearly Gaussian attractor statistics. The large error arises from the presence of a robust transient phase of fat-tailed dynamics in the system (3.4) which is poorly captured by the coarse grained statistics.

Below, we summarize the results illustrated in Figs. 9–12 with the focus on the model error in the Gaussian approximations $p_G(u,\gamma,t)$ and $\pi_G(u,t)$:

(1) For both the Gaussian estimators $p_G(u,\gamma)$, $\pi_G(u)$ and the conditionally Gaussian estimators $p_2(u,\gamma)$, $\pi_2(u)$, there exists a phase of large model errors at intermediate lead times. This phase exists in all the examined regimes of (3.4) irrespective of the initial conditions, and it arises due to a transient highly non-Gaussian fat-tailed dynamical phase in (3.4) which the Gaussian estimators fail to capture.
(2) The trends in the model error evolution for the joint and the marginal densities are similar. This is to be expected based on Proposition 4.1.
(3) The contributions to the model error in the Gaussian estimators $p_G(u,\gamma)$ and $\pi_G(u)$ from the intrinsic information barrier $\mathcal{P}(p,p_2)$ (see Proposition 4.1), and from the error $\mathcal{P}(p_2,p_G)$ due to the fully Gaussian vs conditionally Gaussian approximations depend on the dynamical regime.

(a) The effects of the intrinsic information barrier are the most pronounced in the non-Gaussian Regime I of (3.4) with abundant transient instabilities in u (see Figs. 8 and 11). In this regime, the information barrier dominates the total model errors. In the nearly Gaussian regime, the intrinsic information barrier is negligible except at short times due to the errors in coarse-graining the highly-non-Gaussian initial conditions (see Figs. 7 and 9).
(b) In the highly non-Gaussian Regime I with abundant instabilities and the fat-tailed equilibrium PDFs (see Fig. 8), the differences in the model error between different initial conditions quickly become irrelevant. The intrinsic information barrier dominates the model error, and there is a significant error for long range predictions in both the joint and the marginal coarse-grained densities.
(c) In the non-Gaussian Regime II of (3.4) with large amplitude intermittent instabilities, the intrinsic information barrier dominates the error in the Gaussian estimators at short ranges. At intermediate lead times, the error due to the fully Gaussian vs conditionally Gaussian approximations exceeds the intrinsic barrier. The error at long lead times is significantly smaller than those in Regime I with comparable contributions from $\mathcal{P}(p, p_2)$ and $\mathcal{P}(p_2, p_G)$.
(d) In the nearly Gaussian Regime III of (3.4), the intrinsic information barrier in the Gaussian estimators is small and dominated by the errors in coarse-graining the non-Gaussian initial conditions.

(4) The intrinsic information barriers in the joint density $\mathcal{P}(p, p_2)$ and in the marginal density $\mathcal{P}(\pi, \pi_2)$, are comparable throughout the evolution and almost identical at short lead times.

5.3 Ensemble Prediction with Model Errors Due to Imperfect Dynamics

We focus on the model error which arises through common approximations associated with the ensemble prediction: (i) Errors due to imperfect/simplified dynamics, and (ii) errors due to coarse-graining the statistics of the perfect system which is used for tuning the imperfect models. While the above two approximations are often simultaneously present in applications and are generally difficult to disentangle, it is important to understand the effects of these two contributions in a controlled environment which is developed below.

Similar to the framework used in the previous sections, we consider the dynamics with the structure as in the test models (3.1)–(3.2), where the non-Gaussian perfect system, as in (3.5), is given by

$$\begin{aligned}
&\text{(a)}\quad d\gamma = \big[-a\gamma + b\gamma^2 - c\gamma^3 + f_\gamma(t)\big]dt + (A - B\gamma)dW_C + \sigma_\gamma dW_\gamma,\\
&\text{(b)}\quad du = \big(-\gamma u + f_u(t)\big)dt + \sigma_u dW_u,
\end{aligned}\tag{5.2}$$

with cubic nonlinearity in the damping fluctuations γ. The imperfect non-Gaussian model introduces errors by assuming Gaussian dynamics in the damping fluctuations, as in (3.4),

$$
\begin{aligned}
&\text{(a)}\quad d\gamma^{\mathrm{M}} = \big(-d_\gamma^{\mathrm{M}}(\gamma^{\mathrm{M}} - \hat{\gamma}^{\mathrm{M}}) + f_\gamma^{\mathrm{M}}(t)\big)dt + \sigma_\gamma^{\mathrm{M}} dW_\gamma^{\mathrm{M}},\\
&\text{(b)}\quad du^{\mathrm{M}} = \big(-\gamma^{\mathrm{M}} u^{\mathrm{M}} + f_u^{\mathrm{M}}(t)\big)dt + \sigma_u^{\mathrm{M}} dW_u^{\mathrm{M}}.
\end{aligned}
\tag{5.3}
$$

The imperfect model (5.3) is optimized by tuning its marginal attractor statistics, in either u^{M} or γ^{M} depending on the context, to reproduce the respective true marginal statistics. This is a prototype problem for a number of important issues. Two topical examples are:

(1) Reduced models with a subset of unresolved variables (here γ^{M}) whose statistics is tuned for statistical fidelity in the resolved variables (here u^{M}).

(2) Simplification of parts of the dynamics in complex multi-component models such as the coupled atmosphere-ocean-land models in climate science; in the present toy model setting γ can be regarded as the atmospheric forcing of the ocean dynamics u.

In order to illustrate the framework developed in Sect. 2–4, we compare the model error arising in the optimized imperfect statistics, $p^{\mathrm{M}*}(u,\gamma,t)$ or $\pi^{\mathrm{M}*}(u,t)$, associated with (5.3) with the model error in $p_2(u,\gamma,t)$ or $\pi_2(u,t)$ due to the Gaussian coarse-graining of the conditional density $p(u \mid \gamma, t)$ of the perfect system (5.2) using the CGFPE framework of Sect. 2.

In particular, we show that a small model error can be achieved at medium and long lead times for imperfect predictions of the marginal dynamics $\pi^{\mathrm{M}*}(u)$ using models with tuned unresolved dynamics γ despite a large model error in the joint density $p^{\mathrm{M}*}(u,\gamma)$.

5.3.1 Ensemble Predictions with Imperfect Dynamics and Time-Independent Statistics on the Attractor

We consider the perfect system (5.2) and its model (5.3) with invariant measures at their respective equilibria. This configuration is achieved by assuming constant forcing $f_\gamma{=}0.8220$, $f_u{=}-0.5$, $f_\gamma^{\mathrm{M}}{=}0$, $f_u^{\mathrm{M}} = -0.5$ in both (5.2) and (5.3). We first examine the effects of model errors associated with two distinct ways of optimizing the imperfect model (5.3):

(I) Tuning the marginal equilibrium statistics of the damping fluctuations γ^{M} in (5.3) for fidelity to the true statistics of γ in (5.2).

In order to tune the mean and variance of γ^{M} to coincide with the true moments, we simply set

$$
\hat{\gamma}^{\mathrm{M}} = \langle\gamma\rangle_{\mathrm{eq}}, \qquad \frac{\sigma^2_{\gamma^{\mathrm{M}}}}{2d_\gamma^{\mathrm{M}}} = \mathrm{Var}_{\mathrm{eq}}(\gamma), \tag{5.4}
$$

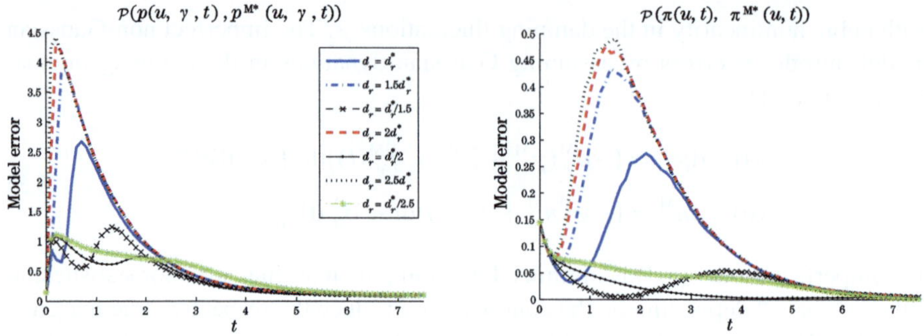

Fig. 13 The ensemble prediction of (5.2) with imperfect models in (5.3); dependence of model error on the decorrelation time in the imperfect model. The model error (4.11) via the relative entropy for the imperfect prediction of the system (5.2) using imperfect models in (5.3) with the correct climatology in γ^{M} but different decorrelation times of the damping fluctuations. Note that in this case underdamped imperfect models have the best medium range prediction skill. The results shown are obtained in the skewed two-state unimodal regime (see Fig. 1) of (5.2), starting from the statistical initial condition $\widetilde{p}_1(u,\gamma)$ (see Sect. 5.1 and Fig. 7)

which leads to a one-parameter family of models in (5.3) with a correct marginal equilibrium density in γ^{M}. Below, we choose the damping d_γ^{M} in (5.3) as the free parameter and study the dependence of model errors in the class of models satisfying (5.4) and parameterized by the damping/decorrelation dime in γ^{M} (see Fig. 13). Note that only one model in this family can match both the equilibrium density $\pi(\gamma)$ and the decorrelation time $\tau_\gamma = \int \mathrm{Corr}_\gamma(\tau)\mathrm{d}\tau$, of the true damping fluctuations in (5.2). For such a model we have, in addition to (5.4),

$$\tau_\gamma^{\mathrm{M}} = \frac{1}{d_{\gamma^{\mathrm{M}}}} = \tau_\gamma. \tag{5.5}$$

Examples of the prediction error in models of (5.3) optimized for equilibrium fidelity in γ^{M} but different dampings $d_\gamma^{\mathrm{M}*}$ are shown in Fig. 13 for the two-state unimodal regime of (5.2) (see Fig. 1). We highlight two important observations here:

(1) Underdamped models of (3.4) optimized for equilibrium fidelity in the damping fluctuations γ^{M} have the smallest error for medium range forecasts (all models are comparable for long range forecasts). These results are similar to those reported recently in [25], where the short and medium range predictive skills of linear models with optimized marginal statistics of the unresolved dynamics were shown to often exceed the skills of models with correct marginal statistics and decorrelation time.

(2) Despite the striking reduction in the model error at intermediate lead times achieved through underdamping the unresolved dynamics in (3.4), caution is needed when tuning imperfect models for short range forecasts or forced response predictions, where the damping, in both the resolved and unresolved dynamics, is relevant for correct system responses (see [25]).

(II) Tuning the marginal equilibrium statistics of the damping fluctuations γ^{M} in (5.3) for fidelity to the true statistics of u in (5.2).

This case corresponds to the situation in which we construct a simplified model of a system with unresolved degrees of freedom (here γ); these stochastically 'superparameterized' unresolved dynamics are then tuned to correctly reproduce the statistical features of the resolved dynamics (here u).

We consider this optimization in the Gaussian framework and optimize the imperfect model (5.3) by tuning the dynamics of the damping fluctuations γ^{M} in order to minimize the lack of information in the imperfect marginal density for the resolved variable, i.e., the optimal imperfect model satisfies

$$\mathcal{P}\big(\pi_G(u), \pi_G^{\mathrm{M}*}(u)\big) = \min_{d_\gamma^{\mathrm{M}}, \sigma_\gamma^{\mathrm{M}}, \widehat{\gamma}^{\mathrm{M}}} \mathcal{P}\big(\pi_G(u), \pi_G^{\mathrm{M}}(u)\big), \tag{5.6}$$

where π_G and π_G^{M} are the Gaussian estimators of the respective marginal densities associated with (5.2) and (5.3), respectively. With the conditional moments of u in the perfect system (5.2), $M_1(\gamma)$ and $M_2(\gamma)$ obtained by solving (2.10) in the CGFPE framework in Sect. 2, the mean and the variance of $p_G(u)$ are given by

$$\overline{u} = \int \mathcal{M}_1(\gamma)\mathrm{d}\gamma, \qquad R_u = \int \mathcal{M}_2(\gamma)\mathrm{d}\gamma - \overline{u}^2, \tag{5.7}$$

respectively. Analogous expressions hold for the mean and the variance of $p_G^{\mathrm{M}}(u)$ which are used in the optimization (5.6).

The two types of model optimization are compared in Fig. 14 for the two-state unimodal regime of (5.2) (see Fig. 1). Both procedures yield comparably good results at long lead times when the model error in the marginal densities in $\pi^{\mathrm{M}*}(u,t)$ is considered. Unsurprisingly, optimizing the marginal dynamics of u^{M} by tuning the dynamics of γ^{M} generally leads to a smaller model error for short and medium range predictions. But the type of the optimization largely depends on the applications.

In Figs. 15, 16, 17, 18, we illustrate the evolution of the model error in the imperfect statistical prediction of (5.2) which is optimized according to the procedure (I) above. Two non-Gaussian regimes of the true system (5.2) illustrated in Fig. 1 are used to analyze the error in imperfect predictions with optimized models in (5.3).

5.3.2 Ensemble Predictions with Imperfect Dynamics and Time-Periodic Statistics on the Attractor

We finish the analysis by considering the dynamics of the perfect system (5.2) and its model (5.3) with time-periodic statistics on the attractor. We focus on the highly non-Gaussian regime of the perfect system (5.2) with the cubic nonlinearity in the damping fluctuations periodic transitions between the nearly Gaussian and highly skewed marginal densities in the damping fluctuations γ which are induced by the simple time-periodic forcing

$$f_\gamma(t) = f_{\gamma,0} + f_{\gamma,1}\sin(\omega t + \phi).$$

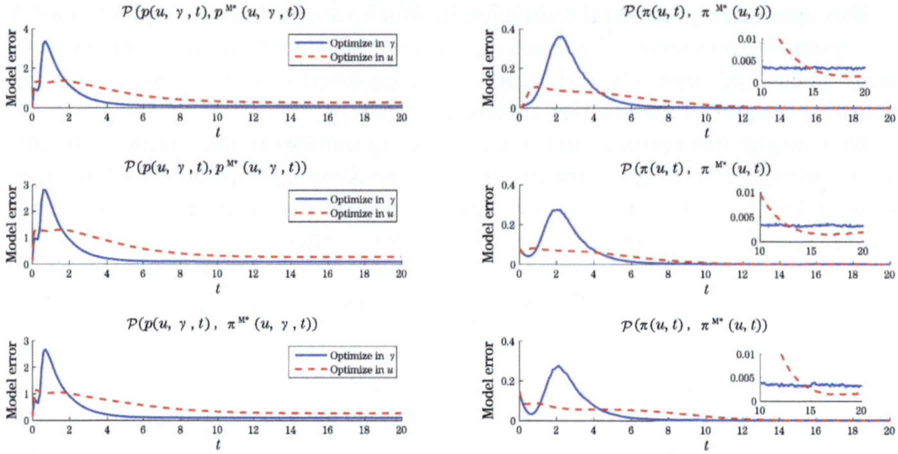

Fig. 14 The ensemble prediction of (5.2) with imperfect models in (5.3); comparison of model errors for different types of model optimization. Evolution of the model error (4.11) via the relative entropy for imperfect models in (5.3) where the imperfect dynamics of the damping fluctuations γ^{M}, is either (I) tuned to correctly reproduce the marginal equilibrium statistics of γ, or (II) tuned to correctly reproduce the marginal equilibrium statistics of u in (5.2). The results shown are obtained for the perfect dynamics in (5.2) in the regime with skewed unimodal statistics and the two-state switching in the path-wise dynamics (see Fig. 1), and for three different statistical initial conditions: (*top*) the initial density $\widetilde{p}_1(u,\gamma)$, (*middle*) the initial density $\widetilde{p}_2(u,\gamma)$, (*bottom*) the initial density $\widetilde{p}_3(u,\gamma)$ (see also Fig. 7 and Sect. 5.1)

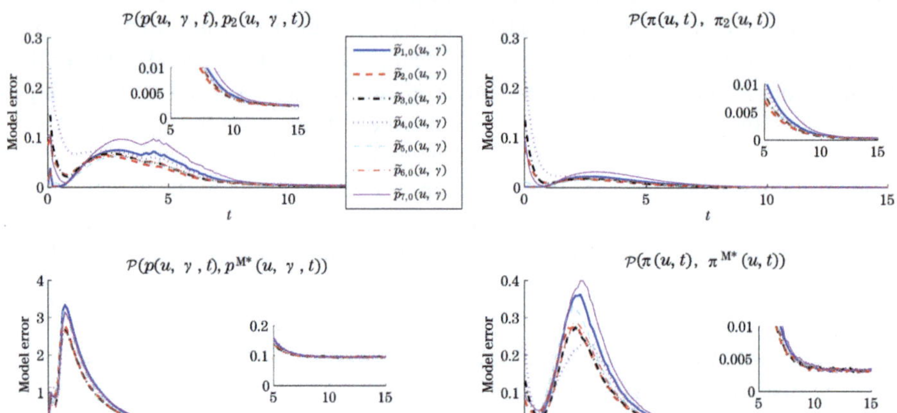

Fig. 15 The ensemble prediction with optimized imperfect dynamics; the perfect system (5.2) with skewed unimodal statistics and the regime switching, imperfect model given by (5.3). Comparison of two types of model errors in ensemble predictions: (*top row*) the model error (4.11) due to coarse-graining the perfect conditional statistics (see Sect. 4), and (*bottom row*) the model error due to imperfect dynamics in (5.3) where γ^{M} is tuned for the correct marginal equilibrium statistics and the correlation time of the damping fluctuations γ in (5.2). The model error via the relative entropy 4.11 is shown for the joint densities (*left column*) and the marginal densities in u (*right column*). The respective initial conditions are shown in Fig. 7

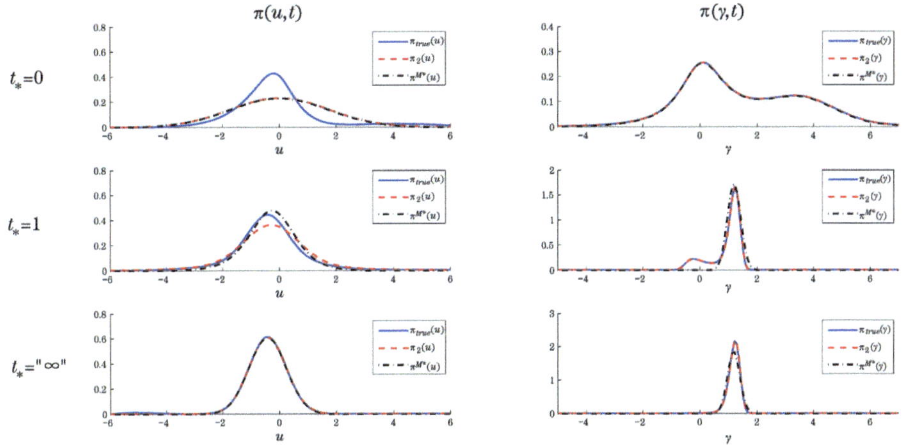

Fig. 16 Three distinct stages in the statistical evolution of the system (5.2) and its imperfect models (5.3) with different contributions to model errors; the example shown corresponds to the evolution from the initial condition $\widetilde{p}_3$ (see Sect. 5.1) in the regime with time-invariant statistics at the equilibrium with unimodal PDFs and the regime switching (see Fig. 1). (*Top*) The initial configuration at $t_* = 0$. (*Middle*) The fat-tailed phase in the true marginal $\pi(u, t)$ corresponding to the large error phase in the coarse-grained and the Gaussian models (see Fig. 15). (*Bottom*) Equilibrium marginal statistics on the attractor with the skewed marginals $\pi(\gamma)$ and $\pi(\gamma^{\mathrm{M}})$ of the damping fluctuations

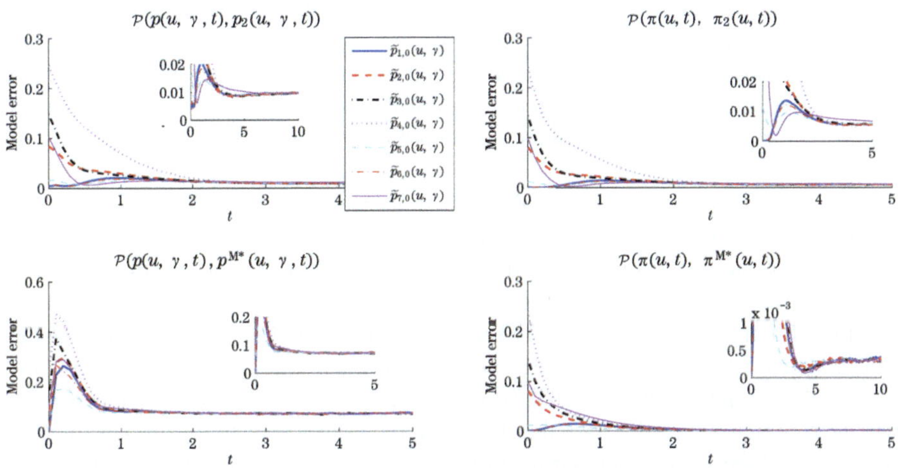

Fig. 17 The ensemble prediction with optimized imperfect dynamics; the perfect system (5.2) with fat-tailed statistic, imperfect model given by (5.3). Comparison of two types of model errors in ensemble predictions: (*Top row*) The model error due to coarse-graining the perfect dynamics (5.2), and (*bottom row*) the model error due to imperfect dynamics (5.3), where γ^{M} is tuned for the correct marginal equilibrium statistics and the correlation time of the damping fluctuations γ in (5.2). The model error via the relative entropy (4.11) is shown for the joint densities (*left column*) and the marginal densities in u (*right column*). The respective initial conditions are shown in Fig. 7.

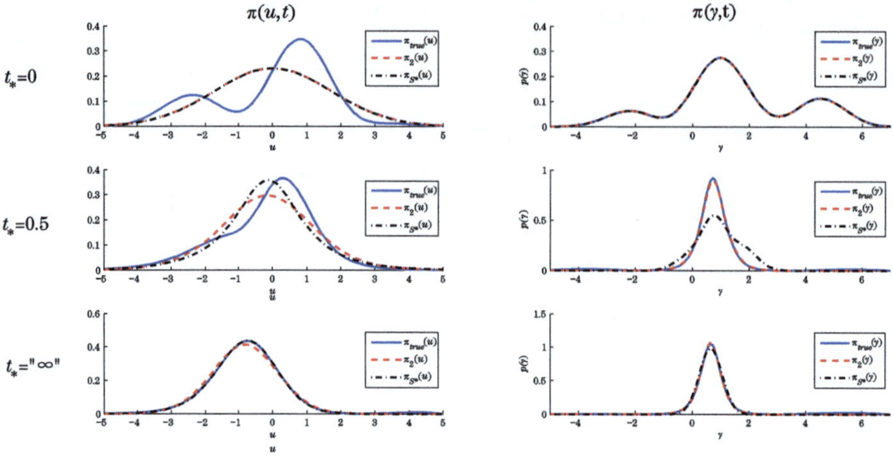

Fig. 18 Three distinct stages in the statistical evolution of the perfect system (5.2) and its imperfect models in (5.3) with different contributions to the model error; the example shown corresponds to the evolution from the initial condition $\widetilde{p}_3$ (see Sect. 5.1) in the regime with time-invariant statistics at the equilibrium and fat-tailed PDFs (see Fig. 1). (*Top*) The initial configuration at $t_* = 0$. (*Middle*) The fat-tailed phase in the true marginal $\pi(u,t)$ corresponding to the large error phase in the coarse-grained and the Gaussian models (see Fig. 15). (*Bottom*) Equilibrium marginal statistics on the attractor with the fat-tailed marginals $\pi(\gamma)$ and $\pi(\gamma^{\mathrm{M}})$ of the damping fluctuations

This regime was previously used in Sect. 3.1.1 to validate the CGFPE framework (see Fig. 2). Similar to the configurations studied with time-independent equilibrium statistics in the previous section, we are interested in the differences between the model error arising in the optimized imperfect dynamics $p^{\mathrm{M}*}(u,\gamma,t)$ and $\pi^{\mathrm{M}*}(u)$, and the error due to coarse-graining the perfect statistics in the densities $p_2(u,\gamma,t), \pi_2(u)$ obtained through the Gaussian approximations of the conditionals $p(u \mid \gamma, t)$.

The issue of tuning the marginal attractor statistics of the damping fluctuations γ^{M} in the imperfect model (5.3) requires more care than in the case with time-independent equilibrium statistics; this is due to the presence of an intrinsic information barrier (see Sect. 4 or [5, 25]) when tuning the statistics of the Gaussian damping fluctuations γ^{M} in (5.3) to the true statistics of (5.2) in γ. Similar to the time-independent case, we aim at tuning the marginal attractor statistics in γ^{M} for best fidelity to the true marginal statistics in γ. However, there exists an information barrier associated with the fact that the attractor variance of the Gaussian fluctuations γ^{M} is always constant regardless of the forcing $f_\gamma^{\mathrm{M}}(t)$. One way to optimize the imperfect statistics of γ is to tune its decorrelation time, and time-averaged mean and variance on the attractor to reproduce the true time-averaged quantities. However, such an approach is clearly insensitive to phase variations of the respective statistical moments. Here, instead, we optimize the imperfect model by first tuning the decorrelation times of γ^{M} and γ and then minimizing the period-averaged relative entropy between the marginal densities for the damping fluctuations, i.e., the

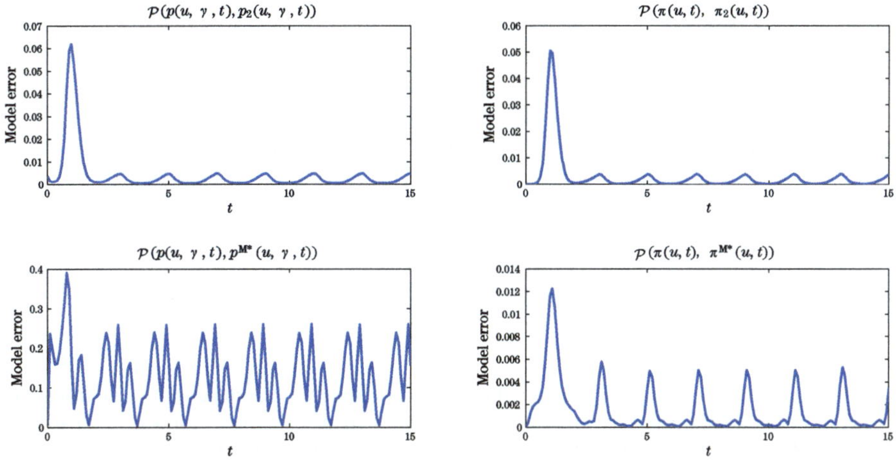

Fig. 19 The model error in imperfect optimized ensemble predictions of non-Gaussian systems with time-periodic statistics; the perfect model (5.2) with attractor statistics nearly Gaussian $\longleftrightarrow$ high skewness in γ (see Fig. 2). Evolution of the model error (4.11) associated with the statistical prediction of (5.2) in the highly non-Gaussian regime with time-periodic statistics using two non-Gaussian models: (*Top row*) Models with a coarse-grained perfect conditional density $p_2(u \mid \gamma)$ (see Sect. 4), and (*bottom row*) models with imperfect dynamics of the damping fluctuations, γ^{M} in (5.3) which are optimized by matching the decorrelation time of γ and minimizing the period-averaged relative entropy (see Sect. 5.3.2 and Sect. 4 for details)

optimized model (5.3) satisfies

$$\overline{\mathcal{P}\big(\pi_G(\gamma,t), \pi_G^{\mathrm{M}*}(\gamma,t)\big)} = \min_{\sigma_\gamma^{\mathrm{M}}, \{f_\gamma^{\mathrm{M}}\}} \overline{\mathcal{P}\big(\pi_G(\gamma,t), \pi_G^{\mathrm{M}}(\gamma,t)\big)}, \tag{5.8}$$

where the overbar denotes the temporal average over one period, and $\{f_\gamma^{\mathrm{M}}\}$ denotes a set of parameters in the forcing f_γ^{M} in (5.3). In the examples below we assume that the form of the forcing f_γ^{M} with the same time dependence on the true one, i.e.,

$$f_\gamma^{\mathrm{M}}(t) = f_{\gamma,0}^{\mathrm{M}} + f_{\gamma,1}^{\mathrm{M}} \sin\big(\omega^{\mathrm{M}} t + \phi^{\mathrm{M}}\big) \quad \text{with } \omega^{\mathrm{M}} = \omega, \quad \phi^{\mathrm{M}} = \phi,$$

so that the optimization in (5.8) is carried out over a three-parameter space $\{\sigma_\gamma^{\mathrm{M}}, f_{\gamma,0}, f_{\gamma,1}\}$ (the optimization in the phase and the frequency are often crucial and interesting, but we skip the discussions for the sake of brevity).

In Figs. 19 and 20, we show the model error for the coarse-grained joint and marginal densities p_2, π_2, and compare them with the model error in the joint and marginal densities associated with the optimized imperfect model (5.3). Here, the parameters used in (5.2) are

$$\begin{aligned} &a = 1, \quad b = 1, \quad c = 1, \quad A = 0.5, \quad B = -0.5, \quad \sigma = 0.5, \quad \sigma_u = 1, \\ &f_u = -0.5, \quad f_{\gamma,0} = 2.5, \quad f_{\gamma,1} = 6.5, \quad \omega = \pi, \quad \phi = -\frac{\pi}{2}. \end{aligned} \tag{5.9}$$

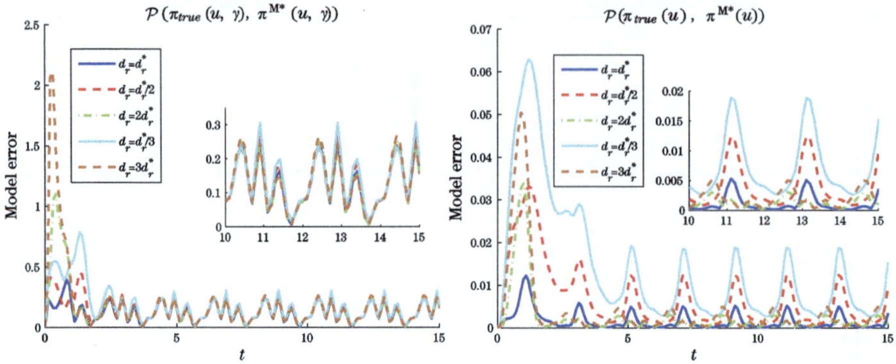

Fig. 20 Dependence on the model error on decorrelation time in imperfect optimized ensemble predictions of non-Gaussian systems with time-periodic statistics; the perfect model (5.2) and its attractor statistics as in Fig. 19. The evolution of the model error (4.11) for imperfect predictions of the true dynamics (5.2) using the models in (5.3) with different decorrelation times of damping fluctuations γ^{M}. $\tau_\gamma = \frac{1}{d_\gamma}$ denotes the decorrelation time of γ in the true dynamics (5.2). For a given decorrelation time $\frac{1}{d_\gamma^{\mathrm{M}}}$, the model (5.3) is optimized in the remaining parameters by minimizing the period-averaged relative entropy $\overline{\mathcal{P}(p(u,\gamma,t), p^{\mathrm{M}*}(u,\gamma,t))}$ (see Sect. 5.3.2 and Sect. 4 for details)

In Fig. 19, the decorrelation time $\tau^{\mathrm{M}*} = \frac{1}{d_\gamma^{\mathrm{M}*}}$ of the damping fluctuations γ^{M} is the same as the one in the true dynamics while the results shown in Fig. 20 illustrate the dependence of the model error in the optimized imperfect model on the decorrelation time (see also Fig. 13 for the configuration with time-independent equilibrium statistics).

The following points summarize the results of Sects. 5.3.1 and 5.3.2:

(1) A small model error can be achieved at medium and long lead times for imperfect predictions of the marginal dynamics $\pi^{\mathrm{M}*}(u)$ using models with tuned unresolved dynamics γ despite a large model error in the joint density $p^{\mathrm{M}*}(u,\gamma)$ (see Figs. 13–15, 17–20).

(2) The error in the coarse-grained densities $p_2(u,\gamma,t)$, $\pi_2(u,t)$ is much smaller than that in the optimized models with imperfect dynamics with $p^{\mathrm{M}*}(u,\gamma,t)$, $\pi^{\mathrm{M}*}(u,t)$ (see Figs. 15–18).

(3) The largest error in the optimized models (3.4) is associated with the presence of transient multimodal phases which can not be captured by the imperfect models in the class (3.4) (see Figs. 15–19).

(4) At long lead times, the model error in the joint density $\mathcal{P}(p(u,\gamma,t), p^{\mathrm{M}*}(u,\gamma,t))$, is largely insensitive to the variation of the damping $d_\gamma^{\mathrm{M}*}$ (see Fig. 20).

(5) The model error in the marginal densities $\pi^{\mathrm{M}*}(u,t)$ of the optimized models has non-trivial dependence on the decorrelation time $\frac{1}{d_\gamma^{\mathrm{M}*}}$ of the damping fluctuations. The overall trend is that underdamped imperfect models have smaller errors in the marginals $\pi^{\mathrm{M}*}(u,t)$ for the constant or slow forcing, while the overdamped imperfect models are better for the strongly varying forcing (see Figs. 13 and 20 for two extreme cases).

6 Concluding Discussion

We consider a class of statistically exactly solvable non-Gaussian test models where the generalized Feynman-Kac formulation developed here reduces the exact behavior of conditional statistical moments to the solution to inhomogeneous Fokker-Planck equations modified by linear lower order coupling and source terms. This procedure is applied to test models with hidden instabilities and is combined with information theory to address two important issues in contemporary statistical predictions of turbulent dynamical systems: The coarse-grained ensemble prediction in a perfect model and the improving long range forecasting in imperfect models. Here, the focus is on studying these model errors in conditionally Gaussian approximations of the highly non-Gaussian test models. In particular, we show that in many turbulent non-Gaussian dynamical regimes, a small model error can be achieved for imperfect medium and long range forecasts of the resolved variables using models with appropriately tuned statistics of the unresolved dynamics. The framework developed here, combining the generalized Feynman-Kac approach with information theory, also allows for identifying dynamical regimes with information barriers and/or transient phases in the non-Gaussian dynamics, where the imperfect models fail to capture the characteristics of the true dynamics. The techniques and models developed here should be useful for quantifying and mitigating the model error in filtering and prediction in a variety of other contexts. These applications will be developed by the authors in the near future.

Appendix A: The Numerical Scheme for Solving the CGFPE System (2.10)

Here, we outline the numerical method for solving the CGFPE system in (2.10) in one spatial dimension. This is achieved by combining the third-order backward differentiation formulas [17] with the method of (see [19]) and the second-order, finite-volume representation for (2.10).

Recall that the CGFPE system consists of a hierarchy of inhomogeneous Fokker-Planck equations for the conditional moments $\mathcal{M}_N(\gamma, t)$ with the forcing terms depending linearly on $\mathcal{M}_N(\gamma, t)$ and inhomogeneities depending linearly on $\mathcal{M}_{N-i}(\gamma, t)$, $i > 1$. Thus, due to the form of (2.10), the linearity of the forcing and inhomogeneities, we outline here the present algorithm applied to the homogeneous Fokker-Planck part of (2.10), written in the conservative form

$$\frac{\partial \pi}{\partial t} = -\frac{\partial}{\partial \gamma}\left[\left(F - \frac{1}{2}G_\gamma\right)\pi - \frac{1}{2}G\pi_\gamma\right], \tag{A.1}$$

where $\pi(\gamma, t) = \int p(u, \gamma, t)\mathrm{d}u$ and $G(\gamma, t) = \widetilde{\sigma}^2(\gamma, t)$. Given the spatial grid with nodes $\gamma_i, i = 1, \ldots, N$, the uniform spacing $\Delta\gamma$, and the approximation

$$Q_i(t) \equiv \frac{1}{\Delta\gamma}\int_{\gamma_{i-\frac{1}{2}}}^{\gamma_{i+\frac{1}{2}}} \pi(\gamma, t)\mathrm{d}\gamma, \tag{A.2}$$

we discretize (A.1) in space through the second-order finite volume formula as

$$\begin{aligned}\frac{\mathrm{d}Q_i}{\mathrm{d}t} = -\frac{1}{\Delta\gamma}\Bigg[&\left(F-\frac{1}{2}G_\gamma\right)_{i+\frac{1}{2}}\left(\frac{9}{16}Q_i+\frac{9}{16}Q_{i+1}-\frac{1}{16}Q_{i-1}-\frac{1}{16}Q_{i+2}\right)\\ &-\left(F-\frac{1}{2}G_\gamma\right)_{i-\frac{1}{2}}\left(\frac{9}{16}Q_{i-1}+\frac{9}{16}Q_i-\frac{1}{16}Q_{i-2}-\frac{1}{16}Q_{i+1}\right)\Bigg]\\ &+\frac{1}{2}\frac{1}{\Delta\gamma}\Bigg[G_{i+\frac{1}{2}}\left(-\frac{9}{8}Q_i+\frac{9}{8}Q_{i+1}+\frac{1}{24}Q_{i-1}-\frac{1}{24}Q_{i+2}\right)\\ &-G_{i-\frac{1}{2}}\left(-\frac{9}{8}Q_{i-1}+\frac{9}{8}Q_i+\frac{1}{24}Q_{i-2}-\frac{1}{24}Q_{i+1}\right)\Bigg].\end{aligned}\tag{A.3}$$

The above expression is obtained by seeking higher order interpolants for $Q^{n+1}_{i+\frac{1}{2}}$ in the standard finite-volume formulation

$$\begin{aligned}\frac{\mathrm{d}Q_i}{\mathrm{d}t} = -\frac{1}{\Delta\gamma}&\left[\left(F-\frac{1}{2}G_\gamma\right)_{i+\frac{1}{2}}Q^{n+1}_{i+\frac{1}{2}}-\left(F-\frac{1}{2}G_\gamma\right)_{i-\frac{1}{2}}Q^{n+1}_{i-\frac{1}{2}}\right]\\ &+\frac{1}{2\Delta\gamma}\left[G_{i+\frac{1}{2}}Q^{n+1}_{i+\frac{1}{2}}-G_{i-\frac{1}{2}}Q^{n+1}_{i-\frac{1}{2}}\right].\end{aligned}\tag{A.4}$$

The second order approximations for $Q^{n+1}_{i+\frac{1}{2}}$ are obtained by determining the coefficients a, b, c, d in the expansion

$$\widetilde{Q}_{i+\frac{1}{2}} = aQ_i + bQ_{i+1} + cQ_{i-1} + dQ_{i+2},$$

such that $\widetilde{Q}_{i+\frac{1}{2}} - Q_{i+\frac{1}{2}}$ is of order $O((\Delta\gamma)^3)$.

The time discretization of (A.1) or (2.10) is obtained by using the three-step backward differentiation formula (BDF3) (see [17]), which belongs to the family of linear multistep methods. In particular, (A.1) is discretized in time as follows

$$Q^{n+3} - \frac{18}{11}Q^{n+2} + \frac{9}{11}Q^{n+1} - \frac{2}{11}Q^n = \frac{6}{11}\Delta t f(Q^{n+3}).\tag{A.5}$$

The above implicit formulation can be solved explicitly due to the linearity of (A.1), where

$$f(Q^{n+3}) = \begin{cases} MQ^{n+3} & \text{for solving } \mathcal{M}_0,\\ MQ^{n+3} + f_{Q_3} & \text{for solving } \mathcal{M}_i \text{ with } i > 1.\end{cases}$$

Thus, (A.5) can be rewritten as

$$Q^{n+3} = \begin{cases} (I - \frac{6}{11}\Delta t M)^{-1}(\frac{18}{11}Q^{n+2} - \frac{9}{11}Q^{n+1} + \frac{2}{11}Q^n) & \text{for solving } \mathcal{M}_0,\\ (I - \frac{6}{11}\Delta t M)^{-1}(\frac{18}{11}Q^{n+2} - \frac{9}{11}Q^{n+1} + \frac{2}{11}Q^n + \frac{6}{11}\Delta t f_{Q_3}) & \\ \quad\text{for solving } \mathcal{M}_i \text{ with } i > 1. & \end{cases}$$

The (local) accuracy of the temporal discretization is $\mathcal{O}((\Delta t)^3)$. Analogous discretization is implemented for solving the inhomogeneous system (2.10).

Appendix B: Expressions for the Initial Densities

Here, we list the formulas used for generating the initial densities $\widetilde{p}_i(u,\gamma)$ introduced in Sect. 5.1. Recall that we chose the initial densities with uncorrelated variables,

$$\widetilde{p}_i(\gamma,u) = \widetilde{\pi}_i(\gamma)\widetilde{\pi}_i(u),$$

where the marginal densities $\widetilde{\pi}_i(\gamma)$ and $\widetilde{\pi}_i(u)$ are given by the mixtures

$$\widetilde{\pi}_i(\gamma) \propto \sum_n R_n(\gamma), \quad \widetilde{\pi}_i(u) \propto \sum_n Q_n(u)$$

with the identical first and second moments chosen as

$$\langle u\rangle = 0, \quad \langle u^2\rangle = 3, \quad \langle\gamma\rangle = 1.5, \quad \langle\gamma^2\rangle = 7.5, \quad \langle\gamma u\rangle = 0.$$

In particular, the seven initial densities in Sect. 5.1 with the same joint second-order statistics are obtained as follows (see Table 4 for the parameters used in (1)–(7)):

(1) Joint density

$$\widetilde{p}_1(u,\gamma) = \frac{1}{2}\big(R_1(\gamma)+R_2(\gamma)\big)Q_1(u),$$

where

$$R_i(\gamma) \propto \exp\left(-\frac{(\gamma-\overline{\gamma}_i)^2}{2\sigma_i^\gamma}\right), \qquad Q_1(u) \propto \exp\left(-\frac{(u-\overline{u}_1)^2}{2\sigma_1^u}\right).$$

(2) Joint density

$$\widetilde{p}_2(u,\gamma) = \frac{1}{4}\big(R_1(\gamma)+R_2(\gamma)\big)\big(Q_1(u)+Q_2(u)\big),$$

where

$$R_i(\gamma) \propto \exp\left(-\frac{(\gamma-\overline{\gamma}_i)^2}{2\sigma_i^\gamma}\right), \qquad Q_i(u) \propto \exp\left(-\frac{(u-\overline{u}_i)^2}{2\sigma_i^u}\right)\big(2+\sin(u)\big).$$

(3) Joint density

$$\widetilde{p}_3(u,\gamma) = \frac{1}{4}\big(R_1(\gamma)+R_2(\gamma)\big)\big(Q_1(u)+Q_2(u)\big),$$

Table 4 The parameters used in (1)–(7)

	$\overline{\gamma}_1$	$\overline{\gamma}_2$	σ_1^γ	σ_2^γ	$\overline{u}_1$	$\overline{u}_2$	σ_1^u	σ_2^u
(1)	0.0000	3.0000	3.0000	3.0000	0.0000		3.0000	
(2)	0.0506	2.9494	2.6492	3.6492	−1.1667	0.5291	2.7234	1.3088
(3)	0.0167	2.9055	2.7649	3.6316	−0.9653	0.8210	2.7216	1.7763
(4)	4.0209	4.1964	21.9235	1.2482	−1.0970	5.0522	2.2703	2.0612
(5)	5.2632		11.1937		1.0204	−1.0203	1.4575	2.4603
(6)	0.0163	2.9064	2.7691	3.6204	0.0000	5.0000	3.0000	2.0000
(7)	1.5000		5.2500		−0.7417	−0.9372	1.3784	2.4465

where

$$R_i(\gamma) \propto \exp\left(-\frac{(\gamma-\overline{\gamma}_i)^2}{2\sigma_i^\gamma}\right)\left(\frac{3}{2}+\sin\left(\frac{\pi\gamma}{2}\right)\right),$$

$$Q_i(u) \propto \exp\left(-\frac{(u-\overline{u}_i)^2}{2\sigma_i^u}\right)\left(\frac{3}{2}+\sin\left(\frac{\pi u}{2}\right)\right).$$

(4) Joint density

$$\widetilde{p}_4(u,\gamma) = \frac{1}{4}\big(R_1(\gamma)+R_2(\gamma)\big)\big(Q_1(u)+Q_2(u)\big),$$

where

$$R_i(\gamma) \propto \exp\left(-\frac{(\gamma-\overline{\gamma}_i)^2}{2\sigma_i^\gamma}\right)\frac{1}{\gamma^2+1}, \quad Q_i(u) \propto \exp\left(-\frac{(u-\overline{u}_i)^2}{2\sigma_i^u}\right)\frac{1}{u^2+1}.$$

(5) Joint density

$$\widetilde{p}_5(u,\gamma) = \frac{1}{2}R_1(\gamma)\big(Q_1(u)+Q_2(u)\big),$$

where

$$R_i(\gamma) \propto \exp\left(-\frac{(\gamma-\overline{\gamma}_i)^2}{2\sigma_i^\gamma}\right)\frac{1}{\gamma^2+1}, \quad Q_i(u) \propto \exp\left(-\frac{(u-\overline{u}_i)^2}{2\sigma_i^u}\right).$$

(6) Joint density

$$\widetilde{p}_6(u,\gamma) = \frac{1}{2}\big(R_1(\gamma)+R_2(\gamma)\big)Q_1(u),$$

where

$$R_i(\gamma) \propto \exp\left(-\frac{(\gamma-\overline{\gamma}_i)^2}{2\sigma_i^\gamma}\right)\left(\frac{3}{2}+\sin\left(\frac{\pi\gamma}{2}\right)\right), \quad Q_1(u) \propto \exp\left(-\frac{(u-\overline{u}_1)^2}{2\sigma_1^u}\right).$$

(7) Joint density

$$\widetilde{p}_7(u,\gamma) = \frac{1}{2} R_1(\gamma)\big(Q_1(u) + Q_2(u)\big),$$

where

$$R_1(\gamma) \propto \exp\left(-\frac{(\gamma - \overline{\gamma}_1)^2}{2\sigma_1^\gamma}\right),$$

$$Q_i(u) \propto \exp\left(-\frac{(u - \overline{u}_i)^2}{2\sigma_i^u}\right)\left(\frac{3}{2} + \sin\left(\frac{\pi u}{2} - 1\right)\right).$$

References

1. Bensoussan, A.: Stochastic Control of Partially Observable Systems. Cambridge University Press, Cambridge (1992)
2. Bourlioux, A., Majda, A.J.: An elementary model for the validation of flamelet approximations in non-premixed turbulent combustion. Combust. Theory Model. **4**(2), 189–210 (2000)
3. Branicki, M., Gershgorin, B., Majda, A.J.: Filtering skill for turbulent signals for a suite of nonlinear and linear Kalman filters. J. Comput. Phys. **231**, 1462–1498 (2012)
4. Branicki, M., Majda, A.J.: Fundamental limitations of polynomial chaos for uncertainty quantification in systems with intermittent instabilities. Commun. Math. Sci. **11**(1) (2012, in press)
5. Branicki, M., Majda, A.J.: Quantifying uncertainty for predictions with model error in non-Gaussian models with intermittency. Nonlinearity **25**, 2543–2578 (2012)
6. Cover, T.A., Thomas, J.A.: Elements of Information Theory, 2nd edn. Wiley-Interscience, Hoboken (2006)
7. Gardiner, C.: Stochastic Methods: A Handbook for the Natural and Social Sciences, 4th edn. Springer Series in Synergetics. Springer, Berlin (2010)
8. Gershgorin, B., Harlim, J., Majda, A.J.: Improving filtering and prediction of spatially extended turbulent systems with model errors through stochastic parameter estimation. J. Comput. Phys. **229**, 32–57 (2010)
9. Gershgorin, B., Harlim, J., Majda, A.J.: Test models for improving filtering with model errors through stochastic parameter estimation. J. Comput. Phys. **229**, 1–31 (2010)
10. Gershgorin, B., Majda, A.J.: A test model for fluctuation-dissipation theorems with time-periodic statistics. Physica D **239**, 1741–1757 (2010)
11. Gershgorin, B., Majda, A.J.: Quantifying uncertainty for climate change and long range forecasting scenarios with model errors. Part I: Gaussian models. J. Climate **25**, 4523–4548 (2012)
12. Gershgorin, B., Majda, A.J.: A nonlinear test model for filtering slow-fast systems. Commun. Math. Sci. **6**, 611–649 (2008)
13. Gershgorin, B., Majda, A.J.: Filtering a nonlinear slow-fast system with strong fast forcing. Commun. Math. Sci. **8**, 67–92 (2009)
14. Gershgorin, B., Majda, A.J.: Filtering a statistically exactly solvable test model for turbulent tracers from partial observations. J. Comput. Phys. **230**, 1602–1638 (2011)
15. Harlim, J., Majda, A.J.: Filtering turbulent sparsely observed geophysical flows. Mon. Weather Rev. **138**(4), 1050–1083 (2010)
16. Hersh, R.: Random evolutions: a survey of results and problems. Rocky Mt. J. Math. **4**(3), 443–477 (1974)
17. Iserles, A.: A First Course in the Numerical Analysis of Differential Equations. Cambridge University Press, Cambridge (1996)

18. Kleeman, R.: Information theory and dynamical system predictability. Entropy **13**, 612–649 (2011)
19. LeVeque, R.: Numerical Methods for Conservation Laws. ETH Lectures in Mathematics Series. Birkhäuser, Basel (1990)
20. Liptser, R.S., Shiryaev, A.N.: Statistics of Random Process, 2nd edn. Springer, New York (2001)
21. Majda, A., Kramer, P.: Simplified models for turbulent diffusion: theory, numerical modeling, and physical phenomena. Phys. Rep. **314**(4), 237–257 (1999)
22. Majda, A.J.: Challenges in climate science and contemporary applied mathematics. Commun. Pure Appl. Math. **65**(7), 920–948 (2012)
23. Majda, A.J., Abramov, R., Gershgorin, B.: High skill in low frequency climate response through fluctuation dissipation theorems despite structural instability. Proc. Natl. Acad. Sci. USA **107**(2), 581–586 (2010)
24. Majda, A.J., Abramov, R.V., Grote, M.J.: Information theory and stochastics for multiscale nonlinear systems. CRM Monograph Series, vol. 25. Am. Math. Soc., Providence (2005)
25. Majda, A.J., Branicki, M.: Lessons in uncertainty quantification for turbulent dynamical systems. Discrete Contin. Dyn. Syst. **32**(9), 3133–3231 (2012)
26. Majda, A.J., Franzke, C., Crommelin, D.: Normal forms for reduced stochastic climate models. Proc. Natl. Acad. Sci. USA **106**(10), 3649–3653 (2009)
27. Majda, A.J., Franzke, C., Fischer, A., Crommelin, D.T.: Distinct metastable atmospheric regimes despite nearly Gaussian statistics: a paradigm model. Proc. Natl. Acad. Sci. USA **103**(22), 8309–8314 (2006)
28. Majda, A.J., Gershgorin, B.: Quantifying uncertainty in climage change science through empirical information theory. Proc. Natl. Acad. Sci. USA **107**(34), 14958–14963 (2010)
29. Majda, A.J., Gershgorin, B.: Improving model fidelity and sensitivity for complex systems through empirical information theory. Proc. Natl. Acad. Sci. USA **108**(5), 10044–10049 (2011)
30. Majda, A.J., Gershgorin, B.: Link between statistical equilibrium fidelity and forecasting skill for complex systems with model error. Proc. Natl. Acad. Sci. USA **108**(31), 12599–12604 (2011)
31. Majda, A.J., Kleeman, R., Cai, D.: A mathematical framework for predictability through relative entropy. Methods Appl. Anal. **9**(3), 425–444 (2002)
32. Majda, A.J., Wang, X.: Nonlinear Dynamics and Statistical Theories for Basic Geophysical Flows. Cambridge University Press, Cambridge (2006)
33. Majda, A.J., Gershgorin, B.: Elementary models for turbulent diffusion with complex physical features: eddy diffusivity, spectrum, and intermittency. Philos. Trans. R. Soc. (2011, in press)
34. Majda, A.J., Harlim, J.: Filtering Complex Turbulent Systems. Cambridge University Press, Cambridge (2012)
35. Majda, A.J., Harlim, J., Gershgorin, B.: Mathematical strategies for filtering turbulent dynamical systems. Discrete Contin. Dyn. Syst. **27**, 441–486 (2010)
36. Risken, H.: The Fokker-Planck Equation: Methods of Solutions and Applications, 2nd edn. Series in Synergetics. Springer, Berlin (1989)

Asymptotic Analysis in a Gas-Solid Combustion Model with Pattern Formation

Claude-Michel Brauner, Lina Hu, and Luca Lorenzi

Abstract The authors consider a free interface problem which stems from a gas-solid model in combustion with pattern formation. A third-order, fully nonlinear, self-consistent equation for the flame front is derived. Asymptotic methods reveal that the interface approaches a solution to the Kuramoto-Sivashinsky equation. Numerical results which illustrate the dynamics are presented.

Keywords Asymptotics · Free interface · Kuramoto-Sivashinsky equation · Pseudo-differential operator · Spectral method

Mathematics Subject Classification 35B40 · 35R35 · 35B35 · 35K55 · 80A25

1 Introduction

Combustion phenomena are particularly important for science and industry, as Lions pointed out in his foreword to the special issue of the CNRS "Images des Mathématiques" in 1996 (see [1]). Flames constitute a complex physical system involving fluid dynamics and multistep chemical kinetics (see, e.g., [9]). In the middle of the 20th century, the Russian School, which included Frank-Kamenetskii and

Project supported by a grant from the Fujian Administration of Foreign Expert Affairs, China (No. SZ2011008).

C.-M. Brauner · L. Hu (✉)
School of Mathematical Sciences, Xiamen University, Xiamen 361005, Fujian, China
e-mail: linahu@stu.xmu.edu.cn

C.-M. Brauner
Institut de Mathématiques de Bordeaux, Université de Bordeaux, 33405 Talence cedex, France
e-mail: claude-michel.brauner@u-bordeaux1.fr

L. Lorenzi
Dipartimento di Matematica e Informatica, Università degli Studi di Parma, Parco Area delle Scienze 53/A, 43124 Parma, Italy
e-mail: luca.lorenzi@unipr.it

P.G. Ciarlet et al. (eds.), *Partial Differential Equations: Theory, Control and Approximation*, DOI 10.1007/978-3-642-41401-5_5,

Zel'dovich, used formal asymptotics based on large activation energy to write simpler descriptions of such a reactive system. Later, the development of systematic asymptotic techniques during the 1960s opened the way towards revealing an underlying simplicity in many combustion processes. Eventually, the full power of asymptotical analysis has been realized by modern singular perturbation theory. Lions was the first one to put these formalities on a rigorous basis in his seminal monograph "Perturbations singulières dans les problèmes aux limites et en contrôle optimal" (see [14]).

In short, the small perturbation parameter in activation-energy asymptotics is the inverse of the normalized activation energy, the Zel'dovich number β. In the limit $\beta \to +\infty$, the flame front reduces to a free interface. The laminar flames of low-Lewis-number premixtures are known to display diffusive-thermal instability responsible for the formation of a non-steady cellular structure (see [24]), when the Lewis number Le (the ratio of thermal and molecular diffusivities) is such that $\mathrm{Le} \lesssim 1$. From an asymptotical viewpoint, one combines the limit of large activation energy with the requirement that $\alpha = \frac{1}{2}\beta(1 - \mathrm{Le})$ remains bounded: in the near equidiffusive flame model (or NEF for short), β^{-1} and $1 - \mathrm{Le}$ are asymptotically of the same order of magnitude (see [22]).

A very challenging problem is the derivation of a single equation for the free interface, which may capture most of the dynamics and, as a consequence, yields a reduction of the effective dimensionality of the system. Asymptotical methods are also the main tool: in a set of conveniently rescaled dependent and independent variables, the flame front is asymptotically represented (see [23]) by a solution to the Kuramoto-Sivashinsky (or K-S for short) equation

$$\Phi_\tau + 4\Phi_{\eta\eta\eta\eta} + \Phi_{\eta\eta} + \frac{1}{2}(\Phi_\eta)^2 = 0. \tag{K-S}$$

This equation has received considerable attention from the mathematical community (see [25]), especially for its ability to generate a cellular structure, pattern formation, and chaotic behavior in an appropriate range of parameters (see [12]). We refer to [2–7] for a rigorous mathematical approach to the derivation of (K-S).

In this paper, we consider a model in gas-solid combustion, proposed in [13]. This model was motivated by the experimental studies of Zik and Moses (see [26]) who observed a striking fingering pattern in flames spreading over thin solid fuels. The phenomenon was interpreted in terms of the diffusive instability similar to that occurring in laminar flames of low-Lewis-number premixtures. As we show below, the gas-solid and premixed gas systems share some common asymptotic features, especially the K-S equation.

The free interface system for the scaled temperature θ, the excess enthalpy S, the prescribed flow intensity U (with $0 < U < 1$), and the moving front $x = \xi(t, y)$, is as follows:

$$U\frac{\partial \theta}{\partial x} = \Delta\theta, \quad x < \xi(t, y), \tag{1.1}$$

$$\theta = 1, \quad x \geq \xi(t, y), \tag{1.2}$$

$$\frac{\partial \theta}{\partial t} + U\frac{\partial S}{\partial x} = \Delta S - \alpha\Delta\theta, \quad x \neq \xi(t, y). \tag{1.3}$$

System (1.1)–(1.3) is coupled with the following jump conditions for the normal derivatives of θ and S:

$$\left[\frac{\partial \theta}{\partial n}\right] = -\exp(S), \qquad \left[\frac{\partial S}{\partial n}\right] = \alpha\left[\frac{\partial \theta}{\partial n}\right]. \tag{1.4}$$

It is not difficult to show that (1.1)–(1.4) admit a planar traveling wave solution with velocity $-V$, where $V = -U\ln U$. Setting $x' = x + Vt$, the traveling wave solution is given by

$$\overline{\theta}(x') = \begin{cases} \exp(Ux'), & x' \leq 0, \\ 1, & x' > 0, \end{cases}$$

$$\overline{S}(x') = \begin{cases} (\alpha - \ln U)Ux'\exp(Ux') + (\ln U)\exp(Ux'), & x' \leq 0, \\ \ln U, & x' > 0. \end{cases}$$

As usual, one fixes the moving front. We set

$$\xi(t, y) = -Vt + \varphi(t, y), \quad x' = x - \xi(t, y),$$

where φ is the perturbation of the planar front. In this new framework, the system (1.1)–(1.3) can be written as follows:

$$U\theta_{x'} = \Delta_\varphi \theta, \quad x' < 0, \tag{1.5}$$

$$\theta = 1, \quad x' \geq 0, \tag{1.6}$$

$$\theta_t + (V - \varphi_t)\theta_{x'} + US_{x'} = \Delta_\varphi S - \alpha\Delta_\varphi\theta, \quad x' \neq 0, \tag{1.7}$$

where

$$\Delta_\varphi = \left(1 + (\varphi_y)^2\right)D_{x'x'} + D_{yy} - \varphi_{yy}D_{x'} - 2\varphi_y D_{x'y}.$$

The front is now fixed at $x' = 0$. The first jump condition in (1.4) is

$$\sqrt{1 + (\varphi_y)^2}\left[\frac{\partial \theta}{\partial x'}\right] = -\exp(S), \tag{1.8}$$

and the second one is

$$\left[\frac{\partial S}{\partial x'}\right] = \alpha\left[\frac{\partial \theta}{\partial x'}\right]. \tag{1.9}$$

We will consider a quasi-steady version of the model, motivated by the fact that, in similar problems, not far from the instability threshold, the respective time derivatives of the temperature and enthalpy (if any) exhibit a relatively small effect on the solution. The dynamics appears to be essentially driven by the front. We can thus introduce a quasi-steady model replacing (1.5)–(1.7) by

$$U\theta_{x'} = \Delta_\varphi \theta, \quad x' < 0,$$

$$\theta = 1, \quad x' \geq 0,$$

$$(V - \varphi_t)\theta_{x'} + U S_{x'} = \Delta_\varphi S - \alpha \Delta_\varphi \theta, \quad x' \neq 0.$$

Next we consider the perturbations of temperature u and enthalpy v,

$$\theta = \overline{\theta} + u, \quad S = \overline{S} + v,$$

and, for simplicity, in the equations satisfied by u, v and φ, we keep only the linear and second-order terms for φ, and the first-order terms for u and v. Writing x instead of x' to avoid a cumbersome notation, some (easy) computations reveal that the triplet (u, v, φ) solves the differential equations

$$Uu_x - \Delta u = (\Delta_\varphi - \Delta)\overline{\theta}, \quad x < 0,$$

$$Vu_x - \Delta(v - \alpha u) + Uv_x - \varphi_t \overline{\theta}_x = (\Delta_\varphi - \Delta)(\overline{S} - \alpha\overline{\theta}), \quad x \neq 0,$$

where $u \equiv 0$ in $[0, +\infty)$, and

$$(\Delta_\varphi - \Delta)\overline{\theta} = \big(U(\varphi_y)^2 - \varphi_{yy}\big)U\mathrm{e}^{Ux},$$

$$(\Delta_\varphi - \Delta)(\overline{S} - \alpha\overline{\theta})$$

$$= \begin{cases} (\varphi_y)^2(\alpha - \ln U)U^2(1 + Ux)\mathrm{e}^{Ux} - \varphi_{yy}(\alpha - \ln U)U^2 x\mathrm{e}^{Ux}, & x < 0, \\ 0, & x > 0. \end{cases}$$

The previous system is endowed with a set of boundary conditions. First, the continuities of θ and S at the front yield the equation

$$u(0^-) = [v] = 0$$

(recall that $u(x) = 0$ for $x \geq 0$). Second, up to the second-order, condition (1.8) gives

$$-U + [u_x] = -\big(1 + (\varphi_y)^2\big)^{-\frac{1}{2}} U\mathrm{e}^{v(0)} \sim -\Big(1 - \frac{1}{2}(\varphi_y)^2\Big)U\Big(1 + v(0) + \frac{1}{2}\big(v(0)\big)^2\Big).$$

By keeping only the first-order for v, we get the condition

$$-u_x(0^-) + Uv(0) = \frac{1}{2}(\varphi_y)^2 U.$$

Finally, the condition $[S_x] = \alpha[\theta_x]$ yields

$$[v_x] = -\alpha u_x(0^-).$$

Summing up, the final system is as follows:

$$\begin{cases} Uu_x - \Delta u = (\Delta_\varphi - \Delta)\overline{\theta}, & x < 0, \\ Vu_x - \Delta(v - \alpha u) + Uv_x - \varphi_t\overline{\theta}_x = (\Delta_\varphi - \Delta)(\overline{S} - \alpha\overline{\theta}), & x \neq 0, \\ u(0^-) = [v] = 0, \\ Uv(0) - u_x(0^-) = \frac{1}{2}(\varphi_y)^2 U, \\ [v_x] = -\alpha u_x(0^-). \end{cases} \tag{1.10}$$

Throughout this paper, we will also use the very convenient notation

$$\gamma = \alpha - \ln U.$$

First, our goal is to derive a self-consistent equation for the front φ,

$$\varphi_t = \mathscr{A}(\varphi) + \mathscr{M}\big((\varphi_y)^2\big), \tag{1.11}$$

where $\mathscr{A}$ is a third-order, pseudo-differential operator, in contrast to the NEF model in gaseous combustion where the corresponding linear operator is of the second-order (see [6]). Another important feature is that the nonlinear term is also of the third-order, which means that Eq. (1.11) is fully nonlinear. Here the spatial domain is a two-dimensional strip $\mathbb{R} \times [-\frac{\ell}{2}, \frac{\ell}{2}]$ with periodic boundary conditions at $\pm\frac{\ell}{2}$.

Second, we define a small parameter $\varepsilon = \gamma - 1$. The main result of this paper states the precise sense in which the front φ approaches a solution to the Kuramoto-Sivashinsky equation when $\varepsilon \to 0$.

Theorem 1.1 *Let $\Phi_0 \in H^m(-\frac{\ell_0}{2}, \frac{\ell_0}{2})$ be a periodic function of period ℓ_0. Further, let Φ be the periodic solution to* (K-S) *(with period ℓ_0) on a fixed time interval $[0, T]$, satisfying the initial condition $\Phi(0, \cdot) = \Phi_0$. Then, if m is large enough, there exists an $\varepsilon_0 = \varepsilon_0(T) \in (0, 1)$ such that, for $0 < \varepsilon \le \varepsilon_0$,* (1.11) *admits a unique classical solution φ on $[0, \frac{T}{\varepsilon^2 U^2}]$, which is periodic with period $\frac{\ell_0}{\sqrt{\varepsilon U}}$ with respect to y, and satisfies*

$$\varphi(0, y) = \varepsilon U^{-1}\Phi_0(y\sqrt{\varepsilon U}), \quad |y| \le \frac{\ell_0}{2\sqrt{\varepsilon U}}.$$

Moreover, there exists a positive constant C such that

$$\left|\varphi(t, y) - \varepsilon U^{-1}\Phi\big(t\varepsilon^2 U^2, y\sqrt{\varepsilon U}\big)\right| \le C\varepsilon^2, \quad 0 \le t \le \frac{T}{\varepsilon^2 U^2}, \quad |y| \le \frac{\ell_0}{2\sqrt{\varepsilon U}}$$

for any $\varepsilon \in (0, \varepsilon_0]$.

This paper is organized as follows. In Sect. 2, we proceed to a formal ansatz in the spirit of [23], defining the rescaled variable $\psi = \varepsilon^{-1} U \varphi$ and expanding $\psi = \psi^0 + \varepsilon\psi^1 + \cdots$. It transpires that ψ^0 verifies (K-S), thanks to an elementary solvability condition.

Section 3 is devoted to the derivation of (1.11), via an explicit computation in the discrete Fourier variable. The asymptotic analysis in the rescaled variables $t = \frac{\tau}{\varepsilon^2 U^2}$, $y = \frac{\eta}{\sqrt{\varepsilon}U}$ is performed in Sect. 4. Since the perturbation in (1.11) is singular as $\varepsilon \to 0$, we turn to the equivalent (at fixed $\varepsilon > 0$) fourth-order, fully nonlinear equation (1.12), whose prima facie limit as $\varepsilon \to 0$ is Eq. (K-S),

$$
\begin{aligned}
&\frac{\partial}{\partial \tau}(\sqrt{I - 4\varepsilon D_{\eta\eta}})\psi \\
&\quad = -4D_{\eta\eta\eta\eta}\psi - D_{\eta\eta}\psi \\
&\qquad + \frac{1}{4}\Big\{(I - 4\varepsilon D_{\eta\eta})^{\frac{3}{2}} - 3(I - 4\varepsilon D_{\eta\eta}) - 4(1+\varepsilon)(\sqrt{I - 4\varepsilon D_{\eta\eta}} - I)\Big\}(D_\eta \psi)^2.
\end{aligned}
\tag{1.12}
$$

We prove a priori estimates, which constitute the key tool to prove the main theorem. Finally, numerical computations which illustrate the dynamics in (1.12) are presented in Sect. 5.

The local existence in time for (1.1)–(1.4) and the stability issue will be addressed in a forthcoming paper, by using the methods of [4, 8] and [15–20].

Notation 1.1 Given a (smooth enough) function $f : (-\frac{\ell}{2}, \frac{\ell}{2}) \to \mathbb{C}$, we denote by $\widehat{f}(k)$ its k-th Fourier coefficient, that is, we write

$$
f(y) = \sum_{k=0}^{+\infty} \widehat{f}(k) w_k(y), \quad y \in \left(-\frac{\ell}{2}, \frac{\ell}{2}\right),
$$

where $\{w_k\}$ is a complete set of (complex valued) eigenfunctions of the operator

$$
D_{yy} : H^2\left(-\frac{\ell}{2}, \frac{\ell}{2}\right) \to L^2\left(-\frac{\ell}{2}, \frac{\ell}{2}\right),
$$

whose eigenvalues are $0, -\frac{4\pi^2}{\ell^2}, -\frac{4\pi^2}{\ell^2}, -\frac{16\pi^2}{\ell^2}, -\frac{16\pi^2}{\ell^2}, -\frac{36\pi^2}{\ell^2}, \ldots$, and we label as $0 = -\lambda_0(\ell) > -\lambda_1(\ell) = -\lambda_2(\ell) > -\lambda_3(\ell) = -\lambda_4(\ell) > \cdots$. Typically, when no confusion may arise, we simply write λ_k instead of $\lambda_k(\ell)$.

For any $s \geq 0$, we denote by $H^s_\sharp$ the usual Sobolev space of order s consisting of ℓ-periodic (generalized) functions, i.e.,

$$
H^s_\sharp = \left\{ u = \sum_{k=0}^{+\infty} \widehat{u}(k) w_k : \sum_{k=0}^{+\infty} \lambda_k^s |\widehat{u}(k)|^2 < +\infty \right\}.
$$

For $s = 0$, we simply write L^2 instead of $H^0_\sharp$, and we denote by $|\cdot|_2$ the usual L^2-norm.

By the notation $\widehat{f}(x,k)$, we mean the k-th Fourier coefficient of the function $f(x,\cdot)$. A similar notation is used for functions which depend also on the time variable.

2 A Formal Ansatz

The aim of this section is to use a formal asymptotic expansion method, in the spirit of [23]. The small perturbation parameter $\varepsilon > 0$ is defined by

$$\alpha = 1 + \ln U + \varepsilon, \quad \text{i.e.,} \quad \gamma = 1 + \varepsilon. \tag{2.1}$$

Accordingly, we now introduce scaled dependent and independent variables

$$t = \frac{\tau}{\varepsilon^2 U^2}, \quad y = \frac{\eta}{\sqrt{\varepsilon} U}, \quad \varphi = \frac{\varepsilon}{U}\psi, \quad u = \varepsilon^2 u_1, \quad v = \varepsilon^2 v_1, \tag{2.2}$$

and the ansatz

$$u_1 = u_1^0 + \varepsilon u_1^1 + \cdots, \quad v_1 = v_1^0 + \varepsilon v_1^1 + \cdots, \quad \psi = \psi^0 + \varepsilon \psi^1 + \cdots.$$

It is easy to rewrite (1.10) in terms of the rescaled variables. At the zeroth order, it comes that

$$\begin{cases} U(u_1^0)_x - (u_1^0)_{xx} = -U^2 e^{Ux}\psi^0_{\eta\eta}, & x < 0, \\ V(u_1^0)_x - (v_1^0)_{xx} + (u_1^0)_{xx} + (\ln U)(u_1^0)_{xx} + U(v_1^0)_x = -U^3 x e^{Ux}\psi^0_{\eta\eta}, & x < 0, \\ u_1^0 = 0, & x \geq 0, \\ (v_1^0)_{xx} - U(v_1^0)_x = 0, & x > 0. \end{cases} \tag{2.3}$$

At $x = 0$, the following conditions should be satisfied:

$$u_1^0(0) = [v_1^0] = 0, \tag{2.4a}$$

$$(u_1^0)_x(0) - U v_1^0(0) = 0, \tag{2.4b}$$

$$[(v_1^0)_x] = -(1 + \ln U)(u_1^0)_x(0). \tag{2.4c}$$

We assume that the functions $x \mapsto e^{-\frac{Ux}{2}} u_1^0(x)$ and $x \mapsto e^{-\frac{Ux}{2}} v_1^0(x)$ are bounded in $(-\infty, 0)$ and $\mathbb{R}$, respectively. Note that (2.3) coupled with conditions (2.4a) and

(2.4b) is uniquely solvable in the unknowns (u_1^0, v_1^0), by taking ψ^0 as a parameter. It turns out that

$$u_1^0 = Uxe^{Ux}\psi^0_{\eta\eta}, \quad x<0,$$
$$v_1^0 = e^{Ux}\psi^0_{\eta\eta} + U(\ln U)xe^{Ux}\psi^0_{\eta\eta} + U^2x^2e^{Ux}\psi^0_{\eta\eta}, \quad x<0,$$
$$u_1^0 = 0, \quad x\ge 0,$$
$$v_1^0 = \psi^0_{\eta\eta}, \quad x\ge 0.$$

One might be tempted to use condition (2.4c) to determine function ψ^0. Unfortunately, whatever ψ^0 is, the triplet (u_1^0, v_1^0, ψ^0) satisfies this condition. As a matter of fact, we are not able to determine uniquely a solution to (2.3)–(2.4c). This situation is not surprising at all in the singular perturbation theory (see [10, 14]). To determine ψ^0, one needs to consider the (linear) problem for the first-order terms in the asymptotic expansion of u_1, v_1 and ψ. As we will show in a while, this problem provides a solvability condition, which is just the missing equation for ψ^0.

The system for (u_1^1, v_1^1, ψ^1) is the following one:

$$\begin{cases} U(u_1^1)_x - (u_1^1)_{xx} - U^2(u_1^0)_{\eta\eta} = (U(\psi^0_\eta)^2 - U\psi^1_{\eta\eta})Ue^{Ux}, \quad x<0,\\ V(u_1^1)_x - (v_1^1)_{xx} - U^2(v_1^0)_{\eta\eta} + (u_1^0)_{xx} + (1+\ln U)((u_1^1)_{xx} \\ \quad + U^2(u_1^0)_{\eta\eta}) + U(v_1^1)_x \\ \quad = U^2\psi^0_\tau e^{Ux} + (\psi^0_\eta)^2U^2e^{Ux} + (\psi^0_\eta)^2U^3xe^{Ux} \\ \qquad - U^3\psi^1_{\eta\eta}xe^{Ux} - U^3\psi^0_{\eta\eta}xe^{Ux}, \quad x<0,\\ (v_1^1)_{xx} + U^2(v_1^0)_{\eta\eta} - U(v_1^1)_x = 0, \quad x>0,\\ u_1^1 = 0, \quad x\ge 0,\\ u_1^1(0) = [v_1^1] = 0,\\ Uv_1^1(0) - (u_1^1)_x(0) = \frac{1}{2}U(\psi^0_\eta)^2,\\ [(v_1^1)_x] = -(1+\ln U)(u_1^1)_x(0) - (u_1^0)_x(0). \end{cases} \tag{2.5}$$

As above, we assume that the functions $x\mapsto e^{-\frac{Ux}{2}}u_1^1(x)$ and $x\mapsto e^{-\frac{Ux}{2}}v_1^1(x)$ are bounded in $(-\infty,0)$ and $\mathbb{R}$, respectively. Using these conditions one can easily show that the more general solutions (u_1^1, v_1^1, ψ^1) to the differential equations and the first boundary condition in (2.5) are given by

$$u_1^1 = Uxe^{Ux}\left(\psi^0_{\eta\eta\eta\eta} - (\psi^0_\eta)^2 + \psi^1_{\eta\eta}\right) - \frac{1}{2}U^2x^2e^{Ux}\psi^0_{\eta\eta\eta\eta}, \quad x<0,$$
$$v_1^1 = v_1^1(0)e^{Ux} + Axe^{Ux} + Bx^2e^{Ux} + Cx^3e^{Ux}, \quad x<0,$$
$$u_1^1 = 0, \quad x\ge 0,$$
$$v_1^1 = v_1^1(0) + Ux\psi^0_{\eta\eta\eta\eta}, \quad x\ge 0,$$

where

$$A = U(\ln U)\left(\psi^1_{\eta\eta} - (\psi^0_\eta)^2 + \psi^0_{\eta\eta\eta\eta}\right) - U\psi^0_\tau - U(\psi^0_\eta)^2 - 3U\psi^0_{\eta\eta\eta\eta},$$
$$B = U^2\psi^0_{\eta\eta} + U^2\psi^1_{\eta\eta} - U^2(\psi^0_\eta)^2 - \frac{1}{2}U^2(\ln U)\psi^0_{\eta\eta\eta\eta} + \frac{3}{2}U^2\psi^0_{\eta\eta\eta\eta},$$
$$C = -\frac{1}{2}U^3\psi^0_{\eta\eta\eta\eta},$$

and $v^1_1(0)$ is an arbitrary parameter. Hence, (u^1_1, v^1_1) depends on ψ^1. To determine both ψ^1 and $v^1_1(0)$, we use the last two boundary conditions which give

$$-U\psi^0_{\eta\eta\eta\eta} + U(\psi^0_\eta)^2 - U\psi^1_{\eta\eta} + Uv^1_1(0) = \frac{U}{2}(\psi^0_\eta)^2 \tag{2.6}$$

and

$$U\psi^1_{\eta\eta} - Uv^1_1(0) = -U\psi^0_\tau - U\psi^0_{\eta\eta} - 5U\psi^0_{\eta\eta\eta\eta}, \tag{2.7}$$

respectively. Obviously, (2.6)–(2.7) is a linear system for $(v^1_1(0), \psi^1_{\eta\eta})$ with the solvability condition

$$\psi^0_\tau + \psi^0_{\eta\eta} + 4\psi^0_{\eta\eta\eta\eta} + \frac{1}{2}(\psi^0_\eta)^2 = 0.$$

Hence, the K-S equation is the missing equation at the zeroth-order, needed to uniquely determine (u^0_1, v^0_1, ψ^0).

3 A Third-Order Fully Nonlinear Pseudo-Differential Equation for the Front

The aim of this section is the derivation of a self-consistent pseudo-differential equation for the front φ. We rewrite (1.10), namely,

$$\begin{cases} Uu_x - \Delta u = Ue^x(U(\varphi_y)^2 - \varphi_{yy}), & x<0,\\ Vu_x - \Delta(v-\alpha u) + Uv_x & \\ \quad = U\varphi_t e^x + (\alpha - \ln U)U^2(1+Ux)e^{Ux}(\varphi_y)^2 & \\ \qquad - U^2(\alpha - \ln U)xe^{Ux}\varphi_{yy}, & x<0,\\ Uv_x - \Delta v = 0, & x>0,\\ u(0) = [v] = 0, & \\ Uv(0) - u_x(0) = \frac{1}{2}U(\varphi_y)^2, & \\ [v_x] = -\alpha u_x(0) & \end{cases} \tag{3.1}$$

in a two-dimensional strip $\mathbb{R} \times [-\frac{\ell}{2}, \frac{\ell}{2}]$, with periodicity in the y variable.

3.1 Computations in the Discrete Fourier Variable

Throughout this subsection, (u, v, φ) is a sufficiently smooth solution to (3.1) such that the functions

$$(x, y) \mapsto e^{-\frac{Ux}{2}} u(t, x, y), \qquad (x, y) \mapsto e^{-\frac{Ux}{2}} v(t, x, y)$$

are bounded in $(-\infty, 0] \times [-\frac{\ell}{2}, \frac{\ell}{2}]$ and $\mathbb{R} \times [-\frac{\ell}{2}, \frac{\ell}{2}]$, respectively.

We start from the first equation in (3.1), namely,

$$Uu_x - \Delta u = \big(U(\varphi_y)^2 - \varphi_{yy}\big)Ue^x, \tag{3.2}$$

and the boundary condition $u(\cdot, 0, \cdot) = 0$. Applying the Fourier transform to both sides of (3.2), we end up with the infinitely many equations

$$U\widehat{u}_x(t,x,k) - \widehat{u}_{xx}(t,x,k) + \lambda_k \widehat{u}(t,x,k) = \big(U\widehat{(\varphi_y)^2}(t,k) + \lambda_k \widehat{\varphi}(t,k)\big)Ue^{Ux}$$

for $k \ge 0$. For notational convenience, we set $\nu_k = \frac{U}{2} + \frac{1}{2}\sqrt{U^2 + 4\lambda_k}$ for any $k \ge 0$.

Since u vanishes at $x = 0$ and tends to 0 as $x \to -\infty$ not slower than $e^{\frac{Ux}{2}}$, the modes $\widehat{u}(\cdot, \cdot, k)$ should enjoy the same properties. Easy computations reveal that

$$\widehat{u}(t,x,0) = -U\widehat{(\varphi_y)^2}(t,0)xe^{Ux}, \quad x \le 0,$$

$$\widehat{u}(t,x,k) = U(\lambda_k)^{-1}\big(U\widehat{(\varphi_y)^2}(t,k) + \lambda_k\widehat{\varphi}(t,k)\big)\big(e^{Ux} - e^{\nu_k x}\big), \quad x \le 0, \quad k \ge 1.$$

Applying the same arguments to the equation for v jointly with the second and the fourth boundary conditions in (3.1), we obtain that the modes $\widehat{v}(\cdot, \cdot, k)$ are given by

$$\begin{aligned}\widehat{v}(t,x,0) = {} & \frac{1}{U}\widehat{\varphi_t}(t,0) + \big(-\gamma U\widehat{(\varphi_y)^2}(t,0) - U(\ln U)\widehat{(\varphi_y)^2}(t,0) - \widehat{\varphi_t}(t,0)\big)xe^{Ux} \\ & - \gamma U^2\widehat{(\varphi_y)^2}(t,0)x^2e^{Ux}, \quad x < 0,\\ \widehat{v}(t,x,0) = {} & \frac{1}{U}\widehat{\varphi_t}(t,0), \quad x > 0\end{aligned}$$

and

$$\begin{aligned}\widehat{v}(t,x,k) &= c_{1,k}e^{\nu_k x} + A_k e^{Ux} + B_k x e^{Ux} + C_k x e^{\nu_k x}, \quad x < 0,\\ \widehat{v}(t,x,k) &= c_{2,k}e^{(U-\nu_k)x}, \quad x \ge 0\end{aligned}$$

for $k \ge 1$, where

$$A_k = \frac{(\alpha+\gamma)U^2}{\lambda_k}\widehat{(\varphi_y)^2}(t,k) + \alpha U\widehat{\varphi}(t,k) + \frac{U}{\lambda_k}\widehat{\varphi_t}(t,k),$$

$$
\begin{aligned}
B_k &= \frac{\gamma U^3}{\lambda_k}\widehat{(\varphi_y)^2}(t,k) + \gamma U^2 \widehat{\varphi}(t,k),\\
C_k &= \frac{\gamma U^3 \nu_k}{\lambda_k(U-2\nu_k)}\widehat{(\varphi_y)^2}(t,k) + \frac{\gamma U^2 \nu_k}{U-2\nu_k}\widehat{\varphi}(t,k),\\
c_{2,k} &= \left(\frac{\gamma U^2(U-\nu_k)}{\lambda_k(U-2\nu_k)} + \frac{\gamma U^3}{\lambda_k(U-2\nu_k)} + \frac{\gamma U^3}{(\nu_k-U)(U-2\nu_k)^2}\right)\widehat{(\varphi_y)^2}(t,k)\\
&\quad + \left(\frac{\gamma U^2}{U-2\nu_k} + \frac{\gamma U^2\nu_k}{(U-2\nu_k)^2}\right)\widehat{\varphi}(t,k) + \frac{U(U-\nu_k)}{\lambda_k(U-2\nu_k)}\widehat{\varphi_t}(t,k),\\
c_{1,k} &= c_{2,k} - A_k.
\end{aligned}
$$

The equation for the front now comes by the last but one boundary condition in (3.1), which we have not used so far, in the Fourier variable, by taking advantage of the formulas for the modes of $\widehat{u}$ and $\widehat{v}$. It turns out that the equation for the front (in Fourier coordinates) is

$$
\widehat{\varphi_t}(t,0) + \frac{1}{2}U\widehat{(\varphi_y)^2}(t,0) = 0,
$$

$$
\begin{aligned}
&\frac{U(U-\nu_k)}{\lambda_k(U-2\nu_k)}\widehat{\varphi_t}(t,k) + \left(\frac{\gamma U^2}{U-2\nu_k} + \frac{\gamma U^2\nu_k}{(U-2\nu_k)^2} + \nu_k - U\right)\widehat{\varphi}(t,k)\\
&\quad + \left(\frac{\gamma U^2(U-\nu_k)}{\lambda_k(U-2\nu_k)} + \frac{\gamma U^3}{\lambda_k(U-2\nu_k)} + \frac{\gamma U^3}{(\nu_k-U)(U-2\nu_k)^2} + \frac{U(\nu_k-U)}{\lambda_k} - \frac{1}{2}\right)\\
&\quad \times \widehat{(\varphi_y)^2}(t,k) = 0,
\end{aligned}
$$

or, even, in the much more compact form

$$
\begin{aligned}
(X_k U)\widehat{\varphi_t}(t,k) &= \frac{1}{4}(U^2 - X_k^2)(X_k^2 - \gamma U^2)\widehat{\varphi}(t,k)\\
&\quad \mid \frac{1}{4}(X_k^3 \quad 3UX_k^2 \quad 4\gamma U^2 X_k \mid 4\gamma U^3)\widehat{(\varphi_y)^2}(t,k)\\
&= \left(-4\lambda_k^2 + (\gamma-1)U^2\lambda_k\right)\widehat{\varphi}(t,k)\\
&\quad + \frac{1}{4}(X_k^3 - 3UX_k^2 - 4\gamma U^2 X_k + 4\gamma U^3)\widehat{(\varphi_y)^2}(t,k) \qquad (3.3)
\end{aligned}
$$

for any $k \geq 0$, if we set

$$
X_k = \sqrt{U^2 + 4\lambda_k}, \quad k \geq 0.
$$

Therefore, we have proved the following proposition.

Proposition 3.1 *Let (u, v, φ) be a sufficiently smooth solution to* (3.1) *such that the functions $(x, y) \mapsto e^{-\frac{Ux}{2}} u(t, x, y)$ and $(x, y) \mapsto e^{-\frac{Ux}{2}} v(t, x, y)$ are bounded in $(-\infty, 0] \times [-\frac{\ell}{2}, \frac{\ell}{2}]$ and $\mathbb{R} \times [-\frac{\ell}{2}, \frac{\ell}{2}]$, respectively. Then, the interface φ solves Eqs.* (3.3) *for any $k \geq 0$.*

3.2 A Fourth-Order Pseudo-Differential Equation for the Front

Let us define the pseudo-differential operators (or Fourier multipliers) $\mathscr{B}, \mathscr{L}$ and $\mathscr{F}$ through their symbols, respectively

$$b_k = X_k U, \qquad l_k = -4\lambda_k^2 + (\gamma - 1)\lambda_k U^2,$$

$$f_k = \frac{1}{4}(X_k^3 - 3UX_k^2 - 4\gamma U^2 X_k + 4\gamma U^3)$$

for any $k \geq 0$. It is easy to see that

$$\mathscr{B} = U\big(U^2 I - 4D_{yy}\big)^{\frac{1}{2}},$$

$$\mathscr{F} = \frac{1}{4}\big(U^2 I - 4D_{yy}\big)^{\frac{3}{2}} - \frac{3}{4}U\big(U^2 I - 4D_{yy}\big) - \gamma U^2\big(\sqrt{U^2 I - 4D_{yy}} - U\big),$$

while the realization of $\mathscr{L}$ in L^2 is the operator

$$L = -4D_{yyyy} - (\gamma - 1)U^2 D_{yy}$$

with $H_\sharp^4$ being a domain.

It follows from Proposition 3.1 that the front φ solves the equation

$$\frac{d}{dt}\mathscr{B}(\varphi) = \mathscr{L}(\varphi) + \mathscr{F}\big((\varphi_y)^2\big). \tag{3.4}$$

The main feature of (3.4) is that the nonlinear part is rather unusual. Actually, it has a fourth-order leading term, as $\mathscr{L}$ does. Therefore, (3.4) is a fully nonlinear equation. More precisely, we have the following result.

Lemma 3.1 *The operators $\mathscr{B}$ and $\mathscr{F}$ admit bounded realizations $B : H_\sharp^1 \to L^2$ and $F : H_\sharp^3 \to L^2$, respectively. Moreover, B is invertible.*

Proof A straightforward asymptotic analysis reveals that

$$b_k \sim 2\sqrt{\lambda_k}U, \quad f_k \sim 2\lambda_k^{\frac{3}{2}}$$

as $k \to +\infty$, from which we deduce that $\mathscr{B}$ and $\mathscr{F}$ admit bounded realizations $B : H_\sharp^1 \to L^2$ and $F : H_\sharp^3 \to L^2$.

Finally, since $b_k \neq 0$ for any $k \geq 0$, it follows that B is invertible. □

3.3 The Third-Order Pseudo-Differential Equation for the Front

In view of Lemma 3.1, we may rewrite (3.4) as

$$\varphi_t = \mathscr{B}^{-1}\mathscr{L}(\varphi) + \mathscr{B}^{-1}\mathscr{F}\big((\varphi_y)^2\big)$$

or, equivalently, as

$$\varphi_t = \mathscr{A}(\varphi) + \mathscr{M}\big((\varphi_y)^2\big). \tag{3.5}$$

We emphasize that (3.5) is a pseudo-differential, fully nonlinear equation of the third-order, since the pseudo-differential operators $\mathscr{A}$ and $\mathscr{M}$ have symbols

$$a_k = \frac{(U^2 - X_k^2)(X_k^2 - \gamma U^2)}{4UX_k} \quad \text{and} \quad m_k = \frac{X_k^3 - 3UX_k^2 - 4\gamma U^2 X_k + 4\gamma U^3}{4UX_k}$$

for $k \geq 0$, respectively. Clearly, any smooth enough solution to (3.4) solves (3.5) as well.

The following result is crucial for the rest of the paper.

Theorem 3.1 *The following properties are satisfied:*

(i) *The realization $A : H_\sharp^3 \to L^2$ of $\mathscr{A}$ is a sectorial operator and the sequence (a_k) constitutes its spectrum $\sigma(A)$. In particular, 0 is a simple eigenvalue of A, and the spectral projection Π associated with 0 is given by*

$$\Pi(\psi) = \frac{1}{\ell}\int_{-\frac{\ell}{2}}^{\frac{\ell}{2}} \psi(y)dy, \quad \psi \in L^2.$$

Finally, $\sigma(A) \setminus \{0\}$ is contained in the left half-plane $\{\lambda \in \mathbb{C} : \mathrm{Re}\lambda < 0\}$ if and only if $\gamma < \gamma_c$.

(ii) *The realization $M : H_\sharp^2 \to L^2$ of the operator $\mathscr{M}$ is bounded.*

Proof (i) Let us split

$$a_k = -\frac{2\lambda_k^{\frac{3}{2}}}{U} + \frac{-4\lambda_k^2 + 4\lambda_k^2\sqrt{\frac{U^2}{4\lambda_k} + 1} - \lambda_k U^2 + \lambda_k\gamma U^2}{U\sqrt{U^2 + 4\lambda_k}} =: -\frac{2\lambda_k^{\frac{3}{2}}}{U} + a_{1,k}$$

for any $k \geq 0$. Since

$$a_{1,k} \sim \frac{1}{4}\sqrt{\lambda_k}(2\gamma - 1)U$$

as $k \to +\infty$, if $\gamma \neq \frac{1}{2}$, we can infer that the realization A of operator $\mathscr{A}$ in $H_\sharp^3$ is well-defined. Moreover, since A splits into the sum of two operators A_0 (whose symbol is $(-2\lambda_k^{\frac{3}{2}}U^{-1})$) and A_1, which is a nice perturbation of A_0 (being a bounded operator in $H^1_\sharp$, which is an intermediate space of class $J_{\frac{1}{3}}$ between L^2 and $H^3_\sharp$), in view of [21, Proposition 2.4.1(i)], it is enough to prove that A_0 is a sectorial operator. But this follows immediately from the general abstract results (see, e.g., [11, Chap. 3]), or a direct computation. Indeed, if λ has the positive real part, then the equation $\lambda u - A_0 u = f$ has the unique solution, for any $f \in L^2$,

$$u = R(\lambda, A_0) f = U \sum_{k=0}^{+\infty} \frac{\widehat{f}(k)}{\lambda U + 2\lambda_k^{\frac{3}{2}}}$$

and

$$|R(\lambda, A_0) f|_2^2 = U^2 \sum_{k=0}^{+\infty} \frac{|\widehat{f}(k)|^2}{|\lambda U + 2\lambda_k^{\frac{3}{2}}|^2} \le \frac{1}{|\lambda|^2} \sum_{k=0}^{+\infty} |\widehat{f}(k)|^2 = \frac{1}{|\lambda|^2} |f|_2^2.$$

Proposition 2.1.1 in [21] yields the sectoriality of A_0.

Next we compute the spectrum of the operator A. Since $H^3_\sharp$ is compactly embedded into L^2, $\sigma(A)$ consists of eigenvalues only. We claim that $\sigma(A)$ consists of the elements of the sequence (a_k). Indeed writing the eigenvalue equation in the Fourier variable, we get the infinitely many equations

$$\lambda \widehat{\psi}(k) - a_k \widehat{\psi}(k) = 0, \quad k \ge 0, \tag{3.6}$$

which should be satisfied by the pair λ (the eigenvalue) and ψ (the eigenfunction). It is clear that this system of infinitely many equations admits a non-identically vanishing solution $(\widehat{\psi}(k))$ if and only if λ equals one of the elements of the sequence. The set equality $\sigma(A) = \{a_k : k \ge 0\}$ is thus proved.

Since the sequence (a_k) converges to $-\infty$ as $k \to +\infty$, all the eigenvalues of A are isolated. In particular, 0 is isolated and, again from formula (3.6), we easily see that the eigenspace associated with the eigenvalue $\lambda = 0$ is one-dimensional. To conclude that $\lambda = 0$ is simple, and in view of [21, Propositions A.1.2 and A.2.1], it suffices to prove that it is a simple pole of the resolvent operator. In such a case, the associated spectral projection is the residual at $\lambda = 0$ of $R(\cdot, A)$.

Clearly, for any $\lambda \notin \sigma(A)$,

$$R(\lambda, A)\zeta = \sum_{k=0}^{+\infty} \frac{1}{\lambda - a_k} \widehat{\zeta}(k) w_k$$

for any $\zeta \in L^2$. Hence,

$$\lambda R(\lambda, A)\zeta = \widehat{\psi}(0) w_0 + \sum_{k=1}^{+\infty} \frac{\lambda}{\lambda - a_k} \widehat{\zeta}(k) w_k =: \Pi\zeta + R_1(\lambda)\zeta.$$

Since $\lambda \ne a_k$ for any $k \ge 1$, and $a_k \to -\infty$ as $k \to +\infty$, there exists a neighborhood of $\lambda = 0$ in which the ratio $\frac{|\lambda|}{|\lambda - a_k|}$ is bounded, uniformly with respect to $k \ge 1$. As a byproduct, in such a neighborhood of $\lambda = 0$, the mapping $\lambda \mapsto R_1(\lambda)$ is bounded with values in $L(L^2)$. This shows that $\lambda = 0$ is a simple pole of operator A.

To conclude the proof of point (i), let us determine the values of γ such that $\sigma(A) \setminus \{0\}$ does not contain nonnegative elements. For this purpose, it suffices to observe that $a_k < 0$ for any $k \geq 1$ if and only if $4\lambda_k + U^2 - \gamma U^2 > 0$ for such k's, which is equivalent to $4\lambda_1 + U^2 - \gamma U^2 > 0$, since (λ_k) is a nondecreasing sequence. Hence, the condition for $\sigma(A) \setminus \{0\}$ be contained in $(-\infty, 0)$, is $\gamma < \gamma_c$, where

$$\gamma_c = 1 + \frac{16\pi^2}{\ell^2 U^2}. \tag{3.7}$$

(ii) As in the proof of Lemma 3.1, it suffices to observe that $m_k \sim \lambda_k U^{-1}$ as $k \to +\infty$. □

The linearized stability principle (see, e.g., [21, Sect. 9.1.1]) and the results in Theorem 3.1 yield the following stability analysis.

Corollary 3.1 *Let γ_c be given by* (3.7).

(a) *If $\gamma < \gamma_c$, then the null solution to* (3.5) *is* (*orbitally*) *stable, with an asymptotic phase, with respect to sufficiently smooth and small perturbations.*
(b) *If $\gamma > \gamma_c$, then the null solution to* (3.5) *is unstable.*

4 Rigorous Asymptotic Derivation of the K-S Equation

The second question that we address is the link between (3.5) and (K-S). As in Sect. 2, we consider the small perturbation parameter $\varepsilon > 0$ defined by

$$\gamma = 1 + \varepsilon$$

(see (2.1)). Moreover, we perform the same change of dependent and independent variables as in (2.2), namely,

$$t - \frac{\tau}{\varepsilon^2 U^2}, \qquad y - \frac{\eta}{\sqrt{\varepsilon} U}, \qquad \varphi - \frac{\varepsilon}{U}\psi.$$

The key-idea is to link the small positive parameter ε and the width of the strip which will blow up as $\varepsilon \to 0$. For fixed $\ell_0 > 0$, we take ℓ of the form,

$$\ell_\varepsilon = \frac{\ell_0}{\sqrt{\varepsilon} U}.$$

Hence γ_c (see (3.7)) converges to 1 as $\varepsilon \to 0$.

In view of Corollary 3.1, in order to avoid a trivial dynamics, we assume that $\gamma_c > 1$. This means that we take the bifurcation parameter ℓ_0 larger than 4π and obtain that $\gamma_c \in (1, 1+\varepsilon)$.

For the new variables, $\mathscr{B}$ is replaced by the operator $\mathscr{B}_\varepsilon = U^2\sqrt{I - 4\varepsilon D_{\eta\eta}}$. Lemma 3.1 applies to this operator and guarantees that, for any fixed $\varepsilon > 0$, the realization $B_\varepsilon : H^1_\sharp \to L^2$ of $\mathscr{B}_\varepsilon$ is bounded. However, the perturbation is clearly singular as $\varepsilon \to 0$, since obviously $B_\varepsilon \to U^2 I$. Therefore, it is hopeless to take the limit $\varepsilon \to 0$ in the third-order equation (3.5). Fortunately, the fourth-order equation (3.4) is more friendly, since, after the division by ε^3 and U^3, it comes that

$$\begin{aligned}
&\frac{\partial}{\partial\tau}(\sqrt{I - 4\varepsilon D_{\eta\eta}})\psi \\
&\quad = -4D_{\eta\eta\eta\eta}\psi - D_{\eta\eta}\psi \\
&\qquad + \frac{1}{4}\big\{(I - 4\varepsilon D_{\eta\eta})^{\frac{3}{2}} - 3(I - 4\varepsilon D_{\eta\eta}) - 4(1+\varepsilon)(\sqrt{I - 4\varepsilon D_{\eta\eta}} - I)\big\}(D_\eta\psi)^2,
\end{aligned} \tag{4.1}$$

which is the perturbed equation that we are going to study, with periodic boundary conditions at $\eta = \pm\frac{\ell_0}{2}$.

Mimicking (3.4), we rewrite (4.1) in the abstract way,

$$\frac{\mathrm{d}}{\mathrm{d}\tau}\mathscr{B}_\varepsilon\psi = \mathscr{L}\psi + \mathscr{F}_\varepsilon\big((\psi_\eta)^2\big), \tag{4.2}$$

where the symbols of the operators $\mathscr{B}_\varepsilon$, $\mathscr{L}$ and $\mathscr{F}_\varepsilon$ are

$$b_{\varepsilon,k} = X_{\varepsilon,k}, \qquad s_k = -\lambda_k(4\lambda_k - 1),$$

$$f_{\varepsilon,k} = \frac{1}{4}\big(X^3_{\varepsilon,k} - 3X^2_{\varepsilon,k} - 4(1+\varepsilon)X_{\varepsilon,k} + 4 + 4\varepsilon\big)$$

for any $k \ge 0$, respectively, and

$$X_{\varepsilon,k} = \sqrt{1 + 4\varepsilon\lambda_k}, \quad k \ge 0.$$

Writing (4.2) in the discrete Fourier variable gives infinitely many equations

$$b_{\varepsilon,k}\widehat{\psi_\tau}(\tau,k) = -\lambda_k(4\lambda_k - 1)\widehat{\psi}(\tau,k) + f_{\varepsilon,k}\widehat{(\psi_\eta)^2}(\tau,k)$$

for any $k \ge 0$. Note that the leading terms (namely, at order 0 in ε) of $b_{\varepsilon,k}$ and $f_{\varepsilon,k}$ are 1 and $-\frac{1}{2}$, respectively.

Fix $T > 0$. For $\Phi_0 \in H^m_\sharp$ ($m \ge 4$), the Cauchy problem

$$\begin{cases}
\Phi_\tau(\tau,\eta) = -4\Phi_{\eta\eta\eta\eta}(\tau,\eta) - \Phi_{\eta\eta}(\tau,\eta) - \frac{1}{2}(\Phi_\eta(\tau,\eta))^2, & \tau \ge 0,\ |\eta| \le \frac{\ell_0}{2}, \\
D^k_\eta\Phi(\tau, -\frac{\ell_0}{2}) = D^k_\eta\Phi(\tau, \frac{\ell_0}{2}), & \tau \ge 0,\ k \le m-1, \\
\Phi(0,\eta) = \Phi_0(\eta), & |\eta| \le \frac{\ell_0}{2}
\end{cases}$$

admits a unique solution $\Phi \in C([0,T]; H^m_\sharp)$ such that $\Phi_\tau \in C([0,T]; H^{m-4}_\sharp)$ (see, e.g., [6, Appendix B]).

Through Φ, we split $\psi = \Phi + \varepsilon\rho_\varepsilon$. For simplicity, we take zero as the initial condition for ρ_ε, and to avoid cumbersome notation, in the sequel, we write ρ for ρ_ε.

If ψ solves (4.2), then

$$\frac{\partial}{\partial\tau}\mathscr{B}_\varepsilon(\rho) + \mathscr{H}_\varepsilon(\Phi_\tau) = \mathscr{L}(\rho) + \mathscr{M}_\varepsilon\big((\Phi_\eta)^2\big) + \varepsilon\mathscr{F}_\varepsilon\big((\rho_\eta)^2\big) + 2\mathscr{F}_\varepsilon(\Phi_\eta\rho_\eta), \quad (4.3)$$

where the symbols of the operators $\mathscr{H}_\varepsilon$ and $\mathscr{M}_\varepsilon$ are

$$h_{\varepsilon,k} = \frac{1}{\varepsilon}(X_{\varepsilon,k} - 1), \qquad m_{\varepsilon,k} = \frac{1}{4\varepsilon}\big(X_{\varepsilon,k}^3 - 3X_{\varepsilon,k}^2 - 4(1+\varepsilon)X_{\varepsilon,k} + 6 + 4\varepsilon\big)$$

for any $k \ge 0$.

Proposition 4.1 *There exists a positive constant C_* such that the following properties are satisfied for any $\varepsilon \in (0, 1]$:*

(a) *For any $s = 2, 3, \ldots$, the operators $\mathscr{B}_\varepsilon$ and $\mathscr{H}_\varepsilon$ admit bounded realizations B_ε and H_ε, respectively, mapping $H^s_\sharp$ into $H^{s-2}_\sharp$. Moreover,*

$$\|B_\varepsilon\|_{L(H^s_\sharp, H^{s-2}_\sharp)} + \|H_\varepsilon\|_{L(H^s_\sharp, H^{s-2}_\sharp)} \le C_*.$$

Finally, the operator B_ε is invertible from $H^s_\sharp$ to $H^{s-2}_\sharp$.

(b) *For any $s = 3, 4, \ldots$, the operators $\mathscr{F}_\varepsilon$ and $\mathscr{M}_\varepsilon$ admit bounded realizations F_ε and M_ε, respectively, mapping H^s into H^{s-3}. Moreover,*

$$\|F_\varepsilon\|_{L(H^s_\sharp, H^{s-3}_\sharp)} + \|M_\varepsilon\|_{L(H^s_\sharp, H^{s-3}_\sharp)} \le C_*.$$

Proof The statement follows from an analysis of the symbols of the operators $\mathscr{B}_\varepsilon$, $\mathscr{F}_\varepsilon$, $\mathscr{H}_\varepsilon$ and $\mathscr{M}_\varepsilon$. Without much effort, one can show that

$$|h_{\varepsilon,k}| \le 4\lambda_k, \qquad |m_{\varepsilon,k}| \le 2\lambda_k^{\frac{3}{2}} + 25\lambda_k$$

for any $k \ge 0$ and any $\varepsilon \in (0, 1]$. These estimates combined with the formulas $0 \ne b_{\varepsilon,k} = \varepsilon h_{\varepsilon,k} + 1$ and $f_k = \varepsilon m_{\varepsilon,k} - \frac{1}{2}$, for any $k \ge 0$ and any $\varepsilon \in (0, 1]$, yield the assertion. □

Instead of studying (3.4), we find it much more convenient to deal with the equation satisfied by $\zeta := \rho_\eta$, i.e.,

$$\frac{\partial}{\partial\tau}\mathscr{B}_\varepsilon(\zeta) + \mathscr{H}_\varepsilon(\Psi_\tau) = \mathscr{L}(\zeta) + \mathscr{M}_\varepsilon\big((\Psi^2)_\eta\big) + \varepsilon\mathscr{F}_\varepsilon\big((\zeta^2)_\eta\big) + 2\mathscr{F}_\varepsilon\big((\Psi\zeta)_\eta\big), \quad (4.4)$$

which we couple with the initial condition $\zeta(0, \cdot) = 0$. Here, $\Psi = \Phi_\eta$.

4.1 A Priori Estimates

For any $n = 0, 1, 2, \ldots$ and any $T > 0$, we set

$$X_n(T) = \{\zeta \in C([0,T]; H_\sharp^{4\vee 2n}) \cap C^1([0,T]; L^2) : \zeta_\tau \in C([0,T]; H_\sharp^{2\vee(n+1)})\},$$

where $a \vee b := \max\{a, b\}$.

For any $\varepsilon > 0$, we introduce in $H_\sharp^{\frac{1}{2}}$ the norm

$$\|\zeta\|^2_{\frac{1}{2},\varepsilon} = \sum_{k=0}^{+\infty} \sqrt{1 + 4\varepsilon\lambda_k} |\widehat{\zeta}(k)|^2, \qquad \zeta \in H_\sharp^{\frac{1}{2}}.$$

Note that, for any fixed $\varepsilon > 0$, $\|\cdot\|_{\frac{1}{2},\varepsilon}$ is a norm, equivalent to the usual norm in $H_\sharp^{\frac{1}{2}}$.

The main result of this subsection is contained in the following theorem, where we set $\Psi_0 = (\Phi_0)_\eta$.

Theorem 4.1 *Fix an integer $n \ge 0$ and $T > 0$. Further, suppose that $\Psi_0 \in H_\sharp^{n+6}$. Then, there exist $\varepsilon_1 = \varepsilon_1(n, T) \in (0, 1)$ and $K_n = K_n(T) > 0$ such that, if $\zeta \in X_n(T_1)$ is a solution on the time interval $[0, T_1]$ to (4.4) for some $T_1 \le T$, then*

$$\sup_{\tau\in[0,T_1]} \|D_\eta^n \zeta(\tau,\cdot)\|^2_{\frac{1}{2},\varepsilon} \le K_n, \tag{4.5}$$

whenever $0 < \varepsilon \le \varepsilon_1$.

Note that the assumptions on Ψ_0 guarantee that $\Psi \in C([0,T]; H_\sharp^{n+4}) \cap C^1([0,T]; H_\sharp^{n+2})$.

The proof of Theorem 4.1 heavily relies on the following lemma.

Lemma 4.1 *Let A_0, c_0, c_1, c_2, c_3, ε, T_0 be positive constants, and let T_1 be such that $0 < T_1 < T_0$. Further, let f_ε and $A_\varepsilon : [0, T_1] \to \mathbb{R}$ be a positive continuous function and a positive continuously differentiable function respectively such that*

$$\begin{cases} A'_\varepsilon(\tau) + (c_0 - \varepsilon^2(A_\varepsilon(\tau))^2) f_\varepsilon(\tau) \le c_1 + c_2 A_\varepsilon(\tau) + c_3\varepsilon(A_\varepsilon(\tau))^2, & \tau \in [0, T_1], \\ A_\varepsilon(0) = 0. \end{cases}$$

Then, there exist $\varepsilon_1 = \varepsilon_1(T_0) \in (0, 1)$ and a constant $K = K(T_0)$ such that $A_\varepsilon(\tau) \le K$ for any $\tau \in [0, T_1]$ and any $\varepsilon \in (0, \varepsilon_1]$.

Proof When f_ε identically vanishes, the proof follows from [2, Lemma 3.1], which shows that we can take

$$\varepsilon_1(T_0) = \frac{3c_2^2}{16 c_1 c_3 (e^{c_2 T_0} - 1)} \quad \text{and} \quad K \le \frac{4 c_1 e^{c_2 T_0}}{3 c_2}.$$

Let us now consider the general case when f_ε does not identically vanish in $[0, T_1]$. We fix

$$\varepsilon_0 = \varepsilon_0(T_0) \le \frac{3c_2^2}{16c_1c_3(\mathrm{e}^{c_2T_0} - 1)}$$

such that $9c_0c_2^2 - 12c_1c_2\mathrm{e}^{c_2T_0}\varepsilon_0 - 16c_1^2\mathrm{e}^{2c_2T_0}\varepsilon_0^2 > 0$, and $\varepsilon \in (0, \varepsilon_0]$. We claim that $c_0 - \varepsilon^2(A_\varepsilon(\tau))^2 > 0$ for any $\tau \in [0, T_1]$.

Let $(0, T_\varepsilon)$ be the largest interval (possibly depending on ε), where $c_0 - \varepsilon A_\varepsilon - \varepsilon^2(A_\varepsilon)^2$ is positive. The existence of this interval is clear, since A_ε vanishes at 0. The positivity of $c_0 - \varepsilon^2(A_\varepsilon)^2$ in $(0, T_\varepsilon)$ shows that $A_\varepsilon' \le c_1 + c_2A_\varepsilon + c_3\varepsilon A_\varepsilon^2$ in such an interval. From the above result we can infer that $A_\varepsilon(\tau) \le \frac{4c_1\mathrm{e}^{c_2T_0}}{3c_2}$ for any $\tau \in [0, T_\varepsilon]$, so that $c_0 - \varepsilon^2(A_\varepsilon(T_\varepsilon))^2 > 0$. By the definition of T_ε, this clearly implies that $T_\varepsilon = T_1$. □

Proof of Theorem 4.1 Throughout the proof, we assume that $T_1 \le T$ is fixed, and ε and τ are arbitrarily fixed in $(0, 1]$ and in $[0, T_1]$, respectively. Moreover, to avoid cumbersome notations, we denote by c almost all the constants appearing in the estimates. Hence, the exact value of c may change from line to line, but we do not need to follow the constants throughout the estimates. We just need to stress how the estimates depend on ε. As a matter of fact, all the c's are independent not only of ε but also of τ, Ψ and ζ. On the contrary, they may depend on n (and, actually, in most cases they do). Finally, we denote by $K(\Psi)$ a constant, which may depend on n and also on Ψ. As above, $K(\Psi)$ may vary from estimate to estimate.

The first step of the proof consists in multiplying both sides of (4.4) by $(-1)^n D_\eta^{2n}\zeta$, and integrating by parts over $(-\frac{\ell_0}{2}, \frac{\ell_0}{2})$. This yields

$$\begin{aligned}
&\int_{-\frac{\ell_0}{2}}^{\frac{\ell_0}{2}} B_\varepsilon\big(\zeta_\tau(\tau,\cdot)\big)(-1)^n D_\eta^{2n}\zeta(\tau,\cdot)\mathrm{d}\eta + 4\int_{\frac{\ell_0}{2}}^{\frac{\ell_0}{2}} |D_\eta^{n+2}\zeta(\tau,\cdot)|^2\mathrm{d}\eta \\
&\quad - \int_{\frac{\ell_0}{2}}^{\frac{\ell_0}{2}} |D_\eta^{n+1}\zeta(\tau,\cdot)|^2\mathrm{d}\eta \\
&= -\int_{-\frac{\ell_0}{2}}^{\frac{\ell_0}{2}} \Big(H_\varepsilon\big(\Psi_\tau(\tau,\cdot)\big) - M_\varepsilon\big((\Psi^2)_\eta(\tau,\cdot)\big)\Big)(-1)^n D_\eta^{2n}\zeta(\tau,\cdot)\mathrm{d}\eta \\
&\quad + \varepsilon\int_{-\frac{\ell_0}{2}}^{\frac{\ell_0}{2}} F_\varepsilon\big((\zeta^2)_\eta(\tau,\cdot)\big)(-1)^n D_\eta^{2n}\zeta(\tau,\cdot)\mathrm{d}\eta \\
&\quad + 2\int_{-\frac{\ell_0}{2}}^{\frac{\ell_0}{2}} F_\varepsilon\big((\Psi\zeta)_\eta(\tau,\cdot)\big)(-1)^n D_\eta^{2n}\zeta(\tau,\cdot)\mathrm{d}\eta.
\end{aligned} \tag{4.6}$$

Using Parseval's formula and the definition of the symbol $b_{\varepsilon,k}$, one can easily show that

$$\int_{-\frac{\ell_0}{2}}^{\frac{\ell_0}{2}} B_\varepsilon\big(\zeta_\tau(\tau,\cdot)\big)(-1)^n D_\eta^{2n}\zeta(\tau,\cdot)\mathrm{d}\eta = \frac{1}{2}\frac{\mathrm{d}}{\mathrm{d}\tau}\|D_\eta^n\zeta(\tau,\cdot)\|_{\frac{1}{2},\varepsilon}^2. \tag{4.7}$$

We now deal with the other terms in (4.6). Integrating n-times by parts and, then, using Poincaré-Wirtinger and Cauchy-Schwarz inequalities, jointly with Proposition 4.1, it is not difficult to show that

$$\left|\int_{-\frac{\ell_0}{2}}^{\frac{\ell_0}{2}} \big(H_\varepsilon(\Psi_\tau(\tau,\eta)) - M_\varepsilon\big((\Psi^2)_\eta(\tau,\cdot)\big)\big)(-1)^n D_\eta^{2n}\zeta(\tau,\cdot)\mathrm{d}\eta\right|$$
$$\leq K(\Psi) + |D_\eta^n\zeta(\tau,\cdot)|_2^2 \tag{4.8}$$

for any $\zeta \in X_n(T_1)$.

Estimating the other two integral terms in the right-hand side of (4.6) demands much more effort. The starting point is the following estimate:

$$\left|\int_{-\frac{\ell_0}{2}}^{\frac{\ell_0}{2}} F_\varepsilon\big(\chi_\eta(\tau,\cdot)\big)(-1)^n D_\eta^{2n}\zeta(\tau,\cdot)\mathrm{d}\eta\right|$$
$$\leq c\varepsilon^{\frac{3}{2}}|D_\eta^{n+2}\chi(\tau,\cdot)|_2|D_\eta^{n+2}\zeta(\tau,\cdot)|_2 + c\varepsilon|D_\eta^{n+2}\chi(\tau,\cdot)|_2|D_\eta^{n+1}\zeta(\tau,\cdot)|_2$$
$$+ c\sqrt{\varepsilon}|D_\eta^n\chi(\tau,\cdot)|_2|D_\eta^{n+2}\zeta(\tau,\cdot)|_2 + c|D_\eta^n\chi(\tau,\cdot)|_2|D_\eta^{n+1}\zeta(\tau,\cdot)|_2, \tag{4.9}$$

which holds for any $\chi \in C([0,T_1]; H_\sharp^{4\vee 2n})$. Such a formula follows by observing that

$$\left|\int_{-\frac{\ell_0}{2}}^{\frac{\ell_0}{2}} F_\varepsilon\big(\chi_\eta(\tau,\cdot)\big)(-1)^n D_\eta^{2n}\zeta(\tau,\cdot)\mathrm{d}\eta\right|$$
$$\leq \sum_{k=0}^{+\infty}\lambda_k^n|f_{\varepsilon,k}||\widehat{\chi_\eta}(\tau,k)||\widehat{\zeta}(\tau,k)|$$
$$\leq c\sum_{k=0}^{+\infty}\lambda_k^n\big(\varepsilon^{\frac{3}{2}}\lambda_k^{\frac{3}{2}} + \varepsilon\lambda_k + \varepsilon^{\frac{1}{2}}\lambda_k^{\frac{1}{2}} + 1\big)|\widehat{\chi_\eta}(\tau,k)||\widehat{\zeta}(\tau,k)|.$$

Then, by using Young inequality, we estimate the terms in the round brackets.

Now, we plug $\chi = \zeta^2$ into (4.9), and use the estimates

$$\left|D_\eta^{n+2}\big(\zeta(\tau,\cdot)\big)^2\right|_2 \leq c\big(|D_\eta^{n+2}\zeta(\tau,\cdot)|_2|D_\eta^n\zeta(\tau,\cdot)|_2 + |D_\eta^{n+1}\zeta(\tau,\cdot)|_2^2\big), \tag{4.10}$$

$$\left|D_\eta^n\big(\zeta(\tau,\cdot)\big)^2\right|_2 \leq c|D_\eta^n\zeta(\tau,\cdot)|_2^2, \tag{4.11}$$

which can be obtained by using the Poincaré-Wirtinger inequality and Leibniz formula, the Cauchy-Schwarz inequality. Again, by using the Poincaré-Wirtinger inequality, we obtain

$$
\begin{aligned}
&\left|\int_{-\frac{\ell_0}{2}}^{\frac{\ell_0}{2}} F_\varepsilon\big((\zeta^2)_\eta(\tau,\cdot)\big)(-1)^n D_\eta^{2n}\zeta(\tau,\cdot)\mathrm{d}\eta\right| \\
&\quad \le c\varepsilon^{\frac{3}{2}}|D_\eta^n\zeta(\tau,\cdot)|_2|D_\eta^{n+2}\zeta(\tau,\cdot)|_2^2 + c\varepsilon^{\frac{3}{2}}|D_\eta^{n+1}\zeta(\tau,\cdot)|_2^2|D_\eta^{n+2}\zeta(\tau,\cdot)|_2 \\
&\qquad + c\varepsilon|D_\eta^n\zeta_\eta(\tau,\cdot)|_2|D_\eta^{n+1}\zeta(\tau,\cdot)|_2|D_\eta^{n+2}\zeta(\tau,\cdot)|_2 + c\varepsilon|D_\eta^{n+1}\zeta(\tau,\cdot)|_2^3 \\
&\qquad + c\varepsilon^{\frac{1}{2}}(1+\varepsilon)|D_\eta^n\zeta(\tau,\cdot)|_2^2|D_\eta^{n+2}\zeta(\tau,\cdot)|_2 + c|D_\eta^n\zeta(\tau,\cdot)|_2^2|D_\eta^{n+1}\zeta(\tau,\cdot)|_2 \\
&\le c\varepsilon^{\frac{3}{2}}|D_\eta^n\zeta(\tau,\cdot)|_2|D_\eta^{n+2}\zeta(\tau,\cdot)|_2^2 + c\varepsilon^{\frac{3}{2}}|D_\eta^{n+1}\zeta(\tau,\cdot)|_2^2|D_\eta^{n+2}\zeta(\tau,\cdot)|_2 \\
&\qquad + c\varepsilon|D_\eta^{n+1}\zeta(\tau,\cdot)|_2^2|D_\eta^{n+2}\zeta(\tau,\cdot)|_2 + c\varepsilon|D_\eta^{n+1}\zeta(\tau,\cdot)|_2^2|D_\eta^{n+2}\zeta(\tau,\cdot)|_2 \\
&\qquad + c\varepsilon^{\frac{1}{2}}|D_\eta^n\zeta(\tau,\cdot)|_2^2|D_\eta^{n+2}\zeta(\tau,\cdot)|_2 + c|D_\eta^n\zeta(\tau,\cdot)|_2^2|D_\eta^{n+2}\zeta(\tau,\cdot)|_2 \\
&\le c\varepsilon^{\frac{3}{2}}|D_\eta^n\zeta(\tau,\cdot)|_2|D_\eta^{n+2}\zeta(\tau,\cdot)|_2^2 + c\varepsilon^2|D_\eta^{n+1}\zeta(\tau,\cdot)|_2^4 + c\varepsilon|D_\eta^{n+2}\zeta(\tau,\cdot)|_2^2 \\
&\qquad + c|D_\eta^n\zeta(\tau,\cdot)|_2^4 + c|D_\eta^{n+2}\zeta(\tau,\cdot)|_2^2 \\
&\le c\varepsilon^{\frac{3}{2}}|D_\eta^n\zeta(\tau,\cdot)|_2|D_\eta^{n+2}\zeta(\tau,\cdot)|_2^2 + c\varepsilon^2|D_\eta^{n+1}\zeta(\tau,\cdot)|_2^4 \\
&\qquad + c|D_\eta^{n+2}\zeta(\tau,\cdot)|_2^2 + c|D_\eta^n\zeta(\tau,\cdot)|_2^4.
\end{aligned}
\tag{4.12}
$$

In the similar way, using the estimate

$$|D_\eta^m(\Psi\zeta)(\tau,\cdot)|_2 \le c|D_\eta^m\zeta(\tau,\cdot)|_2|D_\eta^m\Psi(\tau,\cdot)|_2$$

(with $m \in \{n, n+2\}$) in place of (4.10)–(4.11), from (4.9), we get

$$
\begin{aligned}
&\left|\int_{-\frac{\ell_0}{2}}^{\frac{\ell_0}{2}} F_\varepsilon\big((\Psi\zeta)_\eta(\tau,\cdot)\big)(-1)^n D_\eta^{2n}\zeta(\tau,\cdot)\mathrm{d}\eta\right| \\
&\quad \le c\varepsilon^{\frac{3}{2}}|D_\eta^{n+2}\Psi(\tau,\cdot)|_2|D_\eta^{n+2}\zeta(\tau,\cdot)|_2^2 + c\varepsilon|D_\eta^{n+2}\Psi(\tau,\cdot)|_2|D_\eta^{n+1}\zeta(\tau,\cdot)|_2^2 \\
&\qquad + c\varepsilon|D_\eta^{n+2}\Psi(\tau,\cdot)|_2|D_\eta^{n+2}\zeta(\tau,\cdot)|_2^2 + c|D_\eta^n\Psi(\tau,\cdot)|_2|D_\eta^n\zeta(\tau,\cdot)|_2^2 \\
&\qquad + c\varepsilon|D_\eta^n\Psi(\tau,\cdot)|_2|D_\eta^{n+2}\zeta(\tau,\cdot)|_2^2 + c\delta^{-1}|D_\eta^n\Psi(\tau,\cdot)|_2^2|D_\eta^n\zeta(\tau,\cdot)|_2^2 \\
&\qquad + c\delta|D_\eta^{n+2}\zeta(\tau,\cdot)|_2^2
\end{aligned}
\tag{4.13}
$$

for any $\delta > 0$. We just mention the inequality

$$|D^n\Psi(\tau,\cdot)|_2|D^n\zeta(\tau,\cdot)|_2|D^{n+1}\zeta(\tau,\cdot)|_2$$
$$\le c\delta^{-1}|D^n\Psi(\tau,\cdot)|_2^2|D^n\zeta(\tau,\cdot)|_2^2+\delta|D^{n+1}\zeta(\tau,\cdot)|_2^2,$$

obtained by means of Young and Poincaré-Wirtinger inequalities, which we use to estimate one of the intermediate terms appearing in the proof of (4.13).

Now, taking $c\delta = \frac{5}{2}$, we get the estimate

$$\left|\int_{-\frac{\ell_0}{2}}^{\frac{\ell_0}{2}} F_\varepsilon\big((\Psi\zeta)_\eta(\tau,\cdot)\big)(-1)^n D_\eta^{2n}\zeta(\tau,\cdot)\mathrm{d}\eta\right|$$
$$\le K(\Psi)\big(\varepsilon|D_\eta^{n+2}\zeta(\tau,\cdot)|_2^2+\varepsilon|D_\eta^{n+1}\zeta(\tau,\cdot)|_2^2+|D_\eta^n\zeta(\tau,\cdot)|_2^2\big)+\frac{5}{2}|D_\eta^{n+2}\zeta(\tau,\cdot)|_2^2. \tag{4.14}$$

From (4.6)–(4.8), (4.12), (4.14) and the interpolative inequality

$$|D_\eta^{n+1}\zeta(\tau,\cdot)|_2^2\le|D_\eta^n\zeta(\tau,\cdot)|_2^2+\frac{1}{4}|D_\eta^{n+2}\zeta(\tau,\cdot)|_2^2,$$

we can infer that

$$\frac{1}{2}\frac{\mathrm{d}}{\mathrm{d}\tau}\|D_\eta^n\zeta(\tau,\cdot)\|_{\frac12,\varepsilon}^2+\big(1-\varepsilon K(\Psi)-c\varepsilon^{\frac52}|D_\eta^n\zeta(\tau,\cdot)|_2\big)|D_\eta^{n+2}\zeta(\tau,\cdot)|_2^2$$
$$\le K(\Psi)+K(\Psi)|D_\eta^n\zeta(\tau,\cdot)|_2^2+\varepsilon K(\Psi)|D_\eta^{n+1}\zeta(\tau,\cdot)|_2^2+c\varepsilon|D_\eta^n\zeta(\tau,\cdot)|_2^4$$
$$+c\varepsilon^3|D_\eta^{n+1}\zeta(\tau,\cdot)|_2^4$$
$$\le K(\Psi)+K(\Psi)|D_\eta^n\zeta(\tau,\cdot)|_2^2+\varepsilon K(\Psi)|D_\eta^{n+2}\zeta(\tau,\cdot)|_2^2$$
$$+c\varepsilon|D_\eta^n\zeta(\tau,\cdot)|_2^4+c\varepsilon^3|D_\eta^n\zeta(\tau,\cdot)|_2^2|D_\eta^{n+2}\zeta(\tau,\cdot)|_2^2,$$

which we can rewrite in the form

$$\frac{\mathrm{d}}{\mathrm{d}\tau}\|D_\eta^n\zeta(\tau,\cdot)\|_{\frac12,\varepsilon}^2+\big(2-\varepsilon K(\Psi)-c\varepsilon^2\|D_\eta^n\zeta(\tau,\cdot)\|_{\frac12,\varepsilon}^2\big)|D_\eta^{n+2}\zeta(\tau,\cdot)|_2^2$$
$$\le K(\Psi)+K(\Psi)\|D_\eta^n\zeta(\tau,\cdot)\|_{\frac12,\varepsilon}^2+c\varepsilon\|D_\eta^n\zeta(\tau,\cdot)\|_{\frac12,\varepsilon}^4,$$

by estimating $2\|D_\eta^n\zeta(\tau,\cdot)\|_{\frac12,\varepsilon}\le\|D_\eta^n\zeta(\tau,\cdot)\|_{\frac12,\varepsilon}^2+1$ and recalling that $\varepsilon\in(0,1]$.

Up to replacing $(0,1]$ by a smaller interval $(0,\varepsilon_0]$, we can assume that $\varepsilon K(\Psi)<1$ for any $\varepsilon\in(0,\varepsilon_0]$. Hence, applying Lemma 4.1 with

$$c_0=1,\qquad c_1=K(\Psi),\qquad c_2=K(\Psi),\qquad c_3=c,$$

$$A_\varepsilon(\tau)=\|D^n_\eta\zeta(\tau,\cdot)\|^2_{\frac{1}{2},\varepsilon},\qquad f_\varepsilon(\tau)=|D^{n+2}_\eta\zeta(\tau,\cdot)|_2^2,$$

we complete the proof. □

Now, taking advantage of the previous a priori estimates, which can be extended also to variational solutions ζ_N to (4.4) belonging to the space spanned by the functions $w_1,\dots,w_N$ (with constants independent of $N\in\mathbb{N}$), and using the classical Faedo-Galerkin method, the following result can be proved.

Theorem 4.2 *Fix $T>0$. Then, there exists an $\varepsilon_0(T)>0$ such that, for any $0<\varepsilon\le\varepsilon_0(T)$,* (4.4) *has a unique classical solution ζ on $[0,T]$, vanishing at $\tau=0$.*

4.2 *Proof of Theorem 1.1*

Since the unique solution ζ to (4.4) is the candidate to be the η-derivative of the solution ρ to (4.3), ρ should split into the sum $\rho(\tau,\eta)=(\mathscr{P}(\zeta))(\tau,\eta)+\upsilon(\tau)$ for some scalar valued function υ, where

$$(\mathscr{P}(\zeta))(\tau,\eta)=\int_{-\frac{\ell_0}{2}}^{\eta}\zeta(s)\mathrm{d}s-\frac{1}{2}\int_{-\frac{\ell_0}{2}}^{\frac{\ell_0}{2}}\zeta(s)\left(1-\frac{2s}{\ell_0}\right)\mathrm{d}s.$$

Imposing that ρ in the previous form is a solution to (4.3) and projecting along $\Pi(L^2)$, we see that ρ is a solution to (4.3) if and only if υ solves the following Cauchy problem:

$$\begin{cases}\frac{\mathrm{d}\upsilon}{\mathrm{d}\tau}=-\Pi(H_\varepsilon(\Phi_\tau))-\frac{1}{2}\varepsilon\Pi(\zeta^2)-\Pi(\Phi_\eta\zeta),\\ \upsilon(0)=0.\end{cases}$$

Since this problem has in fact a unique solution, and $\mathscr{P}(\zeta)+\upsilon$ vanishes at $\tau=0$, we conclude that problem (4.3) is uniquely solvable.

To complete the proof, we should show that there exists an $M>0$ such that

$$\sup_{\substack{\tau\in[0,T]\\ \eta\in[-\frac{\ell_0}{2},\frac{\ell_0}{2}]}}|\rho(\tau,\eta)|\le M \tag{4.15}$$

uniformly in $0<\varepsilon\le\varepsilon_0(T)$. Once this estimate is proved, coming back from (4.3) to (1.11), we see that the latter one has a unique classical solution $\varphi:[0,\frac{T}{\varepsilon^2U^2}]\times$

$\mathbb{R} \to \mathbb{R}$, which is periodic (with respect to the spatial variable) with period $\ell_\varepsilon = \frac{\ell_0}{\sqrt{\varepsilon U}}$, and satisfies

$$\varphi(0,\cdot) = \varepsilon U^{-1} \Phi_0(\sqrt{\varepsilon U}\cdot),$$

as well as the estimate

$$\left\| \varphi(t,\cdot) - \varepsilon U^{-1} \Phi\left(t\varepsilon^2 U^2, \cdot\sqrt{\varepsilon U}\right) \right\|_{C([-\frac{\ell_\varepsilon}{2},\frac{\ell_\varepsilon}{2}])} \le \frac{\varepsilon^2 M}{U}, \quad t \in [0, T_\varepsilon],$$

as is claimed.

So, let us prove (4.15). For this purpose, it is enough to use the a priori estimate (4.5) jointly with the Poincaré-Wirtinger inequality to estimate ζ, and to use (4.5) to estimate υ. This completes the proof of Theorem 1.1.

5 Numerical Experiments

In this section, we intend to solve numerically (4.1) for small positive ε and illustrate the convergence to the solution to (K-S).

In order to reformulate (4.1) on the interval $[0, 2\pi]$ with periodic boundary conditions, we set $x = \frac{\eta}{2\widetilde{\ell_0}}$, where $\widetilde{\ell_0} = \frac{\ell_0}{4\pi}$. It comes that

$$\begin{aligned}
&\frac{\partial}{\partial \tau}\left(\sqrt{I - \frac{\varepsilon}{\widetilde{\ell_0}^2} D_{xx}}\right)\psi \\
&= -\frac{1}{4\widetilde{\ell_0}^4} D_{xxxx}\psi - \frac{1}{4\widetilde{\ell_0}^2} D_{xx}\psi + \frac{1}{16\widetilde{\ell_0}^2}\left\{\left(I - \frac{\varepsilon}{\widetilde{\ell_0}^2} D_{xx}\right)^{\frac{3}{2}} - 3\left(I - \frac{\varepsilon}{\widetilde{\ell_0}^2} D_{xx}\right)\right. \\
&\quad \left. - 4(1+\varepsilon)\left(\sqrt{I - \frac{\varepsilon}{\widetilde{\ell_0}^2} D_{xx}} - I\right)\right\}(D_x\psi)^2.
\end{aligned}$$

Next, we define the bifurcation parameter $\beta = 4\widetilde{\ell_0}^2$ as in [7, 12]. After multiplication by β^2, it comes that

$$\begin{aligned}
&\frac{\partial}{\partial \tau}\left(\sqrt{\beta^4 - 4\varepsilon\beta^3 D_{xx}}\right)\psi \\
&= -4D_{xxxx}\psi - \beta D_{xx}\psi + \frac{\beta}{4}\left\{\left(I - \frac{4\varepsilon}{\beta} D_{xx}\right)^{\frac{3}{2}} - 3\left(I - \frac{4\varepsilon}{\beta} D_{xx}\right)\right. \\
&\quad \left. - 4(1+\varepsilon)\left(\sqrt{I - \frac{4\varepsilon}{\beta} D_{xx}} - I\right)\right\}(D_x\psi)^2.
\end{aligned}$$

Finally, we rescale the time by setting $t = \frac{\tau}{\beta^2}$,

$$\frac{\partial}{\partial t}\left(\sqrt{I - \frac{4\varepsilon}{\beta} D_{xx}}\right)\psi$$
$$= -4D_{xxxx}\psi - \beta D_{xx}\psi + \frac{\beta}{4}\left\{\left(I - \frac{4\varepsilon}{\beta}D_{xx}\right)^{\frac{3}{2}} - 3\left(I - \frac{4\varepsilon}{\beta}D_{xx}\right)\right.$$
$$\left. - 4(1+\varepsilon)\left(\sqrt{I - \frac{4\varepsilon}{\beta}D_{xx}} - I\right)\right\}(D_x\psi)^2.$$

By setting $\varepsilon' = \frac{\varepsilon}{\beta}$, with the prime being omitted hereafter, we obtain

$$\frac{\partial}{\partial t}(\sqrt{I - 4\varepsilon D_{xx}})\psi$$
$$= -4D_{xxxx}\psi - \beta D_{xx}\psi$$
$$+ \frac{\beta}{4}\{(I - 4\varepsilon D_{xx})^{\frac{3}{2}} - 3(I - 4\varepsilon D_{xx}) - 4(1+\varepsilon)(\sqrt{I - 4\varepsilon D_{xx}} - I)\}(D_x\psi)^2. \tag{5.1}$$

The initial condition is given by $\psi(0,\cdot) = \psi_0$, where ψ_0 is periodic with period 2π. Note that, in contrast to [7, 12], we do not subtract the drift.

Equation (5.1) in the discrete Fourier variable gives

$$\frac{\partial}{\partial t}(\sqrt{1 + 4\varepsilon k^2})\widehat{\psi}(t,k)$$
$$= -4k^4\widehat{\psi}(t,k) + \beta k^2\widehat{\psi}(t,k)$$
$$+ \frac{\beta}{4}\{(1 + 4\varepsilon k^2)^{\frac{3}{2}} - 3(1 + 4\varepsilon k^2) - 4(1+\varepsilon)(\sqrt{1 + 4\varepsilon k^2} - 1)\}\widehat{(\psi_x)^2}(t,k).$$

We use a backward-Euler scheme for the first-order time derivative to treat implicitly all the linear terms and to treat explicitly the nonlinear terms. The implicit treatment of the fourth- and second-order terms reduces the stability constraint, while the explicit treatment of the nonlinear terms avoids the expensive process of solving nonlinear equations at each time step. For simplicity, in the rest of this section, we use the notation $\widehat{f_k}$ instead of $\widehat{f}(k)$. It comes that

$$(\sqrt{1 + 4\varepsilon k^2})\frac{\widehat{\psi}_k^{n+1} - \widehat{\psi}_k^n}{\Delta t}$$
$$= -4k^4\widehat{\psi}_k^{n+1} + \beta k^2\widehat{\psi}_k^{n+1}$$
$$+ \frac{\beta}{4}\{(1 + 4\varepsilon k^2)^{\frac{3}{2}} - 3(1 + 4\varepsilon k^2) - 4(1+\varepsilon)(\sqrt{1 + 4\varepsilon k^2} - 1)\}\{[(\psi_x)^n]^2\}_k,$$

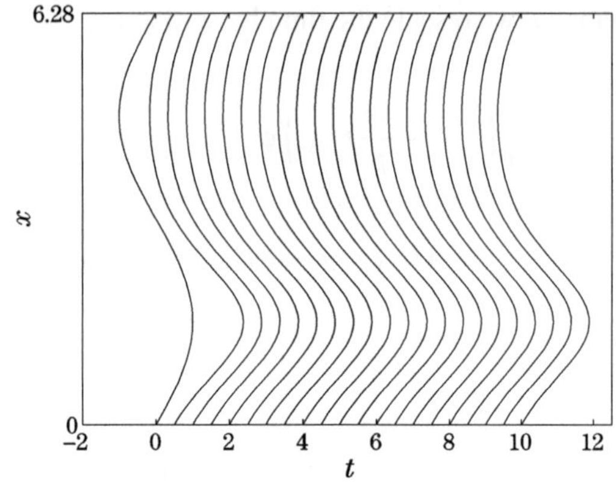

Fig. 1 Front propagation with $\beta = 10$, $\varepsilon = 0.1$ and $\psi_0(x) = \sin(x)$

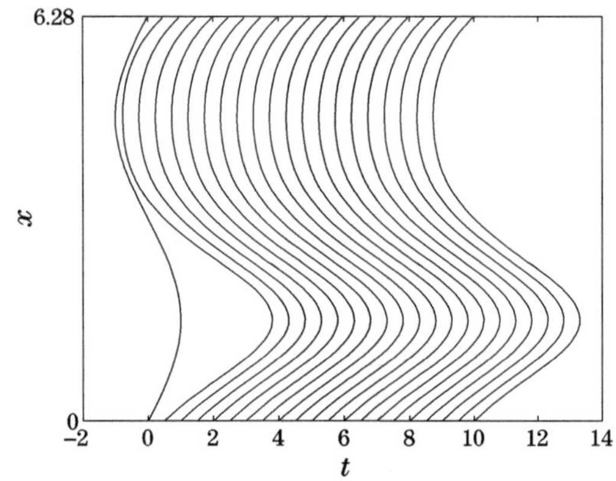

Fig. 2 Front propagation with $\beta = 10$, $\varepsilon = 0.01$ and $\psi_0(x) = \sin(x)$

where $\{(\psi_x)^2\}_k$ represents the k-th Fourier coefficient of $(\psi_x)^2$. This method is of the first order with respect to time. From the previous equation, it is easy to compute the k-th Fourier coefficient $\widehat{\psi}_k^{n+1}$. One gets

$$\widehat{\psi}_k^{n+1} = \left((1+4\varepsilon k^2)^{\frac{1}{2}} + 4k^4\Delta t - \beta k^2 \Delta t\right)^{-1}\Bigg((1+4\varepsilon k^2)^{\frac{1}{2}}\widehat{\psi}_k^n + \frac{\beta\Delta t}{4}\{(1+4\varepsilon k^2)^{\frac{3}{2}} - 3(1+4\varepsilon k^2) - 4(1+\varepsilon)[(1+4\varepsilon k^2)^{\frac{1}{2}} - 1]\} \times \left\{\left[(\psi_x)^n\right]^2\right\}_k\Bigg). \tag{5.2}$$

Practical calculations hold in the spectral space. We use an additional FFT to recover the physical nodal values ψ_j from $\widehat{\psi}_k$, where j stands for the division node in the physical space.

The numerical tests aim at checking the behavior of the solutions to (5.1) for values of ε close to 0, and comparing them to those for the Kuramoto-Sivashinsky equation. In Figs. 1, 2, 3, 4 and 5, 6, 7, 8, we plot consecutive front positions com-

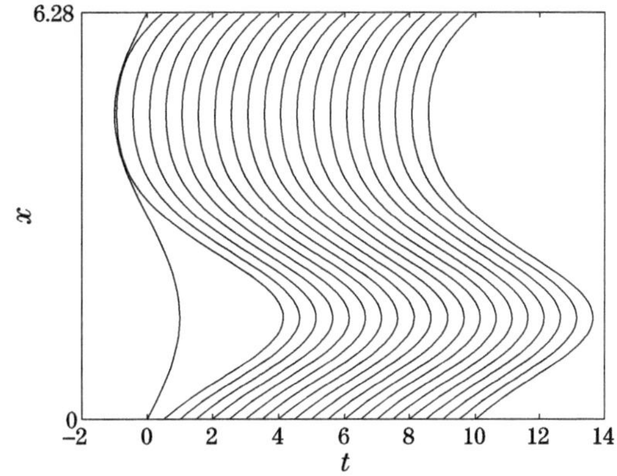

Fig. 3 Front propagation with $\beta = 10$, $\varepsilon = 0.001$ and $\psi_0(x) = \sin(x)$

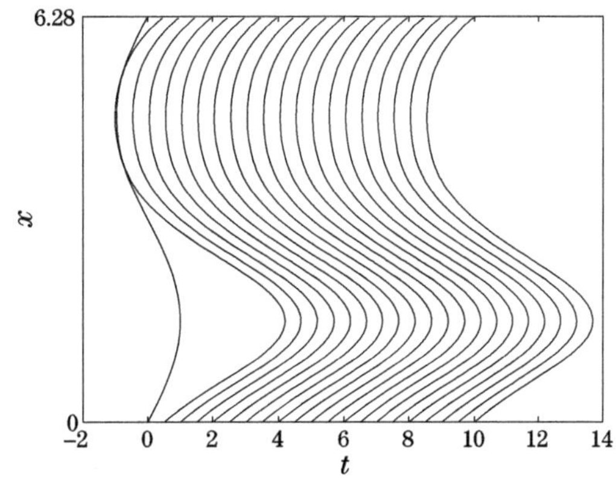

Fig. 4 Front propagation with $\beta = 10$, $\varepsilon = 0$ and $\psi_0(x) = \sin(x)$

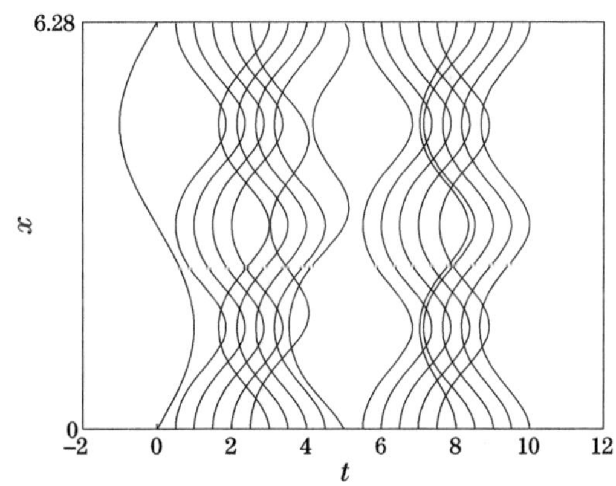

Fig. 5 Front propagation with $\beta = 20$, $\varepsilon = 0.1$ and $\psi_0(x) = \sin(x)$

puted by using (5.2), taking $\beta = 10$ and 20 respectively, and giving to ε the following values: 0.1, 0.01, 0.001 and 0 (which correspond to Eq. (K-S)).

We now investigate the dynamics of (5.1) with respect to the parameter β. For this purpose, we fix $\varepsilon = 0.001$.

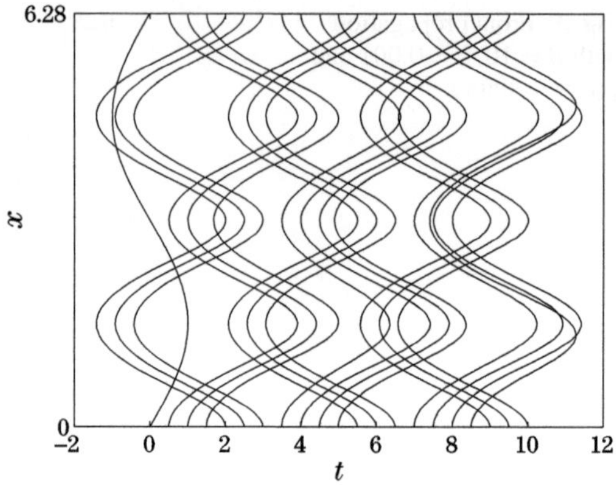

Fig. 6 Front propagation with $\beta = 20$, $\varepsilon = 0.01$ and $\psi_0(x) = \sin(x)$

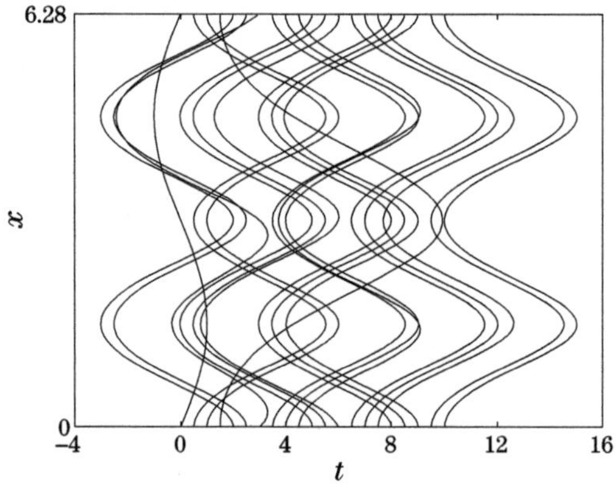

Fig. 7 Front propagation with $\beta = 20$, $\varepsilon = 0.001$ and $\psi_0(x) = \sin(x)$

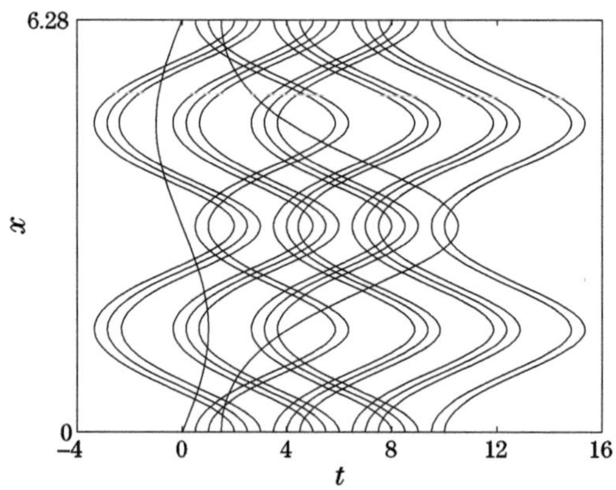

Fig. 8 Front propagation with $\beta = 20$, $\varepsilon = 0$ and $\psi_0(x) = \sin(x)$

The numerical simulations confirm that, as for Eq. (K-S), 0 turns out to be a global attractor for the solution to (5.1), for any $\beta \in [1, 4]$. A non-trivial attractor is expected for larger β's. In Figs. 9, 10, 11, 12, we can see the front evolutions

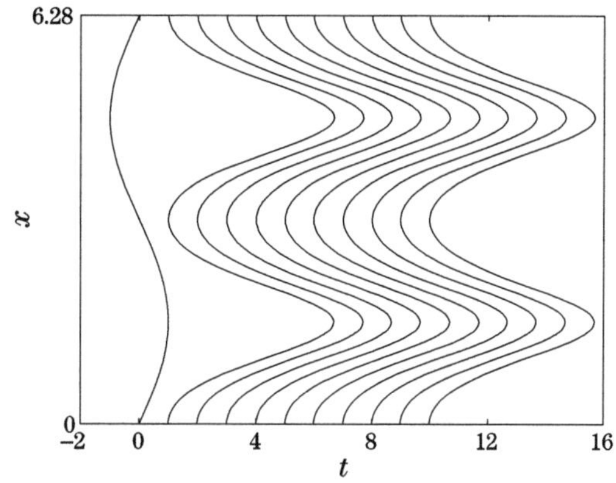

Fig. 9 Front propagation with $\beta = 30$, $\varepsilon = 0.001$ and $\psi_0(x) = \sin(x)$

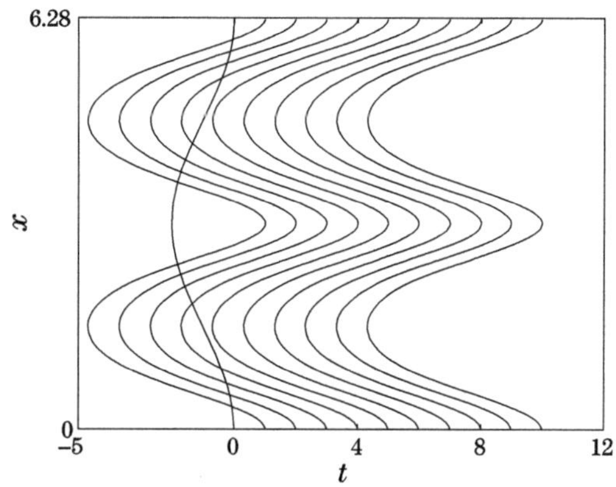

Fig. 10 Front propagation with $\beta = 30$, $\varepsilon = 0.001$ and $\psi_0(x) = \cos(x)$

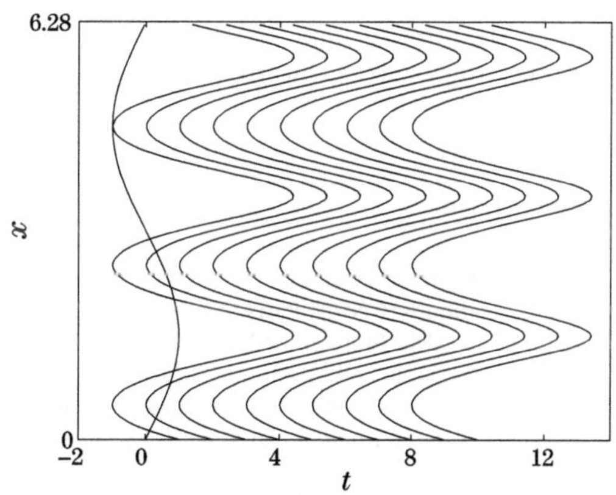

Fig. 11 Front propagation with $\beta = 60$, $\varepsilon = 0.001$ and $\psi_0(x) = \sin(x)$

generated by (5.2) with $\beta = 30, 60$ for two different initial conditions. In all the figures below, the periodic orbit is clearly observed.

Summing up, our numerical tests confirm that (5.1) preserves the same structure as Eq. (K-S). Larger β generates an even richer dynamics (see Fig. 13) where the

Fig. 12 Front propagation with $\beta = 60$, $\varepsilon = 0.001$, and $\psi_0(x) = \cos(x)$

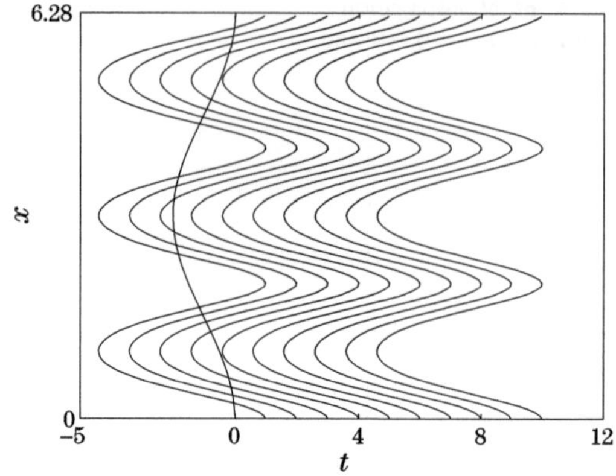

Fig. 13 Front propagation with $\beta = 108$, $\varepsilon = 0.0001$ and $\psi_0(x) = 0.1(\cos(x) + \cos(2x) + \cos(3x))$

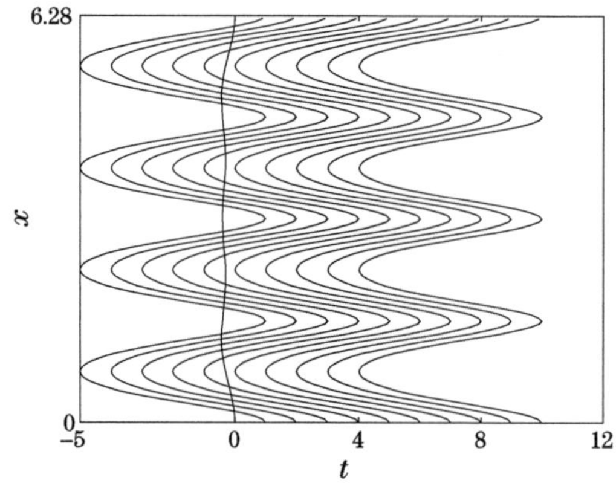

front propagation is captured from a computation with $\beta = 108$. As predicted in [12], the front evolves toward an essentially quadrimodal global attractor.

References

1. Berestycki, H., Brauner, C.-M., Clavin, P., et al.: Modélisation de la Combustion, Images des Mathématiques. CNRS, Paris (1996). Special Issue
2. Brauner, C.-M., Frankel, M.L., Hulshof, J., et al.: On the κ-θ model of cellular flames: existence in the large and asymptotics. Discrete Contin. Dyn. Syst., Ser. S **1**, 27–39 (2008)
3. Brauner, C.-M., Frankel, M.L., Hulshof, J., Sivashinsky, G.I.: Weakly nonlinear asymptotics of the κ-θ model of cellular flames: the Q-S equation. Interfaces Free Bound. **7**, 131–146 (2005)
4. Brauner, C.-M., Hulshof, J., Lorenzi, L.: Stability of the travelling wave in a 2D weakly nonlinear Stefan problem. Kinet. Relat. Models **2**, 109–134 (2009)
5. Brauner, C.-M., Hulshof, J., Lorenzi, L.: Rigorous derivation of the Kuramoto-Sivashinsky equation in a 2D weakly nonlinear Stefan problem. Interfaces Free Bound. **13**, 73–103 (2011)
6. Brauner, C.-M., Hulshof, J., Lorenzi, L., Sivashinsky, G.I.: A fully nonlinear equation for the flame front in a quasi-steady combustion model. Discrete Contin. Dyn. Syst., Ser. A **27**, 1415–1446 (2010)

7. Brauner, C.-M., Lorenzi, L., Sivashinsky, G.I., Xu, C.-J.: On a strongly damped wave equation for the flame front. Chin. Ann. Math. **31B**(6), 819–840 (2010)
8. Brauner, C.-M., Lunardi, A.: Instabilities in a two-dimensional combustion model with free boundary. Arch. Ration. Mech. Anal. **154**, 157–182 (2000)
9. Buckmaster, J.D., Ludford, G.S.S.: Theory of Laminar Flames. Cambridge University Press, Cambridge (1982)
10. Eckhaus, W.: Asymptotic Analysis of Singular Perturbations. Studies in Mathematics and Its Applications, vol. 9. North-Holland, Amsterdam (1979)
11. Haase, M.: The Functional Calculus for Sectorial Operators. Operator Theory: Advances and Applications, vol. 169. Birkhäuser, Basel (2006)
12. Hyman, J.M., Nicolaenko, B.: The Kuramoto-Sivashinsky equation: a bridge between PDEs and dynamical systems. Physica D **18**, 113–126 (1986)
13. Kagan, L., Sivashinsky, G.I.: Pattern formation in flame spread over thin solid fuels. Combust. Theory Model. **12**, 269–281 (2008)
14. Lions, J.-L.: Perturbations Singulières dans les Problèmes aux Limites et en Contrôle Optimal. Lect. Notes in Math., vol. 323. Springer, Berlin (1970)
15. Lorenzi, L.: Regularity and analyticity in a two-dimensional combustion model. Adv. Differ. Equ. **7**, 1343–1376 (2002)
16. Lorenzi, L.: A free boundary problem stemmed from combustion theory. I. Existence, uniqueness and regularity results. J. Math. Anal. Appl. **274**, 505–535 (2002)
17. Lorenzi, L.: A free boundary problem stemmed from combustion theory. II. Stability, instability and bifurcation results. J. Math. Anal. Appl. **275**, 131–160 (2002)
18. Lorenzi, L.: Bifurcation of codimension two in a combustion model. Adv. Math. Sci. Appl. **14**, 483–512 (2004)
19. Lorenzi, L., Lunardi, A.: Stability in a two-dimensional free boundary combustion model. Nonlinear Anal. **53**, 227–276 (2003)
20. Lorenzi, L., Lunardi, A.: Erratum: "Stability in a two-dimensional free boundary combustion model". Nonlinear Anal. **53**(6), 859–860 (2003). Nonlinear Anal. **53**(2), 227–276 (2003). MR1959814
21. Lunardi, A.: Analytic Semigroups and Optimal Regularity in Parabolic Problems. Birkhäuser, Basel (1995)
22. Matkowsky, B.J., Sivashinsky, G.I.: An asymptotic derivation of two models in flame theory associated with the constant density approximation. SIAM J. Appl. Math. **37**, 686–699 (1979)
23. Sivashinsky, G.I.: On flame propagation under conditions of stoichiometry. SIAM J. Appl. Math. **39**, 67–82 (1980)
24. Sivashinsky, G.I.: Instabilities, pattern formation and turbulence in flames. Annu. Rev. Fluid Mech. **15**, 179–199 (1983)
25. Temam, R.: Infinite-Dimensional Dynamical Systems in Mechanics and Physics, 2nd edn. Applied Mathematical Sciences, vol. 68. Springer, New York (1997)
26. Zik, O., Moses, E.: Fingering instability in combustion: an extended view. Phys. Rev. E **60**, 518–531 (1999)

Implicit Sampling, with Application to Data Assimilation

Alexandre J. Chorin, Matthias Morzfeld, and Xuemin Tu

Abstract There are many computational tasks in which it is necessary to sample a given probability density function (or pdf for short), i.e., to use a computer to construct a sequence of independent random vectors x_i $(i = 1, 2, \ldots)$, whose histogram converges to the given pdf. This can be difficult because the sample space can be huge, and more importantly, because the portion of the space where the density is significant, can be very small, so that one may miss it by an ill-designed sampling scheme. Indeed, Markov-chain Monte Carlo, the most widely used sampling scheme, can be thought of as a search algorithm, where one starts at an arbitrary point and one advances step-by-step towards the high probability region of the space. This can be expensive, in particular because one is typically interested in independent samples, while the chain has a memory. The authors present an alternative, in which samples are found by solving an algebraic equation with a random right-hand side rather than by following a chain; each sample is independent of the previous samples. The construction is explained in the context of numerical integration, and it is then applied to data assimilation.

Keywords Importance sampling · Bayesian estimation · Particle filter · Implicit filter · Data assimilation

---

Project supported by the Director, Office of Science, Computational and Technology Research, US Department of Energy (No. DE-AC02-05CH11231) and the National Science Foundation (Nos. DMS-1217065, OCE-0934298).

A.J. Chorin (✉)
Department of Mathematics, University of California, Berkeley, CA 94720, USA
e-mail: chorin@math.berkeley.edu

A.J. Chorin · M. Morzfeld
Lawrence Berkeley National Laboratory, Berkeley, CA 94720, USA

M. Morzfeld
e-mail: mmo@math.lbl.gov

X. Tu
Department of Mathematics, University of Kansas, Lawrence, KS 66045, USA
e-mail: xtu@math.ku.edu

P.G. Ciarlet et al. (eds.), *Partial Differential Equations: Theory, Control and Approximation*, DOI 10.1007/978-3-642-41401-5_6,

Mathematics Subject Classification 11K45 · 34K60 · 62M20 · 65C50 · 93E11

1 Implicit Sampling

Suppose that one wants to evaluate the integral

$$I = \int g(x) f(x) \mathrm{d}x,$$

where x is a vector variable, and $f(x)$ is a probability density function (or pdf for short). If the dimension of x is large, it is natural to do so by Monte Carlo. Write $I = E[g(x)]$, where $E[\cdot]$ denotes an expected value and x is a random variable whose pdf is $f(x)$, $x \sim f(x)$. The integral can then be approximated through the law of large numbers,

$$I \approx I_n = \frac{1}{n} \sum_{j=1}^{n} g(X_j),$$

where the X_i are n independent samples of the pdf f, and the error is proportional to $n^{-\frac{1}{2}}$ (see [1, 2]).

To perform this calculation, one has to find samples X_j of a given pdf f, which is often difficult. One way to proceed is to find an "importance" density f_0, whose support contains the support of f, and which is easier to sample. Write

$$I = \int g(x) \frac{f(x)}{f_0(x)} f_0(x) \mathrm{d}x = E\big[g(x) w(x)\big],$$

where

$$w(x) = \frac{f(x)}{f_0(x)}$$

is a "sampling weight" and $x \sim f_0(x)$. We can approximate this integral through the law of large numbers as above, so that

$$I_n = \frac{1}{n} \sum_{j=1}^{n} g(X_j) w(X_j)$$

converges almost surely to I as $n \to \infty$. One requirement for this to be a practical computing scheme is that the ratio $\frac{f}{f_0}$ be close to a constant, and in particular, that f_0 be large where f is large; otherwise, one wastes one's efforts on samples that contribute little to the result. However, one may not know in advance where f is large—indeed, in the application to data assimilation below, the whole purpose of the computation is to identify the set where f is large.

We now propose a construction that makes it possible to find a suitable importance density under quite general conditions. Write

$$F(x) = -\log f(x),$$

and suppose for the moment that F is convex. Pick a reference variable ξ such that (i) ξ is easy to sample, (ii) its pdf $g(\xi)$ has a maximum at $\xi = 0$, (iii) the logarithm of g is convex, (iv) it is possible to write the variable with pdf f as a function of ξ. It is often convenient to pick ξ as a unit Gaussian variable, $\xi \sim \mathcal{N}(0, I)$, where I is the identity, and $\mathcal{N}(\mu, \Sigma)$ denotes a Gaussian with mean μ and covariance matrix Σ, and we will do so here. This choice does not imply any Gaussianity assumption for the pdf f we wish to sample.

Then proceed as follows: find

$$\phi = \min F,$$

the minimum of F, and pick a sequence of independent samples $\xi \sim \mathcal{N}(0, I)$. For each one, solve the equation

$$F(X) - \phi = \frac{1}{2}\xi^{\mathrm{T}}\xi, \tag{1.1}$$

i.e., equate the logarithm of f, the pdf to be sampled, to the logarithm of the pdf of the reference variable, after subtracting ϕ, the minimum of F. Subtracting ϕ ensures that solutions exist. Pick the solutions so that the map $\xi \to x$ is one-to-one and onto. The resulting samples X are independent, because the samples ξ of the reference density are independent. This is in contrast to Markov-chain Monte Carlo schemes, where the successive samples are dependent. Moreover, under the assumptions on ξ, most of the samples of ξ are close to the origin; the corresponding samples X are near the minimizer of F, and therefore near the mode of f. The minimization of F guides the samples of x to where the probability is high.

It is important to note that this construction can be carried out even if the pdf of x is not known explicitly, as long as one can evaluate f for each value of its argument up to a multiplicative constant. The normalization of f need not be known because a multiplicative factor in f becomes an additive factor in $F = -\log f$, and cancels out when the minimum ϕ is subtracted.

To calculate the sampling weight, note that, on one hand, Eq. (1.1) yields $f(x) = \mathrm{e}^{-\phi}g(\xi)$, where g is the Gaussian $\mathcal{N}(0, I)$. On the other hand, by the change of variable theorem for integrals, the pdf of x is $\frac{g(\xi)}{J}$, where J is the Jacobian of the map $\xi \to x$. The sampling weight is therefore

$$w \propto \mathrm{e}^{-\phi} J.$$

The assumption that F is convex is too strong. Nothing changes if F is merely U-shaped. A function f of a single scalar variable is U-shaped if it has a single minimum ϕ, has no local maxima or inflection points, and tends to ∞ as $|x| \to \infty$. A function of many variables is U-shaped if the intersection of its graph with every vertical plane through its minimum is U-shaped. If F is not U-shaped, the construction above can still be carried out. Often one can write F as a union of U-shaped functions with disjoint supports, and then a compound reference density directs the samples to the various pieces in turn. One can also approximate F by an approximation of its convex hull. For example, one can expand F around its minimizer $m = \operatorname{argmin} F$ (i.e. $F(m) = \phi$),

$$F = \phi + \frac{1}{2}(x - m)^{\mathrm{T}} H(x - m) + \cdots,$$

where a superscript T denotes a transpose, and H is the Hessian of F which may be left over from the minimization that produced ϕ. One defines

$$F_0 = \phi + \frac{1}{2}(x - m)^{\mathrm{T}} H(x - m),$$

and replaces F by F_0 in Eq. (1.1), so that it becomes

$$(x - m)^{\mathrm{T}} H(x - m) = \xi^{\mathrm{T}} \xi,$$

where the left-hand side is now convex. This still maps the neighborhood of the maximum of g onto the neighborhood of the maximum of f. The sampling weight becomes $w \propto \mathrm{e}^{-\phi_0} J$, where $\phi_0 = F(x) - F_0(x)$.

There remains the task of solving Eq. (1.1) and evaluating the Jacobian J. How onerous this task is depends on the problem. Observe that Eq. (1.1) is a single equation while the vector x has many components, so that there are many solutions, but only one is needed. One may, for example, look for a solution in a random direction, reducing the problem of solving Eq. (1.1) to a scalar problem and greatly simplifying the evaluation of J. This is a "random map" implementation of implicit sampling (for details, see [3, 4]).

One may worry that the minimization that produces ϕ may be expensive. However, any successful sampling in a multi-dimensional problem requires a search for high probability areas and, therefore, includes an unacknowledged maximization of a pdf. One may as well do this maximization consciously and bring to bear the tools that make it efficient (see also the comparison with variational methods below).

2 Filtering and Data Assimilation

There are many problems in science and engineering, where one wants to identify the state of a system from an uncertain model supplemented by a stream of noisy and incomplete data. An example of this situation is shown in Fig. 1.

Imagine that a ship sank in the Pacific ocean. Its passengers are floating in a dinghy, and you are the coast guard and want to send a navy ship to the rescue. A model of the currents and winds in the ocean makes it possible to draw possible trajectories, but these are uncertain. A ham radio operator spoke to someone in the dinghy several times, but could not locate it without error. These are the data. The most likely position of the dinghy is somewhere between the trajectories and the observations. Note that the location of the highest probability area is the unknown.

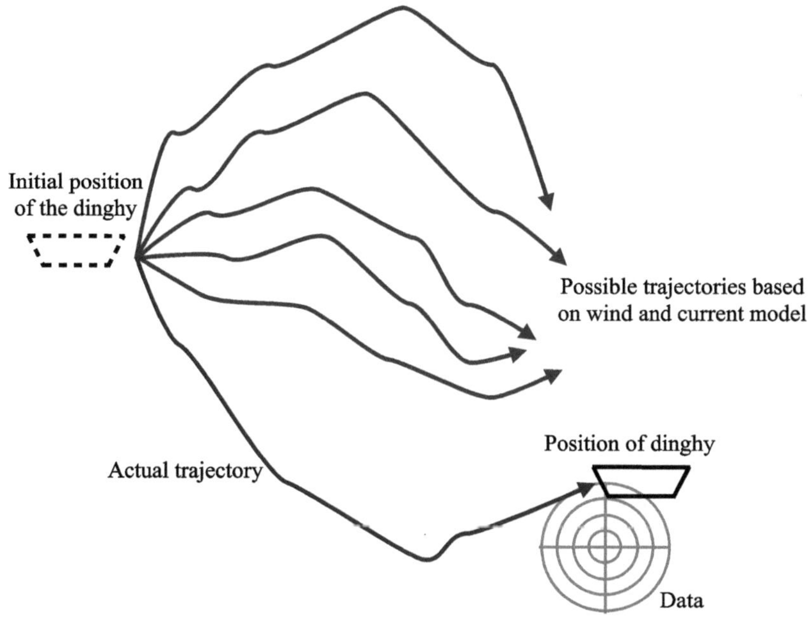

Fig. 1 A dinghy in the Pacific ocean: the floating passengers can be located by combining the information from an uncertain model of the currents and winds with the information from a ham radio operator

In mathematical terms, the model is often a Markov state space model (often a discretization of a stochastic differential equation (or SDE for short) (see [5])) and describes the state sequence $\{x^n; n \in N\}$, where x^n is a real, m-dimensional vector. To simplify notations, we assume here that the noise is additive, so that the model equations are

$$x^n = f^n\left(x^{n-1}\right) + v^{n-1}, \tag{2.1}$$

where f^n is an m-dimensional vector function, and $\{v^{n-1}, n \in N\}$ is a sequence of independent identical distributed (or i.i.d. for short) m-dimensional random vectors which, in many applications, are Gaussian vectors with independent components. One can think of the x^n as values of a process $x(t)$ evaluated at times $n\delta$, where δ is a fixed time increment. The probability density function of the initial state x^0 is assumed to be known.

The model is supplemented by an observation (or measurement) equation, which relates observations $\{b^n; n \in N\}$, where b^n is a real, k-dimensional vector and $k \leq m$, to the states x^n. We assume here that the observation equation is

$$b^n = h^n\left(x^n\right) + z^n, \tag{2.2}$$

where h^n is a k-dimensional, possibly nonlinear, vector function, and $\{z^n, n \in N\}$ is a k-dimensional i.i.d. process, independent of v^n. The model and the observation equations together constitute a hidden Markov state space model. To streamline notation, we denote the state and observation sequences up to time n by

$$x^{0:n} = \{x^0, \ldots, x^n\} \quad \text{and} \quad b^{1:n} = \{b^1, \ldots, b^n\},$$

respectively.

The goal is to estimate the sequence $x^{0:n}$, based on (2.1) and (2.2). This is known as "filtering" or "data assimilation". We compute the estimate by sequential Monte Carlo, i.e., by sampling sequentially from the conditional pdf $p(x^{0:n} \mid b^{1:n})$ (called the target pdf), and using these samples to approximate the conditional mean (the minimum mean square error estimator (see [2])) by the weighted sample mean. We do this by following "particles" (replicas of the system) whose empirical distribution weakly approximates the target density. For simplicity of presentation, we assume that the model equation (2.1) is synchronized with the observations (2.2), i.e. observations b^n are available at every model step (see [3] for an extension to the case where observations are sparse in time). Using Bayes' rule and the Markov property of the model, we obtain the recursion

$$p\left(x^{0:n+1} \mid b^{1:n+1}\right) = \frac{p(x^{0:n} \mid b^{1:n}) p(x^{n+1} \mid x^n) p(b^{n+1} \mid x^{n+1})}{p(b^{n+1} \mid b^{1:n})}. \tag{2.3}$$

At the current time $t = n+1$, the first term in the numerator of the right-hand side of (2.3) is known from the previous steps. The denominator is common to all particles and thus drops out in the importance sampling scheme (where the weights are normalized, so that their sum equals 1). All we have to do is sampling the right-hand side of this expression at every step and for every particle. We do that by implicit sampling, which is indifferent to all the factors on the right-hand side other than $p(x^{n+1} \mid x^n) p(b^{n+1} \mid x^{n+1})$ (see also [3, 4, 6, 7]). The factor $p(x^{n+1} \mid x^n)$ is determined by the model (2.1), while the factor $p(b^{n+1} \mid x^{n+1})$ represents the effect of the observation (2.2). We supplement the sampling by a resampling after each step which equalizes the weights, and gets rid of the factor $p(x^{0:n} \mid b^{1:n})$ and of many of the particles with small weights (see [1, 8] for efficient resampling algorithms).

We claim that the use of implicit sampling in data assimilation makes it possible to improve on what other algorithms can do in this problem. We therefore compare the implicit sampling algorithm with other methods of data assimilation in common use.

3 Comparisons with Other Data Assimilation Algorithms

3.1 *The Standard Bayesian Filter*

Suppose that the observations are highly consistent with the SDE—for example, in Fig. 1, the observations may be somewhere in the middle of the pencil of solutions to the model. There is really no need to explicitly look for the maximum of the pdf, because the model (2.1) already generates samples that are in the high probability region. Therefore, one can set $\phi = \log p(b^{n+1} \mid x^{n+1})$ in Eq. (1.1), and then solving (1.1) is simply sampling a new location determined by the SDE, to which one subsequently assigns a weight determined by the proximity of the sample X to the observation. This sampling scheme is often called the sequential importance sampling with a resampling (or SIR for short) filter. The SIR filter is widely used, and is less expensive than what we propose, but may fail if the data are not close to what the model alone would predict. As the dimension of the vector x increases, the neighborhood of the observations and the pencil of solutions to the SDE occupy an ever decreasing fraction of the available space, so that with SIR, guaranteeing that at least a few samples hit the high probability area requires more and more samples (see [9, 10]). In contrast, with implicit sampling, the observations affect not only the weights of the samples but also their locations. For more on the SIR, see [1, 8, 11–14].

3.2 Optimal Filters

There is a literature on "optimal" particle filters, defined as particle filters in which the variance of the weights of each particular particle (not the variance of all the weights) is zero (see [8, 12, 15]). In general, a filter that is "optimal" in this sense requires a precise knowledge of the normalization of the pdf to be sampled, which is not usually available (see the formulas for the pdf to be sampled, remembering that $\int f \mathrm{d}x = 1$, $\int g \mathrm{d}x = 1$, do not imply that $\int fg \mathrm{d}x = 1$.)

To see why in general the optimal filter can not be implemented without knowing the normalization constants exactly, consider first the problem of sampling a given pdf f, and carry out the following construction (in one dimension for simplicity): let $g(\xi)$ be the pdf of a reference variable ξ. Define $F = -\log f$ as before and find the region of high probability through minimization of F, i.e. compute $m = \operatorname{argmin} F$. To find a sample X, solve the differential equation $f\mathrm{d}x = g\mathrm{d}s$, or

$$\frac{\mathrm{d}x}{\mathrm{d}s} = \frac{g}{f}$$

with the initial condition $x(0) = m$, for $s \in (0, \xi]$. This defines a map $\xi \to x(\xi)$ with $f(x) = g(\xi)J(\xi)$, where $J = |\frac{\mathrm{d}s}{\mathrm{d}x}|$. One can check that the weight is independent of the sample. This sampling scheme fails unless one knows the normalization constant with perfect accuracy, because if one multiplies f in the differential equation by a constant, the resulting samples are not distributed correctly.

In the data assimilation problem one has to sample a different pdf for each particle, so that the application of this sampling scheme yields an "optimal filter" with a zero-variance weight for each particle, provided that one can calculate the normalization constants exactly, which can be done at an acceptable cost only in special cases. In those special cases, the resulting filter coincides with our implicit filter.

The implicit filter avoids the problem of unknown normalization constants by taking logs, converting a harmful unknown multiplicative constant in the pdf into a harmless additive constant.

3.3 The Kalman Filter

If the observation function h is linear, the model (2.1) is linear, the initial data are either constant or Gaussian, and the observation noise z^n in (2.2) is Gaussian, then the pdf we are sampling is Gaussian and is entirely determined by its mean and covariance. It is easy to see that in this case a single particle suffices in the implicit filter, and that one gets the best results by setting $\xi = 0$ in the formulas above. The resulting filter is the Kalman filter (see [16, 17]).

3.4 The Ensemble Kalman Filter

The ensemble Kalman filter (see [18]) estimates a pdf for the SDE by a Monte Carlo solution to a Fokker-Planck equation, extracts from this solution a Gaussian approximation, and then takes the data into account by an (approximate) Kalman filter step. The implicit filter on the other hand can be viewed as a Monte Carlo solution to the Zakai equation (see [19]) for the conditional probability $p(x^{0:n} \mid b^{1:n})$, doing away with the need for an expensive and approximate Kalman step.

3.5 Variational Data Assimilation

There is a significant literature on variational data assimilation methods (see [20–25]), where one makes an estimate by maximizing some objective function of the estimate. Clearly the computation of $\phi = \min F$ above resembles a variational estimate. One can view implicit sampling as a sampling scheme added to a variational estimate. The added cost is small, while the advantages are a better estimate (a least square estimate rather than a maximum likelihood estimate, which is particularly important when the pdf's are not symmetric), and the addition of error estimates, which come naturally with a particle filter but are hard to obtain with a variational estimate. For a thorough discussion, see [26].

4 An Example

As an example, we present a data assimilation calculation for the stochastic Kuramoto-Sivashinksy (or SKS for short) equation presented earlier in [3],

$$u_t + uu_x + u_{xx} + \nu u_{xxxx} = gW(x,t),$$

where $\nu > 0$ is the viscosity, g is a scalar, and $W(x,t)$ is a space-time white noise process. The SKS equation is a chaotic stochastic partial differential equation that has been used to model laminar flames and reaction-diffusion systems (see [27, 28]) and recently, has also been used as a large dimensional test problem for data assimilation algorithms (see [29, 30]).

We consider the m-dimensional Itô-Galerkin approximation of the SKS equation

$$\mathrm{d}U = \big(\mathcal{L}(U) + \mathcal{N}(U)\big)\mathrm{d}t + g\mathrm{d}W_t^m,$$

where U is a finite dimensional column vector whose components are the Fourier coefficients of the solution and W_t^m is a truncated cylindrical Brownian motion (see [31]), obtained from the projection of the noise process $W(x,t)$ onto the Fourier modes. Assuming that the initial conditions $u(x,0)$ are odd with $\widetilde{U}_0(0) = 0$ and that g is imaginary, all Fourier coefficients $U_k(t)$ are imaginary for all $t \geq 0$. Writing $U_k = i\widehat{U}_k$ and subsequently dropping the hat gives

$$\mathcal{L}(U) = \mathrm{diag}\big(\omega_k^2 - \nu\omega_k^4\big)U,$$

$$\big\{\mathcal{N}(U)\big\}_k = -\frac{\omega_k}{2}\sum_{k'=-m}^{m} U_{k'}U_{k-k'},$$

where $\omega_k = \frac{2\pi k}{L}$ $(k = 1, \ldots, m)$, and $\{\mathcal{N}(U)\}_k$ denotes the k-th element of the vector $\mathcal{N}(U)$. We choose a period $L = 16\pi$ and a viscosity $\nu = 0.251$, to obtain SKS equations with 31 linearly unstable modes. This set-up is similar to the SKS equation considered in [30]. With these parameter values there is no steady state as in [29]. We choose zero initial conditions $U(0) = 0$, so that the solution evolves solely due to the effects of the noise. To approximate the SKS equation, we keep $m = 512$ of the Fourier coefficients and use the exponential Euler scheme (see [32]), with time step $\delta = 2^{-12}$ for time discretization (see [3] for details).

We are solving the SKS equations in Fourier variables, but we choose to observe in a physical space (as may be physically reasonable). Specifically, we observe the solution $u(x,t)$ at $\frac{m}{2}$ equidistant locations and at every model step through the nonlinear observation operator $h(x) = x + x^3$. The minimization of F_j was done by using Newton's method (see [33, 34]), initialized by a model run without noise. To obtain samples, we solve the algebraic equation (1.1), which is easy when the functions F_j are nearly diagonal, i.e., when the linearizations around a current state are nearly diagonal matrices. This requires in particular that the variables that are observed coincide with the variables that are evolved by the dynamics. Observing in physical space while computing in Fourier space creates the opposite situation, in which each observation is related to the variables one computes by a dense matrix. This problem was overcome by using the random map algorithm, presented in [3], for solving (1.1).

To test the resulting filter, we generated data by running the model, and then compared the results obtained by the filter with these data. This procedure is called a "twin experiment" and we define, for each twin experiment, the error at time t^n as

$$e^n = \|U_{\mathrm{ref}}^n - U_F^n\|,$$

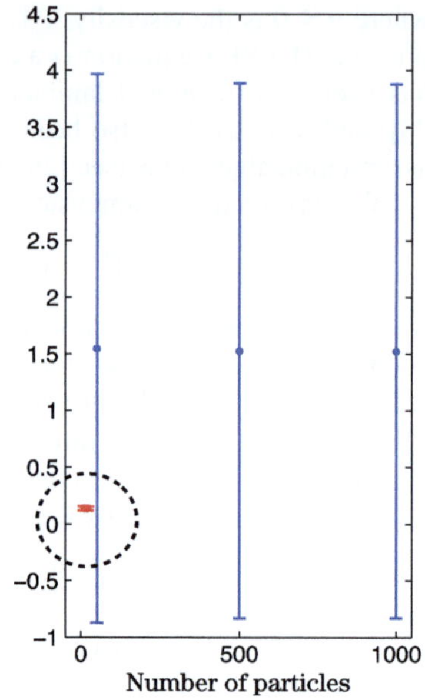

Fig. 2 Filtering results for the SKS equation: the error statistics are shown as a function of the number of particles for the SIR filter (*blue*) and the implicit particle filter (*red*). The error bars represent the mean of the errors and mean of the standard deviations of the errors

where the norm is the Euclidean norm, U^n_{ref} denotes the set of Fourier coefficients of the reference run and U^n_F denotes the reconstruction by the filter, both at the fixed time t^n. The error statistics of 500 twin experiments are shown in Fig. 2.

We observe from Fig. 2 that the implicit particle filter produces accurate state estimates (small errors and small error variances) with a small number of particles. The SIR filter on the other hand requires thousands of particles to achieve a similar accuracy and therefore, is impractical for filtering the SKS equation.

5 Conclusions

We have presented an importance sampling procedure in which the importance density is defined implicitly through a mapping guided by a minimization rather than be given by an explicit formula. This makes it possible to sample effectively a variety of pdfs that are otherwise difficult to work with. In particular, in the data assimilation problem, implicit sampling makes it possible to incorporate the information in the data into the sampling process, so that the target density is sampled efficiently. We are confident that this construction will find wide applicability in the sciences.

References

1. Doucet, A., de Freitas, N., Gordon, N.: Sequential Monte Carlo Methods in Practice. Springer, New York (2001)

2. Chorin, A.J., Hald, O.H.: Stochastic Tools in Mathematics and Science, 2nd edn. Springer, New York (2009)
3. Morzfeld, M., Tu, X., Atkins, E., Chorin, A.J.: A random map implementation of implicit filters. J. Comput. Phys. **231**, 2049–2066 (2012)
4. Morzfeld, M., Chorin, A.J.: Implicit particle filtering for models with partial noise, and an application to geomagnetic data assimilation. Nonlinear Process. Geophys. **19**, 365–382 (2012)
5. Kloeden, P.E., Platen, E.: Numerical Solution of Stochastic Differential Equations, 3rd edn. Springer, New York (1999)
6. Chorin, A.J., Tu, X.: Implicit sampling for particle filters. Proc. Natl. Acad. Sci. USA **106**, 17249–17254 (2009)
7. Chorin, A.J., Morzfeld, M., Tu, X.: Implicit particle filters for data assimilation. Commun. Appl. Math. Comput. Sci. **5**(2), 221–240 (2010)
8. Arulampalam, M.S., Maskell, S., Gordon, N., Clapp, T.: A tutorial on particle filters for online nonlinear/non-Gaussian Bayesian tracking. IEEE Trans. Signal Process. **10**, 197–208 (2002)
9. Bickel, P., Li, B., Bengtsson, T.: Sharp failure rates for the bootstrap particle filter in high dimensions. In: Pushing the Limits of Contemporary Statistics: Contributions in Honor of Jayanta K. Ghosh, pp. 318–329 (2008)
10. Snyder, C.C., Bengtsson, T., Bickel, P., Anderson, J.: Obstacles to high-dimensional particle filtering. Mon. Weather Rev. **136**, 4629–4640 (2008)
11. Gordon, N.J., Salmon, D.J., Smith, A.F.M.: Novel approach to nonlinear/non-Gaussian Bayesian state estimation. IEE Proc., F, Radar Signal Process. **140**, 107–113 (1993)
12. Doucet, A., Godsill, S., Andrieu, C.: On sequential Monte Carlo sampling methods for Bayesian filtering. Stat. Comput. **50**, 174–188 (2000)
13. Del Moral, P.: Feynman-Kac Formulae. Springer, New York (2004)
14. Del Moral, P.: Measure-valued processes and interacting particle systems. Application to nonlinear filtering problems. Ann. Appl. Probab. **8**(2), 438–495 (1998)
15. Zaritskii, V.S., Shimelevich, L.I.: Monte Carlo technique in problems of optimal data processing. Autom. Remote Control **12**, 95–103 (1975)
16. Kalman, R.E.: A new approach to linear filtering and prediction theory. J. Basic Eng. **82**, 35–48 (1960)
17. Kalman, R.E., Bucy, R.S.: New results in linear filtering and prediction theory. J. Basic Eng. **83**, 95–108 (1961)
18. Evensen, G.: Data Assimilation. Springer, New York (2007)
19. Zakai, M.: On the optimal filtering of diffusion processes. Z. Wahrscheinlichkeitstheor. Verw. Geb. **11**, 230–243 (1969)
20. Talagrand, O., Courtier, P.: Variational assimilation of meteorological observations with the adjoint vorticity equation. I. Theory. Q. J. R. Meteorol. Soc. **113**, 1311–1328 (1987)
21. Bennet, A.F., Leslie, L.M., Hagelberg, C.R., Powers, P.E.: A cyclone prediction using a barotropic model initialized by a general inverse method. Mon. Weather Rev. **121**, 1714–1728 (1993)
22. Courtier, P., Thepaut, J.N., Hollingsworth, A.: A strategy for operational implementation of 4D-var, using an incremental appoach. Q. J. R. Meteorol. Soc. **120**, 1367–1387 (1994)
23. Courtier, P.: Dual formulation of four-dimensional variational assimilation. Q. J. R. Meteorol. Soc. **123**, 2449–2461 (1997)
24. Talagrand, O.: Assimilation of observations, an introduction. J. Meteorol. Soc. Jpn. **75**(1), 191–209 (1997)
25. Tremolet, Y.: Accounting for an imperfect model in 4D-var. Q. J. R. Meteorol. Soc. **621**(132), 2483–2504 (2006)
26. Atkins, E., Morzfeld, M., Chorin, A.J.: Implicit particle methods and their connection to variational data assimilation. Mon. Weather Rev. (2013, in press)
27. Kuramoto, Y., Tsuzuki, T.: On the formation of dissipative structures in reaction-diffusion systems. Prog. Theor. Phys. **54**, 687–699 (1975)
28. Sivashinsky, G.: Nonlinear analysis of hydrodynamic instability in laminar flames. Part I. Derivation of basic equations. Acta Astronaut. **4**, 1177–1206 (1977)

29. Chorin, A.J., Krause, P.: Dimensional reduction for a Bayesian filter. Proc. Natl. Acad. Sci. USA **101**, 15013–15017 (2004)
30. Jardak, M., Navon, I.M., Zupanski, M.: Comparison of sequential data assimilation methods for the Kuramoto-Sivashinsky equation. Int. J. Numer. Methods Fluids **62**, 374–402 (2009)
31. Lord, G.J., Rougemont, J.: A numerical scheme for stochastic PDEs with Gevrey regularity. IMA J. Numer. Anal. **24**, 587–604 (2004)
32. Jentzen, A., Kloeden, P.E.: Overcoming the order barrier in the numerical approximation of stochastic partial differential equations with additive space-time noise. Proc. R. Soc. A **465**, 649–667 (2009)
33. Fletcher, R.: Practical Methods of Optimization. Wiley, New York (1987)
34. Nocedal, J., Wright, S.T.: Numerical Optimization, 2nd edn. Springer, New York (2006)

Periodic Homogenization for Inner Boundary Conditions with Equi-valued Surfaces: The Unfolding Approach

Doina Cioranescu, Alain Damlamian, and Tatsien Li

Abstract Making use of the periodic unfolding method, the authors give an elementary proof for the periodic homogenization of the elastic torsion problem of an infinite 3-dimensional rod with a multiply-connected cross section as well as for the general electro-conductivity problem in the presence of many perfect conductors (arising in resistivity well-logging). Both problems fall into the general setting of equi-valued surfaces with corresponding assigned total fluxes. The unfolding method also gives a general corrector result for these problems.

Keywords Periodic homogenization · Elastic torsion · Equi-valued surfaces · Resistivity well-logging · Periodic unfolding method

Mathematics Subject Classification 35B27 · 74Q05 · 74E30 · 74Q15 · 35J25 · 35Q72

Supported by the National Natural Science Foundation of China (No. 11121101) and the National Basic Research Program of China (No. 2013CB834100).

D. Cioranescu (✉)
Laboratoire Jacques-Louis Lions (UMR 7598 du CNRS), Université Pierre et Marie Curie, 4 Place Jussieu, 75005 Paris, France
e-mail: cioran@ann.jussieu.fr

A. Damlamian
Laboratoire d'Analyse et de Mathématiques Appliquées, CNRS UMR 8050 Centre Multidisciplinaire de Créteil, Université Paris-Est, 94010 Créteil Cedex, France
e-mail: damla@u-pec.fr

T. Li
Nonlinear Mathematical Modeling and Methods Laboratory, Shanghai Key Laboratory for Contemporary Applied Mathematics, School of Mathematical Sciences, Fudan University, Shanghai 200433, China
e-mail: dqli@fudan.edu.cn

P.G. Ciarlet et al. (eds.), *Partial Differential Equations: Theory, Control and Approximation*, DOI 10.1007/978-3-642-41401-5_7,

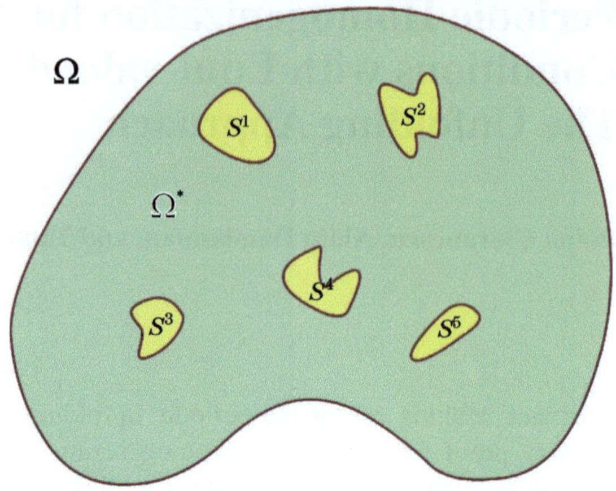

Fig. 1 A 2-dimensional cross-section Ω^* obtained from a bounded open set Ω perforated by a finite number of regular closed subsets (which have a nonempty interior) $S^1, S^2, \dots$

1 Introduction

The periodic unfolding method was introduced in [4] (see also [5]). It gave an elementary proof for the classical periodic homogenization problem, including the case with several micro-scales (a detailed account and proofs can be found in [5]).

In this paper, we show how it can be applied to the periodic homogenization of the general problem of equi-valued surfaces with corresponding assigned total fluxes. Two examples of this type of problems are the elastic torsion problem of an infinite 3-dimensional rod with a multiply-connected cross section (where the equations are set in a 2-dimensional domain) and, in any dimension, the electro-conductivity problem in the presence of many isolated perfect conductors.

In the linear elastic torsion problem (see [13, 15] for the setting of the problem), the material is an infinite cylindrical bar with a 2-dimensional cross-section Ω^* obtained from a bounded open set Ω perforated by a finite number of regular closed subsets (which have a nonempty interior) $S^1, S^2, \dots$ (see Fig. 1). The stress function of the elastic material is shown to be the solution to the following problem:

$$\begin{aligned}
&\varphi \in H_0^1(\Omega) \quad \text{with } f_{|S^j} \text{ (an unknown constant) for each } j,\\
&-\Delta\varphi = 1 \quad \text{in } \Omega^*,\\
&\int_{\partial S^j} \frac{\partial\varphi}{\partial n} d\sigma(x) = |S^j| \quad \text{(the measure of } S^j\text{) for each } j.
\end{aligned} \tag{1.1}$$

The electric conductivity problem arising in resistivity well-logging is set in any dimension with the same type of geometry (Ω, Ω^* and S^j's). The conductivity tensor A can vary with the position in Ω^*, the right-hand side is an L^2 function f

defined on Ω^*, and the total fluxes on the ∂S^j are given numbers g^j.

$$\begin{aligned} &\varphi \in H_0^1(\Omega) \quad \text{with } f_{|S^j} \text{ (an unknown constant) for each } j, \\ &-\operatorname{div}\big(A(x)\nabla\varphi(x)\big) = f \quad \text{in } \Omega^*, \\ &\int_{\partial S^j} \frac{\partial\varphi}{\partial\nu_A} \mathrm{d}\sigma(x) = g^j \quad \text{for each } j. \end{aligned} \tag{1.2}$$

Here we refer to [14, 15] written by Li et al. on the subject. They also include an exposé of the torsion problem.

Here, we consider the periodic homogenization for these problems. We refer to [7] for the first proof of the elastic torsion problem (via extension operators and oscillating test functions), where regularity assumptions are made for the boundary of the inclusions. In [3], this question is addressed, but still with some geometric conditions and by the same use of oscillating test functions. Moreover, some related results can be found in [1, 2, 8, 10–12].

One advantage of the unfolding method is that it requires no regularity for the boundary of the inclusions whatsoever. Actually, there is no need to introduce surface integrals, except if one wants to see the "usual" strong formulation, valid for Lipschitz boundaries. Another advantage of the method is that an immediate consequence is a corrector result which is completely general (without the need for extra regularity).

The plan is as follows. In Sect. 1, we introduce the notations, and set the approximate problem as one which encompasses both the aforementioned problems. Section 2 gives a brief summary of the results of the periodic unfolding method. In Sect. 3, we establish the convergence to the unfolded problem. In Sect. 4, we obtain the homogenized limit. Section 5 is devoted to the convergence of the energy and the construction of correctors. In the last section, we consider variants of the problem, and state the corresponding results.

General Notations (1) In this work, ε indicates the generic element of a bounded subset of $\mathbb{R}_+^*$ in the closure of which 0 lies. Convergence of ε to 0 is understood in this set. Also, c and C denote generic constants, which do not depend upon ε.

(2) As usual, 1_D denotes the characteristic function of the set D.

(3) For a measurable set D in $\mathbb{R}^n$, $|D|$ denotes its Lebesgue measure.

(4) For simplicity, the notation $L^p(\mathcal{O})$ will be used for both scalar and vector-valued functions defined on the set $\mathcal{O}$, since no ambiguity will arise.

2 Setting of the Problem

We use the general framework of [5] and the notations therein.

Let $\mathbf{b} = (b_1, \dots, b_n)$ be a basis of $\mathbb{R}^n$. We denote by

$$\mathcal{G} = \left\{ \xi \in \mathbb{R}^n \mid \xi = \sum_{i=1}^n k_i b_i, (k_1, \dots, k_n) \in \mathbb{Z}^n \right\} \tag{2.1}$$

the group of macroscopic periods for the problem.

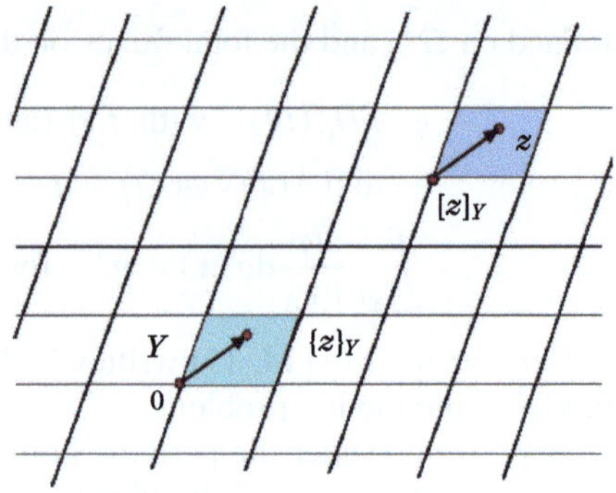

Fig. 2 Definition of $[z]_Y$ and $\{z\}_Y$

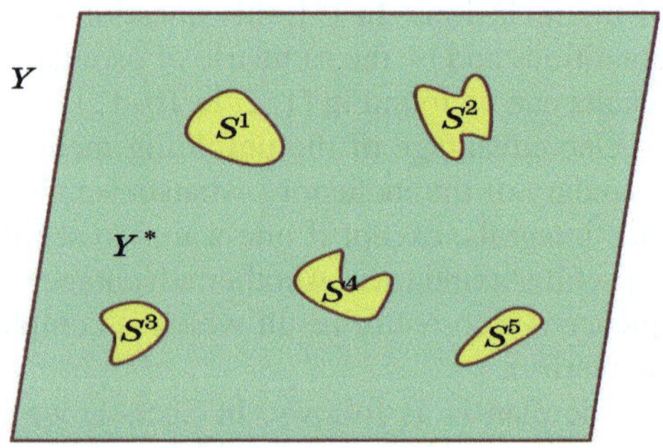

Fig. 3 The sets Y and Y^*

Let Y be the open parallelotope generated by the basis $\mathbf{b}$, i.e.,

$$\left\{ y \in \mathbb{R}^n \mid y = \sum_{i=1}^{n} y_i b_i, (y_1, \ldots, y_n) \in (0,1)^n \right\}. \tag{2.2}$$

More generally, Y can be any bounded connected subset of $\mathbb{R}^n$ with Lipschitz boundary, having the paving property with respect to the group $\mathcal{G}$.

For $z \in \mathbb{R}^n$, $[z]_Y$ denotes the unique (up to a set of measure zero) integer combination $\sum_{j=1}^{n} k_j b_j$ of the periods, such that $z - [z]_Y$ belongs to Y. Now set (see Fig. 2)

$$\{z\}_Y = z - [z]_Y \in Y \quad \text{a.e. for } z \in \mathbb{R}^n.$$

Let S be a given compact subset in Y, which is the finite disjoint union of S^j for $j = 1, \ldots, J$ (with the same property). The only condition on the S^j's is that they are pair-wise separated with the strictly positive measure. For the two examples, we consider that the sets S^j are naturally assumed to be connected (although this is not necessary for the treatment given here). The set $Y \setminus S$ is denoted by Y^* (see Fig. 3).

Let now Ω be an open bounded subset of $\mathbb{R}^n$. We define the "holes" in Ω as follows.

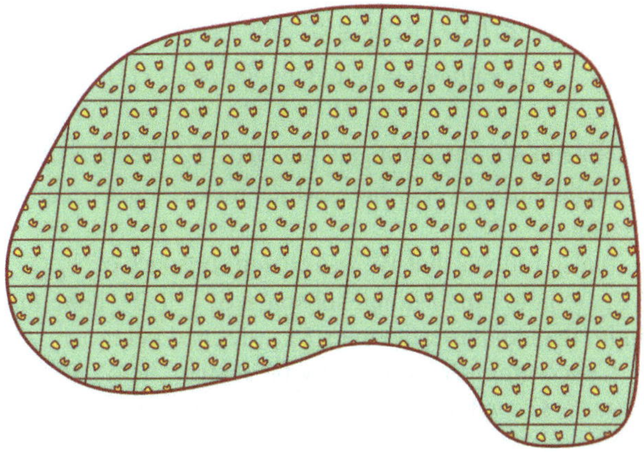

Fig. 4 The sets Ω_ε (in *green*) and S_ε (in *yellow*)

Definition 2.1 (See Fig. 4)

$$
\begin{aligned}
S_\varepsilon &\doteq \left\{x \in \Omega, \left\{\frac{x}{\varepsilon}\right\}_Y \in S\right\},\\
S_\varepsilon^j &\doteq \left\{x \in \Omega, \left\{\frac{x}{\varepsilon}\right\}_Y \in S^j\right\},\\
\Omega_\varepsilon &\doteq \Omega \setminus S_\varepsilon.
\end{aligned}
\tag{2.3}
$$

Remark 2.1 It is well-known that the characteristic function of the sets S_ε^j converges weakly-$*$ to $\frac{|S^j|}{|Y|}$ in $L^\infty(\Omega)$. This is a simple consequence of the properties of the unfolding operator given in the next section.

In this paper, we consider a boundary value problem, which generalizes both cases and does not require consideration of surface integrals. It applies as long as the "inclusions" S_ε^j have the strictly positive measure (so that requiring the restriction of an H_0^1 function to each of them to an arbitrary constant almost everywhere makes sense).

We make no regularity assumption regarding the sets S^j. We only make the natural assumption that they are well separated from each other and from ∂Y in Y (if some are not well separated, then they should be merged into a single one).

Whenever needed, the functions in $H_0^1(\Omega)$ will be extended by 0 in the whole of $\mathbb{R}^n$ (where they belong to $H^1(\mathbb{R}^n)$). Similarly, the functions of $L^2(\Omega)$ will be extended by 0 to the whole of $\mathbb{R}^n$.

We introduce the following two families of subspaces, where $\mathcal{G}_\varepsilon$ denotes the elements $\xi \in \mathcal{G}$, such that the corresponding cell $\varepsilon\xi + \varepsilon Y$ intersects Ω.

Definition 2.2

$$
\begin{aligned}
W_0^\varepsilon \doteq \big\{v \in H_0^1(\Omega);\ &\forall \xi \in \mathcal{G}_\varepsilon \text{ and } j \in \{1, \dots, J\}\ v_{|\varepsilon\xi+\varepsilon S^j} \text{ is a constant function,}\\
&\text{the value of which depends on } (\xi, j)\big\},
\end{aligned}
\tag{2.4}
$$

$$L_\varepsilon \doteq \big\{w \in L^2(\Omega);\ \forall \xi \in \mathcal{G}_\varepsilon \text{ and } j \in \{1, \dots, J\}\ w_{|\varepsilon\xi+\varepsilon S^j} \text{ is a constant function, the value of which depends on } (\xi, j)\big\}. \tag{2.5}$$

Note that a condition, such as $v_{|\varepsilon\xi+\varepsilon S^j}$ being a constant function, is taken in the sense of almost everywhere, which makes sense because each $\varepsilon\xi + \varepsilon S^j$ is of positive measure.

It also follows that W_0^ε is a closed subspace of $H_0^1(\Omega)$. On the other hand, clearly, L_ε is a finite dimensional subspace of $L^2(\Omega)$. Note that L_ε is the image of $L^2(\Omega)$ under the local average map $\mathcal{M}_\varepsilon$ (defined below).

Concerning the conductivity matrix field A^ε, it is assumed to belong to $M(\alpha, \beta, \Omega_\varepsilon)$ which is traditionally defined as follows.

Definition 2.3 Let $\alpha, \beta \in \mathbb{R}$, such that $0 < \alpha < \beta$. $M(\alpha, \beta, \mathcal{O})$ denotes the set of the $n \times n$ matrices $B = B(x)$, $B = (b_{ij})_{1\le i,j\le n} \in (L^\infty(\mathcal{O}))^{n\times n}$, such that

$$\big(B(x)\lambda, \lambda\big) \ge \alpha|\lambda|^2, \quad |B(x)\lambda| \le \beta|\lambda| \quad \text{for any } \lambda \in \mathbb{R}^n \text{ and a.e. on } \mathcal{O}.$$

Let f_ε be given in $L^2(\Omega_\varepsilon)$, and g_ε^j in L^ε for $j = 1, \dots, J$.
The problem we consider is given in the variational form as

$$(\mathrm{P}_\varepsilon)\begin{cases} \text{Find } u_\varepsilon \text{ in } W_0^\varepsilon, \text{ such that } \forall w \in W_0^\varepsilon, \\ \int_{\Omega_\varepsilon} A^\varepsilon \nabla u_\varepsilon \nabla w \mathrm{d}x = \int_{\Omega_\varepsilon} f_\varepsilon w \mathrm{d}x + \varepsilon^n \sum\limits_{\substack{\xi\in\mathcal{G}_\varepsilon \\ j=1,\dots,J}} g_\varepsilon^j{}_{|\varepsilon\xi+\varepsilon Y}\, w_{|\varepsilon\xi+\varepsilon S^j}. \end{cases} \tag{2.6}$$

Using the obvious formula

$$\varepsilon^n \sum_{\xi\in\mathcal{G}} g_\varepsilon^j{}_{|\varepsilon\xi+\varepsilon Y}\, w_{|\varepsilon\xi+\varepsilon S^j} = \frac{1}{|S^j|}\int_{S_\varepsilon^j} g_\varepsilon^j(x) w(x)\mathrm{d}x, \tag{2.7}$$

this problem can be written as

$$(\mathrm{P}_\varepsilon)\begin{cases} \text{Find } u_\varepsilon \text{ in } W_0^\varepsilon, \text{ such that } \forall w \in W_0^\varepsilon, \\ \int_{\Omega_\varepsilon} A^\varepsilon \nabla u_\varepsilon \nabla w \mathrm{d}x = \int_{\Omega} F_\varepsilon w \mathrm{d}x, \end{cases} \tag{2.8}$$

where

$$F_\varepsilon \doteq f_\varepsilon 1_{\Omega_\varepsilon} + \sum_{\substack{\xi\in\mathcal{G}_\varepsilon \\ j=1,\dots,J}} \frac{1}{|S^j|} g_\varepsilon^j(x) 1_{S_\varepsilon^j}.$$

This problem has a unique solution by the Lax-Milgram theorem applied in the space W_0^ε because of Poincaré's inequality in $H_0^1(\Omega)$.

The "strong" formulation of problem (P_ε), assuming at least Lipschitz regularity for the boundaries involved, is

$$\begin{cases} \text{Find } u_\varepsilon \in W_0^\varepsilon, \text{ such that } -\operatorname{div}(A^\varepsilon \nabla u_\varepsilon) = f_\varepsilon \text{ in } \Omega_\varepsilon, \\ \forall \xi \in \mathcal{G}_\varepsilon,\ \forall j = 1, \dots, J,\ \langle A^\varepsilon \nabla u_\varepsilon \cdot n, 1\rangle_{H^{-\frac{1}{2}}, H^{\frac{1}{2}}(\varepsilon\xi + \varepsilon\partial S^j)} = \varepsilon^n g_\varepsilon^j{}_{|\varepsilon\xi+\varepsilon Y}, \end{cases} \tag{2.9}$$

where $n(x)$ is the outward unit normal to $\varepsilon\xi + \varepsilon\partial S^j$. Under the regularity assumption above, it is classical that the duality pairing makes sense, because, since f_ε belongs to $L^2(\Omega_\varepsilon)$, $A^\varepsilon \nabla u_\varepsilon \cdot n$ is an element of $H^{-\frac{1}{2}}(\varepsilon\xi + \varepsilon\partial S^j)$. The pairing is often (somewhat incorrectly) written as

$$\int_{\varepsilon\xi+\varepsilon\partial S^j} A^\varepsilon \nabla u_\varepsilon \cdot n(x) d\sigma(x).$$

Making use of the Lax-Milgram theorem, one gets the following estimate.

Proposition 2.1 *There is a constant C depending only upon α and the Poincaré constant for $H_0^1(\Omega)$ (but not upon ε), such that, for every ε,*

$$|u_\varepsilon|_{H_0^1(\Omega)} \le C|F_\varepsilon|_{L^2(\Omega_\varepsilon)} = C\Big(|f_\varepsilon|^2_{L^2(\Omega_\varepsilon)} + \sum_{\substack{\xi\in\mathcal{G}_\varepsilon \\ j=1,\dots,J}} \varepsilon^2 |g_\varepsilon^j(x)_{|S_\varepsilon^j}|^2\Big)^{\frac{1}{2}}. \tag{2.10}$$

Consequently, assume that the right-hand side of (2.10) is bounded, so is the sequence $\{u_\varepsilon\}$ in $H_0^1(\Omega)$.

The homogenization problem is to investigate the weak convergence of this sequence and the possible problem satisfied by its limit under suitable assumptions on the data.

Remark 2.2 The sequence $\{u_\varepsilon\}$ in $H_0^1(\Omega)$ does not usually converge strongly in $H_0^1(\Omega)$, because, when it does, its limit is the zero function. More generally, if v_ε is in W_0^ε and converges strongly to v_0 in $H_0^1(\Omega)$, then $v_0 = 0$. The proof is elementary. Going to the limit for the product $0 \equiv 1_{S_\varepsilon} \nabla v_\varepsilon$, which, as a consequence of the assumptions, converges weakly to $\frac{|S|}{|Y|}\nabla v_0$, implies that $\nabla v_0 \equiv 0$, and hence the result is obtained.

3 A Brief Summary of the Unfolding Method in Fixed Domains

We recall the following notations used in [5]:

$$\widehat{\Omega}_\varepsilon = \text{interior}\Big\{\bigcup_{\xi\in\Xi_\varepsilon} \varepsilon(\xi + \overline{Y})\Big\}, \quad \Lambda_\varepsilon = \Omega \setminus \widehat{\Omega}_\varepsilon, \tag{3.1}$$

where

$$\Xi_\varepsilon = \big\{\xi \in \mathcal{G}, \varepsilon(\xi + Y) \subset \Omega\big\} \tag{3.2}$$

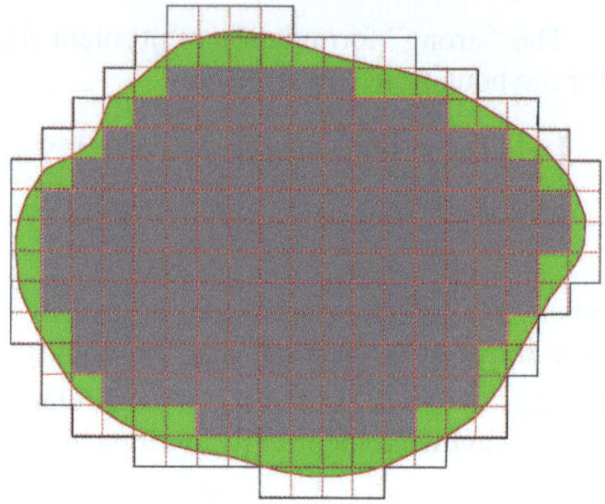

Fig. 5 The sets $\widehat{\Omega}_\varepsilon$ *(in grey)* and Λ_ε *(in green)*

(see Fig. 5). The set $\widehat{\Omega}_\varepsilon$ is the interior of the largest union of $\varepsilon(\xi+\overline{Y})$ cells included in Ω. Here, the set Ξ_ε is slightly smaller than $\mathcal{G}_\varepsilon$ as defined previously.

Definition 3.1 For ϕ Lebesgue-measurable on $\widehat{\Omega}_\varepsilon$, the unfolding operator $\mathcal{T}_\varepsilon$ is defined as follows:

$$\mathcal{T}_\varepsilon(\phi)(x,y)=\begin{cases}\phi(\varepsilon[\frac{x}{\varepsilon}]_Y+\varepsilon y), & \text{a.e. for } (x,y)\in\widehat{\Omega}_\varepsilon\times Y,\\ 0, & \text{a.e. for } (x,y)\in\Lambda_\varepsilon\times Y.\end{cases} \tag{3.3}$$

The properties of the unfolding operator are summarized here.

Theorem 3.1 *Let p belong to* $[1,+\infty)$.

(i) $\mathcal{T}_\varepsilon$ *is linear continuous from* $L^p(\Omega)$ *to* $L^p(\Omega\times Y)$. *Its norm is bounded by* $|Y|^{\frac{1}{p}}$.

(ii) *For every* w *in* $L^1(\Omega)$,

$$\int_\Omega w(x)\mathrm{d}x=\frac{1}{|Y|}\int_{\Omega\times Y}\mathcal{T}_\varepsilon(w)(x,y)\mathrm{d}x\mathrm{d}y+\int_{\Lambda_\varepsilon}w(x)\mathrm{d}x.$$

(iii) *Let* $\{w_\varepsilon\}$ *be a sequence in* $L^p(\Omega)$, *such that* $w_\varepsilon\to w$ *strongly in* $L^p(\Omega)$. *Then*

$$\mathcal{T}_\varepsilon(w_\varepsilon)\to w \quad \textit{strongly in } L^p(\Omega\times Y).$$

(iv) *Let* $\{w_\varepsilon\}$ *be bounded in* $L^p(\Omega)$, *and suppose that the corresponding* $\mathcal{T}_\varepsilon(w_\varepsilon)$ *(which is bounded in* $L^p(\Omega\times Y)$*) converges weakly to* $\widehat{w}$ *in* $L^p(\Omega\times Y)$.

Then

$$w_\varepsilon\rightharpoonup\mathcal{M}_Y(\widehat{w})=\frac{1}{|Y|}\int_Y\widehat{w}(\cdot,y)\mathrm{d}y \quad \textit{weakly in } L^p(\Omega).$$

Here, the operator $\mathcal{M}_Y$ is the average over Y.

Definition 3.2 The operator $\mathcal{M}_\varepsilon \doteq \mathcal{M}_Y \circ \mathcal{T}_\varepsilon$ is the local average operator. It assigns to a function, which is integrable on Ω its local average (associated with the ε-cells $\varepsilon\xi + \varepsilon Y$).

Theorem 3.2 *Let $\{w_\varepsilon\}$ be in $W^{1,p}(\Omega)$ with $p \in (1, +\infty)$, and assume that $\{w_\varepsilon\}$ is a bounded sequence in $W^{1,p}(\Omega)$. Then, there exist a subsequence (still denoted by $\{\varepsilon\}$) and functions w in $W^{1,p}(\Omega)$ and $\widehat{w}$ in $L^p(\Omega; W^{1,p}_{\text{per}}(Y))$ with $\mathcal{M}_Y(\widehat{w}) \equiv 0$, such that*

$$\begin{aligned} \mathcal{T}_\varepsilon(w_\varepsilon) &\rightharpoonup w \quad \text{weakly in } L^p\big(\Omega; W^{1,p}(Y)\big), \\ \mathcal{T}_\varepsilon(\nabla w_\varepsilon) &\rightharpoonup \nabla w + \nabla_y \widehat{w} \quad \text{weakly in } L^p(\Omega \times Y). \end{aligned} \tag{3.4}$$

Furthermore, the sequence $\frac{1}{\varepsilon}(\mathcal{T}_\varepsilon(w_\varepsilon) - \mathcal{M}_\varepsilon(w_\varepsilon))$ converges weakly in $L^p(\Omega; W^{1,p}(Y))$ to $y_M \cdot \nabla w + \widehat{w}$, where $y_M \doteq y - \mathcal{M}_Y(y)$.

Here, $W^{1,p}_{\text{per}}(Y)$ denotes the space of the functions in $W^{1,p}_{\text{loc}}(\mathbb{R}^n)$, which are $\mathcal{G}$-periodic. It is a closed subspace of $W^{1,p}(Y)$, and is endowed with the corresponding norm.

We end this section by recalling the notion of the averaging operator $\mathcal{U}_\varepsilon$. This operator is the adjoint of $\mathcal{T}_\varepsilon$ and maps $L^p(\Omega \times Y)$ into the space $L^p(\Omega)$.

Definition 3.3 For p in $[1, +\infty]$, the averaging operator $\mathcal{U}_\varepsilon : L^p(\Omega \times Y) \mapsto L^p(\Omega)$ is defined as follows:

$$\mathcal{U}_\varepsilon(\Phi)(x) = \begin{cases} \frac{1}{|Y|} \int_Y \Phi(\varepsilon[\frac{x}{\varepsilon}]_Y + \varepsilon z, \{\frac{x}{\varepsilon}\}_Y) \mathrm{d}z, & \text{a.e. for } x \in \widehat{\Omega}_\varepsilon, \\ 0, & \text{a.e. for } x \in \Lambda_\varepsilon. \end{cases}$$

The main properties of $\mathcal{U}_\varepsilon$ are listed in the next proposition.

Proposition 3.1 (Properties of $\mathcal{U}_\varepsilon$) *Suppose that p is in $[1, +\infty)$.*

(i) *The averaging operator is linear and continuous from $L^p(\Omega \times Y)$ to $L^p(\Omega)$, and*

$$\|\mathcal{U}_\varepsilon(\Phi)\|_{L^p(\Omega)} \leq |\, Y \,|^{-\frac{1}{p}} \|\Phi\|_{L^p(\Omega \times Y)}.$$

(ii) *If φ is independent of y, and belongs to $L^p(\Omega)$, then*

$$\mathcal{U}_\varepsilon(\varphi) \to \varphi \quad \text{strongly in } L^p(\Omega).$$

(iii) *Let $\{\Phi_\varepsilon\}$ be a bounded sequence in $L^p(\Omega \times Y)$, such that $\Phi_\varepsilon \rightharpoonup \Phi$ weakly in $L^p(\Omega \times Y)$. Then*

$$\mathcal{U}_\varepsilon(\Phi_\varepsilon) \rightharpoonup \mathcal{M}_Y(\Phi) = \frac{1}{|Y|} \int_Y \Phi(\cdot, y) \mathrm{d}y \quad \text{weakly in } L^p(\Omega).$$

In particular, for every $\Phi \in L^p(\Omega \times Y)$,

$$\mathcal{U}_\varepsilon(\Phi) \rightharpoonup \mathcal{M}_Y(\Phi) \quad \text{weakly in } L^p(\Omega).$$

(iv) *Suppose that* $\{w_\varepsilon\}$ *is a sequence in* $L^p(\Omega)$. *Then, the following assertions are equivalent*:

(a) $\mathcal{T}_\varepsilon(w_\varepsilon) \to \widehat{w}$ *strongly in* $L^p(\Omega \times Y)$ *and* $\int_{\Lambda_\varepsilon} |w_\varepsilon|^p \mathrm{d}x \to 0$,
(b) $w_\varepsilon - \mathcal{U}_\varepsilon(\widehat{w}) \to 0$ *strongly in* $L^p(\Omega)$.

We complete this section with a somewhat unusual convergence property involving the averaging operator $\mathcal{U}_\varepsilon$ which is applied in Corollary 6.3.

Proposition 3.2 *For* $p \in [1, +\infty)$, *suppose that* α *is in* $L^p(\Omega)$ *and* β *in* $L^\infty(\Omega; L^p(Y))$. *Then, the product* $\mathcal{U}_\varepsilon(\alpha)\mathcal{U}_\varepsilon(\beta)$ *belongs to* $L^p(\Omega)$ *and*

$$\mathcal{U}_\varepsilon(\alpha\beta) - \mathcal{U}_\varepsilon(\alpha)\mathcal{U}_\varepsilon(\beta) \to 0 \quad \textit{strongly in } L^p(\Omega). \tag{3.5}$$

4 The Unfolded Limit Problem

In order to use the unfolding operator $\mathcal{T}_\varepsilon$, in S_ε, we extend f_ε by zero and A^ε by αI without changing the notation. This implies that $\mathcal{T}_\varepsilon(f_\varepsilon)|_{\Omega\times S} \equiv 0$, and similarly, $\mathcal{T}_\varepsilon(A^\varepsilon)|_{\widehat{\Omega}_\varepsilon\Omega\times S} \equiv \alpha I$.

Let G_ε be defined as $G_\varepsilon \doteq \sum_{j=1,\dots,J} \frac{1}{|S^j|} g_\varepsilon^j 1_{S_\varepsilon^j}$.

We make the following assumptions concerning the data, for ε converging to 0:

$$(\mathrm{H}) \begin{cases} \mathcal{T}_\varepsilon(A^\varepsilon) \text{ converges in measure (or a.e.) in } \Omega \times Y \text{ to } A^0, \\ \mathcal{T}_\varepsilon(f_\varepsilon) \text{ converges weakly to } f_0 \text{ in } L^2(\Omega \times Y), \\ g_\varepsilon^j \text{ converges weakly to } g_0^j \text{ in } L^2(\Omega) \text{ for } j = 1, \dots, J. \end{cases} \tag{4.1}$$

Note that from the definition of A^ε and f_ε, f_0 vanish on $\Omega \times S$ while $A^0|_{\Omega\times S} \equiv \alpha I$. Since A^ε belongs to $M(\alpha, \beta, \Omega_\varepsilon)$, it follows that A^0 belongs to $M(\alpha, \beta, \Omega \times Y)$.

It follows from the last hypothesis that $\mathcal{T}_\varepsilon(G_\varepsilon)$ converges weakly in $L^2(\Omega \times Y)$ to the function $G_0 \doteq \sum_{j=1,\dots,J} \frac{1}{|S^j|} g_0^j(x) 1_{S^j}(y)$ (it is actually an equivalence). It also implies that $\mathcal{T}_\varepsilon(F_\varepsilon)$ converges weakly in the same space to $F_0 \doteq f_0 + G_0$.

Proposition 4.1 *Under hypothesis* (H), *up to a subsequence (which we still denote by* $\{\varepsilon\}$*), there exist* $u_0 \in H_0^1(\Omega)$ *and* $\widehat{u} \in L^2(\Omega; H^1_{\mathrm{per}}(Y))$, *such that*

$$\begin{aligned}
&\mathcal{T}_\varepsilon(u_\varepsilon) \rightharpoonup u_0 \textit{ weakly in } L^2\big(\Omega; H^1(Y)\big),\\
&\mathcal{T}_\varepsilon(\nabla u_\varepsilon) \rightharpoonup \nabla u_0 + \nabla_y \widehat{u} \textit{ weakly in } L^2(\Omega \times Y),\\
&\frac{1}{\varepsilon}\big(\mathcal{T}_\varepsilon(u_\varepsilon) - \mathcal{M}_\varepsilon(u_\varepsilon)\big) \textit{ converges weakly in } L^2\big(\Omega; H^1(Y)\big) \textit{ to } y_M \cdot \nabla u_0 + \widehat{u},\\
&\eta_\varepsilon \doteq \mathcal{T}_\varepsilon\big(A^\varepsilon\big)\mathcal{T}_\varepsilon(\nabla u_\varepsilon) \rightharpoonup \eta_0 \doteq A^0(\nabla u_0 + \nabla_y \widehat{u}) \textit{ weakly in } L^2(\Omega \times Y),\\
&\eta_0 \textit{ vanishes almost everywhere in } \Omega \times S,\\
&y_M \cdot \nabla u_0 + \widehat{u} \textit{ is independent of } y \textit{ on each } \Omega \times S^j,\ j = 1, \dots, J.
\end{aligned} \tag{4.2}$$

Proof The existence of a subsequence of u_0 and $\widehat{u}$ satisfying the first three conditions follows from Theorem 3.2. The next convergence follows from the fact that convergence in measure (or a.e.) is a multiplier for strong as well as for weak convergence in $L^2(\Omega \times Y)$. The last property follows from the fact that $u_{\varepsilon|\varepsilon\xi+\varepsilon S^j}$ is independent of x, which implies that $\varepsilon^{-1}(\mathcal{T}_\varepsilon(u_\varepsilon) - \mathcal{M}_\varepsilon(u_\varepsilon))|_{\Omega\times S^j}$ is a function only of x. This property is preserved by weak limit, and holds for $y_M \cdot \nabla u_0 + \widehat{u}$. A similar proof implies that η_0 vanishes a.e. on $\Omega \times S$. □

From now on, we use the notation η_ε for $A^\varepsilon(\nabla u_\varepsilon)$ (and η_0 for $A^0(\nabla u_0 + \nabla_y \widehat{u})$), so that $\mathcal{T}_\varepsilon(\eta_\varepsilon)$ converges weakly to η_0.

Definition 4.1 Let $\mathbf{H}^S_{\text{per}}$ denote the subspace of $H^1_{\text{per}}(Y)$, consisting of functions which are constant on each S^j (with independent constants for each j).

Proposition 4.2 *For almost every $x \in \Omega$, and for every $\Phi \in L^2(\Omega; \mathbf{H}^S_{\text{per}})$, the vector field η_0 satisfies*

$$\int_{\Omega\times Y} \eta_0(x, y) \cdot \nabla_y \Phi(y) \mathrm{d}y = 0. \tag{4.3}$$

The corresponding strong formulation, under regularity assumptions, is

$$\begin{cases} -\operatorname{div}_y \eta_0 = 0 \quad \text{in } \mathcal{D}'(Y^*), \\ \langle \eta_0 \cdot n, 1\rangle_{H^{-\frac{1}{2}}, H^{\frac{1}{2}}(\partial S^j)} = 0 \quad \text{for } j = 1, \ldots, J, \\ \text{and periodicity conditions of the normal flux of } \eta_0 \text{ on opposite faces of } \partial Y. \end{cases} \tag{4.4}$$

Proof Let w be fixed in $\mathcal{D}(\Omega)$, ψ in $\mathcal{D}(Y) \cap \mathbf{H}^S_{\text{per}}$ and ϕ in $\mathcal{C}^\infty_{\text{per}}(\overline{Y})$ vanishing on S.
The function v_ε defined as

$$v_\varepsilon(x) \doteq \varepsilon\left(\mathcal{M}_\varepsilon(w)(x)\psi\left(\left\{\frac{x}{\varepsilon}\right\}_Y\right) + w(x)\phi\left(\left\{\frac{x}{\varepsilon}\right\}_Y\right)\right)$$

belongs to the space W^ε_0, since ψ vanishes near ∂Y. Furthermore, it converges to zero uniformly. Using v_ε as a test function in Problem (P_ε) gives, for ε going to zero according to the established subsequence,

$$\int_{\Omega_\varepsilon} A^\varepsilon \nabla u_\varepsilon \, \nabla v_\varepsilon \mathrm{d}x \to 0. \tag{4.5}$$

The gradient of v_ε is given by

$$\nabla v_\varepsilon(x) \equiv \mathcal{M}_\varepsilon(w)(x)\nabla_y\psi\left(\left\{\frac{x}{\varepsilon}\right\}_Y\right) + w(x)\nabla_y\phi\left(\left\{\frac{x}{\varepsilon}\right\}_Y\right) + \varepsilon\nabla w(x)\phi\left(\left\{\frac{x}{\varepsilon}\right\}_Y\right).$$

Consequently,

$$\mathcal{T}_\varepsilon(\nabla v_\varepsilon) = \mathcal{M}_\varepsilon(w)(x)\nabla_y\psi(y) + \mathcal{T}_\varepsilon(w)\nabla_y\phi(y) + \varepsilon\mathcal{T}_\varepsilon(\nabla w)\phi(y).$$

From the fact that $\mathcal{T}_\varepsilon(w)$ and $\mathcal{M}_\varepsilon(w)$ both converge uniformly to w (in $\Omega \times Y$ and Ω, respectively), it follows that

$$\mathcal{T}_\varepsilon(\nabla v_\varepsilon) \to w(x)\nabla_y\big(\psi(y) + \phi(y)\big).$$

Applying Theorem 3.1(ii) to the left-hand side of (4.5), this gives

$$\int_{\Omega_\varepsilon} A^\varepsilon \nabla u_\varepsilon \, \nabla v_\varepsilon \mathrm{d}x = \frac{1}{|Y|}\int_{\Omega \times Y} \mathcal{T}_\varepsilon(\eta_\varepsilon)(x,y)\mathcal{T}_\varepsilon(\nabla v_\varepsilon)\mathrm{d}x\mathrm{d}y \to 0.$$

By Proposition 4.1, this implies

$$\int_{\Omega \times Y} \eta_0(x,y) w(x) \nabla_y\big(\psi(y) + \phi(y)\big)\mathrm{d}x\mathrm{d}y = 0.$$

Now, by a partition of the unity argument, every $\Psi \in \mathcal{C}^\infty_{\text{per}}(\overline{Y}) \cap \mathbf{H}^S_{\text{per}}$ can be written as a sum of a function ψ of the first case and a function ϕ of the second case. Therefore, for every $w \in \mathcal{D}(\Omega)$ and every $\Psi \in \mathcal{C}^\infty_{\text{per}}(\overline{Y}) \cap \mathbf{H}^S_{\text{per}}$, (4.3) is satisfied. Finally, by a totality argument, (4.3) holds for all $\Phi \in L^2(\Omega; \mathbf{H}^S_{\text{per}})$, since η_0 belongs to $L^2(\Omega \times Y)$. □

We now turn to the task of obtaining a relevant formula for $\int_{\Omega \times Y} \eta_0 \nabla w \mathrm{d}x\mathrm{d}y$ for w in $H^1_0(\Omega)$. To this end, we use the following lemma (in [3, Proposition 2.1], a similar argument is used but only to obtain the first statement).

Lemma 4.1 *Let Ψ be in $C^\infty_{\text{per}}(Y)$ with $\Psi \equiv 1$ on S. For every w in $\mathcal{D}(\Omega)$ and every ε, there exists a v_ε in W^ε_0, such that, as $\varepsilon \to 0$,*

$$\begin{cases} v_\varepsilon \text{ converges uniformly to } w \text{ in } \Omega \text{ as well as weakly in } H^1_0(\Omega), \\ \mathcal{T}_\varepsilon(\nabla) \text{ converges strongly in } L^2(\Omega \times Y) \text{ to } \nabla w - \nabla_y((y_M \cdot \nabla w)\Psi(y)). \end{cases} \tag{4.6}$$

Proof It is clear that the function $x \mapsto w(x)(1 - \Psi(\{\frac{x}{\varepsilon}\}_Y) + \mathcal{M}_\varepsilon(w)\Psi(\{\frac{x}{\varepsilon}\}_Y))$ belongs to the space W^ε_0. Furthermore, note that

$$\mathcal{T}_\varepsilon(v_\varepsilon) = \mathcal{T}_\varepsilon(w)\big(1 - \Psi(y)\big) + \mathcal{M}_\varepsilon(w)\Psi(y) \to w(1 - \Psi + \Psi) = w$$

uniformly in $\Omega \times Y$.

Similarly,

$$\begin{aligned} \nabla v_\varepsilon &= \nabla w(1 - \Psi\left(\left\{\frac{\cdot}{\varepsilon}\right\}_Y\right) - \frac{1}{\varepsilon}\big(w - \mathcal{M}_\varepsilon(w)\big)(\nabla_y \Psi\left(\left\{\frac{\cdot}{\varepsilon}\right\}_Y\right), \\ \mathcal{T}_\varepsilon(\nabla v_\varepsilon) &= \mathcal{T}_\varepsilon(\nabla w)\big(1 - \Psi(y)\big) - \frac{1}{\varepsilon}\big(\mathcal{T}_\varepsilon(w) - \mathcal{M}_\varepsilon(w)\big)\nabla_y \Psi(y). \end{aligned} \tag{4.7}$$

We can now show that the latter one converges strongly in $L^2(\Omega \times Y)$.

Indeed, it is enough to show the strong convergence of $\frac{1}{\varepsilon}(\mathcal{T}_\varepsilon(w) - \mathcal{M}_\varepsilon(w))$ in the same space. We claim that it converges to $y_M \cdot \nabla w$. Indeed, set

$$z_\varepsilon \doteq \frac{1}{\varepsilon}\big(\mathcal{T}_\varepsilon(w) - \mathcal{M}_\varepsilon(w)\big) - y_M \cdot \nabla w.$$

Clearly, $\nabla_y z_\varepsilon = \mathcal{T}_\varepsilon(\nabla w) - \nabla w$, which converges strongly to zero in $L^2(\Omega \times Y)$. By the Poincaré-Wirtinger inequality in Y, and since $\mathcal{M}_Y(z_\varepsilon) \equiv 0$, z_ε itself converges to 0 in $L^2(\Omega; H^1(Y))$.

We conclude with the identity

$$\nabla w(\Psi) + (y_M \cdot \nabla w)\nabla_y \Psi = \nabla_y((y_M \cdot \nabla w)\Psi(y)). \qquad \square$$

Choosing such a v_ε as a test function in Problem (P_ε) gives

$$\int_\Omega \eta_\varepsilon \nabla v_\varepsilon \mathrm{d}x = \int_\Omega (f_\varepsilon 1_{\Omega_\varepsilon} + G_\varepsilon) v_\varepsilon \mathrm{d}x. \tag{4.8}$$

Unfolding the right-hand side of this relation gives the convergence

$$\frac{1}{|Y|}\int_{\Omega\times Y} \big(\mathcal{T}_\varepsilon(f_\varepsilon) + \mathcal{T}_\varepsilon(G_\varepsilon)\big)\mathcal{T}_\varepsilon(v_\varepsilon)\mathrm{d}x\mathrm{d}y \to \frac{1}{|Y|}\int_{\Omega\times Y} F_0(x,y)w(x)\mathrm{d}x\mathrm{d}y. \tag{4.9}$$

The left-hand side of (4.8) is also unfolded to obtain the convergence

$$\frac{1}{|Y|}\int_{\Omega\times Y} \mathcal{T}_\varepsilon(\eta_\varepsilon)\mathcal{T}_\varepsilon(\nabla v_\varepsilon)\mathrm{d}x\mathrm{d}y \to \frac{1}{|Y|}\int_{\Omega\times Y} \eta_0\big(\nabla w - \nabla_y\big((y_M \cdot \nabla w)\Psi(y)\big)\big)\mathrm{d}x\mathrm{d}y. \tag{4.10}$$

Regrouping (4.9) and (4.10), we get

$$\begin{aligned}&\frac{1}{|Y|}\int_{\Omega\times Y} \eta_0\big(\nabla w - \nabla_y\big((y_M \cdot \nabla w)\Psi(y)\big)\big)\mathrm{d}x\mathrm{d}y\\ &\quad= \frac{1}{|Y|}\int_{\Omega\times Y} F_0(x,y)w(x)\mathrm{d}x\mathrm{d}y.\end{aligned} \tag{4.11}$$

By a density argument, this still holds for every w in $H_0^1(\Omega)$.

We have proved the following proposition.

Proposition 4.3 *For every* $w \in H_0^1(\Omega)$,

$$\begin{aligned}&\frac{1}{|Y|}\int_{\Omega\times Y} \eta_0\big(\nabla w - \nabla_y\big((y_M \cdot \nabla w)\Psi(y)\big)\big)\mathrm{d}x\mathrm{d}y\\ &\quad= \frac{1}{|Y|}\int_{\Omega\times Y} F_0(x,y)w(x)\mathrm{d}x\mathrm{d}y.\end{aligned} \tag{4.12}$$

Remark 4.1 (1) Because η_0 satisfies (4.3), formula (4.12) does not depend upon the choice of Ψ. Indeed, for such another $\widehat{\Psi}$ for a.e. $x \in \Omega$, by (4.3), one has

$$\int_Y \eta_0 \nabla_y \big((y_M \cdot \nabla w)(\Psi - \widehat{\Psi}) \big) \mathrm{d}y = 0. \tag{4.13}$$

(2) If ∂S is assumed regular, in view of (4.3), the term

$$\frac{1}{|Y|} \int_{\Omega \times Y} \eta_0 \nabla_y \big((y_M \cdot \nabla w) \Psi \big) \mathrm{d}x \mathrm{d}y$$

can be interpreted as

$$\frac{1}{|Y|} \int_\Omega \langle \eta_0 \cdot n, y_M \cdot \nabla w \rangle_{H^{-\frac{1}{2}}, H^{\frac{1}{2}}(\partial S)} \mathrm{d}x,$$

because Ψ is identically 1 on S.

Definition 4.2 Let W be the following space:

$$\begin{aligned} W = \big\{ &(w, \widehat{w});\, w \in H_0^1(\Omega),\, \widehat{w} \in H^1_{\text{per}}(Y),\, \mathcal{M}_Y(\widehat{w}) = 0, \\ &\widehat{w} + (y_M \cdot \nabla w)|_{S^j} \text{ is a constant (depending on } j\text{) for } j = 1, \ldots, J \big\}. \end{aligned} \tag{4.14}$$

It is a closed subspace of $H_0^1(\Omega) \times H^1_{\text{per}}(Y)$, and hence it is a Hilbert space.

We can now state the limit unfolded problem.

Theorem 4.1 *Under Hypothesis* (H), *the whole sequence* $\{u_e\}$ *converges weakly in* $H_0^1(\Omega)$ *to a function* u_0. *There also exists a* $\widehat{u}$ *in* $L^2(\Omega; H^1_{\text{per}}(Y))$, *such that* $(u_0, \widehat{u})$ *is the unique solution to the following problem*:

$$\begin{cases} \text{Find } (u_0, \widehat{u}) \in W, \text{ such that } \forall (w, \widehat{w}) \in W, \\ \frac{1}{|Y|} \int_{\Omega \times Y} A^0 (\nabla u_0 + \nabla_y \widehat{u})(\nabla w + \nabla_y \widehat{w}) \mathrm{d}x \mathrm{d}y \\ \quad = \frac{1}{|Y|} \int_{\Omega \times Y} F_0(x, y) w(x) \mathrm{d}x \mathrm{d}y. \end{cases} \tag{4.15}$$

Proof It has already been established that $(u_0, \widehat{u})$ belongs to the space W. Combining Propositions 4.2 and 4.3 gives that for all $w \in H_0^1(\Omega)$, $\Phi \in L^2(\Omega; \mathbf{H}^S_{\text{per}}(Y))$, the following holds:

$$\begin{aligned} &\frac{1}{|Y|} \int_{\Omega \times Y} A^0 (\nabla u_0 + \nabla_y \widehat{u}) \big(\nabla w + \nabla_y \big(\Phi - (y_M \cdot \nabla w) \Psi \big) \big) \mathrm{d}x \mathrm{d}y \\ &\quad = \frac{1}{|Y|} \int_{\Omega \times Y} F_0(x, y) w(x) \mathrm{d}x \mathrm{d}y. \end{aligned} \tag{4.16}$$

Every element $(w, \widehat{w})$ of W can be written in the form

$$
\begin{aligned}
&\big(w, \Phi - (y_M \cdot \nabla w)\Psi - \mathcal{M}_Y\big(\Phi - (y_M \cdot \nabla w)\Psi\big)\big) \quad \text{with} \\
&\nabla_y \widehat{w} = \nabla_y\big(\Phi - (y_M \cdot \nabla w)\Psi\big),
\end{aligned} \tag{4.17}
$$

with (w, Φ) in $H_0^1(\Omega) \times L^2(\Omega; \mathbf{H}^S_{\text{per}}(Y))$, by setting $\Phi \doteq \widehat{w} + (y_M.\nabla w)\Psi$. This shows that (4.15) is equivalent to (4.16).

To prove the existence and the uniqueness of the solution, we show that the Lax-Milgram theorem applies to (4.15). It is enough to show that the bilinear form on the left-hand side of (4.15) is coercive.

Since A^0 belongs to $M(\alpha, \beta, \Omega \times Y)$, it follows that

$$
\frac{1}{|Y|}\int_{\Omega \times Y} A^0(\nabla u_0 + \nabla_y \widehat{u})(\nabla u_0 + \nabla_y \widehat{u})\mathrm{d}x\mathrm{d}y \geq \frac{\alpha}{|Y|}\|\nabla u_0 + \nabla_y \widehat{u}\|^2_{L^2(\Omega \times Y)}.
$$

But since u_0 is independent of y and $\widehat{u}$ is Y-periodic, the latter one is just

$$
\alpha\left(\|\nabla u_0\|^2_{L^2(\Omega)} + \frac{1}{|Y|}\|\nabla_y \widehat{u}\|^2_{L^2(\Omega \times Y)}\right).
$$

One concludes by using the Poincaré inequality in $H_0^1(\Omega)$ and the Poincaré-Wirtinger inequality in $H^1_{\text{per}}(Y)$ (since $\mathcal{M}_Y(\widehat{u}) = 0$). □

5 The Homogenized Limit Problem

For a given vector $\lambda \in \mathbb{R}^n$, consider the cell-problem for a.e. $x \in \Omega$,

$$
\begin{cases}
\text{Find } \chi_\lambda \in H^1_{\text{per}}(Y), \text{ such that } \mathcal{M}_Y(\chi_\lambda) = 0, \\
(\chi_\lambda + y_M.\lambda)_{|S^j} \text{ is independent of } y \text{ for } j = 1, \dots, J, \\
\int_Y A^0(x, y)(\nabla_y \chi_\lambda(y) + \lambda)\nabla_y \varphi(y)\mathrm{d}y = 0, \ \forall \varphi \in \mathbf{H}^S_{\text{per}}.
\end{cases} \tag{5.1}
$$

This problem itself is not variational. Introducing a fixed function Ψ in $\mathcal{D}(Y)$ with $\Psi_{|S} \equiv 1$ and $\mathcal{M}_Y(y_M \Psi) = 0$, the function $U_\lambda \doteq \chi_\lambda + (y \cdot \lambda)\Psi$ belongs to $\mathbf{H}^S_{\text{per}}$ with $\mathcal{M}_Y(U_\lambda) = 0$, and is the unique solution to the following variational problem in the same space:

$$
\begin{cases}
\text{Find } U_\lambda \in \mathbf{H}^S_{\text{per}}, \text{ such that } \mathcal{M}_Y(U_\lambda) = 0, \\
\int_Y A^0(x, y)(\nabla_y U_\lambda(y))\nabla_y \varphi(y)\mathrm{d}y \\
\quad = \int_Y A^0((\Psi - 1)\lambda + (y_M \cdot \lambda)\nabla_y \Psi)\nabla_y \varphi(y)\mathrm{d}y, \ \forall \varphi \in \mathbf{H}^S_{\text{per}}.
\end{cases} \tag{5.2}
$$

Note that the Lax-Milgram theorem applies to (5.2).

Once U_λ is obtained, set $\chi_\lambda \doteq U_\lambda - (y \cdot \lambda)\Psi$. We now show that it is independent of the choice of Ψ. Indeed, this corresponds to having uniqueness in (5.1). The difference V of two solutions to (5.1) belongs to $\mathbf{H}^S_{\text{per}}$ and satisfies in particular $\int_Y A^0 \nabla_y V \nabla_y V \mathrm{d}y = 0$, which by ellipticity, implies $\nabla_y V \equiv 0$ so that $V = 0$ (since $\mathcal{M}_Y(V) = 0$).

Using U_λ as a test function in (5.2) for a.e. $x \in \Omega$, it is straightforward to see that

$$|\nabla_y \chi_\lambda|_{L^2(Y)} \leq C(\alpha, \beta, Y)|\lambda|, \tag{5.3}$$

with a constant $C(\alpha, \beta, Y)$, which depends only upon α, β and Y.

For simplicity, we write $\chi(\lambda)$ for χ_λ.

Going back to (4.15) with the solution $(u_0, \widehat{u})$, it follows that

$$\widehat{u}(x, y) = \xi(\nabla u_0)\left(= \sum_{i=1}^{n} \frac{\partial u_0}{\partial x_i}(x)\chi_{e_i}(x, y)\right). \tag{5.4}$$

Then, (4.15) becomes

$$\begin{cases} \text{Find } u_0 \in H^1_0(\Omega), \text{ such that } \forall w \in H^1_0(\Omega), \\ \frac{1}{|Y|}\int_{\Omega\times Y} A^0(\nabla u_0 + \nabla_y \chi(\nabla u_0))\nabla w \mathrm{d}x\mathrm{d}y \\ \quad - \frac{1}{|Y|}\int_{\Omega\times Y} A^0(\nabla u_0 + \nabla_y \chi(\nabla u_0))\nabla_y((y_M \cdot \nabla w)\Psi(y))\mathrm{d}x\mathrm{d}y \\ \quad = \frac{1}{|Y|}\int_{\Omega\times Y} F_0(x, y)w(x)\mathrm{d}x\mathrm{d}y. \end{cases} \tag{5.5}$$

Definition 5.1 Set

$$A^{\text{hom}}(\lambda, \mu) \doteq \frac{1}{|Y|}\int_Y A^0(\lambda + \nabla_y \chi_\lambda)(\mu + \nabla_y \chi_\mu)\mathrm{d}y. \tag{5.6}$$

Proposition 5.1 *The homogenized limit problem is*

$$\begin{cases} \textit{Find } u_0 \in H^1_0(\Omega), \textit{ such that } \forall w \in H^1_0(\Omega), \\ \int_\Omega A^{\text{hom}}(\nabla u_0)\nabla w \mathrm{d}x = \int_\Omega \mathcal{M}_Y(F_0) w \mathrm{d}x. \end{cases} \tag{5.7}$$

The matrix field A^{hom} belongs to $M(\alpha, \beta(1 + C(\alpha, \beta, Y)), \Omega)$, so that (5.7) is well-posed.

Proof From (5.5), the homogenized problem (5.7) holds with

$$A^{\text{hom}}(\lambda, \mu) \doteq \frac{1}{|Y|}\int_Y A^0(\lambda + \nabla_y \chi_\lambda)\big(\mu - \nabla_y((y_M \cdot \mu)\Psi)\big)\mathrm{d}y \tag{5.8}$$

for $\lambda, \mu \in \mathbb{R}^n$ and for a.e. $x \in \Omega$. However, by (5.1), since $\chi_\mu + (y_M \cdot \mu)\Psi$ is in $\mathbf{H}^S_{\text{per}}$,

$$\int_Y A^0(\lambda + \nabla_y \chi_\lambda)\nabla_y \chi_\mu \mathrm{d}y = -\int_Y A^0(\lambda + \nabla_y \chi_\lambda)\nabla_y\big((y_M \cdot \mu)\Psi\big)\mathrm{d}y,$$

so that

$$A^{\text{hom}}(\lambda, \mu) \doteq \frac{1}{|Y|} \int_Y A^0(\lambda + \nabla_y \chi_\lambda)(\mu + \nabla_y \chi_\mu)\mathrm{d}y, \tag{5.9}$$

which is formula (5.6).

From the coerciveness of A^0, we get for a.e. $x \in \Omega$

$$A^{\text{hom}}(x)(\lambda)\mu \geq \frac{\alpha}{|Y|}|\lambda + \nabla_y \chi_\lambda|^2_{L^2(Y)}.$$

As before, since χ_λ is Y-periodic, $|\lambda + \nabla_y \chi_\lambda|^2_{L^2(Y)} = |Y||\lambda|^2 + |\nabla_y \chi_\lambda|^2_{L^2(Y)}$, which shows the α-coerciveness of A^{hom}.

Finally, by (5.3), it follows that $|A^{\text{hom}}(\lambda)\mu| \leq (\beta(1+C(\alpha, \beta, Y)))^2|\lambda||\mu|$, which completes the proof. □

6 Convergence of the Energy and Correctors

Proposition 6.1 *Under the hypotheses of the preceding sections, the following holds*:

$$\lim_{\varepsilon \to 0} \int_Y A^\varepsilon(\nabla u_\varepsilon)\nabla u_\varepsilon \mathrm{d}x = \int_Y A^{\text{hom}}(\nabla u_0)\nabla u_0 \mathrm{d}x. \tag{6.1}$$

Proof By the definition of Problem (P_ε),

$$\int_Y A^\varepsilon(\nabla u_\varepsilon)\nabla u_\varepsilon \mathrm{d}x = \int_\Omega F_\varepsilon u_\varepsilon \mathrm{d}x.$$

Using the established convergences, it is straightforward to see that the right-hand side, once unfolded, converges to $\int_\Omega \mathcal{M}_Y(F_0)u_0 \mathrm{d}x$. We conclude by comparing with (5.7). □

Corollary 6.1 *The following strong convergence holds*:

$$\begin{cases} \mathcal{T}_\varepsilon(\nabla u_\varepsilon) \to \nabla u_0 + \nabla_y \widehat{u} \quad \textit{strongly in } L^2(\Omega \times Y), \\ \int_{\Lambda_\varepsilon} |\nabla u_\varepsilon|^2 \mathrm{d}x \to 0. \end{cases} \tag{6.2}$$

Proof By definition of A^{hom},

$$\int_Y A^{\text{hom}}(\nabla u_0)\nabla u_0 \mathrm{d}x = \frac{1}{|Y|} \int_{\Omega \times Y} A^0(\nabla u_0 + \nabla_y \widehat{u})(\nabla u_0 + \nabla_y \widehat{u})\mathrm{d}x\mathrm{d}y.$$

On the other hand, by the coercivity of A^ε,

$$\int_\Omega A^\varepsilon(\nabla u_\varepsilon)\nabla u_\varepsilon \mathrm{d}x \geq \frac{1}{|Y|} \int_{\Omega \times Y^*} \mathcal{T}_\varepsilon\big(A^\varepsilon\big)\mathcal{T}_\varepsilon(\nabla u_\varepsilon)\mathcal{T}_\varepsilon(\nabla u_\varepsilon)\mathrm{d}x\mathrm{d}y + \alpha \int_{\Lambda_\varepsilon} |\nabla u_\varepsilon|^2 \mathrm{d}x. \tag{6.3}$$

Therefore, by (6.1),

$$\limsup_{\varepsilon\to 0}\frac{1}{|Y|}\int_{\Omega\times Y^*}\mathcal{T}_\varepsilon(A^\varepsilon)\mathcal{T}_\varepsilon(\nabla u_\varepsilon)\mathcal{T}_\varepsilon(\nabla u_\varepsilon)\mathrm{d}x\mathrm{d}y$$
$$\leq \frac{1}{|Y|}\int_{\Omega\times Y}A^0(\nabla u_0+\nabla_y\widehat{u})(\nabla u_0+\nabla_y\widehat{u})\mathrm{d}x\mathrm{d}y. \tag{6.4}$$

Since $\mathcal{T}_\varepsilon(A^{\varepsilon)}$ converges a.e. to A^0, and $\mathcal{T}_\varepsilon(\nabla u_\varepsilon)$ converges weakly to $\nabla u_0+\nabla_y\widehat{u}$ in $L^2(\Omega\times Y)$, if follows from [6, Lemma 4.9] that

$$\mathcal{T}_\varepsilon(\nabla u_\varepsilon)\to\nabla u_0+\nabla_y\widehat{u}\quad\text{strongly in } L^2(\Omega\times Y).$$

In turn, this, together with (6.3) implies

$$\int_{\Lambda_\varepsilon}|\nabla u_\varepsilon|^2\mathrm{d}x\to 0.$$

□

Classically in the unfolding method, the convergences of Corollary 6.1 imply the existence of a corrector as follows.

Corollary 6.2 *Under the hypotheses of the preceding sections, as* $\varepsilon\to 0$,

$$|\nabla u_\varepsilon-\nabla u_0-U_\varepsilon(\nabla_y\widehat{u})|_{L^2(\Omega)}\to 0. \tag{6.5}$$

Making use of formula (5.4) for $\widehat{u}$ and Proposition 3.2, we get the following result.

Corollary 6.3 *Under the hypotheses of the preceding sections, as* $\varepsilon\to 0$, *the following strong convergence holds*:

$$\left\|\nabla u_\varepsilon-\nabla u_0-\sum_{i=1}^n\mathcal{U}_\varepsilon\left(\frac{\partial u_0}{\partial x_i}\right)\mathcal{U}_\varepsilon(\nabla_y\chi_i)\right\|_{L^2(\Omega)}\to 0. \tag{6.6}$$

In the case where the matrix field A does not depend on x, the following corrector result holds:

$$\left\|u_\varepsilon-u_0-\varepsilon\sum_{i=1}^n\mathcal{Q}_\varepsilon\left(\frac{\partial u_0}{\partial x_i}\right)\chi_i\left(\left\{\frac{\cdot}{\varepsilon}\right\}_Y\right)\right\|_{H^1(\Omega)}\to 0. \tag{6.7}$$

Proof By construction, for $i=1,\dots,n$, the function χ_i belongs to $L^\infty(\Omega;H^1(Y))$. By (6.5),

$$\|\nabla u_\varepsilon-\mathcal{U}_\varepsilon(\nabla u_0+\nabla_y\widehat{u_0})\|_{L^2(\Omega)}\to 0. \tag{6.8}$$

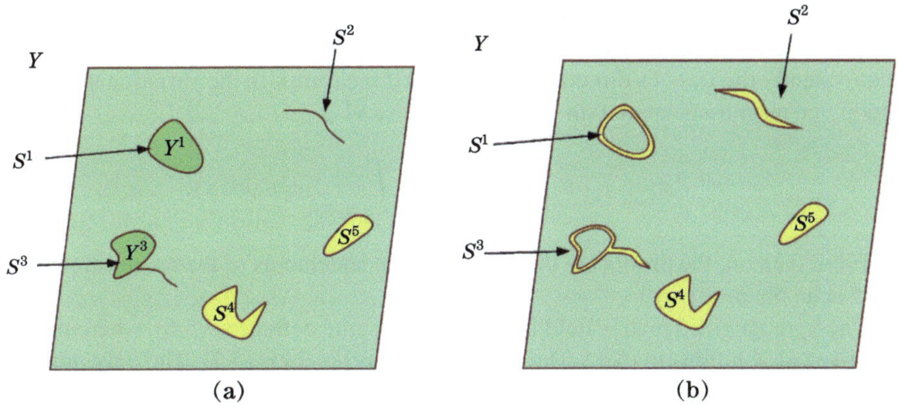

Fig. 6 The various cases of cracks and fissures (S^1, S^2 and S^3) as limits of thick inclusions

By Proposition 3.1(ii), this implies

$$\left\| \nabla u_\varepsilon - \nabla u_0 - \sum_{i=1}^{n} \mathcal{U}_\varepsilon\left(\frac{\partial u_0}{\partial x_i}\nabla_y \chi_i\right)\right\|_{L^2(\Omega)} \to 0. \tag{6.9}$$

Hence (6.6) follows directly from Proposition 3.2.

Convergence (6.7) follows from (6.6) as in [5]. □

7 Other Connected Problems

7.1 Cracks and Fissures

One can consider the case when some of the sets S^j are cracks or fissures, i.e., they are Lipschitz submanifolds of codimension one in Y. The case of a submanifold without boundary corresponds to the case of the boundary of a compact subset Y^j in Y. The case of a submanifold with a boundary corresponds to a crack in Y. A combination of the two can also occur (see Fig. 6(a) for the various cases).

Each of these cases can be seen as limits of thick inclusions (see Fig. 6(b)).

The corresponding conditions for regular cracks (which have two sides) in the strong form are as follows:

$$\begin{cases} \text{The solution } u \text{ is an unknown constant on each fissure } \varepsilon\xi + \varepsilon S^j, \\ \int_{\varepsilon\xi+\varepsilon S^j} [\frac{\partial u}{\partial \nu_A}]_{\varepsilon\xi+\varepsilon S^j} d\sigma(x) = g^j_{\varepsilon}{}_{|\varepsilon\xi+\varepsilon Y}, \text{ a given number,} \end{cases} \tag{7.1}$$

where $[\frac{\partial u}{\partial \nu_A}]_{\varepsilon\xi+\varepsilon S^j}$ denotes the sum of the two outward conormal derivatives from both sides of the crack (it can be considered as the jump of the conormal derivative across the fissure $\varepsilon\xi + \varepsilon S^j$, and hence it is denoted by the notation).

In the definition of the space W_0^ε, there is the requirement that the functions be constant almost everywhere with respect to the surface measure on the fissures (equivalently the $(n-1)$-dimensional Hausdorff measure). In the variational formulation, the term associated with the fissure $\varepsilon\xi + \varepsilon S^j$ is

$$\varepsilon^{n-1} g^j_{\varepsilon\,|\varepsilon\xi+\varepsilon Y} w_{|\varepsilon\xi+\varepsilon S^j} = \frac{1}{|S^j|}\int_{\varepsilon\xi+\varepsilon S^j} g^j_\varepsilon w \mathrm{d}\sigma(x).$$

From then on, the proofs are the same, and the statements of the results are modified in an obvious way.

The homogenized matrix field is given by the same definition (5.6), where the ξ_λ are given as solutions to (5.1). The homogenized problem is (5.7). The only modification is in the definition of the space $\mathbf{H}^S_{\mathrm{per}}$, where the conditions on the fissures are taken in the sense of traces (i.e., almost everywhere for the corresponding surface measures).

7.2 The Global Conductor 1

In the global conductor case, the situation is the same as in the previous cases, but all the conductors are somehow connected, so that the solution takes the same unknown constant value on all of S_ε. The problem is therefore set in the smaller subspace

$$W^\varepsilon_{0c} \doteq \big\{w \in H^1_0(\Omega);\ w|_{S_\varepsilon} \text{ is constant}\big\}, \tag{7.2}$$

and is defined for given A^ε and f_ε as before and for a given real number g_ε as

$$(\widetilde{\mathcal{P}}_\varepsilon)\ \begin{cases} \text{Find } u_\varepsilon \in W^\varepsilon_{0c}, \text{ such that for all } w \in W^\varepsilon_{0c}, \\ \int_\Omega A^\varepsilon \nabla u_\varepsilon \nabla w \mathrm{d}x = \int_\Omega f_\varepsilon w \mathrm{d}x + g_\varepsilon w|_{S_\varepsilon}. \end{cases} \tag{7.3}$$

It is easy to see that if f_ε is bounded in $H^{-1}(\Omega)$, and g_ε is bounded in $\mathbb{R}$, so is u_ε in $H^1_0(\Omega)$. By compactness of the Sobolev embedding, it follows that $\{u_\varepsilon\}$ is compact in $L^2(\Omega)$. Since 1_{S_ε} converges weakly-$*$ in $L^\infty(\Omega)$ to $\theta \doteq \frac{|S|}{|Y|} > 0$, in view of the identity $u_\varepsilon 1_{S_\varepsilon} \equiv C_\varepsilon (\in \mathbb{R})$, which converges weakly in $L^2(\Omega)$, it follows that the whole sequence $\{u_\varepsilon\}$ converges to a constant (namely, $\theta^{-1}\lim C_\varepsilon$). But the only constant in $H^1_0(\Omega)$ is 0. Therefore, u_ε also converges weakly to 0 in $H^1_0(\Omega)$, and C_ε converges to 0.

Here we use the obvious variant of Theorem 3.2, where the sequence $\frac{1}{\varepsilon}(\mathcal{T}_\varepsilon(u_\varepsilon) - C_\varepsilon)$ instead of $\frac{1}{\varepsilon}(\mathcal{T}_\varepsilon(u_\varepsilon) - \mathcal{M}_\varepsilon(u_\varepsilon))$ is used to obtain the limit $\widehat{u}$. This is valid because of the existence of the corresponding Poincaré-Wirtinger inequality in the space

$$H^S(Y) \doteq \big\{\psi \in H^1(Y) \text{ with } \psi_{|S} \equiv 0\big\}. \tag{7.4}$$

Proposition 7.1 *There exists a positive real number C_P, such that for every $\psi \in H^S(Y)$,*

$$|\psi|_{L^2(Y)} \leq C_P |\nabla_y \psi|_{L^2(Y)}. \tag{7.5}$$

Proof This is a straightforward consequence of the existence of a Poincaré-Wirtinger constant C_{PW} for $H^1(Y)$, which implies that for every $\psi \in H^1(Y)$,

$$|\psi - M_Y(\psi)|_{L^2(Y)} \leq C_{PW} |\nabla_y \psi|_{L^2(Y)}.$$

Assuming that ψ vanishes on S and taking the average over S (which is a positive Lebesgue measure) imply $|M_Y(\psi)|_{L^2(Y)} = |M_Y(\psi)||Y|^{\frac{1}{2}} \leq C_{PW}|\nabla_y \psi|_{L^2(Y)}$. Combining with the previous inequality, this implies inequality (7.5) with $C_P = 2C_{PW}$. □

We denote by $H^S_{\text{per}}(Y)$ the subspace of $H^S(Y)$ consisting of its Y-periodic elements.

It then follows that, up to a subsequence, $\frac{1}{\varepsilon}(\mathcal{T}_\varepsilon(u_\varepsilon) - C_\varepsilon)$ converges weakly to some $\widehat{u}$ in $L^2(\Omega; H^1(Y))$. Furthermore, $\widehat{u}$ belongs to $L^2(\Omega; H^S_{\text{per}}(Y))$. Under the same hypothesis on $\mathcal{T}_\varepsilon(A^{\varepsilon)}$ as before, $\mathcal{T}_\varepsilon(A^\varepsilon)\mathcal{T}_\varepsilon(\nabla u_\varepsilon)$ converges weakly to $\eta_0 \doteq A^0 \nabla_y \widehat{u}$ in $L^2(\Omega \times Y)$.

Considering as before a $w \in \mathcal{D}(\Omega)$ and a ϕ in $C^\infty_{\text{per}}(\overline{Y})$ which vanishes on S, one gets

$$\int_{\Omega \times Y} \eta_0(x, y) w(x) \nabla_y \phi(y) \mathrm{d}x \mathrm{d}y = 0.$$

By the same totality argument,

$$\int_{\Omega \times Y} \eta_0(x, y) \cdot \nabla_y \Phi(y) \mathrm{d}y = 0$$

holds for every $\Phi \in L^2(\Omega; H^1_{\text{per}})$, which vanishes on $\Omega \times S$. However, $\widehat{u}$ is itself in $L^2(\Omega; H^1_{\text{per}})$, and therefore, one concludes that $\nabla_y \widehat{u} \equiv 0$, and since $\widehat{u}$ vanishes on $\Omega \times S$, this implies that $\widehat{u}$ itself is 0.

The interesting question is to determine the next term in the expansion of u_ε in powers of ε. This requires more estimates, both for C_e and for $|\nabla u_\varepsilon|_{L^2(\Omega)}$.

If S_ε intersects $\partial\Omega$ on a set of non-zero capacity (which may well happen quite often), then clearly, $C_\varepsilon = 0$ (see Remark 7.3). When this is not the case, we use the well-known Hardy inequality in $H^1_0(\Omega)$, which requires that $\partial\Omega$ be Lipschitz. Denote by $\delta(x)$ the distance of x to $\partial\Omega$, and by Ω^d the set $\{x \in \Omega, \delta(x) < d\}$. The Hardy inequality states that there exists a constant C_H independent of d, such that

$$\forall w \in H^1_0(\Omega),\ \forall d > 0 \text{ and } d \text{ is small enough,} \quad \left|\frac{w}{\delta}\right|_{L^2(\Omega^d)} \leq C_H |\nabla w|_{L^2(\Omega^d)}. \tag{7.6}$$

Choosing $d = d_\varepsilon$ so that Ω^{d_ε} contains the boundary layer Λ_ε as well as all neighboring ε-cells, and applying the Hardy inequality to u_ε, one gets

$$\frac{|C_\varepsilon|}{d_\varepsilon}|S_\varepsilon \cap \Omega^{d_\varepsilon}|^{\frac{1}{2}} \leq \left|\frac{u_\varepsilon}{\delta}\right|_{L^2(\Omega^d)} \leq C_H|\nabla u_\varepsilon|_{L^2(\Omega^{d_\varepsilon})}. \tag{7.7}$$

Since $\partial\Omega$ is Lipschitz, $|\Omega^{d_\varepsilon}|$ is of order d_ε, and d_ε is of order ε, it follows that $|S_\varepsilon \cap \Omega^{d_\varepsilon}|$, which is of order $\theta|\Omega^{d_\varepsilon}|$, is itself of order ε. This implies that

$$|C_\varepsilon| \leq c\varepsilon^{\frac{1}{2}}|\nabla u_\varepsilon|_{L^2(\Omega^{d_\varepsilon})}. \tag{7.8}$$

A similar computation gives

$$|u_\varepsilon|_{L^2(\Omega^{d_\varepsilon})} \leq c\varepsilon|\nabla u_\varepsilon|_{L^2(\Omega^{d_\varepsilon})}. \tag{7.9}$$

We return to the first estimate of $|\nabla u_\varepsilon|_{L^2(\Omega)}$, letting F_ε denote $f_\varepsilon 1_{\Omega_\varepsilon} + g_\varepsilon 1_{S_\varepsilon}$ and using the unfolding formula as follows:

$$\alpha|\nabla u_\varepsilon|^2_{L^2(\Omega)} \leq \int_{\Lambda_\varepsilon} F_\varepsilon u_\varepsilon \mathrm{d}x + \int_{\Omega\setminus\Lambda_\varepsilon} F_\varepsilon(u_\varepsilon - C_\varepsilon)\mathrm{d}x + C_\varepsilon \int_{\Omega\setminus\Lambda_\varepsilon} F_\varepsilon \mathrm{d}x. \tag{7.10}$$

By the Proposition 7.1, it follows that

$$\begin{aligned}|u_\varepsilon - C_\varepsilon|_{L^2(\Omega\setminus\Lambda_\varepsilon)} &= |\mathcal{T}_\varepsilon(u_\varepsilon) - C_\varepsilon|_{L^2(\Omega\times Y)} \leq C|\nabla_y \mathcal{T}_\varepsilon(u_\varepsilon)|_{L^2(\Omega\times Y)}\\ &= C\varepsilon|\mathcal{T}_\varepsilon(\nabla u_\varepsilon)|_{L^2(\Omega\times Y)} = C\varepsilon|\nabla u_\varepsilon|_{L^2(\Omega)}.\end{aligned} \tag{7.11}$$

Since Ω^{d_ε} contains Λ_ε, combining (7.8)–(7.9) and (7.11) with (7.10) gives

$$\alpha|\nabla u_\varepsilon|_{L^2(\Omega)} \leq c\left(\varepsilon|f_\varepsilon|_{L^2(\Omega)} + \varepsilon^{\frac{1}{2}}\left|\int_{\Omega\setminus\Lambda_\varepsilon} F_\varepsilon \mathrm{d}x\right|\right). \tag{7.12}$$

Suppose that $|F_\varepsilon|_{L^2(\Omega)}$ is bounded. Then, by (7.12), $|\nabla u_\varepsilon|_{L^2(\Omega)}$ is an $O(\varepsilon^{\frac{1}{2}})$, so that by (7.8), C_ε is an $O(\varepsilon)$. Together with (7.9) and (7.11), this shows that $|u_\varepsilon|_{L^2(\Omega)}$ is also an $O(\varepsilon)$. Inequality (7.11) also implies that $|u_\varepsilon - C_\varepsilon|_{L^2_{\mathrm{loc}}(\Omega)}$ is an $O(\varepsilon^{\frac{3}{2}})$. These estimates do not suffice to obtain the next term in the expansion if C_ε is not zero. We need the extra hypothesis

$$(\mathrm{H}_0) \text{ For the values of } \varepsilon, \text{ such that } C_\varepsilon \neq 0, \quad \left|\int_{\Omega\setminus\Lambda_\varepsilon} F_\varepsilon \mathrm{d}x\right| \text{ is an } O\left(\varepsilon^{\frac{1}{2}}\right). \tag{7.13}$$

Proposition 7.2 *Under* (H_0), $|u_\varepsilon|_{H^1_0(\Omega)}$ *is an* $O(\varepsilon)$. *Furthermore,* $|u_\varepsilon|_{L^2(\Omega)}$ *and* C_ε *are* $O(\varepsilon^{\frac{3}{2}})$ *and* $|u_\varepsilon - C_\varepsilon|_{L^2_{\mathrm{loc}}(\Omega)}$ *is an* $O(\varepsilon^2)$.

Proof This is a consequence of (7.12), and then of (7.8)–(7.9) and (7.11). □

Remark 7.1 Since $\partial\Omega$ is assumed Lipschitz, it follows that $|\int_{\Lambda_\varepsilon} F_\varepsilon \mathrm{d}x|$ is bounded above by $|F_\varepsilon|_{L^2(\Omega)}|\Lambda_\varepsilon|^{\frac{1}{2}}$, which is itself an $O(\varepsilon^{\frac{1}{2}})$. Consequently, in the condition (H_0) above, $|\int_{\Omega\setminus\Lambda_\varepsilon} F_\varepsilon \mathrm{d}x|$ can be replaced by $|\int_\Omega F_\varepsilon \mathrm{d}x|$.

One can now apply the variant of Theorem 3.2 to the sequence $U_\varepsilon \doteq \frac{u_\varepsilon}{\varepsilon}$. Up to a subsequence, there exist two functions U_0 in $H_0^1(\Omega)$ and $\widehat{U}$ in $L^2(\Omega; H^S_{\mathrm{per}}(Y))$, such that

$$
\begin{aligned}
&U_\varepsilon \text{ converges weakly to } U_0 \text{ in } H_0^1(\Omega),\\
&\mathcal{T}_\varepsilon(\nabla U_\varepsilon) \text{ converges weakly to } \nabla U_0 + \nabla_y \widehat{U} \text{ in } L^2(\Omega \times Y),\\
&\frac{1}{\varepsilon}\big(\mathcal{T}_\varepsilon(U_\varepsilon) - \varepsilon^{-1}C_\varepsilon\big) \text{ converges weakly to } y_M \cdot \nabla U_0 + \widehat{U} \text{ in } L^2\big(\Omega; H^S(Y)\big).
\end{aligned}
\tag{7.14}
$$

Because of Proposition 7.2, a simplification $U_0 \equiv 0$ occurs.

Consequently, $\frac{1}{\varepsilon}(\mathcal{T}_\varepsilon(U_\varepsilon) - \varepsilon^{-1}C_\varepsilon)$ converges weakly to $\widehat{U}$ in $L^2(\Omega; H^S(Y))$.

We complete the assumptions with

$$
(\widehat{\mathrm{H}}) \quad \begin{cases} \mathcal{T}_\varepsilon(A^\varepsilon) \text{ converges in measure (or a.e.) in } \Omega \times Y \text{ to } A^0,\\ \mathcal{T}_\varepsilon(f_\varepsilon) \text{ converges weakly to } f_0 \text{ in } L^2(\Omega \times Y^*),\\ g_\varepsilon \text{ converges to } g_0 \text{ in } \mathbb{R}. \end{cases}
$$

Note that under the condition of Remark 7.1, $\int_{\Omega\times Y^*} f_0(x,y)\mathrm{d}x\mathrm{d}y + g_0|\Omega||S| = 0$.

Theorem 7.1 *Under assumptions* (H_0) *and* $(\widehat{\mathrm{H}})$, $\mathcal{T}_\varepsilon(\frac{1}{\varepsilon^2}(u_\varepsilon - C_\varepsilon))$ *converges weakly in* $L^2(\Omega; H^S(Y))$ *to* $\widehat{U}$, *which is the unique solution of the following variational problem*:

$$
\begin{cases} \widehat{U} \in L^2(\Omega; H^S_{\mathrm{per}}(Y)), \ \forall \Psi \in L^2(\Omega; H^S_{\mathrm{per}}(Y)),\\ \int_{\Omega\times Y} A^0(x,y)\nabla_y \widehat{U}(x,y)\nabla_y \Psi(x,y)\mathrm{d}x\mathrm{d}y = \int_{\Omega\times Y} f_0(x,y)\Psi(x,y)\mathrm{d}x\mathrm{d}y. \end{cases}
\tag{7.15}
$$

Proof Consider a ψ in $C^\infty_{\mathrm{per}}(\overline{Y})$, which vanishes on S. Again let ϕ be in $\mathcal{D}(\Omega)$. Then, $w_\varepsilon(x) \doteq \varepsilon\phi(x)\psi(\{\frac{x}{\varepsilon}\}_Y)$ is in the space W^ε_{0c}, so it is an acceptable test function. As in the previous computation, this gives at the limit

$$
\int_{\Omega\times Y} A^0(x,y)\nabla_y \widehat{U}(x,y)\phi(x)\nabla_y\psi(y)\mathrm{d}x\mathrm{d}y = \int_{\Omega\times Y} f_0(x,y)\phi(x)\psi(y)\mathrm{d}x\mathrm{d}y.
$$

By the usual totality argument, this implies (7.15). The existence and uniqueness of $\widehat{U}$ follow from the application of the Lax-Milgram theorem. □

The strong formulation of Problem (7.15) is as follows: $\widehat{U} \in L^2(\Omega; H^S_{\rm per}(Y))$ and

$$-\operatorname{div}_y\big(A^0(x,y)\nabla_y \widehat{U}(x,y)\big) = f_0(x,y) \quad \text{for a.e. } (x,y) \in \Omega \times Y^*.$$

The result obtained here can be seen as a sort of expansion of order 2 in ε.

From the weak convergence of $\mathcal{T}_\varepsilon(\frac{1}{\varepsilon^2}(u_\varepsilon - C_\varepsilon))$ to $\widehat{U}$ in $L^2(\Omega; H^S(Y))$, it follows that $\frac{1}{\varepsilon^2}(u_\varepsilon - C_\varepsilon)$ converges weakly to $\mathcal{M}_Y(\widehat{U})$ in $L^2(\Omega)$.

These are not completely satisfactory, because we do not know if, in general, C_ε is of order ε^2 itself.

Remark 7.2 If one assumes a stronger condition than (H_0), namely, $|\int_\Omega F_\varepsilon dx|$ is an $o(\varepsilon^{\frac{1}{2}})$, then one can show the convergence of the energy, which implies that

$$\mathcal{T}_\varepsilon\left(\frac{\nabla u_\varepsilon}{\varepsilon}\right) \text{ converges strongly in } L^2(\Omega \times Y) \text{ to } \nabla_y \widehat{U}. \tag{7.16}$$

In turn, this gives the following corrector result:

$$\nabla u_\varepsilon = \varepsilon U_\varepsilon(\nabla_y \widehat{U}) + o_{L^2(\Omega)}(\varepsilon).$$

There remains a boundary layer, if one wants a corrector for u_ε itself, as well as the open question of the behavior of $\frac{C_\varepsilon}{\varepsilon^2}$.

Remark 7.3 In the case where S_ε intersects $\partial\Omega$ in a set of non-zero capacity, it follows that C_ε is 0. Assuming that this is the case for all ε's of the sequence, this problem reduces to that of a homogeneous Dirichlet condition in S_ε. The corresponding problem was originally studied for the Stokes system by Tartar in the appendix of [18] and for the Laplace equation by Lions [16]. The nonlinear case was later studied in [9].

7.3 The Global Conductor 2

The presentation of the previous section is not necessarily realistic because of the unpredictability of the fact that S_ε intersects the boundary of Ω. It may be more realistic to consider that the conductor is restricted to a compact subset $\overline{\Omega_0}$ of Ω, i.e., $S^0_\varepsilon \doteq S_\varepsilon \cap \overline{\Omega_0}$. We make this hypothesis in this section. We shall also assume that $\partial\Omega_0$ is a null set for the Lebesgue measure, and denote $\Omega \setminus \overline{\Omega_0}$ by Ω_1 (see Fig. 7).

In this case, the variational space for the original problem is

$$W^{S_\varepsilon}_{0c} \doteq \big\{w \in H^1_0(\Omega);\ w_{|S^0_\varepsilon} \text{ is a constant}\big\}. \tag{7.17}$$

Fig. 7 The global conductor

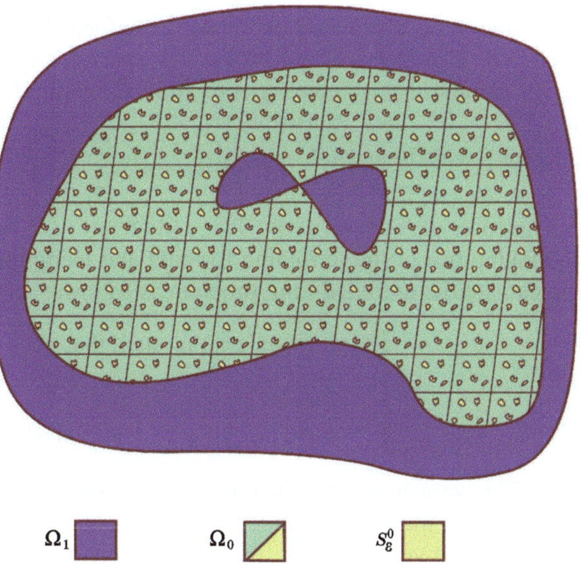

The variational formulation is for given A^ε and f_ε as before and for a given real number g_ε as

$$(\widetilde{\mathcal{P}}_\varepsilon)\ \begin{cases} \text{Find } u_\varepsilon \in W_{0c}^{S_\varepsilon}, \text{ such that for all } w \in W_{0c}^{S_\varepsilon}, \\ \int_\Omega A^\varepsilon \nabla u_\varepsilon \nabla w dx = \int_\Omega f_\varepsilon w dx + g_\varepsilon w|_{S_\varepsilon^0}. \end{cases} \tag{7.18}$$

It is easy to see that if f_ε is bounded in $L^2(\Omega)$ (actually $H^{-1}(\Omega)$ is enough!), and g_ε is bounded in $\mathbb{R}$, so is u_ε in $H_0^1(\Omega)$. By compactness of the Sobolev embedding, it follows that $\{u_\varepsilon\}$ is compact in $L^2(\Omega)$. Since $1_{S_\varepsilon^0}$ converges weakly-$*$ in $L^\infty(\Omega)$ to $\theta \doteq \frac{|S|}{|Y|} 1_{\overline{\Omega}_0} > 0$, and in view of the identity $u_\varepsilon 1_{S_\varepsilon^0} \equiv C_\varepsilon \in \mathbb{R}$, which converges weakly in $L^2(\Omega)$, it follows that the whole sequence $\{u_\varepsilon 1_{\overline{\Omega}_0}\}$ converges to a constant (namely, $\frac{|Y|}{|S|} \lim C_\varepsilon$). Consequently, every weak limit point of $\{u_\varepsilon\}$ is constant on $\overline{\Omega}_0$.

By Theorem 3.2, it follows that, up to a subsequence, u_ε converges weakly to some u_0 in $H_0^1(\Omega)$ and $\frac{1}{\varepsilon}(\mathcal{T}_\varepsilon(u_\varepsilon) - \mathcal{M}_Y(u_\varepsilon))$ converges weakly to some $\widehat{u}$ in $L^2(\Omega; H^1(Y))$. Furthermore, $\widehat{u}$ belongs to $L^2(\Omega; H^1_{\mathrm{per}}(Y))$ and $\mathcal{M}_Y(\widehat{u}) = 0$. Also note that, as before, $(y_M \cdot \nabla u_0 + \widehat{u})_{|\Omega_0 \times S}$ is a constant. Since u_0 is a constant on Ω_0, this reduces to that $\widehat{u}_{|\Omega_0 \times S}$ is a constant.

Let $\mathcal{H}$ denote the subspace of $H_0^1(\Omega)$, consisting of the functions, which are constant a.e. in $\overline{\Omega}_0$.

Now, the pair $(u_0, \widehat{u})$ belongs to the space

$$\begin{aligned} \mathcal{W} = \big\{ & (w, \widehat{w});\ w \in \mathcal{H}, \widehat{w} \in L^2\big(\Omega; H^1_{\mathrm{per}}(Y)\big), \\ & \mathcal{M}_Y(\widehat{w}) = 0 \text{ and } \widehat{w}_{|\overline{\Omega}_0 \times S} \text{ is a constant}\big\}. \end{aligned} \tag{7.19}$$

Under the same hypothesis on $\mathcal{T}_\varepsilon(A^{\varepsilon)}$ as before, $\mathcal{T}_\varepsilon(A^\varepsilon)\mathcal{T}_\varepsilon(\nabla u_\varepsilon)$ converges weakly to $\eta_0 \doteq A^0\nabla_y\widehat{u}$ in $L^2(\Omega \times Y)$ (of course, η_0 vanishes on $\overline{\Omega}_0 \times S$).

We assume that g_ε converges to g_0 in $\mathbb{R}$, while $\mathcal{T}_\varepsilon(f_\varepsilon)$ converges weakly to f_0 in $L^2(\Omega \times Y)$ (they vanish on $\overline{\Omega}_0 \times S$). Consequently, $F_\varepsilon \doteq \mathcal{T}_\varepsilon(f_\varepsilon) + g_\varepsilon 1_{\overline{\Omega}_0\times S}$ converges weakly to $F_0 \doteq f_0 + g_0 1_{\overline{\Omega}_0\times S}$ in $L^2(\Omega \times Y)$.

Now let $(w, \widehat{w})$ be an element of $\mathcal{W} \cap (\mathcal{D}(\Omega) \times \mathcal{D}(\Omega; C^1(\overline{Y})))$, and set as the test function $\varphi_\varepsilon(x) \doteq w(x) + \varepsilon\widehat{w}(x, \{\frac{x}{\varepsilon}\}_Y)$.

Making use of the unfolding formula and using the standard density argument, one can get the unfolded limit problem as follows:

$$\begin{cases} \text{Find } (u_0, \widehat{u}) \in \mathcal{W}, \text{ such that } \forall (w, \widehat{w}) \in W, \\ \frac{1}{|Y|}\int_{\Omega\times Y} A^0(\nabla u_0 + \nabla_y\widehat{u})(\nabla w + \nabla_y\widehat{w})dxdy \\ \quad = \frac{1}{|Y|}\int_{\Omega\times Y} F_0(x, y)w(x)dxdy. \end{cases} \tag{7.20}$$

Since ∇u_0 vanishes in Ω_0, this implies that $\widehat{u}$ is also zero on $\Omega_0 \times Y$.

Therefore, the left-hand side in (7.20) is computed only on $\Omega_1 \times Y$.

For a given vector $\lambda \in \mathbb{R}^n$, the cell-problem is defined for a.e. $x \in \Omega_1$ as

$$\begin{cases} \text{Find } \chi_\lambda \in H^1_{\text{per}}(Y), \text{ such that } \mathcal{M}_Y(\chi_\lambda) = 0, \\ \int_Y A^0(x, y)(\lambda + \nabla_y\chi_\lambda(y))\nabla_y\varphi(y)dy = 0, \quad \forall\varphi \in \mathbf{H}^S_{\text{per}}. \end{cases} \tag{7.21}$$

This is actually the "standard" corrector for periodic homogenization in Ω_1. The corresponding homogenized matrix is the standard one, namely,

$$A^{\text{hom}}(\lambda, \mu) \doteq \frac{1}{|Y|}\int_Y A^0(\lambda + \nabla_y\chi_\lambda)(\mu + \nabla_y\chi_\mu)dy. \tag{7.22}$$

Proposition 7.3 *The homogenized limit problem takes the following form:*

$$\begin{cases} \textit{Find } u_0 \in \mathcal{H}, \textit{ such that } \forall w \in \mathcal{H}, \\ \int_{\Omega_1} A^{\text{hom}}(\nabla u_0)\nabla w dx = \int_\Omega \mathcal{M}_Y(F_0)w dx. \end{cases} \tag{7.23}$$

The strong formulation can be given here, and if $\partial\Omega_0$ is Lipschitz as a problem on Ω_1,

$$\begin{cases} -\operatorname{div}(A^{\text{hom}}\nabla u_0) = \mathcal{M}_Y(f_0) \text{ in } \Omega_1, \\ u_{0|\partial\Omega_0} \text{ is an unknown constant}, \\ \int_{\partial\Omega_0} \frac{\partial u_0}{\partial\nu_{A^{\text{hom}}}} d\sigma(x) = \int_{\Omega_0} \mathcal{M}_Y(F_0)dx. \end{cases} \tag{7.24}$$

Remark 7.4 (1) The convergence of the energy holds in this situation, which implies the existence of a corrector.

(2) This result holds in the more general situation, where the sequence of matrix fields A^ε H-converges to A^{hom} in Ω_1 (for the definition and properties of H-convergence, see [17]).

Acknowledgements It was during a visit of the second author at the ISFMA that the work on this paper was started. He expresses his thanks for ISFMA's generous support.

References

1. Bellieud, M.: Torsion effects in elastic composites with high contrast. SIAM J. Math. Anal. **41**(6), 2514–2553 (2009/2010)
2. Braides, A., Garroni, A.: Homogenization of nonlinear periodic media with stiff and soft inclusions. Math. Models Methods Appl. Sci. **5**(4), 543–564 (1995)
3. Briane, M.: Homogenization of the torsion problem and the Neumann problem in nonregular periodically perforated domains. Math. Models Methods Appl. Sci. **7**(6), 847–870 (1997)
4. Cioranescu, D., Damlamian, A., Griso, G.: Periodic unfolding and homogenization. C. R. Acad. Sci. Paris, Ser. I **335**, 99–104 (2002)
5. Cioranescu, D., Damlamian, A., Griso, G.: The periodic unfolding method in homogenization. SIAM J. Math. Anal. **40**(4), 1585–1620 (2008)
6. Cioranescu, D., Damlamian, A., Donato, P., et al.: The periodic unfolding method in domains with holes. SIAM J. Math. Anal. **44**(2), 718–760 (2012)
7. Cioranescu, D., Paulin, J.S.J.: Homogenization in open sets with holes. J. Math. Anal. Appl. **71**, 590–607 (1979)
8. Cioranescu, D., Paulin, J.S.J.: Homogenization of Reticulated Structures. Applied Mathematical Sciences, vol. 136. Springer, New York (1999)
9. Donato, P., Picard, C.: Convergence of Dirichlet problems for monotone operators in a class of porous media. Ric. Mat. **49**, 245–268 (2000)
10. Gianni, G.D.M., Murat, F.: Asymptotic behaviour and correctors for Dirichlet problems in perforated domains with homogeneous monotone operators. Ann. Sc. Norm. Super. Pisa, Cl. Sci. **24**(2), 239–290 (1997)
11. De Arcangelis, R., Gaudiello, A., Paderni, G.: Some cases of homogenization of linearly coercive gradients constrained variational problems. Math. Models Methods Appl. Sci. **6**, 901–940 (1996)
12. Griso, G.: Error estimate and unfolding for periodic homogenization. Asymptot. Anal. **40**, 269–286 (2004)
13. Lanchon, H.: Torsion éastoplastique d'un arbre cylindrique de section simplement ou multiplement connexe. J. Méc. **13**, 267–320 (1974)
14. Li, T.T., Tan, Y.J.: Mathematical problems and methods in resistivity well-loggings. Surv. Math. Ind. **5**(3), 133–167 (1995)
15. Li, T.T., Zheng, S.M., Tan, Y.Y., Shen, W.X.: Boundary Value Problems with Equivalued Surface and Resistivity Well-Logging. Pitman Research Notes in Mathematics Series, vol. 382. Longman, Harlow (1998)
16. Lions, J.L.: Some Methods in the Mathematical Analysis of Systems and Their Control. Science Press/Gordon & Breach, Beijing/New York (1981)
17. Murat, F., Tartar, L.: H-Convergence. In: Kohn, R.V. (ed.) Topics in the Mathematical Modelling of Composite Materials. Progr. Nonlinear Differential Equations Appl., vol. 31, pp. 21–43. Birkhäser, Boston (1997)
18. Sanchez-Palencia, E.: Nonhomogeneous Media and Vibration Theory. Lecture Notes in Physics, vol. 127. Springer, New York (1980)

Global Null Controllability of the 1-Dimensional Nonlinear Slow Diffusion Equation

Jean-Michel Coron, Jesús Ildefonso Díaz, Abdelmalek Drici, and Tommaso Mingazzini

Abstract The authors prove the global null controllability for the 1-dimensional nonlinear slow diffusion equation by using both a boundary and an internal control. They assume that the internal control is only time dependent. The proof relies on the return method in combination with some local controllability results for nondegenerate equations and rescaling techniques.

Keywords Nonlinear control · Nonlinear slow diffusion equation · Porous medium equation

Mathematics Subject Classification 35L65 · 35L567

Project supported by the ITN FIRST of the Seventh Framework Programme of the European Community (No. 238702), the ERC advanced grant 266907 (CPDENL) of the 7th Research Framework Programme (FP7), DGISPI of Spain (ProjectMTM2011-26119) and the Research Group MOMAT (No. 910480) supported by UCM.

J.-M. Coron (✉)
UMR 7598 Laboratoire Jacques-Louis Lions, Institut Universitaire de France and Université Pierre et Marie Curie (Paris 6), 4, place Jussieu, 75252 Paris cedex 5, France
e-mail: coron@ann.jussieu.fr

J.I. Díaz
Instituto de Matemática Interdisiplinar and Dpto. de Matemática Aplicada, Universidad Complutense de Madrid, Plaza de las Ciencias 3, 28040 Madrid, Spain
e-mail: diaz.racefyn@insde.es

A. Drici
UMR 7598 Laboratoire Jacques-Louis Lions, Université Pierre et Marie Curie (Paris 6), 4, place Jussieu, 75252 Paris cedex 5, France
e-mail: drici@ann.jussieu.fr

T. Mingazzini
Dpto. de Matemática Aplicada, Universidad Complutense de Madrid, Plaza de las Ciencias 3, 28040 Madrid, Spain
e-mail: tommaso.mingazzini@gmail.com

P.G. Ciarlet et al. (eds.), *Partial Differential Equations: Theory, Control and Approximation*, DOI 10.1007/978-3-642-41401-5_8,

1 Introduction

We study the null controllability of the 1-dimensional nonlinear slow diffusion equation, sometimes referred to as the Porous Media Equation (or PME for short), using both internal and boundary controls. The methods we used need such a combination of controls due to the degenerate nature of this quasilinear parabolic equation.

The PME belongs to the more general family of nonlinear diffusion equations of the form

$$y_t - \Delta\phi(y) = f, \tag{1.1}$$

where ϕ is a continuous nondecreasing function with $\phi(0) = 0$. For the PME, the constitutive law is precisely given by

$$\phi(y) = |y|^{m-1}y \tag{1.2}$$

with $m \geq 1$.

This family of equations arises in many different frameworks and, depending on the nature of ϕ, it models different diffusion processes, mainly grouped into three categories: "slow diffusion", "fast diffusion" and linear processes.

The "slow diffusion" case is characterized by a finite speed of propagation and the formation of free boundaries, while the "fast diffusion" one is characterized by a finite extinction time, which means that the solution becomes identically zero after a finite time.

If one neglects the source term, i.e., $f \equiv 0$, and imposes the constraint of non-negativeness to the solutions (which is fundamental in all the applications where y represents for example a density), then one can precisely characterize these phenomena. In fact, it was shown in [12] that the solution of the homogeneous Dirichlet problem associated to (1.1) on a bounded open set Ω of $\mathbb{R}^N$ satisfies a finite extinction time if and only if

$$\int_0^1 \frac{\mathrm{d}s}{\phi(s)} < +\infty,$$

which corresponds to the case $m \in (0, 1)$ for constitutive laws given by (1.2). On the contrary, if

$$\int_0^1 \frac{\mathrm{d}s}{\phi(s)} = +\infty, \tag{1.3}$$

(which is the case for $m \geq 1$) then, for any initial datum $y_0 \in H^{-1}(\Omega) \cap L^1(\Omega)$ with $(-\Delta)^{-1}y_0 \in L^\infty(\Omega)$, there is a kind of "retention property". This means that, if $y_0(x) > 0$ on a positively measured subset $\Omega' \subset \Omega$, then $y(\cdot, t) > 0$ on Ω' for any $t > 0$. In addition to (1.3), if ϕ satisfies

$$\int_0^1 \frac{\phi'(s)\mathrm{d}s}{s} < +\infty,$$

(i.e., $m > 1$ in the case of (1.2)) then the solution enjoys a finite speed of propagation and generates a free boundary given by that of its support ($\partial\{y > 0\}$).

Most typical applications of "slow diffusion" are as follows: Nonlinear heat propagation, groundwater filtration and the flow of an ideal gas in a homogeneous porous medium. With regard to the "fast diffusion", it rather finds a paradigmatic application to the flow in plasma physics. Many results and references can by found in the monographs [2, 23].

As already said, the aim of this paper is to show how a combined action of boundary controls and a spatially homogeneous internal control may allow the global extinction of the solution (the so-called global null controllability) in any prescribed temporal horizon $T > 0$. We shall prove the global null controllability for the following two control problems:

$$\mathrm{P}_{DD}\begin{cases} y_t - (y^m)_{xx} = u(t)\chi_I(t), & (x,t) \in (0,1) \times (0,T), \\ y(0,t) = v_0(t)\chi_I(t), & t \in (0,T), \\ y(1,t) = v_1(t)\chi_I(t), & t \in (0,T), \\ y(x,0) = y_0(x), & x \in (0,1), \end{cases} \tag{1.4}$$

and

$$\mathrm{P}_{DN}\begin{cases} y_t - (y^m)_{xx} = u(t)\chi_I(t), & (x,t) \in (0,1) \times (0,T), \\ (y^m)_x(0,t) = 0, & t \in (0,T), \\ y(1,t) = v_1(t)\chi_I(t), & t \in (0,T), \\ y(x,0) = y_0(x), & x \in (0,1), \end{cases} \tag{1.5}$$

where $I := (t_1, T)$ with $t_1 \in (0,T)$, $m \geq 1$ and χ_I is the characteristic function of I. In both problems, y represents the state variable and $U_{DN} := (u\chi_I, 0, v_1\chi_I)$, respectively $U_{DD} := (u\chi_I, v_0\chi_I, v_1\chi_I)$, is the control variable. The function y^m should be more properly written in form (1.2), but as we shall impose the constraint $y \geq 0$, it makes no real difference.

We emphasize the fact that the internal control $u(t)$ has the property to be independent of the space variable x and that all the controls are active only on a part of the time interval. Moreover, as we shall show later, the systems are null controllable in arbitrarily fixed time, and then the localized form of the control $u(t)\chi_I(t)$ (the same for the boundary controls) on a subinterval of $[0,T]$ is more an emphatic difficulty than a real difficulty. It serves mostly to underline that the controls are not active in the first time lapse. In the same way, it could be possible to take a control interval $(\underline{t}, \overline{t})$ with $\underline{t}, \overline{t} \in (0,T)$ or, even more generally, three different intervals, one for each control v_0, v_1, u, such that the intersection of the three is not empty.

The main results of this paper are contained in the following statement.

Theorem 1.1 *Let* $m \in [1, +\infty)$.

(i) *For any initial data* $y_0 \in H^{-1}(0,1)$ *such that* $y_0 \geq 0$ *and any time* $T > 0$, *there exist controls* $v_0(t), v_1(t)$ *and* $u(t)$ *with* $v_0(t)\chi_I(t), v_1(t)\chi_I(t) \in H^1(0,T)$,

$v_0, v_1 \geq 0$ and $u \in L^\infty(0,T)$ such that the solution y of P_{DD} satisfies $y \geq 0$ on $(0,1)\times(0,T)$, and $y(\cdot,T)\equiv 0$ on $(0,1)$.

(ii) *For any initial data $y_0 \in H^{-1}(0,1)$ such that $y_0 \geq 0$ and any time $T > 0$, there exist controls $v_1(t)$ and $u(t)$ with $v_1(t)\chi_I(t) \in H^1(0,T)$, $v_1 \geq 0$ and $u \in L^\infty(0,T)$ such that the solution y of P_{DN} satisfies $y \geq 0$ on $(0,1)\times(0,T)$, and $y(\cdot,T)\equiv 0$ on $(0,1)$.*

Notice that since $H^{-1}(0,1) = (H_0^1(0,1))'$ and $H_0^1(0,1) \subset C([0,1])$, we have $H^{-1}(0,1) \supset \mathcal{M}(0,1)$, where $\mathcal{M}(0,1)$ is the set of bounded Borel measures on $(0,1)$; for instance, the initial datum can be a Dirac mass distribution at a point in $(0,1)$. As said before in the case of "slow diffusion" ($m > 1$), the solution may present a free boundary given by the boundary of its support (whenever the support of y_0 is strictly smaller than $[0,1]$). Nevertheless, our strategy is built in order to prevent such a situation. Indeed, on the set of points (x,t) where y vanishes (i.e., on the points $(x,t) \in (0,1)\times(0,T)\setminus \mathrm{supp}(y)$), the diffusion operator is not differentiable at $y \equiv 0$, and so some linearization methods which work quite well for second order semilinear parabolic problems (see, e.g., [13, 17, 19, 20]) can not be applied directly. Moreover, the evanescent viscosity perturbation with some higher order terms only gives some controllability results for suitable functions ϕ, as the ones of the Stefan problem (see [13–15]).

Here we follow a different approach which is mainly based on the so-called return method introduced in [9, 10] (see [11, Chap. 6] for information on this method). More precisely, we shall prove first the null controllability of problem (1.4) by applying an idea appeared in [8] (for the controllability of the Burgers equation). In the second step, we shall show, using some symmetry arguments, that the same result holds for (1.5).

Our version of the return method consists in choosing a suitable parametrized family of trajectories $\frac{a(t)}{\varepsilon}$, which is independent of the space variable, going from the initial state $y \equiv 0$ to the final state $y \equiv 0$. We shall use the controls to reach one of such trajectories, no matter which one, in some positive time smaller than the final T. Once we fix a partition of the form $0 < t_1 < t_2 < t_3 < T$, we shall choose a function $a(t)$ satisfying the following properties:

(i) $a \in C^2([0,T])$;
(ii) $a(t) = 0$, $0 \leq t \leq t_1$ and $t = T$;
(iii) $a(t) > 0$, $t \in (t_1, T)$;
(iv) $a(t) = 1$, $t_2 \leq t \leq t_3$.

Then, the solution y of problem P_{DD} can be written as a perturbation of the explicit solution $\frac{a(t)}{\varepsilon}$ of the same equation with the control $U := (\frac{a(t)}{\varepsilon}, \frac{a(t)}{\varepsilon}, \frac{a(t)}{\varepsilon})$ in the following way:

$$y(x,t) = \left(\frac{a(t)}{\varepsilon} + z(x,t)\right). \tag{1.6}$$

Now, our aim is to find controls such that $z(\cdot,t_3) \equiv 0$, which means that we have controlled our solution $y(\cdot,t)$ to the state $\frac{1}{\varepsilon}$ at time $t = t_3$; this will be done by using

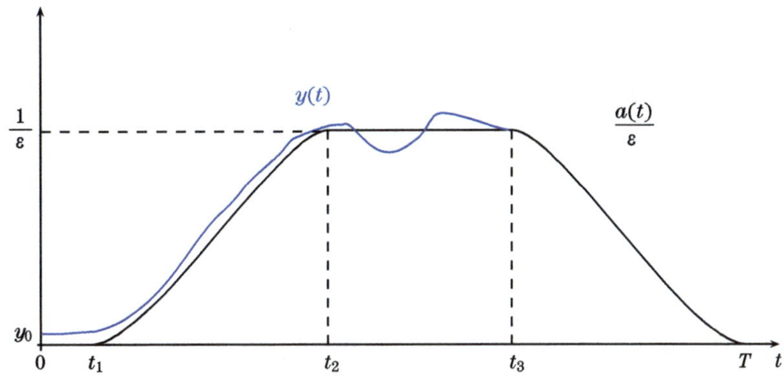

Fig. 1 Solution profile

a slight modification of a result in [4]. On the final time interval (t_3, T), we shall use the same trajectory $y(\cdot, t) \equiv \frac{a(t)}{\varepsilon}$ to reach the final state $y(\cdot, T) \equiv 0$. An ideal representation of the trajectory can be seen in Fig. 1.

One can see that the central core of our procedure is to drive the initial state to a constant state in a finite time thanks to the use of a boundary and internal control which only depends on the time variable.

On the first interval $(0, t_1)$, we shall not make any use of the controls. So we let the solution $y(t) := y(\cdot, t)$ regularize itself from an initial state in $H^{-1}(0, 1)$ to a smoother one in $H_0^1(0, 1)$ for $t = t_1$. Then, as the degenerate character of the diffusion operator neglects the diffusion effects outside the support of the state, we move $y(t)$ away from the zero state by asking $z(t) := z(\cdot, t)$ to be nonnegative at least on the interval (t_1, t_2). With this trick, the solution $y(t)$ will be far enough from zero. On the interval (t_2, t_3) the states $y(t)$ will be kept strictly positive even if the internal control $u(t)$ will be allowed to take negative values.

As already mentioned concerning the local retention property, we point out that the presence of the control $u(t)$ is fundamental for the global null controllability. To be more precise, notice that if we assume $u(t) \equiv 0$ then we can find initial states which can not be steered to zero at time T just with some nonnegative boundary controls. As a matter of fact, one can use the well known family of Barenblatt solutions (see [3, 23]) (also known as ZKB solutions) to show it. Indeed, if we introduce the parameters

$$\alpha = \frac{1}{m+1}, \qquad k = \frac{m-1}{2m(m+1)}, \qquad \tau \ll 1,$$

and choose C such that $(\frac{C}{k})^{\frac{1}{2}}(T+\tau)^{\alpha} < \frac{1}{2}$, then the function

$$y_m(x,t) = (t+\tau)^{-\alpha}\left(C - k|x - \frac{1}{2}|^2(t+\tau)^{-2\alpha}\right)_+^{\frac{1}{m-1}}$$

is a solution of system (1.4) with $u = 0$, $v_0 = v_1 = 0$ and $y_m(\cdot, T) \neq 0$. Any other solution of system (1.4) with the same initial datum and $v_0, v_1 \geq 0$ would be a supersolution of y_m, which implies that $y_m(\cdot, 0)$ can not be connected with $y(\cdot, T) \equiv 0$.

Remark 1.1 It would be very interesting to know if, in the case of the problem P_{DD}, one could take $v_1 = 0$ in Theorem 1.1 as it has been done in [22] for a viscous Burgers' control system.

2 Well-Posedness of the Cauchy Problem

For the existence theory of problem (1.4), we refer to [1, 5–7, 21, 23]; in particular, we shall use a frame similar to the ones in [1, 6]. More precisely, we adopt the following definition.

Definition 2.1 Let $(v_0, v_1) \in L^\infty(0, T)^2$ and $v_D = (1 - x)v_0(t) + xv_1(t)$ and let $u \in L^\infty(0, T)$. Assume that $y_0 \in H^{-1}(0, 1)$. We say that y is a weak solution of

$$\mathrm{P}_{DD} \begin{cases} y_t - (|y|^{m-1}y)_{xx} = u(t), & (x, t) \in (0, 1) \times (0, T), \\ y(0, t) = v_0(t), & t \in (0, T), \\ y(1, t) = v_1(t), & t \in (0, T), \\ y(x, 0) = y_0(x), & x \in (0, 1), \end{cases} \tag{2.1}$$

if

$$y \in C^0\big([0, T]; H^{-1}(0, 1)\big) \quad \text{and} \quad y(0) = y_0, \quad \text{in } H^{-1}(0, 1), \tag{2.2}$$

$$y \in L^\infty\big(\tau, T; L^1(0, 1)\big), \quad \forall \tau \in (0, T], \tag{2.3}$$

$$\partial_t y \in L^2\big(\tau, T; H^{-1}(0, 1)\big), \quad \forall \tau \in (0, T], \tag{2.4}$$

$$|y|^{m-1}y \in |v_D|^{m-1}v_D + L^2\big(\tau, T; H_0^1(0, 1)\big), \quad \forall \tau \in (0, T], \tag{2.5}$$

and for every $\tau \in (0, T]$, $\xi \in L^2(0, T; H_0^1(0, 1))$,

$$\int_\tau^T \langle \partial_t y, \xi \rangle \mathrm{d}t + \int_\tau^T \int_0^1 \big(|y|^{m-1}y\big)_x \xi_x \mathrm{d}x\mathrm{d}t = \int_\tau^T \int_0^1 u\xi \mathrm{d}x\mathrm{d}t, \tag{2.6}$$

where the symbol $\langle \cdot, \cdot \rangle$ stands for the dual pairing between $H^{-1}(0, 1)$ and $H_0^1(0, 1)$.

Remark 2.1 We have changed the definition of weak solution given in [1] in order to handle the case where y_0 is only in $H^{-1}(0, 1)$, instead of $y_0 \in L^{m+1}(0, 1)$ as assumed in [1].

The modifications to extend the previous definition to the case of problem P_{ND} are straightforward (see [1]). For instance, the extension to the interior of the boundary datum can be taken now as $v_D = (c_1 + c_2 x^2) v_1(t)$.

With this definition, one has the following proposition.

Proposition 2.1 *The boundary-value problem* (1.4) *has at most one weak solution.*

The proof of Proposition 2.1 is the same as in [1, Theorem 2.4] due to the regularizing effect required in Definition 2.1 (see also [5]).

The next two propositions follow from results which can be found in [1, Theorems 1.7 and 2.4] and [7].

Proposition 2.2 *Suppose that* $(v_0, v_1) \in H^1(0,T)^2$ *and vanishes in a neighbourhood of* $t = 0$, *then there exists one and only one weak solution of problem* (1.4).

Proposition 2.3 *Suppose that* $(v_0, v_1) \in H^1(0,T)^2$ *and that* $y_0 \in L^{m+1}$, *then there exists one and only one weak solution* y *of problem* (1.4). *Moreover, this solution satisfies*

$$y \in L^\infty\big(0,T; L^1(0,1)\big), \tag{2.7}$$

$$\partial_t y \in L^2\big(0,T; H^{-1}(0,1)\big), \tag{2.8}$$

$$|y|^{m-1} y \in |v_D|^{m-1} v_D + L^2\big(0,T; H_0^1(0,1)\big). \tag{2.9}$$

Now, we emphasize that the solution of problem P_{DD} enjoys an additional semigroup property (we will need it to construct the final trajectory), which directly follows from Definition 2.1, Propositions 2.2 and 2.3.

Lemma 2.1 (Matching) *Suppose that* y_1, *respectively* y_2, *is a weak solution of* (1.4) *on the interval* $(0, T_1)$, *respectively* (T_1, T), *with* $y_2(T_1) = y_1(T_1) \in L^2(0,1)$. *If we denote*

$$y(t) = \begin{cases} y_1(t), & t \in (0, T_1), \\ y_2(t), & t \in (T_1, T), \end{cases}$$

then y *is a weak solution of* (1.4) *in the interval* $(0, T)$.

3 Proof of the Main Theorem: First Step

In the interval $(0, t_1]$ the solution with no control evolves as in [7], hence $0 \le y^m(t) \in H_0^1(0,1)$ for all $t \in (0, t_1]$. Due to the inclusion $H_0^1(0,1) \subset L^\infty(0,1)$, we get that $y_1(x) := y(x, t_1)$ is a bounded function. We call the solution on the first interval y^0, i.e.,

$$y_{|(0,t_1)} = y^0. \tag{3.1}$$

In order to be able to apply the null controllability result in [4] to the function $z(x,t)$, given in the decomposition (1.6), on the interval (t_2, t_3) we need the H^1-norm of $z(t_2)$ to be small enough. We want to find some estimates of the solution z of

$$\begin{cases} z_t - (m(\frac{a(t)}{\varepsilon} + z)^{m-1} z_x)_x = 0, & (x,t) \in (0,1) \times (t_1, t_2), \\ z_x(t,0) = z_x(t,1) = 0, & t \in (t_1, t_2), \\ z(x,0) = y_1(x), & x \in (0,1). \end{cases} \tag{3.2}$$

For the existence, regularity and comparison results for this problem, we refer to [18], where the equation is recast in the form $(|Y|^{\frac{1}{m}} \operatorname{sign}(Y))_t - Y_{xx} = \frac{a'}{\varepsilon}$. From the maximum principle, we deduce that $y_1 \in L^\infty(0,1)$ and $y_1 \geq 0$ imply that $z \in L^\infty((0,1) \times (t_1,t_2))$ and $z \geq 0$. In fact, we have $0 \leq z \leq M$, where $M := \|y_1\|_{L^\infty(0,1)}$ is a solution of the state equation of (3.2), and in particular a super solution of (3.2).

To study the behaviour of z, we will actually make use of rescaling.

3.1 Small Initial Data and a priori Estimates

For $\delta > 0$, we define $\tilde{z} := \delta z$. Then $\tilde{z}$ satisfies

$$\begin{cases} \tilde{z}_t - (m(\frac{a(t)}{\varepsilon} + \frac{1}{\delta}\tilde{z})^{m-1} \tilde{z}_x)_x = 0, & (x,t) \in (0,1) \times (t_1, t_2), \\ \tilde{z}_x(t,0) = \tilde{z}_x(t,1) = 0, & t \in (t_1, t_2), \\ \tilde{z}(x,0) = \delta y_1, & x \in (0,1). \end{cases} \tag{3.3}$$

After collecting the factor $\frac{1}{\varepsilon}$ and rescaling the time $\tau := \frac{t}{\varepsilon^{m-1}}$, we get

$$\tilde{z}_t - \left(m \left(a(\tau) + \frac{\varepsilon}{\delta} \tilde{z} \right)^{m-1} \tilde{z}_x \right)_x = 0.$$

Choosing $\delta := \varepsilon^{1-\alpha}$ with $0 < \alpha < 1$, the system can be written in the following form:

$$\begin{cases} \tilde{z}_\tau - (m(a(\tau) + \varepsilon^\alpha \tilde{z})^{m-1} \tilde{z}_x)_x = 0, & (x,\tau) \in (0,1) \times (\tau_1, \tau_2), \\ \tilde{z}_x(\tau,0) = \tilde{z}_x(\tau,1) = 0, & \tau \in (\tau_1, \tau_2), \\ \tilde{z}(x,0) = \varepsilon^{1-\alpha} y_1, & x \in (0,1), \end{cases} \tag{3.4}$$

where $\tau := \frac{t}{\varepsilon^{m-1}}$. For simplicity, we take $\alpha = \frac{1}{2}$.

Thus, the null controllability of system (3.2) is reduced to the null controllability of system (3.4). As we can see, the initial datum in (3.4) are now depending on ε and tend to 0 as $\varepsilon \to 0$.

3.2 H^1-Estimate

We recall that, according to regularity theory for linear parabolic equations with bounded coefficients, $\tilde{z}(t) \in H^2(0, 1)$ for $t > 0$ (see, e.g., [16, pp. 360–364]). Multiplying by $\tilde{z}_{xx}$ the first equation of (3.4) and integrating on $x \in (0, 1)$, we get

$$\int_0^1 \tilde{z}_\tau \tilde{z}_{xx} \mathrm{d}x = \int_0^1 \left(m\left(a(\tau) + \sqrt{\varepsilon}\tilde{z}\right)^{m-1} \tilde{z}_x\right)_x \tilde{z}_{xx} \mathrm{d}x.$$

Then, integrating by parts and using the boundary condition in (3.4), we are led to

$$\begin{aligned}\frac{1}{2m}\frac{\mathrm{d}}{\mathrm{d}\tau}\int_0^1 \tilde{z}_x^2 \mathrm{d}x &= -\int_0^1 \left(a(\tau) + \sqrt{\varepsilon}\tilde{z}\right)^{m-1} \tilde{z}_{xx}^2 \mathrm{d}x \\ &\quad - \frac{(m-1)}{3}\sqrt{\varepsilon}\int_0^1 \left(a(\tau) + \sqrt{\varepsilon}\tilde{z}\right)^{m-2} \left(\tilde{z}_x^3\right)_x \mathrm{d}x \\ &= -\int_0^1 \left(a(\tau) + \sqrt{\varepsilon}\tilde{z}\right)^{m-1} \tilde{z}_{xx}^2 \mathrm{d}x \\ &\quad + \frac{(m-1)(m-2)}{3}\varepsilon\int_0^1 \left(a(\tau) + \sqrt{\varepsilon}\tilde{z}\right)^{m-3} \tilde{z}_x^4 \mathrm{d}x.\end{aligned}$$

We denote by

$$\begin{aligned}IT_1 &:= -\int_0^1 \left(a(\tau) + \sqrt{\varepsilon}\tilde{z}\right)^{m-1} \tilde{z}_{xx}^2 \mathrm{d}x, \\ IT_2 &:= \frac{(m-1)(m-2)}{3}\varepsilon\int_0^1 \left(a(\tau) + \sqrt{\varepsilon}\tilde{z}\right)^{m-3} \tilde{z}_x^4 \mathrm{d}x.\end{aligned}$$

We observe that $IT_1 \le 0$. Let us look at the term IT_2. For $m \in (1, 2)$, we have that $IT_2 \le 0$. Otherwise,

$$IT_2 \le \frac{(m-1)(m-2)}{3}\left(a(\tau) + \sqrt{\varepsilon}\|\tilde{z}\|_\infty\right)^{m-3} \varepsilon\int_0^1 \tilde{z}_x^4 \mathrm{d}x.$$

The fact that the L^∞-norm of $\tilde{z}$ is finite comes from that $\tilde{z} = \delta z$ and that the supremum of z is bounded, as already pointed out. We now use a well-known Gagliardo-Nirenberg's inequality in the case of a bounded interval.

Lemma 3.1 *Suppose $z \in L^\infty(0, 1)$ with $z_{xx} \in L^2(0, 1)$ and either $z(0) = z(1) = 0$ or $z_x(0) = z_x(1) = 0$. Then*

$$\|z_x\|_{L^4} \le \sqrt{3}\|z_{xx}\|_{L^2}^{\frac{1}{2}}\|z\|_{L^\infty}^{\frac{1}{2}}.$$

Proof Integrating by parts and using the boundary conditions, we obtain

$$\int_0^1 z_x^4 \mathrm{d}x = \int_0^1 z_x^3 z_x \mathrm{d}x = -3 \int_0^1 z_x^2 z_{xx} z \mathrm{d}x.$$

Then, using Cauchy-Schwarz's inequality, we get

$$\|z_x\|_{L^4}^4 \leq 3 \|z_x\|_{L^4}^2 \|z\|_{L^\infty} \|z_{xx}\|_{L^2},$$

and the result follows immediately. □

Setting $C' := C\|\tilde{z}\|_{L^\infty}^2$ and considering that $\|\tilde{z}_x\|_{L^4}^4 \leq C'\|\tilde{z}_{xx}\|_{L^2}^2$, we have

$$\begin{aligned}
\frac{1}{2m}\frac{\mathrm{d}}{\mathrm{d}\tau}\int_0^1 \tilde{z}_x^2 \mathrm{d}x &\leq -\int_0^1 \left(a(\tau)+\sqrt{\varepsilon}\tilde{z}\right)^{m-1} \tilde{z}_{xx}^2 \mathrm{d}x \\
&\quad + \frac{(m-1)(m-2)}{3}\left(a(\tau)+\sqrt{\varepsilon}\|\tilde{z}\|_\infty\right)^{m-3} \varepsilon \int_0^1 \tilde{z}_x^4 \mathrm{d}x \\
&\leq -\left(a(\tau)\right)^{m-1} \int_0^1 \tilde{z}_{xx}^2 \mathrm{d}x \\
&\quad + C' \frac{(m-1)(m-2)}{3}\left(a(\tau)+\sqrt{\varepsilon}\|\tilde{z}\|_\infty\right)^{m-3} \varepsilon \int_0^1 \tilde{z}_{xx}^2 \mathrm{d}x \\
&= C''(m,\tau,\varepsilon) \int_0^1 \tilde{z}_{xx}^2 \mathrm{d}x,
\end{aligned}$$

where

$$C''(m,\tau,\varepsilon) := \left(C' \frac{(m-1)(m-2)}{3}\left(a(\tau)+\sqrt{\varepsilon}\|\tilde{z}\|_\infty\right)^{m-3} \varepsilon - \left(a(\tau)\right)^{m-1} \right).$$

For $\tau > 0$, we have

$$C''(m,\tau,\varepsilon) < 0,$$

if ε is small enough.

From these estimates, we deduce that the H^1-norm is non-increasing in the interval (τ_1, τ_2). Hence, for all $\rho \geq 0$, we can choose ε small enough to get $\|\tilde{z}(\tau_2)\|_{H^1(0,1)} \leq \varepsilon \|y_1\|_{H^1(0,1)} \leq \rho$.

4 The End of the Proof of the Main Theorem

Now, we go back to problem (3.4) but with Dirichlet boundary conditions and initial data $\tilde{z}(\tau_2)$. We apply an extension method that can be found for instance in [19,

Chap. 2]. It consists in extending the space domain from (0, 1) to $E := (-d, 1+d)$ and inserting a sparse control in ω, a nonempty open interval whose closure in $\mathbb{R}$ is included in $(-d, 0)$. We look at the following system:

$$\begin{cases} w_t - (m(1+\sqrt{\varepsilon} w)^{m-1} w_x)_x = \chi_\omega \tilde{u}, & (x,\tau) \in Q', \\ w(-d,\tau) = w(1+d,\tau) = 0, & \tau \in (\tau_2, \tau_3), \\ w(x,\tau_2) = w_2(x), & x \in E, \end{cases} \tag{4.1}$$

where $Q' := E \times (\tau_2, \tau_3)$ and $\tau_3 := \frac{t_3}{\varepsilon^{m-1}}$. The function $w_2 \in H_0^1(E) \cap H^2(E)$ is an extension of $\tilde{z}(\tau_2)$ to E which does not increase the H^1-norm, i.e., $\|w_2\|_{H^1(E)} \leq k\|\tilde{z}(\tau_2)\|_{H^1(0,1)} \leq \sqrt{\varepsilon} k \|y_1\|_{H^1(0,1)}$, for some $k > 0$ independent of $\tilde{z}(\tau_2)$.

Proposition 4.1 *There exists a $\rho > 0$ such that, for any initial datum w_2 with $\|w_2\|_{H^1} \leq \rho$ and for any ε sufficiently small, system* (4.1) *is null controllable, i.e., there exists a $\tilde{u} \in L^2(Q')$ such that $w(\tau_3) = 0$.*

Sketch of the Proof It is substantially the same as in [4]. We just have to choose ρ sufficiently small such that the solution of the control problem satisfies, for suitable value of ε, $\|w\|_{L^\infty} < \frac{1}{\sqrt{\varepsilon}}$. □

Remark 4.1 Note that, combining the results in [4] and [16, pp. 360–364], the solution of (4.1) satisfies $w(0,\cdot), w(1,\cdot) \in H^1(\tau_2, \tau_3)$.

Proof of Theorem 1.1 We consider the function

$$y(\cdot,t) = \begin{cases} y^0(\cdot,t), & t \in (0,t_1), \\ \frac{a(t)}{\varepsilon} + z(\cdot,t) = \frac{a(t)}{\varepsilon} + \frac{\tilde{z}(\cdot,t)}{\sqrt{\varepsilon}}, & t \in (t_1,t_2), \\ \frac{a(t)}{\varepsilon} + \frac{w(\cdot,t)}{\sqrt{\varepsilon}}, & t \in (t_2,t_3), \\ \frac{a(t)}{\varepsilon}, & t \in (t_3,T), \end{cases} \tag{4.2}$$

which is a solution of system (1.4) with controls given by

$$u(t) := \frac{a'(t)}{\varepsilon}, \quad t \in (0,T), \tag{4.3}$$

$$v_0(t) := \begin{cases} 0, & t \in (0,t_1), \\ \frac{a(t)}{\varepsilon} + \frac{\tilde{z}(0,t)}{\sqrt{\varepsilon}}, & t \in (t_1,t_2), \\ \frac{a(t)}{\varepsilon} + \frac{w(0,t)}{\sqrt{\varepsilon}}, & t \in (t_2,t_3), \\ \frac{a(t)}{\varepsilon}, & t \in (t_3,T), \end{cases} \tag{4.4}$$

$$v_1(t) := \begin{cases} 0, & t \in (0, t_1), \\ \frac{a(t)}{\varepsilon} + \frac{\tilde{z}(1,t)}{\sqrt{\varepsilon}}, & t \in (t_1, t_2), \\ \frac{a(t)}{\varepsilon} + \frac{w(1,t)}{\sqrt{\varepsilon}}, & t \in (t_2, t_3), \\ \frac{a(t)}{\varepsilon}, & t \in (t_3, T). \end{cases} \tag{4.5}$$

The function satisfies $y \in C([0, T]; H^{-1}(0, 1))$, and, as one can check using the improved regularity of the solution when it is strictly positive, $(v_1, v_2) \in H^1(0, T)^2$. Combining Propositions 2.2–2.3 and Lemma 2.1, it is easy to see that the function given by (4.2) is the solution in the interval $(0, T)$ of problem (1.4) with nonhomogeneous term (4.3) and boundary conditions given by (4.4)–(4.5).

To conclude, we have from construction that $y(\cdot, T) \equiv 0$.

The proof of part (ii) follows the common argument of extension by symmetry. First, one notices that, using the smoothing property of (1.5) when $u \equiv 0$ and $v_1 \equiv 0$, we may assume that y_0 is in $L^2(0, 1)$. Then, we consider the auxiliary problem

$$\mathrm{P}^s_{DD} \begin{cases} y_t - (y^m)_{xx} = \tilde{u}(t)\chi_I(t), & (x, t) \in (-1, 1) \times (0, T), \\ y(-1, t) = v_0(t)\chi_I(t), & t \in (0, T), \\ y(1, t) = v_1(t)\chi_I(t), & t \in (0, T), \\ y(x, 0) = \tilde{y}_0(x), & x \in (-1, 1) \end{cases} \tag{4.6}$$

with $\tilde{y}_0 \in L^2(-1, 1)$ defined by

$$\tilde{y}_0(x) = y_0(x), \quad \tilde{y}_0(-x) = y_0(x), \quad \forall x \in (0, 1), \tag{4.7}$$

and with $v_0(t) = v_1(t)$. We apply the arguments of part (i) to P^s_{DD} with $(0, 1)$ replaced by $(-1, 1)$ and adjusting the formulation of (4.1) in such a way that the control region ω is now symmetric with respect to $x = 0$. Then, as we will show later, the restriction of the solution of P^s_{DD} to the space interval $(0, 1)$ is the sought trajectory for system P_{DN}. □

Lemma 4.1 *Let ω be a nonempty open subset of $[-1 - d, 1 + d] \setminus [-1, 1]$ which is symmetric with respect to (w.r.t.) $x = 0$. Then, if w_2 is symmetric w.r.t. $x = 0$, we can find a control u_s, symmetric w.r.t. $x = 0$, such that the solution w of system* (4.1) *satisfies*

(1) *w is symmetric w.r.t. $x = 0$,*
(2) *$w(\cdot, \tau_3) = 0$.*

Proof The proof follows almost straightforwardly from [4, Theorems 4.1–4.2]. We just have to minimize the functional which appears in [4, Theorems 4.1] in the space of L^2 functions which are symmetric w.r.t. $x = 0$.

The symmetry of the initial value implies, as a consequence, the symmetry of the solution w.

To conclude the proof of part (ii) of Theorem 1.1, we note that as the solution $y(\cdot, t)$ of (4.6) belongs to $H^2(-1, 1)$ for all $t \in (0, T)$, we see that $y_x(0, t) = 0$ for all $t \in (0, T)$ and so, the conclusion is a direct consequence of part (i). □

References

1. Alt, H.W., Luckhaus, S.: Quasilinear elliptic-parabolic differential equations. Math. Z. **183**(3), 311–341 (1983)
2. Antontsev, S.N., Díaz, J.I., Shmarev, S.: Energy methods for free boundary problems. In: Applications to Nonlinear PDEs and Fluid Mechanics. Progress in Nonlinear Differential Equations and Their Applications, vol. 48. Birkhäuser Boston, Cambridge (2002)
3. Barenblatt, G.I.: On some unsteady motions of a liquid and gas in a porous medium. Prikl. Mat. Meh. **16**, 67–68 (1952)
4. Beceanu, M.: Local exact controllability of the diffusion equation in one dimension. Abstr. Appl. Anal. **14**, 711–793 (2003)
5. Brezis, H.: Propriétés régularisantes de certains semi-groupes non linéaires. Isr. J. Math. **9**, 513–534 (1971)
6. Brézis, H.: Monotonicity methods in Hilbert spaces and some applications to nonlinear partial differential equations. In: Contributions to Non-linear Functional Analysis. Proc. Sympos., Math. Res., pp. 101–156. Center, Univ. Wisconsin, Madison, Wis (1971). Academic Press, New York, 1971
7. Brézis, H.: Opérateurs Maximaux Monotones et Semi-groupes de Contractions dans les Espaces de Hilbert. North Holland, Amsterdam (1973)
8. Chapouly, M.: Global controllability of nonviscous and viscous Burgers-type equations. SIAM J. Control Optim. **48**(3), 1567–1599 (2009)
9. Coron, J.M.: Global asymptotic stabilization for controllable systems without drift. Math. Control Signals Syst. **5**(3), 295–312 (1992)
10. Coron, J.M.: On the controllability of 2-D incompressible perfect fluids. J. Math. Pures Appl. **75**(2), 155–188 (1996)
11. Coron, J.M.: Control and Nonlinearity. Mathematical Surveys and Monographs, vol. 136. AMS, Providence (2007)
12. Díaz, G., Díaz, J.I.: Finite extinction time for a class of nonlinear parabolic equations. Commun. Partial Differ. Equ. **4**(11), 1213–1231 (1979)
13. Díaz, J.I., Ramos, Á.M.: Positive and negative approximate controllability results for semilinear parabolic equations. Rev. R. Acad. Cienc. Exactas Fís. Nat. Madr. **89**(1–2), 11–30 (1995)
14. Díaz, J.I., Ramos, Á.M.: Approximate controllability and obstruction phenomena for quasilinear diffusion equations. In: Bristeau, M.-O., Etgen, G., Fitzgibbon, W., et al. (eds.) Computational Science for the 21st Century, pp. 698–707. Wiley, Chichester (1997)
15. Díaz, J.I., Ramos, Á.M.: Un método de viscosidad para la controlabilidad aproximada de ciertas ecuaciones parabólicas cuasilineales. In: Caraballo, T., et al. (eds.) Actas de Jornada Científica en Homenaje al Prof. A. Valle Sánchez, pp. 133–151. Universidad de Sevilla (1997)
16. Evans, L.C.: Partial Differential Equations, 2nd edn. Graduate Studies in Mathematics, vol. 19. AMS, Providence (2010)
17. Fabre, C., Puel, J.P., Zuazua, E.: Approximate controllability of the semilinear heat equation. Proc. R. Soc. Edinb. A **125**(1), 31–61 (1995)
18. Filo, J.: A nonlinear diffusion equation with nonlinear boundary conditions: method of lines. Math. Slovaca **38**(3), 273–296 (1988)
19. Fursikov, A.V., Imanuvilov, O.Y.: Controllability of evolution equations. Lecture Notes Series, vol. 34. Seoul National University Research Institute of Mathematics Global Analysis Research Center, Seoul (1996)

20. Henry, J.: Etude de la contrôlabilité de certains équations paraboliques, Thèse d' État, Université de Paris VI, Paris (1978)
21. Lions, J.L.: Quelques Méthodes de Résolution des Probl'emes aux Limites non Linéaires. Dunod, Paris (1969)
22. Marbach, F.: Fast global null controllability for a viscous Burgers equation despite the presence of a boundary layer (2013). Preprint. arXiv:1301.2987v1
23. Vázquez, J.L.: The Porous Medium Equation. Oxford Mathematical Monographs, Mathematical Theory. Clarendon/Oxford University Press, Oxford (2007)

Sharp Interpolation Inequalities on the Sphere: New Methods and Consequences

Jean Dolbeault, Maria J. Esteban, Michal Kowalczyk, and Michael Loss

Abstract This paper is devoted to various considerations on a family of sharp interpolation inequalities on the sphere, which in dimension greater than 1 interpolate between Poincaré, logarithmic Sobolev and critical Sobolev (Onofri in dimension two) inequalities. The connection between optimal constants and spectral properties of the Laplace-Beltrami operator on the sphere is emphasized. The authors address a series of related observations and give proofs based on symmetrization and the ultraspherical setting.

Keywords Inequality · Interpolation · Gagliardo-Nirenberg inequality · Logarithmic Sobolev inequality · Heat equation

Mathematics Subject Classification 26D10 · 46E35 · 58E35

Project supported by ANR grants CBDif and NoNAP, the ECOS project (No. C11E07), the Chilean research grants Fondecyt (No. 1090103), Fondo Basal CMM-Chile, Project Anillo ACT-125 CAPDE and the National Science Foundation (No. DMS-0901304).

J. Dolbeault (✉) · M.J. Esteban
Ceremade, Université Paris-Dauphine, Place de Lattre de Tassigny, 75775 Paris Cedex 16, France
e-mail: dolbeaul@ceremade.dauphine.fr

M.J. Esteban
e-mail: esteban@ceremade.dauphine.fr

M. Kowalczyk
Departamento de Ingeniería Matemática and Centro de Modelamiento Matemático (UMI 2807 CNRS), Universidad de Chile, Casilla 170 Correo 3, Santiago, Chile
e-mail: kowalczy@dim.uchile.cl

M. Loss
Georgia Institute of Technology, Skiles Building, Atlanta, GA 30332-0160, USA
e-mail: loss@math.gatech.edu

P.G. Ciarlet et al. (eds.), *Partial Differential Equations: Theory, Control and Approximation*, DOI 10.1007/978-3-642-41401-5_9,

1 Introduction

The following interpolation inequality holds on the sphere:

$$\frac{p-2}{d}\int_{\mathbb{S}^d}|\nabla u|^2 \mathrm{d}\mu+\int_{\mathbb{S}^d}|u|^2 \mathrm{d}\mu \geq\left(\int_{\mathbb{S}^d}|u|^p \mathrm{d}\mu\right)^{\frac{2}{p}}, \quad \forall u \in \mathrm{H}^1\left(\mathbb{S}^d, \mathrm{d}\mu\right) \quad (1.1)$$

for any $p \in (2, 2^*]$ with $2^* = \frac{2d}{d-2}$ if $d \geq 3$, and for any $p \in (2, \infty)$ if $d = 2$. In (1.1), $\mathrm{d}\mu$ is the uniform probability measure on the d-dimensional sphere, that is, the measure induced by Lebesgue's measure on $\mathbb{S}^d \subset \mathbb{R}^{d+1}$, up to a normalization factor such that $\mu(\mathbb{S}^d) = 1$.

Such an inequality was established by Bidaut-Véron and Véron [21] in the more general context of compact manifolds with uniformly positive Ricci curvature. Their method is based on the Bochner-Lichnerowicz-Weitzenböck formula and the study of the set of solutions to an elliptic equation, which is seen as a bifurcation problem and contains the Euler-Lagrange equation associated to the optimality case in (1.1). Later, in [12], Beckner gave an alternative proof based on Legendre's duality, the Funk-Hecke formula, proved in [27, 31], and the expression of some optimal constants found by Lieb [33]. Bakry, Bentaleb and Fahlaoui in a series of papers based on the carré du champ method and mostly devoted to the ultraspherical operator showed a result which turns out to give yet another proof, which is anyway very close to the method of [21]. Their computations allow to slightly extend the range of the parameter p (see [7, 8, 14–20] and [34, 37] for earlier related works).

In all computations based on the Bochner-Lichnerowicz-Weitzenböck formula, the choice of exponents in the computations appears somewhat mysterious. The seed for such computations can be found in [28]. Our purpose is on one hand to give alternative proofs, at least for some ranges of the parameter p, which do not rely on such a technical choice. On the other hand, we also aim at simplifying the existing proofs (see Sect. 3.2).

Inequality (1.1) is remarkable for several reasons as follows:

(1) It is optimal in the sense that 1 is the optimal constant. By Hölder's inequality, we know that $\|u\|_{\mathrm{L}^2(\mathbb{S}^d)} \leq \|u\|_{\mathrm{L}^p(\mathbb{S}^d)}$, so that the equality case can only be achieved by functions, which are constants a.e. Of course, the main issue is to prove that the constant $\frac{p-2}{d}$ is optimal, which is one of the classical issues of the so-called A-B problem, for which we primarily refer to [30].

(2) If $d \geq 3$, the case $p = 2^*$ corresponds to the Sobolev's inequality. Using the stereographic projection as in [33], we easily recover Sobolev's inequality in the Euclidean space $\mathbb{R}^d$ with the optimal constant and obtain a simple characterization of the extremal functions found by Aubin and Talenti [5, 36, 37].

(3) In the limit $p \to 2$, one obtains the logarithmic Sobolev inequality on the sphere, while by taking $p \to \infty$ if $d = 2$, one recovers Onofri's inequality (see [25] and Corollary 2.1).

Exponents are not restricted to $p > 2$. Consider indeed the functional

$$\mathcal{Q}_p[u] := \frac{p-2}{d} \frac{\int_{\mathbb{S}^d} |\nabla u|^2 d\mu}{\left(\int_{\mathbb{S}^d} |u|^p d\mu\right)^{\frac{2}{p}} - \int_{\mathbb{S}^d} |u|^2 d\mu}$$

for $p \in [1,2) \cup (2,2^*]$ if $d \geq 3$, or $p \in [1,2) \cup (2,\infty)$ if $d = 2$, and

$$\mathcal{Q}_2[u] := \frac{2}{d} \frac{\int_{\mathbb{S}^d} |\nabla u|^2 d\mu}{\int_{\mathbb{S}^d} |u|^2 \log\left(\frac{|u|^2}{\int_{\mathbb{S}^d} |u|^2 d\mu}\right) d\mu}$$

for any $d \geq 1$. Because $d\mu$ is a probability measure, $\left(\int_{\mathbb{S}^d} |u|^p d\mu\right)^{\frac{2}{p}} - \int_{\mathbb{S}^d} |u|^2 d\mu$ is nonnegative if $p > 2$, nonpositive if $p \in [1,2)$, and equal to zero if and only if u is constant a.e. Denote by $\mathcal{A}$ the set of $\mathrm{H}^1(\mathbb{S}^d, d\mu)$ functions, which are not a.e. constants. Consider the infimum

$$\mathcal{I}_p := \inf_{u \in \mathcal{A}} \mathcal{Q}_p[u]. \tag{1.2}$$

With these notations, we can state a slight result more general than the one of (1.1), which goes as follows and also covers the range $p \in [1,2]$.

Theorem 1.1 *With the above notations, $\mathcal{I}_p = 1$ for any $p \in [1,2^*]$ if $d \geq 3$, or any $p \in [1,\infty)$ if $d = 1, 2$.*

As already explained above, in the case $(2,2^*]$, the above theorem was proved first in [21, Corollary 6.2], and then in [12], by using the previous results of Lieb [33] and the Funk-Hecke formula (see [27, 31]). The case $p = 2$ was covered in [12]. The whole range $p \in [1,2^*]$ was covered in the case of the ultraspherical operator in [19, 20]. Here we give alternative proofs for various ranges of p, which are less technical and interesting in themselves, as well as some extensions.

Notice that the case $p = 1$ can be written as

$$\int_{\mathbb{S}^d} |\nabla u|^2 d\mu \geq d \left[\int_{\mathbb{S}^d} |u|^2 d\mu - \left(\int_{\mathbb{S}^d} |u| d\mu \right)^2 \right], \quad \forall u \in \mathrm{H}^1(\mathbb{S}^d, d\mu),$$

which is equivalent to the usual Poincaré inequality

$$\int_{\mathbb{S}^d} |\nabla u|^2 d\mu \geq d \int_{\mathbb{S}^d} |u - \overline{u}|^2 d\mu, \quad \forall u \in \mathrm{H}^1(\mathbb{S}^d, d\mu) \quad \text{with } \overline{u} = \int_{\mathbb{S}^d} u \, d\mu.$$

See Remark 2.1 for more details. The case $p = 2$ provides the logarithmic Sobolev inequality on the sphere. It holds as a consequence of the inequality for $p \neq 2$ (see Corollary 1.1).

For $p \neq 2$, the existence of a minimizer of

$$u \mapsto \int_{\mathbb{S}^d} |\nabla u|^2 d\mu + \frac{d\mathcal{I}_p}{p-2} \left[\|u\|^2_{\mathrm{L}^2(\mathbb{S}^d)} - \|u\|^2_{\mathrm{L}^p(\mathbb{S}^d)} \right]$$

in $\{u \in \mathrm{H}^1(\mathbb{S}^d, d\mu) : \int_{\mathbb{S}^d} |u|^p d\mu = 1\}$ is easily achieved by variational methods, and will be taken for granted. The compactness for either $p \in [1, 2)$ or $2 < p < 2^*$ is indeed classical, while the case $p = 2^*$, $d \geq 3$ can be studied by concentration-compactness methods. If a function $u \in \mathrm{H}^1(\mathbb{S}^d, d\mu)$ is optimal for (1.1) with $p \neq 2$, then it solves the Euler-Lagrange equation

$$-\Delta_{\mathbb{S}^d} u = \frac{d\mathcal{I}_p}{p-2} \left[\|u\|_{\mathrm{L}^p(\mathbb{S}^d)}^{2-p} u^{p-1} - u \right], \tag{1.3}$$

where $\Delta_{\mathbb{S}^d}$ denotes the Laplace-Beltrami operator on the sphere $\mathbb{S}^d$.

In any case, it is possible to normalize the $\mathrm{L}^p(\mathbb{S}^d)$-norm of u to 1 without restriction because of the zero homogeneity of $\mathcal{Q}_p$. It turns out that the optimality case is achieved by the constant function, with value $u \equiv 1$ if we assume $\int_{\mathbb{S}^d} |u|^p d\mu = 1$, in which case the inequality degenerates because both sides are equal to 0. This explains why the dimension d shows up here: the sequence $(u_n)_{n \in \mathbb{N}}$, satisfying

$$u_n(x) = 1 + \frac{1}{n} v(x)$$

with $v \in \mathrm{H}^1(\mathbb{S}^d, d\mu)$, such that $\int_{\mathbb{S}^d} v \, d\mu = 0$, is indeed minimizing if and only if

$$\int_{\mathbb{S}^d} |\nabla v|^2 d\mu \geq d \int_{\mathbb{S}^d} |v|^2 d\mu,$$

and the equality case is achieved if v is an optimal function for the above Poincaré inequality, i.e., a function associated to the first non-zero eigenvalue of the Laplace-Beltrami operator $-\Delta_{\mathbb{S}^d}$ on the sphere $\mathbb{S}^d$. Up to a rotation, this means

$$v(\xi) = \xi_d, \quad \forall \xi = (\xi_0, \xi_1, \ldots, \xi_d) \in \mathbb{S}^d \subset \mathbb{R}^{d+1},$$

since $-\Delta_{\mathbb{S}^d} v = d v$. Recall that the corresponding eigenspace of $-\Delta_{\mathbb{S}^d}$ is d-dimensional and is generated by the composition of v with an arbitrary rotation.

Remark 1.1 Some related results can be found in [4, 6, 13, 22, 26, 32, 35].

1.1 The Logarithmic Sobolev Inequality

As the first classical consequence of (1.2), we have a logarithmic Sobolev inequality. This result is rather classical. Related forms of the result can be found, for instance, in [9] or in [3].

Corollary 1.1 *Let $d \geq 1$. For any $u \in \mathrm{H}^1(\mathbb{S}^d, d\mu) \setminus \{0\}$, we have*

$$\int_{\mathbb{S}^d} |u|^2 \log\left(\frac{|u|^2}{\int_{\mathbb{S}^d} |u|^2 d\mu} \right) d\mu \leq \frac{2}{d} \int_{\mathbb{S}^d} |\nabla u|^2 d\mu.$$

Moreover, the constant $\frac{2}{d}$ is sharp.

Proof The inequality is achieved by taking the limit as $p \to 2$ in (1.2). To see that the constant $\frac{2}{d}$ is sharp, we can observe that

$$\lim_{\varepsilon\to 0}\int_{\mathbb{S}^d}|1+\varepsilon v|^2\log\left(\frac{|1+\varepsilon v|^2}{\int_{\mathbb{S}^d}|1+\varepsilon v|^2 d\mu}\right)d\mu = 2\int_{\mathbb{S}^d}|v-\overline{v}|^2 d\mu$$

with $\overline{v} = \int_{\mathbb{S}^d} v\,d\mu$. The result follows by taking $v(\xi) = \xi_d$. □

2 Extensions

2.1 *Onofri's Inequality*

In the case of dimension $d = 2$, (1.1) holds for any $p > 2$, and we recover Onofri's inequality by taking the limit $p \to \infty$. This result is standard in the literature (see for instance [12]). For completeness, let us give a statement and a short proof.

Corollary 2.1 *Let $d = 1$ or $d = 2$. For any $v \in \mathrm{H}^1(\mathbb{S}^d, d\mu)$, we have (Fig. 1)*

$$\int_{\mathbb{S}^d} e^{v-\overline{v}} d\mu \le e^{\frac{1}{2d}\int_{\mathbb{S}^d}|\nabla v|^2 d\mu},$$

where $\overline{v} = \int_{\mathbb{S}^d} v\,d\mu$ is the average of v. Moreover, the constant $\frac{1}{2d}$ in the right-hand side is sharp.

Proof In dimension $d = 1$ or $d = 2$, (1.1) holds for any $p > 2$. Take $u = 1 + \frac{v}{p}$ and consider the limit as $p \to \infty$. We observe that

$$\int_{\mathbb{S}^d}|\nabla u|^2 d\mu = \frac{1}{p^2}\int_{\mathbb{S}^d}|\nabla v|^2 d\mu \quad \text{and} \quad \lim_{p\to\infty}\int_{\mathbb{S}^d}|u|^p d\mu = \int_{\mathbb{S}^d} e^v d\mu,$$

so that

$$\left(\int_{\mathbb{S}^d}|u|^p d\mu\right)^{\frac{2}{p}} - 1 \sim \frac{2}{p}\log\left(\int_{\mathbb{S}^d} e^v d\mu\right) \quad \text{and} \quad \int_{\mathbb{S}^d}|u|^2 d\mu - 1 \sim \frac{2}{p}\int_{\mathbb{S}^d} v\,d\mu.$$

The conclusion holds by passing to the limit $p \to \infty$ in (1.1). Optimality is once more achieved by considering $v = \varepsilon v_1$, $v_1(\xi) = \xi_d$, $d = 1$ and Taylor expanding both sides of the inequality in terms of $\varepsilon > 0$ small enough. Notice indeed that $-\Delta_{\mathbb{S}^d} v_1 = \lambda_1 v_1$ with $\lambda_1 = d$, so that

$$\|\nabla u\|^2_{\mathrm{L}^2(\mathbb{S}^d)} = \varepsilon^2\|\nabla v_1\|^2_{\mathrm{L}^2(\mathbb{S}^d)} = \varepsilon^2 d\,\|v_1\|^2_{\mathrm{L}^2(\mathbb{S}^d)},$$

$\int_{\mathbb{S}^d} v_1 d\mu = \overline{v}_1 = 0$, and

$$\int_{\mathbb{S}^d} e^{v-\overline{v}} d\mu - 1 \sim \frac{\varepsilon^2}{2}\int_{\mathbb{S}^d}|v-\overline{v}|^2 d\mu = \frac{1}{2}\varepsilon^2\|v_1\|^2_{\mathrm{L}^2(\mathbb{S}^d)}.$$

□

2.2 Interpolation and a Spectral Approach for $p \in (1, 2)$

In [10], Beckner gave a method to prove interpolation inequalities between the logarithmic Sobolev and the Poincaré inequalities in the case of a Gaussian measure. Here we shall prove that the method extends to the case of the sphere and therefore provides another family of interpolating inequalities, in a new range: $p \in [1, 2)$, again with optimal constants. For further considerations on inequalities that interpolate between the Poincaré and the logarithmic Sobolev inequalities, we refer to [1, 2, 9, 10, 23, 24, 27, 33] and the references therein.

Our purpose is to extend (1.1) written as

$$\frac{1}{d}\int_{\mathbb{S}^d}|\nabla u|^2 d\mu \geq \frac{\left(\int_{\mathbb{S}^d}|u|^p d\mu\right)^{\frac{2}{p}} - \int_{\mathbb{S}^d}|u|^2 d\mu}{p-2}, \quad \forall u \in \mathrm{H}^1(\mathbb{S}^d, d\mu) \tag{2.1}$$

to the case $p \in [1, 2)$. Let us start with a remark.

Remark 2.1 At least for any nonnegative function v, using the fact that μ is a probability measure on $\mathbb{S}^d$, we may notice that

$$\int_{\mathbb{S}^d}|v-\overline{v}|^2 d\mu = \int_{\mathbb{S}^d}|v|^2 d\mu - \left(\int_{\mathbb{S}^d} v\, d\mu\right)^2$$

can be rewritten as

$$\int_{\mathbb{S}^d}|v-\overline{v}|^2 d\mu = \frac{\int_{\mathbb{S}^d}|v|^2 d\mu - \left(\int_{\mathbb{S}^d}|v|^p d\mu\right)^{\frac{2}{p}}}{2-p}$$

for $p = 1$. Hence this extends (1.1) to the case $q = 1$. However, as already noticed for instance in [1], the inequality

$$\int_{\mathbb{S}^d}|v|^2 d\mu - \left(\int_{\mathbb{S}^d}|v| d\mu\right)^2 \leq \frac{1}{d}\int_{\mathbb{S}^d}|\nabla v|^2 d\mu$$

also means that, for any $c \in \mathbb{R}$,

$$\int_{\mathbb{S}^d}|v+c|^2 d\mu - \left(\int_{\mathbb{S}^d}|v+c| d\mu\right)^2 \leq \frac{1}{d}\int_{\mathbb{S}^d}|\nabla v|^2 d\mu .$$

If v is bounded from below a.e. with respect to μ and $c > -\operatorname{ess\,inf}_\mu v$, so that $v + c > 0$ μ a.e., and the left-hand side is

$$\int_{\mathbb{S}^d}|v+c|^2 d\mu - \left(\int_{\mathbb{S}^d}|v+c| d\mu\right)^2$$

$$= c^2 + 2c\int_{\mathbb{S}^d} v\, d\mu + \int_{\mathbb{S}^d}|v|^2 d\mu - \left(c + \int_{\mathbb{S}^d} v\, d\mu\right)^2 = \int_{\mathbb{S}^d}|v-\overline{v}|^2 d\mu ,$$

so that the inequality is the usual Poincaré inequality. By density, we recover that (2.1) written for $p = 1$ exactly amounts to Poincaré inequality written not only for $|v|$, but also for any $v \in \mathrm{H}^1(\mathbb{S}^d, d\mu)$.

Next, using the method introduced by Beckner [10] in the case of a Gaussian measure, we are in the position to prove (2.1) for any $p \in (1, 2)$, knowing that the inequality holds for $p = 1$ and $p = 2$.

Proposition 2.1 *Inequality* (2.1) *holds for any* $p \in (1, 2)$ *and any* $d \geq 1$. *Moreover,* d *is the optimal constant.*

Proof Optimality can be checked by Taylor expanding $u = 1 + \varepsilon v$ at order two in terms of $\varepsilon > 0$ as in the case $p = 2$ (the logarithmic Sobolev inequality). To establish the inequality itself, we may proceed in two steps.

Step 1 (Nelson's Hypercontractivity Result) Although the result can be established by direct methods, we follow here the strategy of Gross [29], which proves the equivalence of the optimal hypercontractivity result and the optimal logarithmic Sobolev inequality.

Consider the heat equation of $\mathbb{S}^d$, namely,

$$\frac{\partial f}{\partial t} = \Delta_{\mathbb{S}^d} f$$

with the initial data $f(t = 0, \cdot) = u \in L^{\frac{2}{p}}(\mathbb{S}^d)$ for some $p \in (1, 2]$, and let $F(t) := \|f(t, \cdot)\|_{\mathrm{L}^{p(t)}(\mathbb{S}^d)}$. The key computation goes as follows:

$$\frac{F'}{F} = \frac{\mathrm{d}}{\mathrm{d}t} \log F(t) = \frac{\mathrm{d}}{\mathrm{d}t}\left[\frac{1}{p(t)} \log\left(\int_{\mathbb{S}^d} |f(t, \cdot)|^{p(t)} \mathrm{d}\mu\right)\right]$$
$$= \frac{p'}{p^2 F^p}\left[\int_{\mathbb{S}^d} v^2 \log\left(\frac{v^2}{\int_{\mathbb{S}^d} v^2 \mathrm{d}\mu}\right) \mathrm{d}\mu + 4\frac{p-1}{p'} \int_{\mathbb{S}^d} |\nabla v|^2 \mathrm{d}\mu\right]$$

with $v := |f|^{\frac{p(t)}{2}}$. Assuming that $4\frac{p-1}{p'} = \frac{2}{d}$, that is,

$$\frac{p'}{p-1} = 2d,$$

we find that

$$\log\left(\frac{p(t) - 1}{p - 1}\right) = 2dt,$$

if we require that $p(0) = p < 2$. Let $t_* > 0$ satisfy $p(t_*) = 2$. As a consequence of the above computation, we have

$$\|f(t_*, \cdot)\|_{\mathrm{L}^2(\mathbb{S}^d)} \leq \|u\|_{\mathrm{L}^{\frac{2}{p}}(\mathbb{S}^d)}, \quad \text{if } \frac{1}{p-1} = \mathrm{e}^{2dt_*}. \tag{2.2}$$

Step 2 (Spectral Decomposition) Let $u = \sum_{k\in\mathbb{N}} u_k$ be a decomposition of the initial datum on the eigenspaces of $-\Delta_{\mathbb{S}^d}$, and denote by $\lambda_k = k(d+k-1)$ the ordered sequence of the eigenvalues: $-\Delta_{\mathbb{S}^d} u_k = \lambda_k u_k$ (see for instance [20]). Let $a_k = \|u_k\|^2_{\mathrm{L}^2(\mathbb{S}^d)}$. As a straightforward consequence of this decomposition, we know that $\|u\|^2_{\mathrm{L}^2(\mathbb{S}^d)} = \sum_{k\in\mathbb{N}} a_k$, $\|\nabla u\|^2_{\mathrm{L}^2(\mathbb{S}^d)} = \sum_{k\in\mathbb{N}} \lambda_k a_k$ and

$$\|f(t_*,\cdot)\|^2_{\mathrm{L}^2(\mathbb{S}^d)} = \sum_{k\in\mathbb{N}} a_k e^{-2\lambda_k t_*}.$$

Using (2.2), it follows that

$$\frac{\left(\int_{\mathbb{S}^d} |u|^p d\mu\right)^{\frac{2}{p}} - \int_{\mathbb{S}^d} |u|^2 d\mu}{p-2} \le \frac{\left(\int_{\mathbb{S}^d} |u|^2 d\mu\right) - \int_{\mathbb{S}^d} |f(t_*,\cdot)|^2 d\mu}{2-p}$$

$$= \frac{1}{2-p} \sum_{k\in\mathbb{N}^*} \lambda_k a_k \frac{1-e^{-2\lambda_k t_*}}{\lambda_k}.$$

Notice that $\lambda_0 = 0$ so that the term corresponding to $k=0$ can be omitted in the series. Since $\lambda \mapsto \frac{1-e^{-2\lambda t_*}}{\lambda}$ is decreasing, we can bound $\frac{1-e^{-2\lambda_k t_*}}{\lambda_k}$ from above by $\frac{1-e^{-2\lambda_1 t_*}}{\lambda_1}$ for any $k \ge 1$. This proves that

$$\frac{\left(\int_{\mathbb{S}^d} |u|^p d\mu\right)^{\frac{2}{p}} - \int_{\mathbb{S}^d} |u|^2 d\mu}{p-2} \le \frac{1-e^{-2\lambda_1 t_*}}{(2-p)\lambda_1} \sum_{k\in\mathbb{N}^*} \lambda_k a_k = \frac{1-e^{-2\lambda_1 t_*}}{(2-p)\lambda_1} \|\nabla u\|^2_{\mathrm{L}^2(\mathbb{S}^d)}.$$

The conclusion follows easily if we notice that $\lambda_1 = d$ and $e^{-2\lambda_1 t_*} = p-1$, so that

$$\frac{1-e^{-2\lambda_1 t_*}}{(2-p)\lambda_1} = \frac{1}{d}.$$

The optimality of this constant can be checked as in the case $p > 2$ by a Taylor expansion of $u = 1 + \varepsilon v$ at order two in terms of $\varepsilon > 0$ small enough. □

3 Symmetrization and the Ultraspherical Framework

3.1 A Reduction to the Ultraspherical Framework

We denote by $(\xi_0, \xi_1, \ldots, \xi_d)$ the coordinates of an arbitrary point $\xi \in \mathbb{S}^d$ with $\sum_{i=0}^d |\xi_i|^2 = 1$. The following symmetry result is a kind of folklore in the literature, and we can see [5, 11, 33] for various related results.

Lemma 3.1 *Up to a rotation, any minimizer of* (1.2) *depends only on* ξ_d.

Proof Let u be a minimizer for $\mathcal{Q}_p$. By writing u in (1.1) in spherical coordinates $\theta \in [0, \pi], \varphi_1, \varphi_2, \ldots, \varphi_{d-1} \in [0, 2\pi)$ and using decreasing rearrangements (see, for instance, [24]), it is not difficult to prove that among optimal functions, there is one which depends only on θ. Moreover, the equality in the rearrangement inequality means that u has to depend on only one coordinate, i.e., $\xi_d = \sin\theta$.

Let us observe that the problem on the sphere can be reduced to a problem involving the ultraspherical operator as follows:

(1) Using Lemma 3.1, we know that (1.1) is equivalent to

$$\frac{p-2}{d}\int_0^\pi |v'(\theta)|^2 d\sigma + \int_0^\pi |v(\theta)|^2 d\sigma \geq \left(\int_0^\pi |v(\theta)|^p d\sigma\right)^{\frac{2}{p}}$$

for any function $v \in \mathrm{H}^1([0, \pi], d\sigma)$, where

$$d\sigma(\theta) := \frac{(\sin\theta)^{d-1}}{Z_d} d\theta \quad \text{with } Z_d := \sqrt{\pi}\,\frac{\Gamma(\frac{d}{2})}{\Gamma(\frac{d+1}{2})}.$$

(2) The change of variables $x = \cos\theta$ and $v(\theta) = f(x)$ allows to rewrite the inequality as

$$\frac{p-2}{d}\int_{-1}^1 |f'|^2 \nu\, d\nu_d + \int_{-1}^1 |f|^2 d\nu_d \geq \left(\int_{-1}^1 |f|^p d\nu_d\right)^{\frac{2}{p}},$$

where $d\nu_d$ is the probability measure defined by

$$\nu_d(x)dx = d\nu_d(x) := Z_d^{-1}\nu^{\frac{d}{2}-1}dx \quad \text{with } \nu(x) := 1 - x^2, \quad Z_d = \sqrt{\pi}\,\frac{\Gamma(\frac{d}{2})}{\Gamma(\frac{d+1}{2})}.$$

We also want to prove the result in the case $p < 2$, to obtain the counterpart of Theorem 1.1 in the ultraspherical setting. On $[-1, 1]$, consider the probability measure $d\nu_d$, and define

$$\nu(x) := 1 - x^2,$$

so that $d\nu_d = Z_d^{-1}\nu^{\frac{d}{2}-1}dx$. We consider the space $\mathrm{L}^2((-1, 1), d\nu_d)$ with the scalar product

$$\langle f_1, f_2\rangle = \int_{-1}^1 f_1 f_2\, d\nu_d,$$

and use the notation

$$\|f\|_p = \left(\int_{-1}^1 f^p d\nu_d\right)^{\frac{1}{p}}.$$

On $L^2((-1,1),d\nu_d)$, we define the self-adjoint ultraspherical operator by

$$\mathcal{L}f := \left(1-x^2\right)f'' - d\,x\,f' = \nu f'' + \frac{d}{2}\nu' f',$$

which satisfies the identity

$$\langle f_1, \mathcal{L}f_2\rangle = -\int_{-1}^{1} f_1' f_2' \,\nu\, d\nu_d.$$

Then the result goes as follows. □

Proposition 3.1 *Let $p \in [1, 2^*]$, $d \geq 1$. Then we have*

$$-\langle f, \mathcal{L}f\rangle = \int_{-1}^{1} |f'|^2 \nu\, d\nu_d \geq d\,\frac{\|f\|_p^2 - \|f\|_2^2}{p-2}, \quad \forall f \in H^1\big([-1,1], d\nu_d\big), \tag{3.1}$$

if $p \neq 2$; and

$$-\langle f, \mathcal{L}f\rangle = \frac{d}{2}\int_{-1}^{1} |f|^2 \log\left(\frac{|f|^2}{\|f\|_2^2}\right) d\nu_d,$$

if $p = 2$.

We may notice that the proof in [21] requires $d \geq 2$, while the case $d = 1$ is also covered in [12]. In [20], the restriction $d \geq 2$ was removed by Bentaleb et al. Our proof is inspired by [21] and also [14, 17], but it is a simplification (in the particular case of the ultraspherical operator) in the sense that only integration by parts and elementary estimates are used.

3.2 A Proof of Proposition 3.1

Let us start with some preliminary observations. The operator $\mathcal{L}$ does not commute with the derivation, but we have the relation

$$\left[\frac{\partial}{\partial x}, \mathcal{L}\right] u = (\mathcal{L}u)' - \mathcal{L}u' = -2x u'' - d u'.$$

As a consequence, we obtain

$$\langle \mathcal{L}u, \mathcal{L}u\rangle = -\int_{-1}^{1} u' (\mathcal{L}u)' \nu\, d\nu_d = -\int_{-1}^{1} u' \mathcal{L}u' \nu\, d\nu_d + \int_{-1}^{1} u'\big(2xu'' + du'\big)\nu\, d\nu_d,$$

$$\langle \mathcal{L}u, \mathcal{L}u\rangle = \int_{-1}^{1} |u''|^2 \nu^2 d\nu_d - d\langle u, \mathcal{L}u\rangle$$

and

$$\int_{-1}^{1}(\mathcal{L}u)^2 \mathrm{d}\nu_d = \langle \mathcal{L}u, \mathcal{L}u\rangle = \int_{-1}^{1}|u''|^2\nu^2 \mathrm{d}\nu_d + d\int_{-1}^{1}|u'|^2\nu \mathrm{d}\nu_d. \tag{3.2}$$

On the other hand, a few integrations by parts show that

$$\left\langle \frac{|u'|^2}{u}\nu\mathcal{L}u\right\rangle = \frac{d}{d+2}\int_{-1}^{1}\frac{|u'|^4}{u^2}\nu^2 \mathrm{d}\nu_d - 2\frac{d-1}{d+2}\int_{-1}^{1}\frac{|u'|^2u''}{u}\nu^2 \mathrm{d}\nu_d, \tag{3.3}$$

where we have used the fact that $\nu\nu'\nu_d = \frac{2}{d+2}(\nu^2\nu_d)'$.

Let $p \in (1,2)\cup(2,2^*)$. In $H^1([-1,1],\mathrm{d}\nu_d)$, now consider a minimizer f for the functional

$$f \mapsto \int_{-1}^{1}|f'|^2\nu \mathrm{d}\nu_d - d\frac{\|f\|_p^2 - \|f\|_2^2}{p-2} =: \mathcal{G}[f],$$

made of the difference of the two sides in (3.1). The existence of such a minimizer can be proved by classical minimization and compactness arguments. Up to a multiplication by a constant, f satisfies the Euler-Lagrange equation

$$-\frac{p-2}{\mathrm{d}}\mathcal{L}f + f = f^{p-1}.$$

Let β be a real number to be fixed later and define u by $f = u^\beta$, such that

$$\mathcal{L}f = \beta u^{\beta-1}\left(\mathcal{L}u + (\beta-1)\frac{|u'|^2}{u}\nu\right).$$

Then u is a solution to

$$-\mathcal{L}u - (\beta-1)\frac{|u'|^2}{u}\nu + \lambda u = \lambda u^{1+\beta(p-2)} \quad \text{with } \lambda := \frac{d}{(p-2)\beta}.$$

If we multiply the equation for u by $\frac{|u'|^2}{u}\nu$ and integrate, we get

$$-\int_{-1}^{1}\mathcal{L}u\frac{|u'|^2}{u}\nu \mathrm{d}\nu_d - (\beta-1)\int_{-1}^{1}\frac{|u'|^4}{u^2}\nu^2 \mathrm{d}\nu_d + \lambda\int_{-1}^{1}|u'|^2\nu \mathrm{d}\nu_d$$
$$= \lambda\int_{-1}^{1}u^{\beta(p-2)}|u'|^2\nu \mathrm{d}\nu_d.$$

If we multiply the equation for u by $-\mathcal{L}u$ and integrate, we get

$$\int_{-1}^{1}(\mathcal{L}u)^2 \mathrm{d}\nu_d + (\beta-1)\int_{-1}^{1}\mathcal{L}u\frac{|u'|^2}{u}\nu \mathrm{d}\nu_d + \lambda\int_{-1}^{1}|u'|^2\nu \mathrm{d}\nu_d$$
$$= (\lambda+d)\int_{-1}^{1}u^{\beta(p-2)}|u'|^2\nu \mathrm{d}\nu_d.$$

Collecting terms, we find that

$$\int_{-1}^{1}(\mathcal{L}u)^2 \mathrm{d}\nu_d+\left(\beta+\frac{d}{\lambda}\right)\int_{-1}^{1}\mathcal{L}u\frac{|u'|^2}{u}\nu \mathrm{d}\nu_d+(\beta-1)\left(1+\frac{d}{\lambda}\right)\int_{-1}^{1}\frac{|u'|^4}{u^2}\nu^2 \mathrm{d}\nu_d$$
$$-d\int_{-1}^{1}|u'|^2\nu \mathrm{d}\nu_d=0.$$

Using (3.2)–(3.3), we get

$$\int_{-1}^{1}|u''|^2\nu^2 \mathrm{d}\nu_d+\left(\beta+\frac{d}{\lambda}\right)\left[\frac{d}{d+2}\int_{-1}^{1}\frac{|u'|^4}{u^2}\nu^2 \mathrm{d}\nu_d-2\frac{d-1}{d+2}\int_{-1}^{1}\frac{|u'|^2u''}{u}\nu^2 \mathrm{d}\nu_d\right]$$
$$+(\beta-1)\left(1+\frac{d}{\lambda}\right)\int_{-1}^{1}\frac{|u'|^4}{u^2}\nu^2 \mathrm{d}\nu_d=0,$$

that is,

$$\mathsf{a}\int_{-1}^{1}|u''|^2\nu^2 \mathrm{d}\nu_d+2\mathsf{b}\int_{-1}^{1}\frac{|u'|^2u''}{u}\nu^2 \mathrm{d}\nu_d+\mathsf{c}\int_{-1}^{1}\frac{|u'|^4}{u^2}\nu^2 \mathrm{d}\nu_d=0, \tag{3.4}$$

where

$$\mathsf{a}=1,$$
$$\mathsf{b}=-\left(\beta+\frac{d}{\lambda}\right)\frac{d-1}{d+2},$$
$$\mathsf{c}=\left(\beta+\frac{d}{\lambda}\right)\frac{d}{d+2}+(\beta-1)\left(1+\frac{d}{\lambda}\right).$$

Using $\frac{d}{\lambda}=(p-2)\beta$, we observe that the reduced discriminant

$$\delta=\mathsf{b}^2-\mathsf{ac}<0$$

can be written as

$$\delta=A\beta^2+B\beta+1 \quad \text{with } A=(p-1)^2\frac{(d-1)^2}{(d+2)^2}-p+2 \text{ and } B=p-3-\frac{d(p-1)}{d+2}.$$

If $p<2^*$, B^2-4A is positive, and therefore it is possible to find β, such that $\delta<0$.

Hence, if $p<2^*$, we have shown that $\mathcal{G}[f]$ is positive unless the three integrals in (3.4) are equal to 0, that is, u is constant. It follows that $\mathcal{G}[f]=0$, which proves (3.1) if $p\in(1,2)\cup(2,2^*)$. The cases $p=1$, $p=2$ (see Corollary 1.1) and $p=2^*$ can be proved as limit cases. This completes the proof of Proposition 3.1.

4 A Proof Based on a Flow in the Ultraspherical Setting

Inequality (3.1) can be rewritten for $g = f^p$, i.e., $f = g^\alpha$ with $\alpha = \frac{1}{p}$, as

$$-\langle f, \mathcal{L} f\rangle = -\langle g^\alpha, \mathcal{L} g^\alpha\rangle =: \mathcal{I}[g] \geq d\,\frac{\|g\|_1^{2\alpha} - \|g^{2\alpha}\|_1}{p-2} =: \mathcal{F}[g].$$

4.1 Flow

Consider the flow associated to $\mathcal{L}$, that is,

$$\frac{\partial g}{\partial t} = \mathcal{L} g, \tag{4.1}$$

and observe that

$$\frac{\mathrm{d}}{\mathrm{d}t}\|g\|_1 = 0, \quad \frac{\mathrm{d}}{\mathrm{d}t}\|g^{2\alpha}\|_1 = -2(p-2)\langle f, \mathcal{L} f\rangle = 2(p-2)\int_{-1}^{1} |f'|^2 \nu\, \mathrm{d}\nu_d,$$

which finally gives

$$\frac{\mathrm{d}}{\mathrm{d}t}\mathcal{F}\big[g(t,\cdot)\big] = -\frac{d}{p-2}\frac{\mathrm{d}}{\mathrm{d}t}\|g^{2\alpha}\|_1 = -2d\,\mathcal{I}\big[g(t,\cdot)\big].$$

4.2 Method

If (3.1) holds, then

$$\frac{\mathrm{d}}{\mathrm{d}t}\mathcal{F}\big[g(t,\cdot)\big] \leq -2d\,\mathcal{F}\big[g(t,\cdot)\big], \tag{4.2}$$

and thus we prove

$$\mathcal{F}\big[g(t,\cdot)\big] \leq \mathcal{F}\big[g(0,\cdot)\big]\mathrm{e}^{-2dt}, \quad \forall t \geq 0.$$

This estimate is actually equivalent to (3.1) as shown by estimating $\frac{\mathrm{d}}{\mathrm{d}t}\mathcal{F}[g(t,\cdot)]$ at $t = 0$.

The method based on the Bakry-Emery approach amounts to establishing first that

$$\frac{\mathrm{d}}{\mathrm{d}t}\mathcal{I}\big[g(t,\cdot)\big] \leq -2d\,\mathcal{I}\big[g(t,\cdot)\big] \tag{4.3}$$

and proving (4.2) by integrating the estimates on $t \in [0, \infty)$. Since

$$\frac{\mathrm{d}}{\mathrm{d}t}\Big(\mathcal{F}\big[g(t,\cdot)\big] - \mathcal{I}\big[g(t,\cdot)\big]\Big) \geq 0$$

and $\lim_{t\to\infty}(\mathcal{F}[g(t,\cdot)] - \mathcal{I}[g(t,\cdot)]) = 0$, this means that

$$\mathcal{F}\big[g(t,\cdot)\big] - \mathcal{I}\big[g(t,\cdot)\big] \le 0, \quad \forall t \ge 0,$$

which is precisely (3.1) written for $f(t,\cdot)$ for any $t \ge 0$ and in particular for any initial value $f(0,\cdot)$.

The equation for $g = f^p$ can be rewritten in terms of f as

$$\frac{\partial f}{\partial t} = \mathcal{L} f + (p-1)\frac{|f'|^2}{f}\nu.$$

Hence, we have

$$-\frac{1}{2}\frac{\mathrm{d}}{\mathrm{d}t}\int_{-1}^{1} |f'|^2 \nu \mathrm{d}\nu_d = \frac{1}{2}\frac{\mathrm{d}}{\mathrm{d}t}\langle f, \mathcal{L} f\rangle = \langle \mathcal{L} f, \mathcal{L} f\rangle + (p-1)\Big\langle \frac{|f'|^2}{f}\nu, \mathcal{L} f\Big\rangle.$$

4.3 An Inequality for the Fisher Information

Instead of proving (3.1), we will established the following stronger inequality, for any $p \in (2, 2^\sharp]$, where $2^\sharp := \frac{2d^2+1}{(d-1)^2}$:

$$\langle \mathcal{L} f, \mathcal{L} f\rangle + (p-1)\Big\langle \frac{|f'|^2}{f}\nu, \mathcal{L} f\Big\rangle + d\langle f, \mathcal{L} f\rangle \ge 0. \tag{4.4}$$

Notice that (3.1) holds under the restriction $p \in (2, 2^\sharp]$, which is stronger than $p \in (2, 2^*]$. We do not know whether the exponent $2^\sharp$ in (4.4) is sharp or not.

4.4 Proof of (4.4)

Using (3.2)–(3.3) with $u = f$, we find that

$$\frac{\mathrm{d}}{\mathrm{d}t}\int_{-1}^{1} |f'|^2 \nu \mathrm{d}\nu_d + 2d\int_{-1}^{1} |f'|^2 \nu \mathrm{d}\nu_d$$
$$= -2\int_{-1}^{1}\left(|f''|^2 + (p-1)\frac{d}{d+2}\frac{|f'|^4}{f^2} - 2(p-1)\frac{d-1}{d+2}\frac{|f'|^2 f''}{f}\right)\nu^2 \mathrm{d}\nu_d.$$

The right-hand side is nonpositive, if

$$|f''|^2 + (p-1)\frac{d}{d+2}\frac{|f'|^4}{f^2} - 2(p-1)\frac{d-1}{d+2}\frac{|f'|^2 f''}{f}$$

is pointwise nonnegative, which is granted if

$$\left[(p-1)\frac{d-1}{d+2}\right]^2 \le (p-1)\frac{d}{d+2},$$

a condition which is exactly equivalent to $p \le 2^\sharp$.

4.5 An Improved Inequality

For any $p \in (2, 2^\sharp)$, we can write that

$$|f''|^2 + (p-1)\frac{d}{d+2}\frac{|f'|^4}{f^2} - 2(p-1)\frac{d-1}{d+2}\frac{|f'|^2 f''}{f}$$
$$= \alpha|f''|^2 + \frac{p-1}{d+2}\left|\frac{d-1}{\sqrt{d}} f'' - \sqrt{d}\frac{|f'|^2}{f}\right|^2 \ge \alpha|f''|^2,$$

where

$$\alpha := 1 - (p-1)\frac{(d-1)^2}{d(d+2)}$$

is positive. Now, using the Poincaré inequality

$$\int_{-1}^{1} |f''|^2 \mathrm{d}\nu_{d+4} \ge (d+2)\int_{-1}^{1} |f' - \overline{f'}|^2 \mathrm{d}\nu_{d+2},$$

where

$$\overline{f'} := \int_{-1}^{1} f' \mathrm{d}\nu_{d+2} = -d\int_{-1}^{1} x f \mathrm{d}\nu_d,$$

we obtain an improved form of (4.4), namely,

$$\langle \mathcal{L}f, \mathcal{L}f\rangle + (p-1)\left\langle \frac{|f'|^2}{f}\nu, \mathcal{L}f\right\rangle + \left[d + \alpha(d+2)\right]\langle f, \mathcal{L}f\rangle \ge 0,$$

if we can guarantee that $\overline{f'} \equiv 0$ along the evolution determined by (4.1). This is the case if we assume that $f(x) = f(-x)$ for any $x \in [-1, 1]$. Under this condition, we find that

$$\int_{-1}^{1} |f'|^2 \nu \mathrm{d}\nu_d \ge \left[d + \alpha(d+2)\right]\frac{\|f\|_p^2 - \|f\|_2^2}{p-2}.$$

As a consequence, we also have

$$\int_{\mathbb{S}^d} |\nabla u|^2 \mathrm{d}\mu + \int_{\mathbb{S}^d} |u|^2 \mathrm{d}\mu \ge \frac{d + \alpha(d+2)}{p-2}\left(\int_{\mathbb{S}^d} |u|^p \mathrm{d}\mu\right)^{\frac{2}{p}}$$

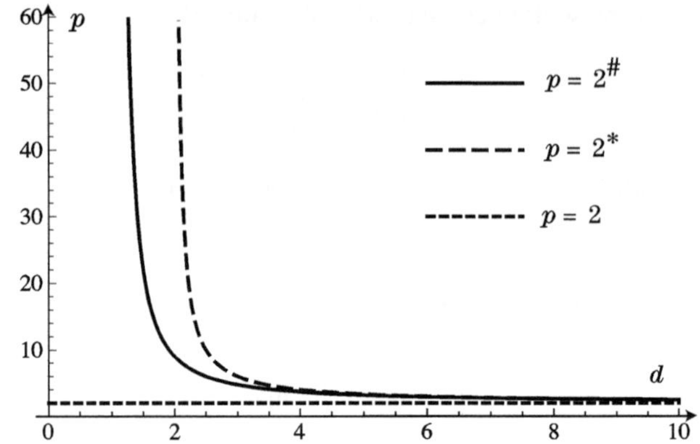

Fig. 1 Plot of $d \mapsto 2^\sharp = \frac{2d^2+1}{(d-1)^2}$ and $d \mapsto 2^* = \frac{2d}{d-2}$

for any $u \in \mathrm{H}^1(\mathbb{S}^d, d\mu)$, such that, using spherical coordinates,

$$u(\theta, \varphi_1, \varphi_2, \ldots, \varphi_{d-1}) = u(\pi - \theta, \varphi_1, \varphi_2, \ldots, \varphi_{d-1}),$$
$$\forall (\theta, \varphi_1, \varphi_2, \ldots, \varphi_{d-1}) \in [0, \pi] \times [0, 2\pi)^{d-1}.$$

4.6 One More Remark

The computation is exactly the same if $p \in (1, 2)$, and henceforth we also prove the result in such a case. The case $p = 1$ is the limit case corresponding to the Poincaré inequality

$$\int_{-1}^{1} |f'|^2 d\nu_{d+2} \geq d \left(\int_{-1}^{1} |f|^2 d\nu_d - | \int_{-1}^{1} f d\nu_d |^2 \right)$$

and arises as a straightforward consequence of the spectral properties of $\mathcal{L}$. The case $p = 2$ is achieved as a limiting case. It gives rise to the logarithmic Sobolev inequality (see, for instance, [34]).

4.7 Limitation of the Method

The limitation $p \leq 2^\sharp$ comes from the pointwise condition

$$h := |f''|^2 + (p-1) \frac{d}{d+2} \frac{|f'|^4}{f^2} - 2(p-1) \frac{d-1}{d+2} \frac{|f'|^2 f''}{f} \geq 0.$$

Can we find special test functions f, such that this quantity can be made negative? Which are admissible, such that $h\nu^2$ is integrable? Notice that at $p = 2^\sharp$, we have that $f(x) = |x|^{1-d}$, such that $h \equiv 0$, but such a function or functions obtained by slightly changing the exponent, are not admissible for larger values of p.

By proving that there is contraction of $\mathcal{I}$ along the flow, we look for a condition which is stronger than one of asking that there is contraction of $\mathcal{F}$ along the flow. It is therefore possible that the limitation $p \leq 2^\sharp$ is intrinsic to the method.

References

1. Arnold, A., Bartier, J.P., Dolbeault, J.: Interpolation between logarithmic Sobolev and Poincaré inequalities. Commun. Math. Sci. **5**, 971–979 (2007)
2. Arnold, A., Dolbeault, J.: Refined convex Sobolev inequalities. J. Funct. Anal. **225**, 337–351 (2005)
3. Arnold, A., Markowich, P., Toscani, G., et al.: On convex Sobolev inequalities and the rate of convergence to equilibrium for Fokker-Planck type equations. Commun. Partial Differ. Equ. **26**, 43–100 (2001)
4. Aubin, T.: Problèmes isopérimétriques et espaces de Sobolev. J. Differ. Geom. **11**, 573–598 (1976)
5. Baernstein, A., Taylor, B.A.: Spherical rearrangements, subharmonic functions, and $*$-functions in n-space. Duke Math. J. **43**, 245–268 (1976)
6. Bakry, D.: Une suite d'inégalités remarquables pour les opérateurs ultrasphériques. C. R. Math. Acad. Sci. **318**, 161–164 (1994)
7. Bakry, D., Bentaleb, A.: Extension of Bochner-Lichnérowicz formula on spheres. Ann. Fac. Sci. Toulouse **14**(6), 161–183 (2005)
8. Bakry, D., Émery, M.: Hypercontractivité de semi-groupes de diffusion. C. R. Math. Acad. Sci. **299**, 775–778 (1984)
9. Bakry, D., Émery, M.: In: Diffusions Hypercontractives, Séminaire de Probabilités, XIX, 1983/1984. Lecture Notes in Math., vol. 1123, pp. 177–206. Springer, Berlin (1985)
10. Beckner, W.: A generalized Poincaré inequality for Gaussian measures. Proc. Am. Math. Soc. **105**, 397–400 (1989)
11. Beckner, W.: Sobolev inequalities, the Poisson semigroup, and analysis on the sphere S^n. Proc. Natl. Acad. Sci. USA **89**, 4816–4819 (1992)
12. Beckner, W.: Sharp Sobolev inequalities on the sphere and the Moser-Trudinger inequality. Ann. Math. **138**(2), 213–242 (1993)
13. Bentaleb, A.: Développement de la moyenne d'une fonction pour la mesure ultrasphérique. C. R. Math. Acad. Sci. **317**, 781–784 (1993)
14. Bentaleb, A.: Inégalité de Sobolev pour l'opérateur ultrasphérique. C. R. Math. Acad. Sci. **317**, 187–190 (1993)
15. Bentaleb, A.: Sur l'hypercontractivité des Semi-groupes Ultrasphériques, Séminaire de Probabilités, XXXIII. Lecture Notes in Math., vol. 1709, pp. 410–414. Springer, Berlin (1999)
16. Bentaleb, A.: L'hypercontractivité des semi-groupes de Gegenbauer multidimensionnels—famille d'iné- galités sur le cercle. Int. J. Math. Game Theory Algebr. **12**, 259–273 (2002)
17. Bentaleb, A.: In: Sur les Fonctions Extrémales des Inégalités de Sobolev des Opérateurs de Diffusion, Séminaire de Probabilités, XXXVI. Lecture Notes in Math., vol. 1801, pp. 230–250. Springer, Berlin (2003)
18. Bentaleb, A., Fahlaoui, S.: Integral inequalities related to the Tchebychev semigroup. Semigroup Forum **79**, 473–479 (2009)
19. Bentaleb, A., Fahlaoui, S.: A family of integral inequalities on the circle $\mathbf{S}^1$. Proc. Jpn. Acad., Ser. A, Math. Sci. **86**, 55–59 (2010)
20. Berger, M., Gauduchon, P., Mazet, E.: Le Spectre d'une Variété Riemannienne. Lecture Notes in Mathematics, vol. 194. Springer, Berlin (1971)
21. Bidaut-Véron, M.F., Véron, L.: Nonlinear elliptic equations on compact Riemannian manifolds and asymptotics of Emden equations. Invent. Math. **106**, 489–539 (1991)

22. Bolley, F., Gentil, I.: Phi-entropy inequalities and Fokker-Planck equations. In: Progress in Analysis and Its Applications, pp. 463–469. World Scientific, Hackensack (2010)
23. Bolley, F., Gentil, I.: Phi-entropy inequalities for diffusion semigroups. J. Math. Pures Appl. **93**(9), 449–473 (2010)
24. Brock, F.: A general rearrangement inequality à la Hardy-Littlewood. J. Inequal. Appl. **5**, 309–320 (2000)
25. Carlen, E., Loss, M.: Computing symmetries, the logarithmic HLS inequality and Onofri's inequality on S^n. Geom. Funct. Anal. **2**, 90–104 (1992)
26. Chafaï, D.: Entropies, convexity, and functional inequalities: on Φ-entropies and Φ-Sobolev inequalities. J. Math. Kyoto Univ. **44**, 325–363 (2004)
27. Funk, P.: Beiträge zur Theorie der Kegelfunktionen. Math. Ann. **77**, 136–162 (1915)
28. Gidas, B., Spruck, J.: Global and local behavior of positive solutions of nonlinear elliptic equations. Commun. Pure Appl. Math. **34**, 525–598 (1981)
29. Gross, L.: Logarithmic Sobolev inequalities. Am. J. Math. **97**, 1061–1083 (1975)
30. Hebey, E.: Nonlinear analysis on manifolds: Sobolev spaces and inequalities. Courant Lecture Notes in Mathematics, vol. 5. New York University Courant Institute of Mathematical Sciences, New York (1999)
31. Hecke, E.: Über orthogonal-invariante Integralgleichungen. Math. Ann. **78**, 398–404 (1917)
32. Latała, R., Oleszkiewicz, K.: In: Between Sobolev and Poincaré, Geometric Aspects of Functional Analysis. Lecture Notes in Math., vol. 1745, pp. 147–168. Springer, Berlin (2000)
33. Lieb, E.H.: Sharp constants in the Hardy-Littlewood-Sobolev and related inequalities. Ann. Math. **118**(2), 349–374 (1983)
34. Mueller, C.E., Weissler, F.B.: Hypercontractivity for the heat semigroup for ultraspherical polynomials and on the n-sphere. J. Funct. Anal. **48**, 252–283 (1982)
35. Rosen, G.: Minimum value for c in the Sobolev inequality $\phi^3\| \leq c\nabla\phi\|^3$. SIAM J. Appl. Math. **21**, 30–32 (1971)
36. Talenti, G.: Best constant in Sobolev inequality. Ann. Mat. Pura Appl. **110**(4), 353–372 (1976)
37. Weissler, F.B.: Logarithmic Sobolev inequalities and hypercontractive estimates on the circle. J. Funct. Anal. **37**, 218–234 (1980)

On the Numerical Solution to a Nonlinear Wave Equation Associated with the First Painlevé Equation: An Operator-Splitting Approach

Roland Glowinski and Annalisa Quaini

Abstract The main goal of this article is to discuss the numerical solution to a nonlinear wave equation associated with the first of the celebrated Painlevé transcendent ordinary differential equations. In order to solve numerically the above equation, whose solutions blow up in finite time, the authors advocate a numerical methodology based on the Strang's symmetrized operator-splitting scheme. With this approach, one can decouple nonlinearity and differential operators, leading to the alternate solution at every time step of the equation as follows: (i) The first Painlevé ordinary differential equation, (ii) a linear wave equation with a constant coefficient. Assuming that the space dimension is two, the authors consider a fully discrete variant of the above scheme, where the space-time discretization of the linear wave equation sub-steps is achieved via a Galerkin/finite element space approximation combined with a second order accurate centered time discretization scheme. To handle the nonlinear sub-steps, a second order accurate centered explicit time discretization scheme with adaptively variable time step is used, in order to follow accurately the fast dynamic of the solution before it blows up. The results of numerical experiments are presented for different coefficients and boundary conditions. They show that the above methodology is robust and describes fairly accurately the evolution of a rather "violent" phenomenon.

Keywords Painlevé equation · Nonlinear wave equation · Blow-up solution · Operator-splitting

Mathematics Subject Classification 35L70 · 65N06

R. Glowinski (✉) · A. Quaini
Department of Mathematics, University of Houston, 4800 Calhoun Rd, Houston, TX 77204, USA
e-mail: roland@math.uh.edu

A. Quaini
e-mail: quaini@math.uh.edu

R. Glowinski
Laboratoire Jacques-Louis Lions, University Pierre et Marie Curie, 4, Place Jussieu, 75252 Paris Cedex 05, France

P.G. Ciarlet et al. (eds.), *Partial Differential Equations: Theory, Control and Approximation*, DOI 10.1007/978-3-642-41401-5_10,

1 Introduction

Although discovered from purely mathematical considerations, the six Painlevé "transcendent" ordinary differential equations arise in a variety of important physical applications (from plasma physics to quantum gravity), motivating the Painlevé project presented in [1], whose goal is to explore the various aspects of the six Painlevé equations. There is an abundant literature concerning the Painlevé equations (see [2–4] and the references therein). Surprisingly, very few of the related publications are of numerical nature, with notable exceptions being [4, 5], which also contain additional references on the numerical solution to the Painlevé equations. Our goal in this article is, in some sense, more modest, since it is to associate with the first Painlevé equation

$$\frac{d^2 y}{dt^2} = 6y^2 + t, \tag{1.1}$$

and the following nonlinear wave equation:

$$\frac{\partial^2 u}{\partial t^2} - c^2 \nabla^2 u = 6u^2 + t \quad \text{in } \Omega \times (0, T_{\max}), \tag{1.2}$$

and to discuss the numerical solution to (1.2). Actually, we are going to consider the numerical solution to two initial/boundary value problems associated with (1.2), namely, we supplement (1.2) with initial conditions and pure homogeneous Dirichlet boundary conditions (resp. mixed Dirichlet-Sommerfeld boundary conditions), that is

$$\begin{cases} u = 0 & \text{on } \partial\Omega \times (0, T_{\max}), \\ u(0) = u_0, & \frac{\partial u}{\partial t}(0) = u_1, \end{cases} \tag{1.3}$$

resp.

$$\begin{cases} u = 0 & \text{on } \Gamma_0 \times (0, T_{\max}), \\ \frac{1}{c}\frac{\partial u}{\partial t} + \frac{\partial u}{\partial n} = 0 & \text{on } \Gamma_1 \times (0, T_{\max}), \\ u(0) = u_0, & \frac{\partial u}{\partial t}(0) = u_1. \end{cases} \tag{1.4}$$

In (1.2)–(1.4), we have

(i) $c\,(> 0)$ is the speed of the propagation of the linear wave solutions to the equation

$$\frac{\partial^2 u}{\partial t^2} - c^2 \nabla^2 u = 0.$$

(ii) Ω is a bounded domain of $\mathbb{R}^d$, and $\partial\Omega$ is its boundary.
(iii) Γ_0 and Γ_1 are two disjoint non-empty subsets of $\partial\Omega$ satisfying $\Gamma_0 \cup \Gamma_1 = \partial\Omega$.
(iv) $\phi(t)$ denotes the function $x \to \phi(x, t)$.

The two problems under consideration are of multi-physics (reaction-propagation type) and multi-time scales nature. Thus, it makes sense to apply an operator-splitting method for the solutions to (1.2), (1.3) and (1.2), (1.4), in order to decouple the nonlinearity and differential operators and to treat the resulting sub-initial value problems with appropriate (and necessarily variable) time discretization sub-steps. Among the available operator-splitting methods, we chose the Strang's symmetrized operator-splitting scheme (introduced in [6]), because it provides a good compromise between accuracy and robustness as shown in [7–9] (and references therein).

The article is structured as follows. In Sect. 2, we discuss the time discretization of the problems (1.2), (1.3) and (1.2), (1.4) by the Strang's symmetrized scheme. In Sects. 3 and 4, we discuss the solution to the initial value subproblems originating from the splitting, and the discussion includes the finite element approximation of the linear wave steps and the adaptive in time solution to the nonlinear ODE steps. In Sect. 5, we present the results of numerical experiments validating the numerical methodology discussed in the previous sections. In this section, we also investigate the influence of c and of the boundary conditions on the behavior of the solutions.

Remark 1.1 Strictly speaking, it is the solutions to the Painlevé equations which are transcendent, not the equations themselves.

Remark 1.2 This article is dedicated to J. L. Lions. We would like to mention that one can find, in Chap. 1 of his book celebrated in 1969 (see [10]), a discussion and further references on the existence and the non-existence of solutions to the following nonlinear wave problem:

$$\begin{cases} \frac{\partial^2 u}{\partial t^2} - \nabla^2 u = u^2 & \text{in } \Omega \times (0, T_{\max}), \\ u = 0 & \text{on } \partial\Omega \times (0, T_{\max}), \\ u(0) = u_0, & \frac{\partial u}{\partial t}(0) = u_1, \end{cases} \tag{1.5}$$

which is a related and simpler variant of problem (1.2), (1.3). The numerical methods discussed in this article can easily handle problem (1.5).

Remark 1.3 The numerical methodology discussed here can be applied more or less easily to other nonlinear wave equations of the following type:

$$\frac{\partial^2 u}{\partial t^2} - c^2\nabla^2 u = f\left(u, \frac{\partial u}{\partial t}, x, t\right).$$

2 Application of the Strang's Symmetrized Operator-Splitting Scheme to the Solution to Problems (1.2), (1.3) and (1.2), (1.4)

2.1 A Brief Discussion of the Strang's Operator-Splitting Scheme

Although the Strang's symmetrized scheme is quite well-known, it may be useful to present briefly this scheme before applying it to the solution to problems (1.2), (1.3) and (1.2), (1.4). Our presentation follows closely the ones in [7, Chap. 6] and [11].

Let us consider thus the following non-autonomous abstract initial value problem (taking place in a Banach space, for example):

$$\begin{cases} \frac{d\phi}{dt} + A(\phi, t) + B(\phi, t) = 0 \quad \text{in } (0, T_{\max}), \\ \phi(0) = \phi_0, \end{cases} \tag{2.1}$$

where the operators A and B can be nonlinear and even multivalued (in which case one has to replace $=0$ by $\ni 0$ in (2.1)). Let Δt be a time-step (fixed, for simplicity), and let us denote $(n+\alpha)\Delta t$ by $t^{n+\alpha}$. When applied to the time discretization of (2.1), the basic Strang's symmetrized scheme reads as follows:

Step 1 Set

$$\phi^0 = \phi_0. \tag{2.2}$$

For $n \geq 0$, ϕ^n being known, compute ϕ^{n+1} as below.

Step 2 Set $\phi^{n+\frac{1}{2}} = \phi(t^{n+\frac{1}{2}})$, where ϕ is the solution to

$$\begin{cases} \frac{d\phi}{dt} + A(\phi, t) = 0 \quad \text{in } (t^n, t^{n+\frac{1}{2}}), \\ \phi(t^n) = \phi^n. \end{cases} \tag{2.3}$$

Step 3 Set $\widehat{\phi}^{n+\frac{1}{2}} = \phi(\Delta t)$, where ϕ is the solution to

$$\begin{cases} \frac{d\phi}{dt} + B(\phi, t^{n+\frac{1}{2}}) = 0 \quad \text{in } (0, \Delta t), \\ \phi(0) = \phi^{n+\frac{1}{2}}. \end{cases} \tag{2.4}$$

Step 4 Set $\phi^{n+1} = \phi(t^{n+1})$, where ϕ is the solution to

$$\begin{cases} \frac{d\phi}{dt} + A(\phi, t) = 0 \quad \text{in } (t^{n+\frac{1}{2}}, t^{n+1}), \\ \phi(t^{n+\frac{1}{2}}) = \widehat{\phi}^{n+\frac{1}{2}}. \end{cases} \tag{2.5}$$

If the operators A and B are smooth functions of their arguments, the above scheme is second order accurate. In addition to [6–9, 11], useful information about the operator-splitting solution to partial differential equations can be found in [12–16] (and references therein).

2.2 Application to the Solution to the Nonlinear Wave Problem (1.2), (1.3)

In order to apply the symmetrized scheme to the solution to (1.2), (1.3), we reformulate the above problem as a first order in time system by introducing the function

$p = \frac{\partial u}{\partial t}$. We obtain that

$$\begin{cases} \frac{\partial u}{\partial t} - p = 0 & \text{in } \Omega \times (0, T_{\max}), \\ \frac{\partial p}{\partial t} - c^2 \nabla^2 u = 6u^2 + t & \text{in } \Omega \times (0, T_{\max}) \end{cases} \tag{2.6}$$

with boundary and initial conditions

$$\begin{cases} u = 0 & \text{on } \partial\Omega \times (0, T_{\max}), \\ u(0) = u_0, & p(0) = u_1. \end{cases} \tag{2.7}$$

Clearly, (2.6), (2.7) is equivalent to (1.2), (1.3).

With Δt as in Sect. 2.1, we introduce $\alpha, \beta \in (0, 1)$ such that $\alpha + \beta = 1$. Applying scheme (2.2)–(2.5) to the solution of (2.6), (2.7), we obtain the following:

Step 1 Set

$$u^0 = u_0, \quad p^0 = u_1. \tag{2.8}$$

For $n \geq 0$, $\{u^n, p^n\}$ being known, compute $\{u^{n+1}, p^{n+1}\}$ as below.

Step 2 Set $u^{n+\frac{1}{2}} = u(t^{n+\frac{1}{2}})$, $p^{n+\frac{1}{2}} = p(t^{n+\frac{1}{2}})$, where $\{u, p\}$ is the solution to

$$\begin{cases} \frac{\partial u}{\partial t} - \alpha p = 0 & \text{in } \Omega \times (t^n, t^{n+\frac{1}{2}}), \\ \frac{\partial p}{\partial t} = 6u^2 + t & \text{in } \Omega \times (t^n, t^{n+\frac{1}{2}}), \\ u(t^n) = u^n, & p(t^n) = p^n. \end{cases} \tag{2.9}$$

Step 3 Set $\widehat{u}^{n+\frac{1}{2}} = u(\Delta t)$, $\widehat{p}^{n+\frac{1}{2}} = p(\Delta t)$, where $\{u, p\}$ is the solution to

$$\begin{cases} \frac{\partial u}{\partial t} - \beta p = 0 & \text{in } \Omega \times (0, \Delta t), \\ \frac{\partial p}{\partial t} - c^2 \nabla^2 u = 0 & \text{in } \Omega \times (0, \Delta t), \\ u = 0 & \text{on } \partial\Omega \times (0, \Delta t), \\ u(0) = u^{n+\frac{1}{2}}, & p(0) = p^{n+\frac{1}{2}}. \end{cases} \tag{2.10}$$

Step 4 Set $u^{n+1} = u(t^{n+1})$, $p^{n+1} = p(t^{n+1})$, where $\{u, p\}$ is the solution to

$$\begin{cases} \frac{\partial u}{\partial t} - \alpha p = 0 & \text{in } \Omega \times (t^{n+\frac{1}{2}}, t^{n+1}), \\ \frac{\partial p}{\partial t} = 6u^2 + t & \text{in } \Omega \times (t^{n+\frac{1}{2}}, t^{n+1}), \\ u(t^{n+\frac{1}{2}}) = \widehat{u}^{n+\frac{1}{2}}, & p(t^{n+\frac{1}{2}}) = \widehat{p}^{n+\frac{1}{2}}. \end{cases} \tag{2.11}$$

By the partial elimination of p, (2.8)–(2.11) reduces to the following:

Step 1 As in (2.8).

For $n \geq 0$, $\{u^n, p^n\}$ being known, compute $\{u^{n+1}, p^{n+1}\}$ as below.

Step 2 Set $u^{n+\frac{1}{2}} = u(t^{n+\frac{1}{2}})$, $p^{n+\frac{1}{2}} = \frac{1}{\alpha}\frac{\partial u}{\partial t}(t^{n+\frac{1}{2}})$, where u is the solution to

$$\begin{cases} \frac{\partial^2 u}{\partial t^2} = \alpha(6u^2 + t) & \text{in } \Omega \times (t^n, t^{n+\frac{1}{2}}), \\ u(t^n) = u^n, & \frac{\partial u}{\partial t}(t^n) = \alpha p^n. \end{cases} \tag{2.12}$$

Step 3 Set $\widehat{u}^{n+\frac{1}{2}} = u(\Delta t)$, $\widehat{p}^{n+\frac{1}{2}} = \frac{1}{\beta}\frac{\partial u}{\partial t}(\Delta t)$, where u is the solution to

$$\begin{cases} \frac{\partial^2 u}{\partial t^2} - \beta c^2 \nabla^2 u = 0 & \text{in } \Omega \times (0, \Delta t), \\ u = 0 & \text{on } \partial\Omega \times (0, \Delta t), \\ u(0) = u^{n+\frac{1}{2}}, & \frac{\partial u}{\partial t}(0) = \beta p^{n+\frac{1}{2}}. \end{cases} \tag{2.13}$$

Step 4 Set $u^{n+1} = u(t^{n+1})$, $p^{n+1} = \frac{1}{\alpha}\frac{\partial u}{\partial t}(t^{n+1})$, where u is the solution to

$$\begin{cases} \frac{\partial^2 u}{\partial t^2} = \alpha(6u^2 + t) & \text{in } \Omega \times (t^{n+\frac{1}{2}}, t^{n+1}), \\ u(t^{n+\frac{1}{2}}) = \widehat{u}^{n+\frac{1}{2}}, & \frac{\partial u}{\partial t}(t^{n+\frac{1}{2}}) = \alpha \widehat{p}^{n+\frac{1}{2}}. \end{cases} \tag{2.14}$$

2.3 Application to the Solution to the Nonlinear Wave Problem (1.2), (1.4)

Proceeding as in Sect. 2.2, we introduce $p = \frac{\partial u}{\partial t}$ in order to reformulate (1.2), (1.4) as the first order in time system. We obtain the system (2.6) supplemented with the following boundary and initial conditions:

$$\begin{cases} u(0) = 0 & \text{on } \Gamma_0 \times (0, T_{\max}), \\ \frac{p}{c} + \frac{\partial u}{\partial n} = 0 & \text{on } \Gamma_1 \times (0, T_{\max}), \\ u(0) = u_0, & p(0) = u_1. \end{cases} \tag{2.15}$$

Applying scheme (2.2)–(2.5) to the solution to the equivalent problem (2.6), (2.15), we obtain the following:

Step 1 As in (2.8).
For $n \geq 0$, $\{u^n, p^n\}$ being known, compute $\{u^{n+1}, p^{n+1}\}$ as below.

Step 2 As in (2.12).

Step 3 Set $\widehat{u}^{n+\frac{1}{2}} = u(\Delta t)$, $\widehat{p}^{n+\frac{1}{2}} = \frac{1}{\beta}\frac{\partial u}{\partial t}(\Delta t)$, where u is the solution to

$$\begin{cases} \frac{\partial^2 u}{\partial t^2} - \beta c^2 \nabla^2 u = 0 & \text{in } \Omega \times (0, \Delta t), \\ u = 0 & \text{on } \Gamma_0 \times (0, \Delta t), \\ \frac{1}{\beta c}\frac{\partial u}{\partial t} + \frac{\partial u}{\partial n} = 0 & \text{on } \Gamma_1 \times (0, \Delta t), \\ u(0) = u^{n+\frac{1}{2}}, & \frac{\partial u}{\partial t}(0) = \beta p^{n+\frac{1}{2}}. \end{cases} \tag{2.16}$$

Step 4 As in (2.14).

3 On the Numerical Solution to the Sub-initial Value Problems (2.13) and (2.16)

3.1 Some Generalities

Since problem (2.13) is the particular case of (2.16) corresponding to $\Gamma_1 = \emptyset$, we are going to consider the second problem only. This linear wave problem is a particular case of

$$\begin{cases} \frac{\partial^2 \phi}{\partial t^2} - \beta c^2 \nabla^2 \phi = 0 & \text{in } \Omega \times (t_0, t_f), \\ \phi = 0 & \text{on } \Gamma_0 \times (t_0, t_f), \\ \frac{1}{\beta c}\frac{\partial \phi}{\partial t} + \frac{\partial \phi}{\partial n} = 0 & \text{on } \Gamma_1 \times (t_0, t_f), \\ \phi(t_0) = \phi_0, & \frac{\partial \phi}{\partial t}(t_0) = \phi_1. \end{cases} \tag{3.1}$$

Assuming that ϕ_0 and ϕ_1 have enough regularity, a variational (weak) formulation of problem (3.1) is given by the following: Find $\phi(t) \in V_0$, a.e. on (t_0, t_f), such that

$$\begin{cases} \langle \frac{\partial^2 \phi}{\partial t^2}, \theta \rangle + \beta c^2 \int_\Omega \nabla \phi \cdot \nabla \theta \, dx + c \int_{\Gamma_1} \frac{\partial \phi}{\partial t} \theta \, d\Gamma = 0, & \forall \theta \in V_0, \\ \phi(t_0) = \phi_0, & \frac{\partial \phi}{\partial t}(t_0) = \phi_1, \end{cases} \tag{3.2}$$

where

(i) V_0 is the Sobolev space defined by

$$V_0 = \{\theta \mid \theta \in H^1(\Omega),\ \theta = 0 \text{ on } \Gamma_0\}, \tag{3.3}$$

(ii) $\langle \cdot, \cdot \rangle$ is the duality pairing between V_0' (the dual of V_0) and V_0, coinciding with the canonical inner product of $L^2(\Omega)$ if the first argument is smooth enough,
(iii) $dx = dx_1 \cdots dx_d$.

3.2 A Finite Element Method for the Space Discretization of the Linear Wave Problem (3.1)

From now on, we are going to assume that Ω is a bounded polygonal domain of $\mathbb{R}^2$. Let $\mathcal{T}_h$ be a classical finite element triangulation of Ω, as considered in [17, Appendix 1] and related references therein. We approximate the space V_0 in (3.3) by

$$V_{0h} = \{\theta \mid \theta \in C^0(\overline{\Omega}),\ \theta|_{\Gamma_0} = 0,\ \theta|_K \in \mathbb{P}_1,\ \forall K \in \mathcal{T}_h\}, \tag{3.4}$$

where $\mathbb{P}_1$ is the space of the polynomials of two variables of degree ≤ 1. If $\Gamma_1 \neq \emptyset$, the points at the interface of Γ_0 and Γ_1 have to be (for consistency reasons) vertices of $\mathcal{T}_h$, at which any element of V_{0h} has to vanish. It is natural to approximate the wave problem (3.2) as follows: Find $\phi_h(t) \in V_{0h}$, a.e. on $(t_0, t_f]$, such that

$$\begin{cases} \int_\Omega \frac{\partial^2 \phi_h}{\partial t^2}\theta \mathrm{d}x + \beta c^2 \int_\Omega \nabla\phi_h \cdot \nabla\theta \mathrm{d}x + c\int_{\Gamma_1} \frac{\partial \phi_h}{\partial t}\theta \mathrm{d}\Gamma = 0, \quad \forall \theta \in V_{0h}, \\ \phi_h(t_0) = \phi_{0h}, \qquad \frac{\partial \phi_h}{\partial t}(t_0) = \phi_{1h}, \end{cases} \tag{3.5}$$

where ϕ_{0h} and ϕ_{1h} belong to V_{0h}, and approximate ϕ_0 and ϕ_1, respectively.

In order to formulate (3.5) as a second order in time system of linear ordinary differential equations, we introduce first the set $\Sigma_{0h} = \{P_j\}_{j=1}^{N_{0h}}$ of the vertices of $\mathcal{T}_h$, which do not belong to $\overline{\Gamma}_0$, and associate with it the following basis of V_{0h}:

$$\mathcal{B}_{0h} = \{w_j\}_{j=1}^{N_{0h}},$$

where the basis function w_j is defined by

$$w_j \in V_{0h}, \quad w_j(P_j) = 1, \quad w_j(P_k) = 0, \quad \forall k \in \{1, \dots, N_{0h}\}, \quad k \neq j.$$

Expanding the solution ϕ_h to (3.5) over the above basis, we obtain

$$\phi_h(t) = \sum_{j=1}^{N_{0h}} \phi_h(P_j, t) w_j.$$

Denoting $\phi_h(P_j, t)$ by $\phi_j(t)$ and the N_{0h}-dimensional vector $\{\phi_j(t)\}_{j=1}^{N_{0h}}$ by $\mathbf{\Phi}_h(t)$, we can easily show that the approximated problem (3.5) is equivalent to the following ordinary differential system:

$$\begin{cases} \mathbf{M}_h \ddot{\mathbf{\Phi}}_h + \beta c^2 \mathbf{A}_h \mathbf{\Phi}_h + c\mathbf{C}_h \dot{\mathbf{\Phi}}_h = \mathbf{0} \quad \text{on } (t_0, t_f), \\ \mathbf{\Phi}_h(t_0) = \mathbf{\Phi}_{0h} \ (= (\phi_{0h}(P_j))_{j=1}^{N_{0h}}), \quad \dot{\mathbf{\Phi}}_h(t_0) = \mathbf{\Phi}_{1h} \ (= (\phi_{1h}(P_j))_{j=1}^{N_{0h}}), \end{cases} \tag{3.6}$$

where the mass matrix $\mathbf{M}_h$, the stiffness matrix $\mathbf{A}_h$, and the damping matrix $\mathbf{C}_h$ are defined by

$$\mathbf{M}_h = (m_{ij})_{1 \le i,j \le N_{0h}} \quad \text{with } m_{ij} = \int_\Omega w_i w_j \mathrm{d}x,$$

$$\mathbf{A}_h = (a_{ij})_{1 \le i,j \le N_{0h}} \quad \text{with } a_{ij} = \int_\Omega \nabla w_i \cdot \nabla w_j \mathrm{d}x,$$

$$\mathbf{C}_h = (c_{ij})_{1 \le i,j \le N_{0h}} \quad \text{with } c_{ij} = \int_{\Gamma_1} w_i w_j \mathrm{d}\Gamma,$$

respectively.

The matrices $\mathbf{M}_h$ and $\mathbf{A}_h$ are sparse and positive definite, while matrix $\mathbf{C}_h$ is "very" sparse and positive semi-definite. Indeed, if P_i and P_j are not neighbors, i.e., they are not vertices of a same triangle of $\mathcal{T}_h$, we have $m_{ij} = 0$, $a_{ij} = 0$ and $c_{ij} = 0$. All these matrix entries can be computed exactly, using, for example, the two-dimensional Simpson's rule for the m_{ij} and the one-dimensional Simpson's rule for the c_{ij}. Since ∇w_i and ∇w_j are piecewise constant, computing a_{ij} is (relatively) easy (see [7, Chap. 5] for more details on these calculations).

Remark 3.1 Using the trapezoidal rule, instead of Simpson's one, to compute the m_{ij} and c_{ij} brings simplification as follows: The resulting $\mathbf{M}_h$ and $\mathbf{C}_h$ will be diagonal matrices, retaining the positivity properties of their Simpson's counterparts. The drawback is some accuracy loss associated with this simplification.

3.3 A Centered Second Order Finite Difference Scheme for the Time Discretization of the Initial Value Problem (3.6)

Let Q be a positive integer (≥ 3, in practice). We associate with Q a time discretization step $\tau = \frac{t_f - t_0}{Q}$. After dropping the subscript h, a classical time discretization scheme for problem (3.6) reads as: Set

$$\mathbf{\Phi}^0 = \mathbf{\Phi}_0, \qquad \mathbf{\Phi}^1 - \mathbf{\Phi}^{-1} = 2\tau\mathbf{\Phi}_1, \tag{3.7}$$

then for $q = 0, \ldots, Q$, compute $\mathbf{\Phi}^{q+1}$ by

$$\mathbf{M}(\mathbf{\Phi}^{q+1} + \mathbf{\Phi}^{q-1} - 2\mathbf{\Phi}^q) + \beta c^2\tau^2\mathbf{A}\mathbf{\Phi}^q + c\frac{\tau}{2}\mathbf{C}(\mathbf{\Phi}^{q+1} - \mathbf{\Phi}^{q-1}) = \mathbf{0}. \tag{3.8}$$

It follows from [7, Chap. 6] that the above second order accurate scheme is stable if the following condition holds:

$$\tau < \frac{2}{c\sqrt{\beta\lambda_{\max}}}, \tag{3.9}$$

where $\lambda_{\max}$ is the largest eigenvalue of $\mathbf{M}^{-1}\mathbf{A}$.

Remark 3.2 To obtain $\mathbf{\Phi}^{q+1}$ from (3.8), one has to solve a linear system associated with the symmetric positive definite matrix

$$\mathbf{M} + \frac{\tau}{2}c\mathbf{C}. \tag{3.10}$$

If the above matrix is diagonal from the use of the trapezoidal rule (see Remark 3.1), computing $\mathbf{\Phi}^{q+1}$ is particularly easy and the time discretization scheme (3.8) is fully explicit. Otherwise, scheme (3.8) is not explicit, strictly speaking. However, since matrix (3.10) is well conditioned, a conjugate gradient algorithm with diagonal preconditioning will have a very fast convergence, particularly if one uses $\mathbf{\Phi}^q$ to initialize the computation of $\mathbf{\Phi}^{q+1}$.

Remark 3.3 In order to initialize the discrete analogue of the initial value problem (2.14), we will use

$$\mathbf{\Phi}^Q \quad \text{and} \quad \frac{\alpha}{\beta}\frac{\mathbf{\Phi}^{Q+1} - \mathbf{\Phi}^{Q-1}}{2\tau}. \tag{3.11}$$

Remark 3.4 As the solution to the nonlinear wave problem under consideration gets closer to blow-up, the norms of the corresponding initial data in (3.7) will go to infinity. In order to off-set (partly, at least) the effect of round-off errors, we suggest the following normalization strategy:

(1) Denote by $\|\phi_{0h}\|_{0h}$ and $\|\phi_{1h}\|_{0h}$ the respective approximations of

$$\left(\int_\Omega |\phi_{0h}|^2 \mathrm{d}x\right)^{\frac{1}{2}} \quad \text{and} \quad \left(\int_\Omega |\phi_{1h}|^2 \mathrm{d}x\right)^{\frac{1}{2}}$$

obtained by the trapezoidal rule.

(2) Divide by $\max[1, \sqrt{\|\phi_{0h}\|_{0h}^2 + \|\phi_{1h}\|_{0h}^2}]$ the initial data $\mathbf{\Phi}_0$ and $\mathbf{\Phi}_1$ in (3.7).

(3) Apply the scheme (3.8) with normalized initial data to compute $\mathbf{\Phi}^{Q-1}$, $\mathbf{\Phi}^Q$ and $\mathbf{\Phi}^{Q+1}$.

(4) Prepare the initial data for the following nonlinear sub-step by multiplying (3.11) by the normalization factor $\max[1, \sqrt{\|\phi_{0h}\|_{0h}^2 + \|\phi_{1h}\|_{0h}^2}]$.

4 On the Numerical Solution to the Sub-initial Value Problems (2.12) and (2.14)

4.1 Generalities

From $n = 0$ until blow-up, we have to solve the initial value sub-problems (2.12) and (2.14) for almost every point of Ω. Following what we discussed in Sect. 3 (whose notation we keep), for the solution to the linear wave equation subproblems, we will consider only those nonlinear initial value sub-problems associated with the N_{0h} vertices of $\mathcal{T}_h$ not located on $\overline{\Gamma}_0$. Each of these sub-problem is of the following type:

$$\begin{cases} \frac{\mathrm{d}^2\phi}{\mathrm{d}t^2} = \alpha(6\phi^2 + t) & \text{in } (t_0, t_f), \\ \phi(t_0) = \phi_0, & \frac{\mathrm{d}\phi}{\mathrm{d}t}(t_0) = \phi_1 \end{cases} \tag{4.1}$$

with the initial data for (4.1) as in algorithms (2.8), (2.12), (2.16) and (2.14), after space discretization. A time discretization scheme of (4.1) with automatic adjustment of the time step will be discussed in the following section.

4.2 A Centered Scheme for the Time Discretization of (4.1)

Let M be a positive integer (> 2 in practice). With M, we associate a time discretization step $\sigma = \frac{t_f - t_0}{M}$. For the time discretization of the initial value problem (4.1), we suggest the following nonlinear variant of (3.8): Set

$$\phi^0 = \phi_0, \quad \phi^1 - \phi^{-1} = 2\sigma\phi_1,$$

then for $m = 0, \ldots, M$, compute ϕ^{m+1} by

$$\phi^{m+1} + \phi^{m-1} - 2\phi^m = \alpha\sigma^2\big(6|\phi^m|^2 + t^m\big) \tag{4.2}$$

with $t^m = t^0 + m\sigma$.

Considering the blowing-up properties of the solutions to the nonlinear wave problems (1.2), (1.3) and (1.2), (1.4), we expect that at one point in time, the solution to problem (4.1) will start growing very fast before becoming infinite. In order to track such a behavior, we have to decrease σ in (4.2), until the solution reaches some threshold at which we decide to stop computing (for the computational experiments reported in Sect. 5, we stop computing beyond 10^4). A practical method for the adaptation of the time step σ is described below.

4.3 On the Dynamical Adaptation of the Time Step σ

The starting point of our adaptive strategy will be the following observation: If ϕ is the solution to (4.1), at a time t before blow-up and for σ sufficiently small, we have (Taylor's expansion)

$$\begin{aligned}\phi(t+\sigma) &= \phi(t) + \sigma\dot{\phi}(t) + \frac{\sigma^2}{2}\ddot{\phi}(t) + \frac{\sigma^3}{6}\dddot{\phi}(t+\theta\sigma)\\ &= \phi(t) + \sigma\dot{\phi}(t) + \frac{\sigma^2}{2}\alpha\big(6|\phi(t)|^2 + t\big)\\ &\quad + \sigma^3\alpha\left(2\phi(t+\theta\sigma)\dot{\phi}(t+\theta\sigma) + \frac{1}{6}\right)\end{aligned} \tag{4.3}$$

with $0 < \theta < 1$. Suppose that we drop the σ^3-term in the above expansion, and that we approximate by finite differences the resulted truncated expansion at $t = t^m$. Then we obtain

$$\phi^{m+1} = \phi^m + \sigma\frac{\phi^{m+1} - \phi^{m-1}}{2\sigma} + \frac{\sigma^2}{2}\alpha\big(6|\phi^m|^2 + t^m\big),$$

which is the explicit scheme (4.2). Moreover, from the expansion (4.3), we can derive the following estimate of the relative error at $t = t^{m+1}$:

$$E^{m+1} = \sigma^3 \alpha \frac{|(\phi^{m+1} + \phi^m)\frac{(\phi^{m+1} - \phi^m)}{\sigma}| + \frac{1}{6}}{\max[1, |\phi^{m+1}|]}.$$

Another possible estimator would be

$$\sigma^3 \alpha \frac{|(\phi^{m+1} + \phi^m)\frac{(\phi^{m+1} - \phi^m)}{\sigma}| + \frac{1}{6}}{\max[1, \frac{1}{2}|\phi^m + \phi^{m+1}|]}.$$

In order to adapt σ by using E^{m+1}, we may proceed as follows: If ϕ^{m+1} obtained from the scheme (4.2) verifies

$$E^{m+1} \leq \text{tol}, \tag{4.4}$$

keep integrating with σ as a time discretization step. If criterion (4.4) is not verified, we have two possible situations, one for $m = 0$ and one for $m \geq 1$. If $m = 0$:

– Divide σ by 2 as many times as necessary to have

$$E^1 \leq \frac{\text{tol}}{5}. \tag{4.5}$$

Each time σ is divided by 2, double M accordingly.

– Still calling σ the first time step for which (4.5) holds after successive divisions by 2, apply scheme (4.2) to the solution to (4.1), with the new σ and the associated M.

If $m \geq 1$:

– Go to $t = t^{m-\frac{1}{2}} = t_0 + (m - \frac{1}{2})\sigma$.
– $t^{m-\frac{1}{2}} \to t_0$, $\frac{\phi^{m-1}+\phi^m}{2} \to \phi_0$, $\frac{\phi^m - \phi^{m-1}}{\sigma} \to \phi_1$.
– $\sigma \to \frac{\sigma}{2}$.
– $2(M - m) + 1 \to M$.
– Apply scheme (4.2) on the new interval (t_0, t_f). If criterion (4.4) is not verified, then proceed as in the case of $m = 0$.

For the numerical results reported in Sect. 5, we use tol $= 10^{-4}$.

Remark 4.1 In order to initialize the discrete analogues of the initial value problems (2.12), (2.13), we will use

$$\phi^M, \quad \frac{\phi^{M+1} - \phi^{M-1}}{2\sigma}$$

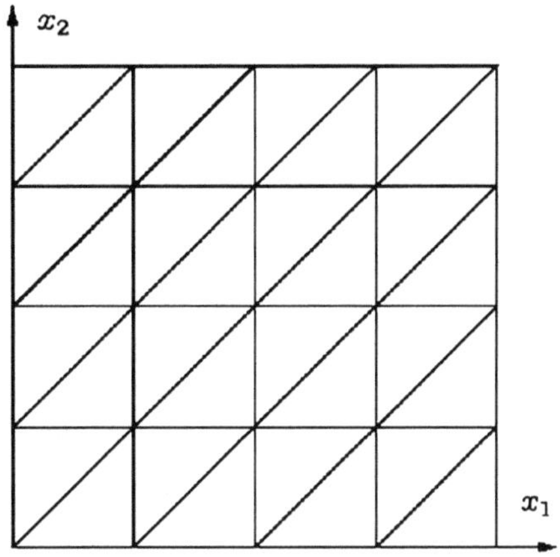

Fig. 1 A uniform triangulation of Ω

and

$$\phi^M, \quad \frac{\beta}{\alpha}\frac{\phi^{M+1} - \phi^{M-1}}{2\sigma},$$

respectively.

5 Numerical Experiments

5.1 Generalities

In this section, we are going to report on the results of numerical experiments concerning the solutions to the nonlinear wave problems (1.2), (1.3) and (1.2), (1.4). The role of these experiments is two-fold as follows: (i) Validate the numerical methodology discussed in Sects. 2–4, (ii) investigate how c and the boundary conditions influence the solutions.

For both problems, we took $\Omega = (0, 1)^2$. For the problem (1.2), (1.4), we took $\Gamma_1 = \{\{x_1, x_2\}, x_1 = 1, 0 < x_2 < 1\}$. The simplicity of the geometry suggests the use of finite differences for the space discretization. Actually, the finite difference schemes which we employ can be obtained via the finite element approximation discussed in Sect. 3, combined with the trapezoidal rule to compute the mass matrix $\mathbf{M}_h$ and the damping matrix $\mathbf{C}_h$. This requires that the triangulations which we employ are uniform like the one depicted in Fig. 1.

5.2 Numerical Experiments for the Nonlinear Wave Problem (1.2), (1.3)

Using well-known notation, we assume that the directional space discretization steps Δx_1 and Δx_2 are equal, and we denote by h their common value. We also assume that $h = \frac{1}{I+1}$, where I is a positive integer. For $0 \le i, j \le I + 1$, we denote by M_{ij} the point $\{ih, jh\}$ and $u_{ij}(t) \simeq u(M_{ij}, t)$. Using finite differences, we obtain the following continuous in time, discrete in space analogue of the problem (1.2), (1.3):

$$\begin{cases} u_{ij}(0) = u_0(M_{ij}), \quad 0 \le i, j \le I+1, \\ \dot{u}_{ij}(0) = u_1(M_{ij}), \quad 1 \le i, j \le I, \\ \ddot{u}_{ij}(t) + (\frac{c}{h})^2(4u_{ij} - u_{i+1j} - u_{i-1j} - u_{ij+1} - u_{ij-1})(t) \\ \quad = 6|u_{ij}(t)|^2 + t \quad \text{on } (0, T_{\max}),\ 1 \le i, j \le I, \\ u_{kl}(t) = 0 \quad \text{on } (0, T_{\max}) \text{ if } M_{kl} \in \partial\Omega. \end{cases} \tag{5.1}$$

In (5.1), we assume that u_0 (resp. u_1) belongs to $C^0(\overline{\Omega}) \cap H_0^1(\Omega)$ (resp. $C^0(\overline{\Omega})$).

The application of the discrete analogue of the operator-splitting scheme (2.8), (2.12)–(2.14) to problem (5.1) leads to the solution at each time step of:

(i) a discrete linear wave problem of the following type:

$$\begin{cases} \phi_{ij}(t_0) = \phi_0(M_{ij}), \quad 0 \le i, j \le I+1, \\ \dot{\phi}_{ij}(t_0) = \phi_1(M_{ij}), \quad 1 \le i, j \le I, \\ \ddot{\phi}_{ij}(t) + \beta(\frac{c}{h})^2(4\phi_{ij} - \phi_{i+1j} - \phi_{i-1j} \\ \quad - \phi_{ij+1} - \phi_{ij-1})(t) = 0 \quad \text{on } (t_0, t_f),\ 1 \le i, j \le I, \\ \phi_{kl}(t) = 0 \quad \text{on } (t_0, t_f) \text{ if } M_{kl} \in \partial\Omega. \end{cases} \tag{5.2}$$

(ii) $2I^2$ nonlinear initial value problems (2 for each interior grid point M_{ij}) like (4.1).

The numerical solution of the problem (4.1) has been addressed in Sects. 4.2 and 4.3. Concerning problem (5.2), it follows from Sect. 3 that its time discrete analogue reads as follows: Set

$$\phi_{ij}^0 = \phi_0(M_{ij}), \quad 0 \le i, j \le I+1 \quad \text{and} \quad \phi_{ij}^1 - \phi_{ij}^{-1} = 2\tau\phi_1(M_{ij}), \quad 1 \le i, j \le I,$$

then, for $q = 0, \ldots, Q$, $1 \le i, j \le I$, we have

$$\begin{cases} \phi_{ij}^{q+1} + \phi_{ij}^{q-1} - 2\phi_{ij}^q + \beta(\frac{\tau}{h}c)^2(4\phi_{ij}^q - \phi_{i+1j}^q - \phi_{i-1j}^q - \phi_{ij+1}^q - \phi_{ij-1}^q) = 0, \\ \phi_{kl}^{q+1} = 0 \quad \text{if } M_{kl} \in \partial\Omega \end{cases} \tag{5.3}$$

with $\tau = \frac{t_f - t_0}{Q}$. In the particular case of scheme (5.3), the stability condition (3.9) takes the following form:

$$\tau < \frac{h}{c\sqrt{2\beta}}. \tag{5.4}$$

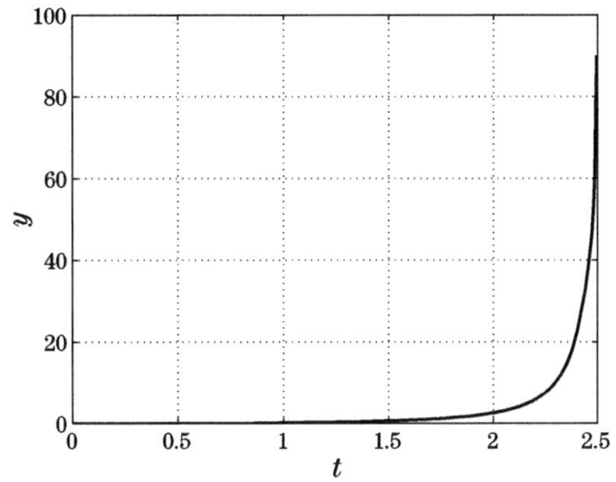

Fig. 2 Case $c = 0$: results obtained by our methodology

For the numerical results presented below, we took

(i) $u_0 = 0$ and $u_1 = 0$.
(ii) c ranging from 0 to 1.5.
(iii) $\alpha = \beta = \frac{1}{2}$.
(iv) $Q = 3$.
(v) For $h = \frac{1}{100}$: $\Delta t = 10^{-2}$ for $c \in [0, 0.6]$; $\Delta t = 8 \times 10^{-3}$ for $c = 0.7, 0.8$; $\Delta t = 5 \times 10^{-3}$ for $c = 0.9, 1, 1.25$; $\Delta t = 10^{-3}$ for $c = 1.5$.
(vi) For $h = \frac{1}{150}$: $\Delta t = 6 \times 10^{-3}$ for $c \in [0, 0.6]$; $\Delta t = 4 \times 10^{-3}$ for $c = 0.7, 0.8$; $\Delta t = 3 \times 10^{-3}$ for $c = 0.9, 1, 1.25$; $\Delta t = 6 \times 10^{-4}$ for $c = 1.5$.

We initialize with $M = 3$ (see Sect. 4.2), and then adapt M following the procedure described in Sect. 4.3.

We consider that the blow-up time is reached as soon as the maximum value of the discrete solution reaches 10^4. Let us remark that the numerical results obtained with $h = \frac{1}{100}$ and $h = \frac{1}{150}$ (and the respective associated values of Δt) are essentially identical.

In Fig. 2, we report the results obtained by our methodology when $c = 0$. They compare quite well with the results reported by Wikipedia.[1]

In Fig. 3, we visualize for $c = 0.8$ and $t \in [0, 14.4]$ (the blow-up time being close to $T_{\max} \simeq 15.512$) the evolution of the computed approximations of the functions

$$u_{\ln} = \operatorname{sgn}(u) \ln(1 + |u|) \quad \text{and} \quad p_{\ln} = \operatorname{sgn}(p) \ln(1 + |p|) \tag{5.5}$$

restricted to the segment $\{\{x_1, x_2\},\ 0 \leq x_1 \leq 1,\ x_2 = \frac{1}{2}\}$. The oscillatory behavior of the solution appears clearly in Fig. 3(b). In Fig. 4, we report the graph of the computed approximations of u and p for $c = 0.8$ at $t = 15.512$, very close to the blow-up time.

[1] http://en.wikipedia.org/wiki/Painlevé_transcendents.

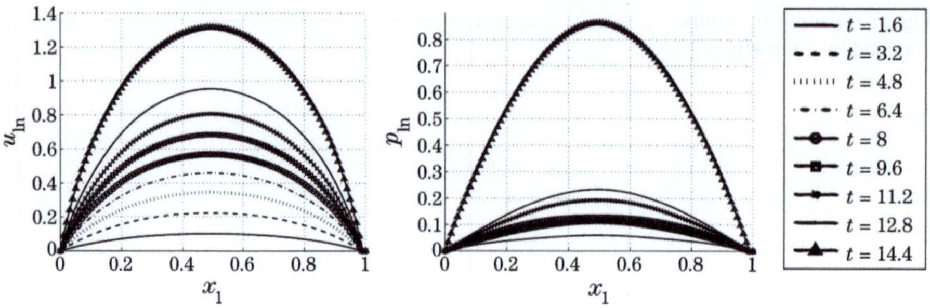

(a) evolution of u_{ln}. (b) evolution of p_{ln}. (c) caption.

Fig. 3 Case $c = 0.8$, pure Dirichlet boundary conditions: evolution of quantities (**a**) u_{ln} and (**b**) p_{ln}. The caption in (**c**) is common to (**a**) and (**b**)

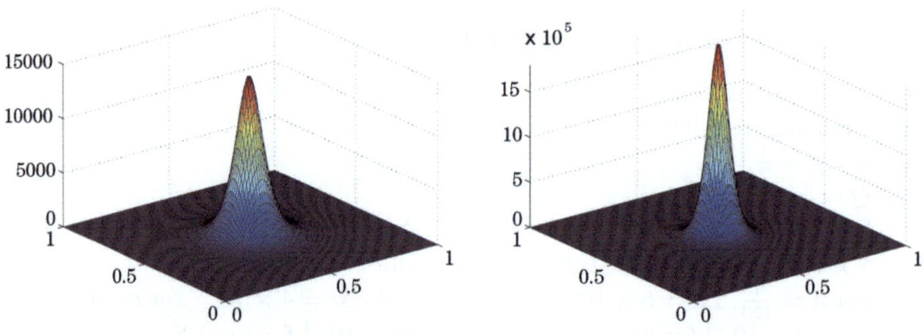

(a) u (b) p

Fig. 4 Case $c = 0.8$, pure Dirichlet boundary conditions: computed approximations for (**a**) u and (**b**) p at $t = 15.512$

In Fig. 5, we show for $c = 1$ the approximated evolution for $t \in [0, 35.03]$ of the function

$$t \to \max_{\{x_{,1},x_2\}\in\Omega} u(x_1, x_2, t). \tag{5.6}$$

The computed maximum value is always achieved at $\{0.5, 0.5\}$. The explosive nature of the solution is obvious from this figure.

In order to better understand the evolution of the function (5.6), we analyze its restriction to the time interval $[0, 28]$ in both the time and frequency domains (see Fig. 6). Actually, concerning the frequency domain, we specially analyze the modulation of the above function, that is the signal obtained after subtracting its convex component from the function in (5.6). Figure 6(b) indicates that the modulation observed in Fig. 6(a) is quasi-monochromatic, with $f \simeq 0.9$ Hz.

Finally, Fig. 7 reports the variation of the blow-up time of the approximated solution as a function of c. As mentioned above, the results obtained with $h = \frac{1}{100}$ and $h = \frac{1}{150}$ match very accurately.

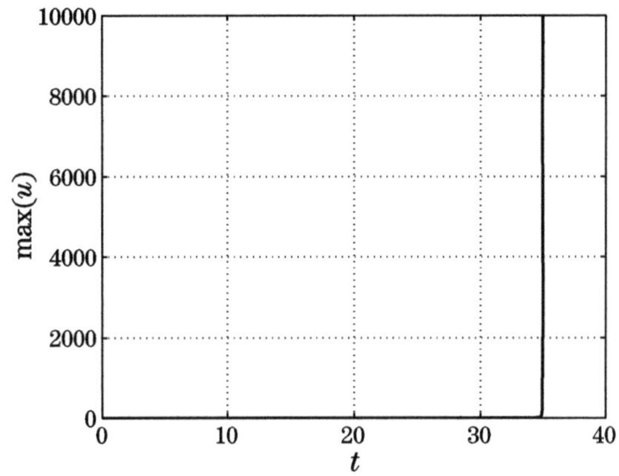

Fig. 5 Case $c = 1$, pure Dirichlet boundary conditions: Evolution of the computed approximation of the function in (5.6) for $t \in [0, 35.03]$

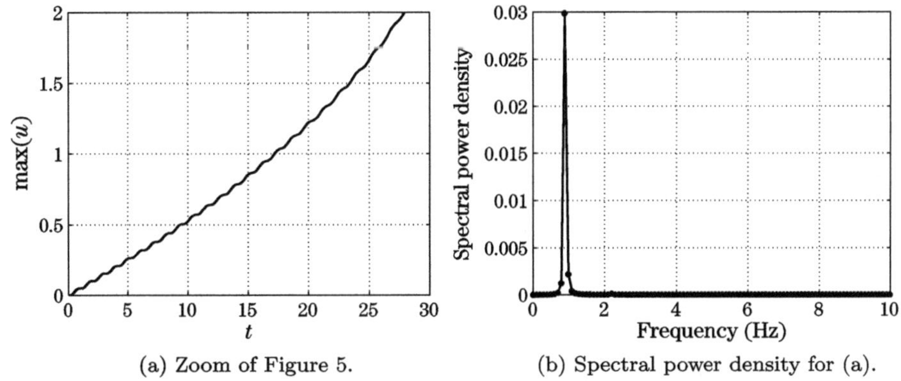

Fig. 6 Case $c = 1$, pure Dirichlet boundary conditions: (**a**) evolution of the computed approximation of the function in (5.6) for $t \in [0, 28]$, (**b**) spectrum of the modulation

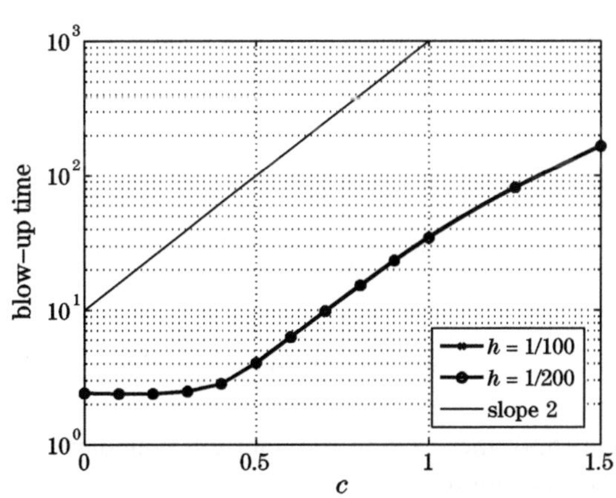

Fig. 7 The blow-up time as a function of c (semi-log scale)

5.3 Numerical Experiments for the Nonlinear Wave Problem (1.2), (1.4)

The time discretization by operator-splitting of the nonlinear wave problem (1.2), (1.4) has been discussed in Sect. 2.3, where we showed that at each time step, we have to solve two nonlinear initial value problems such as (4.1) and one linear wave problem such as (3.1).

The simplicity of the geometry of this test problem (see Sect. 5.1) suggests the use of finite differences for the space discretization. Using the notation in Sect. 5.2, at each time step, we will have to solve $2I(I+1)$ initial value problem such as (4.1), that is two for each grid point M_{ij} $(1 \leq i \leq I+1, 1 \leq j \leq I)$. The solution method discussed in Sect. 4 is still valid. By discretizing problem (3.1) with finite differences, we obtain the following scheme:

$$\phi_{ij}^0 = \phi_0(M_{ij}), \quad 0 \leq i, j \leq I+1,$$

$$\phi_{ij}^1 - \phi_{ij}^{-1} = 2\tau\phi_1(M_{ij}), \quad 1 \leq i \leq I+1, 1 \leq j \leq I,$$

then, for $q = 0, \ldots, Q$, $1 \leq i \leq I+1, 1 \leq j \leq I$,

$$\begin{cases} \phi_{ij}^{q+1} + \phi_{ij}^{q-1} - 2\phi_{ij}^q + \beta(\frac{\tau}{h}c)^2(4\phi_{ij}^q - \phi_{i+1j}^q - \phi_{i-1j}^q - \phi_{ij+1}^q - \phi_{ij-1}^q) = 0, \\ \phi_{kl}^{q+1} = 0 \quad \text{if } M_{kl} \in \Gamma_0, \\ \frac{1}{\beta c}\frac{\phi_{I+1l}^{q+1} - \phi_{I+1l}^{q-1}}{2\tau} + \frac{\phi_{I+2l}^q - \phi_{Il}^q}{2h} = 0, \quad 1 \leq l \leq I, \end{cases} \tag{5.7}$$

where $\tau = \frac{t_f - t_0}{Q}$ and the "ghost" value ϕ_{I+2l}^q is introduced to impose the Sommerfeld condition at the discrete level. Upon elimination of ϕ_{I+2l}^q, we can derive a more practical formulation of the fully discrete problem, namely, for $q = 0, \ldots, Q$, $1 \leq i \leq I$, $1 \leq j \leq I$, instead of (5.7), we have

$$\begin{cases} \phi_{ij}^{q+1} + \phi_{ij}^{q-1} - 2\phi_{ij}^q + \beta(\frac{\tau}{h}c)^2(4\phi_{ij}^q - \phi_{i+1j}^q - \phi_{i-1j}^q - \phi_{ij+1}^q - \phi_{ij-1}^q) = 0, \\ \phi_{kl}^{q+1} = 0, \quad \text{if } M_{kl} \in \Gamma_0 \end{cases} \tag{5.8}$$

and for $q = 0, \ldots, Q$, $i = I+1$, $1 \leq j \leq I$,

$$\left(1 + \frac{\tau}{h}c\right)\phi_{I+1j}^{q+1} + \left(1 - \frac{\tau}{h}c\right)\phi_{I+1j}^{q-1} - 2\phi_{I+1j}^q$$
$$+ \beta\left(\frac{\tau}{h}c\right)^2(4\phi_{I+1j}^q - 2\phi_{Ij}^q - \phi_{I+1j+1}^q - \phi_{I+1j-1}^q) = 0. \tag{5.9}$$

Via (5.9), the discrete Sommerfeld boundary condition is included in the discrete wave equation.

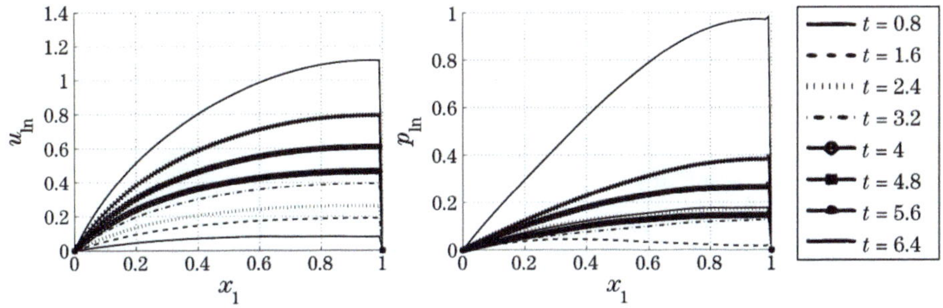

(a) evolution of u_{ln}. (b) evolution of p_{ln}. (c) caption.

Fig. 8 Case $c = 0.8$, mixed Dirichlet-Sommerfeld boundary conditions: evolution of quantities (**a**) u_{ln} and (**b**) p_{ln}. The caption in (**c**) is common to (**a**) and (**b**)

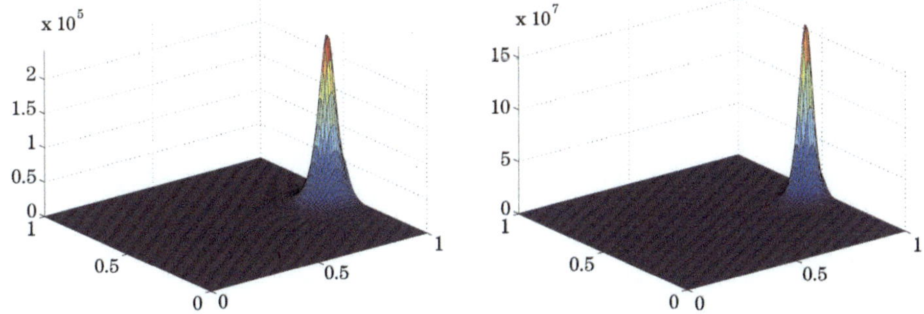

(a) u (b) p

Fig. 9 Case $c = 0.8$, mixed Dirichlet-Sommerfeld boundary conditions: computed approximations for (**a**) u and (**b**) p at $t = 7.432$

We chose the same values for u_0, u_1, c, α, β, Q, h and Δt as in Sect. 5.2. Once again, the results obtained with $h = \frac{1}{100}$ and $h = \frac{1}{150}$ match very accurately.

In Fig. 8, we visualize for $c = 0.8$ and $t \in [0, 6.4]$ (the blow-up time being close to $T_{\max} \simeq 7.432$) the evolution of the computed approximations of the quantities in (5.5) restricted to the segment $\{\{x_1, x_2\},\ 0 \leq x_1 \leq 1,\ x_2 = \frac{1}{2}\}$. These results (and the ones below) show that the blow-up occurs sooner than in the pure Dirichlet boundary condition case. In Fig. 9, we report the graph of the computed approximations of u and p for $c = 0.8$ at $t = 7.432$, very close to the blow-up time.

Figure 10 reports the graph of the computed approximations of u and p for $c = 0.3$ at $t = 2.44$, very close to the blow-up time. Figures 9 and 10 show that for c sufficiently small (resp. large), the blow-up takes place inside Ω (resp. on Γ_1).

In Fig. 11(a), for $c = 1$, we report the approximated evolution of the function in (5.6) for $t \in [0, 15.135]$. In order to have a better view of the expected modulation of the above function, we report in Fig. 11(b) its evolution for $t \in [0, 13.5]$. These figures show the dramatic growth of the solution as t gets closer to $T_{\max}$.

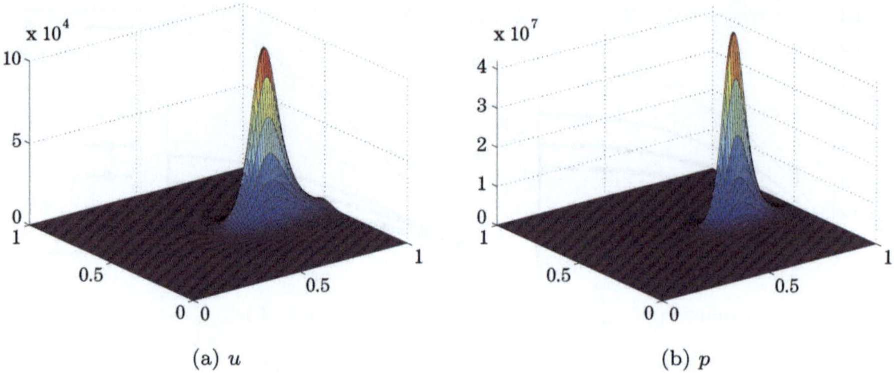

Fig. 10 Case $c = 0.3$, mixed Dirichlet-Sommerfeld boundary conditions: computed approximations for (**a**) u and (**b**) p at $t = 2.44$

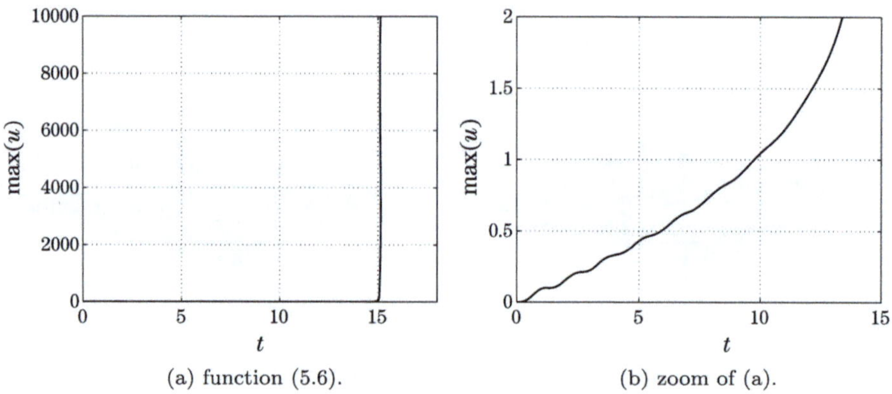

Fig. 11 Case $c = 1$, mixed Dirichlet-Sommerfeld boundary conditions: (**a**) evolution of the computed approximation of the function in (5.6) for $t \in [0, 15.135]$, (**b**) zoomed view for $t \in [0, 13.5]$.

Finally, we report in Fig. 12 the variation versus c of the blow-up time for both the pure Dirichlet and the mixed Dirichlet-Sommerfeld boundary conditions. It is interesting to observe how the presence of a boundary condition with (rather) good transparency properties decreases significantly the blow-up time, everything else being the same. Also, the above figure provides a strong evidence of the very good matching of the approximate solutions obtained for $h = \frac{1}{100}$ and $h = \frac{1}{150}$ (and the related time discretization steps).

6 Further Comments and Conclusions

The methods discussed in this article can be generalized to the coupling of the linear wave equation with other Painlevé equations, or other nonlinearities, such as

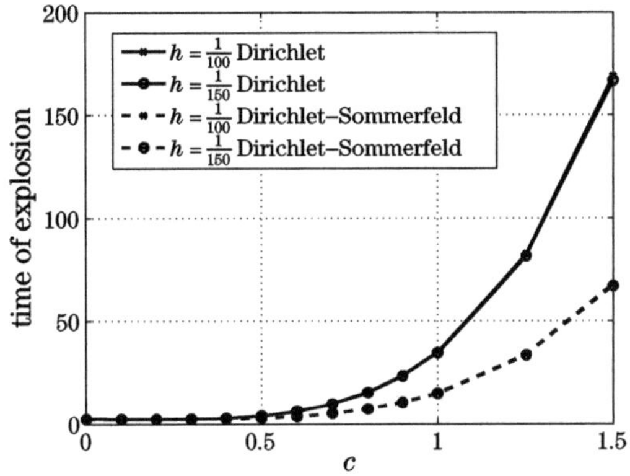

Fig. 12 The blow-up time as a function of c for both the pure Dirichlet and the mixed Dirichlet-Sommerfeld boundary conditions

$v \to e^v$. Actually, this claim is already validated by the results of numerical experiments, which we are performing with these other models. Another generalization under investigation is the application of the methods discussed here to the numerical solution to those nonlinear wave equations of the Euler-Poisson-Darboux type discussed in [18]. This application will require a 5-stage splitting scheme, instead of the 3-stage one, which we employed in this article.

We would like to conclude with the following two comments:

(1) When it goes to the numerical simulation of multi-physics phenomena, there are two possible approaches, namely, the monolithic (that is, un-split) methods and the operator-splitting methods. We think that the splitting methods discussed in this article are better suited for the solution to problems (1.2), (1.3) and (1.2), (1.4) than the monolithic ones.

(2) The splitting methods discussed in this article have good parallelization properties that we intend to investigate in the near future.

Acknowledgements The authors would like to thank the founders of the Painlevé Project. Their one page article in the Notices of the AMS [1] was the motivation of the work reported in this article. The authors thank Dr. Simone Paulotto for his help on the spectral analysis of some results.

References

1. Bornemann, F., Clarkson, P., Deift, P., et al.: A request: The Painlevé Project. Not. Am. Math. Soc. **57**(11), 1938 (2010)
2. Jimbo, M.: Monodromy problem and the boundary condition for some Painlevé equations. Publ. Res. Inst. Math. Sci., Ser. A **18**(3), 1137–1161 (1982)
3. Wong, R., Zhang, H.Y.: On the connection formulas of the fourth Painlevé transcendent. Anal. Appl. **7**(4), 419–448 (2009)
4. Clarkson, P.A.: Painlevé transcendents. In: Olver, F.W.J., Lozier, D.W., Boisvert, R.F., Clark, C.W. (eds.) NIST Handbook of Mathematical Functions, pp. 723–740. Cambridge University Press, Cambridge (2010)
5. Fornberg, B., Weideman, J.A.C.: A numerical methodology for the Painlevé equations. J. Comput. Phys. **230**(15), 5957–5973 (2011)

6. Strang, G.: On the construction and comparison of difference schemes. SIAM J. Numer. Anal. **5**(3), 506–517 (1968)
7. Glowinski, R.: Finite element methods for incompressible viscous flow. In: Ciarlet, P.G., Lions, J.L. (eds.) Handbook of Numerical Analysis, vol. IX, pp. 3–1176. North-Holland, Amsterdam (2003)
8. Bokil, V.A., Glowinski, R.: An operator-splitting scheme with a distributed Lagrange multiplier based fictitious domain method for wave propagation problems. J. Comput. Phys. **205**(1), 242–268 (2005)
9. Glowinski, R., Shiau, L., Sheppard, M.: Numerical methods for a class of nonlinear integro-differential equations. Calcolo **50**, 17–33 (2013)
10. Lions, J.L.: Quelques Méthodes de Résolution des Problèmes aux Limites Non Linéaires. Dunod, Paris (1969)
11. Glowinski, R., Dean, E.J., Guidoboni, G., et al.: Application of operator-splitting methods to the direct numerical simulation of particulate and free-surface flows and to the numerical solution to the two-dimensional elliptic Monge-Ampère equation. Jpn. J. Ind. Appl. Math. **25**(1), 1–63 (2008)
12. Chorin, A.J., Hughes, T.J.R., McCracken, M.F., et al.: Product formulas and numerical algorithms. Commun. Pure Appl. Math. **31**, 205–256 (1978)
13. Beale, J.T., Majda, A.: Rates of convergence for viscous splitting of the Navier-Stokes equations. Math. Comput. **37**(156), 243–259 (1981)
14. Leveque, R.J., Oliger, J.: Numerical methods based on additive splittings for hyperbolic partial differential equations. Math. Comput. **40**(162), 469–497 (1983)
15. Marchuk, G.I.: Splitting and alternating direction method. In: Ciarlet, P.G., Lions, J.L. (eds.) Handbook of Numerical Analysis, vol. I, pp. 197–462. North-Holland, Amsterdam (1990)
16. Temam, R.: Navier-Stokes Equations: Theory and Numerical Analysis. AMS, Providence (2001)
17. Glowinski, R.: Numerical Methods for Nonlinear Variational Problems. Springer, New York (1984)
18. Keller, J.B.: On solutions of nonlinear wave equations. Commun. Pure Appl. Math. **10**(4), 523–530 (1957)

MsFEM à la Crouzeix-Raviart for Highly Oscillatory Elliptic Problems

Claude Le Bris, Frédéric Legoll, and Alexei Lozinski

Abstract We introduce and analyze a multiscale finite element type method (MsFEM) in the vein of the classical Crouzeix-Raviart finite element method that is specifically adapted for highly oscillatory elliptic problems. We illustrate numerically the efficiency of the approach and compare it with several variants of MsFEM.

Keywords Homogenization · Finite elements · Galerkin methods · Highly oscillatory PDE

Mathematics Subject Classification 35B27 · 65M60 · 65M12

The work of the first two authors is partially supported by ONR under Grant (No. N00014-12-1-0383) and EOARD under Grant (No. FA8655-10-C-4002).

C. Le Bris (✉) · F. Legoll
École Nationale des Ponts et Chaussées, 6 et 8 avenue Blaise Pascal, 77455 Marne-La-Vallée Cedex 2, France
e-mail: lebris@cermics.enpc.fr

F. Legoll
e-mail: legoll@lami.enpc.fr

C. Le Bris · F. Legoll
MICMAC project-team, INRIA Rocquencourt, 78153 Le Chesnay Cedex, France

A. Lozinski
Institut de Mathématiques de Toulouse, Université Paul Sabatier, 118 route de Narbonne, 31062 Toulouse Cedex 9, France
e-mail: alexei.lozinski@univ-fcomte.fr

Present address:
A. Lozinski
Laboratoire de Mathématiques CNRS UMR 6623, Université de Franche-Comté, 16 route de Gray, 25030 Besançon Cedex, France

P.G. Ciarlet et al. (eds.), *Partial Differential Equations: Theory, Control and Approximation*, DOI 10.1007/978-3-642-41401-5_11,

1 Introduction

Let $\Omega \subset \mathbb{R}^d$ be a bounded domain and $f \in L^2(\Omega)$ (more regularity on the right-hand side will be needed later on). We consider the problem

$$-\operatorname{div}\left[A_\varepsilon(x)\nabla u^\varepsilon\right] = f \quad \text{in } \Omega, \quad u^\varepsilon = 0 \quad \text{on } \partial\Omega, \tag{1.1}$$

where A_ε is a highly oscillatory, uniformly elliptic and bounded matrix. To fix the ideas (and this will in fact be a necessary assumption for the analysis which we provide below to hold true), one might think of $A_\varepsilon(x) = A_{\text{per}}(\frac{x}{\varepsilon})$, where A_{per} is $\mathbb{Z}^d$ periodic. The approach which we introduce here to address problem (1.1) is a multiscale finite element type method (henceforth abbreviated as MsFEM). As any such method, our approach is not restricted to the periodic setting. Only our analysis is. Likewise, we will assume for simplicity of our analysis that the matrices A_ε which we manipulate are symmetric matrices.

Our purpose is to propose and study a specific multiscale finite element method for the problem (1.1), where the Galerkin approximation space is constructed from ideas similar to those by Crouzeix and Raviart in their construction of a classical FEM space [14].

Recall that the general idea of MsFEM approaches is to construct an approximation space by using precomputed, local functions, that are solutions to the equation under consideration with simple (typically vanishing) right-hand sides. This is in contrast to standard finite element approaches, where the approximation space is based on generic functions, namely piecewise polynomials. To construct our specific multiscale finite element method for the problem (1.1), we recall the classical work of Crouzeix and Raviart [14]. We preserve the main feature of their nonconforming FEM space, i.e., that the continuity across the edges of the mesh is enforced only in a weak sense by requiring that the average of the jump vanishes on each edge. As shown in Sect. 2.1 below, this "weak" continuity condition leads to some natural boundary conditions for the multiscale basis functions.

Our motivation for the introduction of such finite element functions stems from our wish to address several specific multiscale problems, most of them in a nonperiodic setting, for which implementing flexible boundary conditions on each mesh element is of particular interest. A prototypical situation is that of a perforated medium, where inclusions are not periodically located and where the accuracy of the numerical solution is extremely sensitive to an appropriate choice of values of the finite element basis functions on the boundaries of elements when the latter intersect inclusions. The Crouzeix-Raviart type elements which we construct then provide an advantageous flexibility. Additionally, when the problem under consideration is not (as (1.1) above) a simple scalar elliptic Poisson problem but a Stokes type problem, it is well-known that the Crouzeix-Raviart approach also allows—in the classical setting—for directly encoding the incompressibility constraint in the finite element space. This property will be preserved for the approach which we introduce here in the multiscale context. We will not proceed further in this direction and refer the interested reader to our forthcoming publication (see [25]) for more details on this topic and related issues.

Of course, our approach is not the only possible one to address the category of problems we consider. Sensitivity of the numerical solution upon the choice of boundary condition set for the multiscale finite element basis functions is a classical issue. Formally, it may be easily understood. In a one-dimensional situation (see for instance [26] for a formalization of this argument), the error committed by using a multiscale finite element type approach entirely comes from the error committed in the bulk of each element, because it is easy to make the numerical solution agree with the exact solution on nodes. In dimensions greater than one, however, it is impossible to match the finite dimensional approximation on the boundary of elements with the exact, infinite dimensional trace of the exact solution on this boundary. A second source of numerical error thus follows from this. And the derivation of variants of MsFEM type approaches can be seen as the quest to solve the issue of inappropriate boundary conditions on the boundaries of mesh elements.

Many tracks, each of which leads to a specific variant of the general approach, have been followed to address the issue. The simplest choice (see [21, 22]) is to use linear boundary conditions, as in the standard P1 finite element method. This yields a multiscale finite element space consisting of continuous functions. The use of nonconforming finite elements is an attractive alternative, leading to more accurate and more flexible variants of the method. The work [12] uses Raviart-Thomas finite elements for a mixed formulation of a highly oscillatory elliptic problem similar to that considered in the present article. Many contributions such as [1, 2, 5, 7] present variants and follow-up of this work. For non-mixed formulations, we mention the well-known oversampling method (giving birth to nonconforming finite elements, see [16, 20, 21]). We also mention the work [11], where a variant of the classical MsFEM approach (i.e., without oversampling) is presented. Basis functions also satisfy Dirichlet linear boundary conditions on the boundaries of the finite elements, but continuity across the edges is only enforced at the midpoint of the edges, as in the approach suggested by Crouzeix and Raviart [14]. Note that this approach, although also inspired by the work [14], differs from ours in the sense that we do not impose any Dirichlet boundary conditions when constructing the basis functions (see Sect. 2.1 below for more details).

In the context of an HMM-type method, we mention the works [3, 4] for the computation of an approximation of the coarse scale solution. An excellent review of many of the existing approaches is presented in [6], and for the general development of MsFEM (as of 2009) we refer to [15].

Our purpose here is to propose yet another possibility, which may be useful in specific contexts. Results for problems of type (1.1), although good, will not be spectacularly good. However, the ingredients which we employ here to analyze the approach and the structure of our proof will be very useful when studying the same Crouzeix-Raviart type approach for a specific setting of particular interest: the case for perforated domains. In that case, we will show in [25] how extremely efficient our approach is.

Our article is articulated as follows. We outline our approach in Sect. 2 and state the corresponding error estimate, for the periodic setting, in Sect. 3 (Theorem 3.2). The subsequent two sections are devoted to the proof of the main error estimate. We recall some elementary facts and tools of numerical analysis in Sect. 4, and turn to the actual proof of Theorem 3.2 in Sect. 5. Section 6 presents some numerical comparisons between the approach which we introduce here and some existing MsFEM type approaches.

2 Presentation of Our MsFEM Approach

Throughout this article, we assume that the ambient dimension is $d = 2$ or $d = 3$ and that Ω is a polygonal (resp. polyhedral) domain. We define a mesh $\mathcal{T}_H$ on Ω, i.e., a decomposition of Ω into polygons (resp. polyhedra) each of diameter at most H, and denote $\mathcal{E}_H$ the set of all the internal edges (or faces) of $\mathcal{T}_H$. We assume that the mesh does not have any hanging nodes. Otherwise stated, each internal edge (resp. face) is shared by exactly two elements of the mesh. In addition, $\mathcal{T}_H$ is assumed to be a regular mesh in the following sense: for any mesh element $T \in \mathcal{T}_H$, there exists a smooth one-to-one and onto mapping $K : \overline{T} \to T$, where $\overline{T} \subset \mathbb{R}^d$ is the reference element (a polygon, resp. a polyhedron, of fixed unit diameter) and $\|\nabla K\|_{L^\infty} \leq CH$, $\|\nabla K^{-1}\|_{L^\infty} \leq CH^{-1}$, C being some universal constant independent of T, to which we will refer as the regularity parameter of the mesh. To avoid some technical complications, we also assume that the mapping K corresponding to each $T \in \mathcal{T}_H$ is affine on every edge (resp. face) of $\partial\overline{T}$. In the following and to fix the ideas, we will have in mind the two-dimensional situation and a mesh consisting of triangles, which satisfies the minimum angle condition to ensure that the mesh is regular in the sense defined above (see [10, Sect. 4.4]). We will repeatedly use the notations and terminologies (triangle, edge, ...) of this setting, although the analysis carries over to quadrangles if $d = 2$, or to tetrahedra and parallelepipeda if $d = 3$.

The bottom line of our multiscale finite element method à la Crouzeix-Raviart is, as for the classical version of the method, to require the continuity of the (here highly oscillatory) finite element basis functions only in the sense of averages on the edges, rather than to require the continuity at the nodes (which is for instance the case in the oversampling variant of the MsFEM). In doing so, we expect more flexibility, and therefore better approximation properties in delicate cases.

2.1 Construction of the MsFEM Basis Functions

Functional Spaces We introduce the functional space

$$W_H = \Big\{ u \in L^2(\Omega) \text{ such that } u|_T \in H^1(T) \text{ for any } T \in \mathcal{T}_H,$$
$$\int_e [\![u]\!] = 0 \text{ for all } e \in \mathcal{E}_H \text{ and } u = 0 \text{ on } \partial\Omega \Big\},$$

where $[\![u]\!]$ denotes the jump of u over an edge. We next introduce its subspace

$$W_H^0 = \left\{ u \in W_H \text{ such that } \int_e u = 0 \text{ for all } e \in \mathcal{E}_H \right\}$$

and define the MsFEM space à la Crouzeix-Raviart

$$V_H = \left\{ u \in W_H \text{ such that } a_H(u, v) = 0 \text{ for all } v \in W_H^0 \right\}$$

as the orthogonal complement of W_H^0 in W_H, where by orthogonality we mean orthogonality for the scalar product defined by

$$a_H(u, v) = \sum_{T \in \mathcal{T}_H} \int_T (\nabla v)^{\mathrm{T}} A_\varepsilon(x) \nabla u dx. \tag{2.1}$$

We recall that for simplicity we assume all matrices are symmetric.

Notation For any $u \in W_H$, we henceforth denote by

$$\|u\|_E := \sqrt{a_H(u, u)}$$

the energy norm associated with the form a_H.

"Strong" Form To get a more intuitive grasp on the space V_H, we note that any function $u \in V_H$ satisfies, on any element $T \in \mathcal{T}_H$,

$$\int_T (\nabla v)^{\mathrm{T}} A_\varepsilon \nabla u = 0 \quad \text{for all } v \in H^1(T) \quad \text{s.t.} \quad \int_{\Gamma_i} v = 0 \quad \text{for all } i = 1, \ldots, N_\Gamma,$$

where Γ_i (with $i = 1, \ldots, N_\Gamma$) are the N_Γ edges composing the boundary of T (note that, if $\Gamma_i \subset \partial\Omega$, the condition $\int_{\Gamma_i} v = 0$ is replaced by $v = 0$ on Γ_i; this is a convention which we will use throughout our article without explicitly mentioning it). This can be rewritten as

$$\int_T (\nabla v)^{\mathrm{T}} A_\varepsilon \nabla u = \sum_{i=1}^{N_\Gamma} \lambda_i \int_{\Gamma_i} v \quad \text{for all } v \in H^1(T)$$

for some scalar constants $\lambda_1, \ldots, \lambda_{N_\Gamma}$. Hence, the restriction of any $u \in V_H$ to T is a solution to the boundary value problem

$$-\operatorname{div}\left[A_\varepsilon(x) \nabla u\right] = 0 \quad \text{in } T, \qquad n \cdot A_\varepsilon \nabla u = \lambda_i \quad \text{on each } \Gamma_i.$$

The flux along each edge interior to Ω is therefore a constant. This of course defines u only up to an additive constant, which is fixed by the "continuity" condition

$$\int_e [\![u]\!] = 0 \quad \text{for all } e \in \mathcal{E}_H \quad \text{and} \quad u = 0 \quad \text{on } \partial\Omega. \tag{2.2}$$

Remark 2.1 Observe that, in the case $A_\varepsilon = \mathrm{Id}$, we recover the classical nonconforming finite element spaces:

(1) Crouzeix-Raviart element (see [14]) on any triangular mesh: on each T, $u|_T \in \mathrm{Span}\{1, x, y\}$.
(2) Rannacher-Turek element (see [29]) on any rectangular Cartesian mesh: on each T, $u|_T \in \mathrm{Span}\{1, x, y, x^2 - y^2\}$.

Basis Functions We can associate the basis functions of V_H with the internal edges of the mesh as follows. Let e be such an edge and let T_1 and T_2 be the two mesh elements that share that edge e. The basis function ϕ_e associated to e, the support of which is $T_1 \cup T_2$, is constructed as follows. Let us denote the edges composing the boundary of T_k ($k = 1$ or 2) by Γ_i^k (with $i = 1, \dots, N_\Gamma$), and without loss of generality suppose that $\Gamma_1^1 = \Gamma_1^2 = e$. On each T_k, the function ϕ_e is the unique solution in $H^1(T_k)$ to

$$\begin{aligned} -\operatorname{div}\big[A_\varepsilon(x)\nabla\phi_e\big] &= 0 && \text{in } T_k, \\ \int_{\Gamma_i^k} \phi_e &= \delta_{i1} && \text{for } i = 1, \dots, N_\Gamma, \\ n \cdot A_\varepsilon \nabla\phi_e &= \lambda_i^k && \text{on } \Gamma_i^k,\ i = 1, \dots, N_\Gamma, \end{aligned}$$

where δ_{i1} is the Kronecker symbol. Note that, for the edge $\Gamma_1^1 = \Gamma_1^2 = e$ shared by the two elements, the value of the flux may be different from one side of the edge to the other one: λ_1^1 may be different from λ_1^2. The existence and the uniqueness of ϕ_e follow from standard analysis arguments.

Decomposition Property A specific decomposition property based on the above finite element spaces will be useful in the sequel. Consider some function $u \in W_H$, and introduce $v_H \in V_H$ such that, for any element $T \in \mathcal{T}_H$, we have $v_H \in H^1(T)$, and

$$\begin{aligned} -\operatorname{div}\big[A_\varepsilon(x)\nabla v_H\big] &= 0 && \text{in } T, \\ \int_{\Gamma_i} v_H &= \int_{\Gamma_i} u && \text{for } i = 1, \dots, N_\Gamma, \\ n \cdot A_\varepsilon \nabla v_H &= \lambda_i && \text{on } \Gamma_i,\ i = 1, \dots, N_\Gamma. \end{aligned}$$

Consider now $v^0 = u - v_H \in W_H$. We see that, for any edge e,

$$\int_e v^0 = \int_e u - \int_e v_H = 0,$$

and thus $v^0 \in W_H^0$. We can hence decompose (in a unique way) any function $u \in W_H$ as the sum $u = v_H + v^0$, with $v_H \in V_H$ and $v^0 \in W_H^0$.

2.2 Definition of the Numerical Approximation

Using the finite element spaces introduced above, we now define the MsFEM approximation of the solution u^ε to (1.1) as the solution $u_H \in V_H$ to

$$a_H(u_H, v) = \int_\Omega f v \quad \text{for any } v \in V_H, \tag{2.3}$$

where a_H is defined by (2.1). Note that (2.3) is a nonconforming approximation of (1.1), since $V_H \not\subset H_0^1(\Omega)$.

The problem (2.3) is well-posed. Indeed, it is finite dimensional so that it suffices to prove that $f = 0$ implies $u_H = 0$. But $f = 0$ implies, taking $v = u_H$ in (2.3) and using the coercivity of A_ε, that $\nabla u_H = 0$ on every $T \in \mathcal{T}_H$. The continuity condition (2.2) then shows that $u_H = 0$ on Ω.

3 Main Result

The main purpose of our article is to present the numerical analysis of the method outlined in the previous section. To this end, we need to restrict the setting of the approach (stated above for, and indeed applicable to, general matrices A_ε) to the periodic setting. The essential reason for this restriction is that, in the process of the proof of our main error estimate (Theorem 3.2), we need to use an accurate description of the asymptotic behavior (as $\varepsilon \to 0$) of the oscillatory solution u^ε. Schematically speaking, our error estimate is established using a triangle inequality of the form

$$\|u^\varepsilon - u_H\| \leq \|u^\varepsilon - u^{\varepsilon,1}\| + \|u^{\varepsilon,1} - u_H\|,$$

where $u^{\varepsilon,1}$ is an accurate description of the exact solution u^ε to (1.1), for ε small. Such an accurate description is not available in the completely general setting where the method is applicable. In the periodic setting, however, we do have such a description at our disposal. It is provided by the two-scale expansion of the homogenized solution to the problem. This is the reason why we restrict ourselves to this setting. Some other specific settings could perhaps allow for the same type of analysis, but we will not proceed in this direction. On the other hand, in the present state of our understanding of the problem and to the best of our knowledge of the existing literature, we are not aware of any strategy of proof that could accommodate the fully general oscillatory setting.

Periodic Homogenization We henceforth assume that, in (1.1),

$$A_\varepsilon(x) = A_{\text{per}}\left(\frac{x}{\varepsilon}\right), \tag{3.1}$$

where A_{per} is $\mathbb{Z}^d$ periodic (and of course bounded and uniformly elliptic). It is then well-known (see the classical textbooks [8, 13, 24], and also [17] for a general, numerically oriented presentation) that the solution u^ε to (1.1) converges, weakly in $H^1(\Omega)$ and strongly in $L^2(\Omega)$, to the solution $u^\star$ to

$$-\operatorname{div}\left(A^\star_{\mathrm{per}} \nabla u^\star\right) = f \quad \text{in } \Omega, \quad u^\star = 0 \quad \text{on } \partial\Omega, \tag{3.2}$$

with the homogenized matrix given by, for any $1 \le i, j \le d$,

$$\left(A^\star_{\mathrm{per}}\right)_{ij} = \int_{(0,1)^d} \left(e_i + \nabla w_{e_i}(y)\right)^{\mathrm{T}} A_{\mathrm{per}}(y)\left(e_j + \nabla w_{e_j}(y)\right) \mathrm{d}y,$$

where, for any $p \in \mathbb{R}^d$, w_p is the unique (up to the addition of a constant) solution to the corrector problem associated to the periodic matrix A_{per}:

$$-\operatorname{div}\left[A_{\mathrm{per}}(p + \nabla w_p)\right] = 0, \quad w_p \text{ is } \mathbb{Z}^d\text{-periodic.} \tag{3.3}$$

The corrector functions allow to compute the homogenized matrix, and to obtain a convergence result in the H^1 strong norm. Indeed, introduce

$$u^{\varepsilon,1}(x) = u^\star(x) + \varepsilon \sum_{i=1}^d w_{e_i}\left(\frac{x}{\varepsilon}\right) \frac{\partial u^\star}{\partial x_i}(x). \tag{3.4}$$

Then, we have the following proposition.

Proposition 3.1 *Suppose that the dimension is $d > 1$, that the solution $u^\star$ to* (3.2) *belongs to $W^{2,\infty}(\Omega)$ and that, for any $p \in \mathbb{R}^d$, the corrector w_p solution to* (3.3) *belongs to $W^{1,\infty}(\mathbb{R}^d)$. Then*

$$\|u^\varepsilon - u^{\varepsilon,1}\|_{H^1(\Omega)} \le C\sqrt{\varepsilon}\|\nabla u^\star\|_{W^{1,\infty}(\Omega)} \tag{3.5}$$

for a constant C independent of ε and $u^\star$.

We refer to [24, p. 28] for a proof of this result. Note that, in dimension $d = 1$, the rate of convergence of $u^\varepsilon - u^{\varepsilon,1}$ to 0 is even better.

Error Estimate We are now in a position to state our main result.

Theorem 3.2 *Let u^ε be the solution to* (1.1) *for a matrix A_ε given by* (3.1). *We furthermore assume that*

$$A_{\mathrm{per}} \text{ is Hölder continuous} \tag{3.6}$$

and that the solution $u^\star$ to (3.2) *belongs to $C^2(\overline{\Omega})$. Let u_H be the solution to* (2.3). *We have*

$$\|u^\varepsilon - u_H\|_E \le CH\|f\|_{L^2(\Omega)} + C\left(\sqrt{\varepsilon} + H + \sqrt{\frac{\varepsilon}{H}}\right)\|\nabla u^\star\|_{C^1(\overline{\Omega})}, \tag{3.7}$$

where the constant C is independent of H, ε, f and $u^\star$.

Two remarks are in order, first on the necessity of our assumption (3.6), and next on the comparison with other, well established variants of MsFEM.

Remark 3.3 (On the regularity of A_{per}) We recall that, under assumption (3.6), the solution w_p to (3.3) (with, say, zero mean) satisfies, for any $p \in \mathbb{R}^d$,

$$w_p \in C^{1,\delta}(\mathbb{R}^d) \quad \text{for some } \delta > 0. \tag{3.8}$$

We refer to [19, Theorem 8.22 and Corollary 8.36]. Thus, assumption (3.6) implies that $w_p \in W^{1,\infty}(\mathbb{R}^d)$, which in turn is a useful assumption in Proposition 3.1. The regularity (3.8) is also a useful ingredient in the proof of Theorem 3.2 (see (5.11) and (5.14)).

Remark 3.4 (Comparison with other approaches) It is useful to compare our error estimate (3.7) with similar estimates for some existing MsFEM-type approaches in the literature. The classical MsFEM from [22] (by "classical", we mean the method using basis functions satisfying linear boundary conditions on each element) yields an exactly similar majoration in terms of $\sqrt{\varepsilon} + H + \sqrt{\frac{\varepsilon}{H}}$. It is claimed in [22] that the same majoration also holds for the MsFEM-O variant. This variant (in the form presented in [22]) is restricted to the two-dimensional setting. It uses boundary conditions provided by the solution to the oscillatory ordinary differential equation obtained by taking the trace of the original equation (1.1) on the edge considered.

The famous variant of MsFEM using oversampling (see [16, 21]) gives a slightly better estimation, in terms of $\sqrt{\varepsilon} + H + \frac{\varepsilon}{H}$. The best estimation which we are aware of is obtained by using a Petrov-Galerkin variant of MsFEM with oversampling (see [23]). It bounds the error from above by $\sqrt{\varepsilon} + H + \varepsilon$, but this only holds in the regime $\frac{\varepsilon}{H} \leq C^{te}$ and for a sufficiently (possibly prohibitively) large oversampling ratio. All these comparisons show that the method which we present here is guaranteed to be accurate, although not spectacularly accurate, for the equation (1.1) considered. An actually much better behavior will be observed in practice, in particular for the case of a perforated domain that we study in [25].

A comparison with other, related but slightly different in spirit approaches, can also be of interest. The approaches [27, 28] yield an error estimate better than that obtained with the oversampling variant of MsFEM. The computational cost is however larger, owing to the large size of the oversampling domain employed.

4 Some Classical Ingredients for Our Analysis

Before we get to the proof of our main result, Theorem 3.2, we first need to collect here some standard results. These include trace theorems, Poincaré-type inequalities, error estimates for nonconforming finite elements and eventually convergences of oscillating functions. With a view to next using these results for our proof, we actually need not only to recall them but also, for some of them, to make explicit the dependency of the constants appearing in the various estimates upon the size of the domain (which will be taken, in practice, as an element of the mesh, of diameter H). Of course, these results are standard, and their proof is recalled here only for the sake of completeness.

First we recall the definition, borrowed from [18, Definition B.30], of the $H^{1/2}$ space.

Definition 4.1 For any open domain $\omega \subset \mathbb{R}^n$ and any $u \in L^2(\omega)$, we define the norm

$$\|u\|^2_{H^{1/2}(\omega)} := \|u\|^2_{L^2(\omega)} + |u|^2_{H^{1/2}(\omega)},$$

where

$$|u|^2_{H^{1/2}(\omega)} := \int_\omega \int_\omega \frac{|u(x) - u(y)|^2}{|x - y|^{n+1}} \mathrm{d}x\mathrm{d}y,$$

and define the space

$$H^{1/2}(\omega) := \left\{u \in L^2(\omega),\ \|u\|_{H^{1/2}(\omega)} < \infty\right\}.$$

4.1 Reference Element

We first work on the reference element $\overline{T}$, with edges $\overline{e} \subset \partial\overline{T}$ (we recall that our terminology and notation suggest that, to fix the ideas, we have in mind triangles in two dimensions). By the standard trace theorem, we know that there exists C, such that

$$\forall v \in H^1(\overline{T}), \quad \forall \overline{e} \subset \partial\overline{T}, \quad \|v\|_{H^{1/2}(\overline{e})} \leq C\|v\|_{H^1(\overline{T})}. \tag{4.1}$$

In addition, we have the following result.

Lemma 4.2 *There exists C (depending only on the reference mesh element), such that*

$$\forall v \in H^1(\overline{T}) \quad \textit{with} \quad \int_{\overline{e}} v = 0 \quad \textit{for some } \overline{e} \subset \partial\overline{T}, \quad \|v\|_{H^1(\overline{T})} \leq C\|\nabla v\|_{L^2(\overline{T})}. \tag{4.2}$$

The proof follows from the following result (see [18, Lemma A.38]).

Lemma 4.3 (Petree-Tartar) *Let X, Y and Z be three Banach spaces. Let $A \in \mathcal{L}(X, Y)$ be an injective operator and let $T \in \mathcal{L}(X, Z)$ be a compact operator. If there exists $c > 0$, such that $c\|x\|_X \leq \|Ax\|_Y + \|Tx\|_Z$, then* $\mathrm{Im}(A)$ *is closed. Equivalently, there exists $\alpha > 0$, such that*

$$\forall x \in X, \quad \alpha\|x\|_X \leq \|Ax\|_Y.$$

Proof of Lemma 4.2 Consider an edge $\overline{e} \subset \partial \overline{T}$. We apply Lemma 4.3 with $Z = L^2(\overline{T})$, $Y = (L^2(\overline{T}))^d$,

$$X = \left\{ v \in H^1(\overline{T}) \text{ with } \int_{\overline{e}} v = 0 \right\}$$

equipped with the norm $H^1(\overline{T})$, $Av = \nabla v$ (which is indeed injective on X), and $Tv = v$ (which is indeed compact from X to Z). Lemma 4.3 readily yields the bound (4.2) after taking the maximum over all edges $\overline{e}$. □

4.2 Finite Element of Size H

We will repeatedly use the following Poincaré inequality.

Lemma 4.4 *There exists C (depending only on the regularity of the mesh) independent of H such that, for any $T \in \mathcal{T}_H$,*

$$\forall v \in H^1(T) \quad \text{with} \int_e v = 0 \quad \text{for some } e \subset \partial T, \quad \|v\|_{L^2(T)} \le CH \|\nabla v\|_{L^2(T)}. \tag{4.3}$$

Proof To convey the idea of the proof in a simple case, we first assume that the actual mesh element T considered is homothetic to the reference mesh element $\overline{T}$ with a ratio H. We introduce $v_H(x) = v(Hx)$ defined on the reference element. We hence have $v(x) = v_H(\frac{x}{H})$. Thus,

$$\|v\|^2_{L^2(T)} = \int_T v^2(x) \mathrm{d}x = \int_T v_H^2 \left(\frac{x}{H} \right) \mathrm{d}x = H^d \int_{\overline{T}} v_H^2(y) \mathrm{d}y$$

and

$$\|\nabla v\|^2_{L^2(T)} = \int_T |\nabla v(x)|^2 \mathrm{d}x = H^{-2} \int_T \left| \nabla v_H \left(\frac{x}{H} \right) \right|^2 \mathrm{d}x = H^{d-2} \int_{\overline{T}} |\nabla v_H(y)|^2 \mathrm{d}y.$$

We now use Lemma 4.2, and conclude that

$$\|v\|^2_{L^2(T)} = H^d \|v_H\|^2_{L^2(\overline{T})} \le CH^d \|\nabla v_H\|^2_{L^2(\overline{T})} = CH^2 \|\nabla v\|^2_{L^2(T)},$$

which is (4.3) in this simple case. To obtain (4.3) in full generality, we have to slightly adapt the above argument. We shall use, here and throughout the proof of the subsequent lemmas, the notation $A \sim B$ when the two quantities A and B satisfy $c_1 A \le B \le c_2 A$ with the constants c_1 and c_2 depending only on the regularity parameter of the mesh. Let us recall that for all $T \in \mathcal{T}_H$, there exists a smooth one-to-one and onto mapping $K : \overline{T} \to T$ satisfying $\|\nabla K\|_{L^\infty} \le CH$ and

$\|\nabla K^{-1}\|_{L^\infty} \leq C H^{-1}$. We now introduce $v_H(x) = v(K(x))$ defined on the reference element. We hence have

$$\|v\|^2_{L^2(T)} = \int_T v^2(x)\mathrm{d}x = \int_T v_H^2\big(K^{-1}(x)\big)\mathrm{d}x \sim H^d \int_{\overline{T}} v_H^2(y)\mathrm{d}y$$

and

$$\begin{aligned}\|\nabla v\|^2_{L^2(T)} &= \int_T |\nabla v(x)|^2\mathrm{d}x \sim H^{-2} \int_T \left|\nabla v_H\big(K^{-1}(x)\big)\right|^2\mathrm{d}x \\ &\sim H^{d-2} \int_{\overline{T}} |\nabla v_H(y)|^2\mathrm{d}y.\end{aligned}$$

Using Lemma 4.2 (note that $\int_{\overline{e}} v_H(y)\mathrm{d}y = 0$ since the mapping K is affine on the edges, hence, is of constant Jacobian on $\overline{e}$), we obtain

$$\|v\|^2_{L^2(T)} \sim H^d \|v_H\|^2_{L^2(\overline{T})} \leq C H^d \|\nabla v_H\|^2_{L^2(\overline{T})} \leq C H^2 \|\nabla v\|^2_{L^2(T)},$$

which is the bound (4.3). □

We also have the following trace results.

Lemma 4.5 *There exists C (depending only on the regularity of the mesh) such that, for any $T \in \mathcal{T}_H$ and any edge $e \subset \partial T$, we have*

$$\forall v \in H^1(T), \quad \|v\|^2_{L^2(e)} \leq C\big(H^{-1}\|v\|^2_{L^2(T)} + H\|\nabla v\|^2_{L^2(T)}\big). \tag{4.4}$$

Under the additional assumption that $\int_e v = 0$, we have

$$\|v\|^2_{L^2(e)} \leq C H \|\nabla v\|^2_{L^2(T)}. \tag{4.5}$$

If $\int_e v = 0$ and $H \leq 1$, then

$$\|v\|^2_{H^{1/2}(e)} \leq C \|\nabla v\|^2_{L^2(T)}. \tag{4.6}$$

These bounds are classical results (see [10, p. 282]). We provide here a proof for the sake of completeness.

Proof of Lemma 4.5 We proceed as in the proof of Lemma 4.4 and use the same notation. We use $v_H(x) = v(K(x))$ defined on the reference element. We have

$$\begin{aligned}\|v\|^2_{L^2(e)} &= \int_e v^2(x)\mathrm{d}x = \int_e v_H^2\big(K^{-1}(x)\big)\mathrm{d}x \sim H^{d-1} \int_{\overline{e}} v_H^2(y)\mathrm{d}y \\ &= H^{d-1} \|v_H\|^2_{L^2(\overline{e})}.\end{aligned}$$

By a standard trace inequality, we obtain

$$\begin{aligned}\|v\|^2_{L^2(e)} &\le CH^{d-1}\left(\|v_H\|^2_{L^2(\overline{T})} + \|\nabla v_H\|^2_{L^2(\overline{T})}\right)\\ &\le CH^{d-1}\left(\frac{1}{H^d}\|v\|^2_{L^2(T)} + \frac{1}{H^{d-2}}\|\nabla v\|^2_{L^2(T)}\right),\end{aligned}$$

where we have used some ingredients of the proof of Lemma 4.4. This shows that (4.4) holds.

We now turn to (4.5):

$$\begin{aligned}\|v\|^2_{L^2(e)} \sim H^{d-1}\|v_H\|^2_{L^2(\overline{e})} \le CH^{d-1}\|v_H\|^2_{H^1(\overline{T})} &\le CH^{d-1}\|\nabla v_H\|^2_{L^2(\overline{T})}\\ &\le CH\|\nabla v\|^2_{L^2(T)},\end{aligned}$$

where we have used (4.1)–(4.2). This proves (4.5).

We eventually establish (4.6). We first observe, using Definition 4.1 with the domain $\omega \equiv e \subset \mathbb{R}^{d-1}$, that

$$\begin{aligned}|v|^2_{H^{1/2}(e)} &= \int_e\int_e \frac{|v(x)-v(y)|^2}{|x-y|^d}\mathrm{d}x\mathrm{d}y\\ &\sim \frac{1}{H^d}\int_e\int_e \frac{|v_H(K^{-1}(x)) - v_H(K^{-1}(y))|^2}{|K^{-1}(x)-K^{-1}(y)|^d}\mathrm{d}x\mathrm{d}y\\ &\sim H^{d-2}\int_{\overline{e}}\int_{\overline{e}} \frac{|v_H(x)-v_H(y)|^2}{|x-y|^d}\mathrm{d}x\mathrm{d}y\\ &\sim H^{d-2}|v_H|^2_{H^{1/2}(\overline{e})}.\end{aligned}$$

Hence, using (4.1)–(4.2) and since $H \le 1$,

$$\begin{aligned}\|v\|^2_{H^{1/2}(e)} = \|v\|^2_{L^2(e)} + |v|^2_{H^{1/2}(e)} &\sim H^{d-1}\|v_H\|^2_{L^2(\overline{e})} + H^{d-2}|v_H|^2_{H^{1/2}(\overline{e})}\\ &\le CH^{d-2}\|v_H\|^2_{H^{1/2}(\overline{e})} \le CH^{d-2}\|v_H\|^2_{H^1(\overline{T})}\\ &\le CH^{d-2}\|\nabla v_H\|^2_{L^2(\overline{T})} \sim C\|\nabla v\|^2_{L^2(T)}.\end{aligned}$$

This proves (4.6) and concludes the proof of Lemma 4.5. □

The following result is a direct consequence of (4.5) and (4.6).

Corollary 4.6 *Consider an edge $e \in \mathcal{E}_H$, and let $T_e \subset \mathcal{T}_H$ denote all the triangles sharing this edge. There exists C (depending only on the regularity of the mesh), such that*

$$\forall v \in W_H, \quad \|[\![v]\!]\|^2_{L^2(e)} \le CH\sum_{T\in T_e}\|\nabla v\|^2_{L^2(T)}. \tag{4.7}$$

If $H \leq 1$, then

$$\forall v \in W_H, \quad \left\| [\![v]\!] \right\|^2_{H^{1/2}(e)} \leq C \sum_{T \in T_e} \|\nabla v\|^2_{L^2(T)}. \tag{4.8}$$

Proof We introduce $c_e = |e|^{-1} \int_e v$, which is well-defined since $\int_e [\![v]\!] = 0$. On each side of the edge, the function $v - c_e$ has zero average on that edge. Hence, using (4.5), we have

$$\begin{aligned}
\left\| [\![v]\!] \right\|^2_{L^2(e)} &= \left\| [\![v - c_e]\!] \right\|^2_{L^2(e)} = \|(v_1 - c_e) - (v_2 - c_e)\|^2_{L^2(e)} \\
&\leq 2\|v_1 - c_e\|^2_{L^2(e)} + 2\|v_2 - c_e\|^2_{L^2(e)} \\
&\leq CH\big(\|\nabla v_1\|^2_{L^2(T_1)} + \|\nabla v_2\|^2_{L^2(T_2)}\big) \\
&= CH \sum_{T \in T_e} \|\nabla v\|^2_{L^2(T)},
\end{aligned}$$

where we have used the notation $v_1 = v|_{T_1}$. The proof of (4.8) follows a similar pattern, using (4.6). □

4.3 Error Estimate for Nonconforming FEM

The error estimate which we establish in Sect. 5 is essentially based on a Céa-type (or Strang-type) lemma extended to nonconforming finite element methods. We state this standard estimate in the actual context we work in (but again emphasize that it is of course completely general in nature).

Lemma 4.7 (See [10, Lemma 10.1.7]) *Let u^ε be the solution to* (1.1) *and u_H be the solution to* (2.3). *Then*

$$\|u^\varepsilon - u_H\|_E \leq \inf_{v \in V_H} \|u^\varepsilon - v\|_E + \sup_{v \in V_H \setminus \{0\}} \frac{|a_H(u^\varepsilon - u_H, v)|}{\|v\|_E}. \tag{4.9}$$

The first term in (4.9) is the usual best approximation error already present in the classical Céa Lemma. This term measures how accurately the space V_H (or, in general, any approximation space) approximates the exact solution u^ε. The second term of (4.9) measures how the nonconforming setting affects the result. This term would vanish if V_H were a subset of $H^1_0(\Omega)$.

4.4 Integrals of Oscillatory Functions

We shall also need the following result.

Lemma 4.8 *Let $e \in \mathcal{E}_H$, T_1 and T_2 be the two elements adjacent to e and $\tau \in \mathbb{R}^d$, $|\tau| \leq 1$, be a vector tangent (i.e., parallel) to e. Then, for any function $u \in H^1(T_1 \cup T_2)$, any $v \in W_H$ and any $J \in C^1(\mathbb{R}^d)$, we have*

$$\left| \int_e u(x) [\![v(x)]\!] \tau \cdot \nabla J\left(\frac{x}{\varepsilon}\right) \right|$$
$$\leq C\sqrt{\frac{\varepsilon}{H}} \|J\|_{C^1(\mathbb{R}^d)} \sum_{T=T_1,T_2} |v|_{H^1(T)} (\|u\|_{L^2(T)} + H|u|_{H^1(T)}) \tag{4.10}$$

with a constant C which depends only on the regularity of the mesh.

As will be clear from the proof below, the fact that we consider in the above left-hand side the jump of v, rather than v itself, is not essential. A similar estimate holds for the quantity $\int_e u(x)v(x)\tau \cdot \nabla J(\frac{x}{\varepsilon})$, where u and v are any functions of regularity $H^1(T)$ for some $T \in \mathcal{T}_H$ and e is an edge of ∂T.

Proof of Lemma 4.8 Let c_e be the average of v over e and denote $v_j = v|_{T_j}$. Since $[\![v]\!] = (v_1 - c_e) - (v_2 - c_e)$, we obviously have

$$\left| \int_e u(x) [\![v(x)]\!] \tau \cdot \nabla J\left(\frac{x}{\varepsilon}\right) \right| \leq \sum_{j=1}^{2} \left| \int_e u(x) \big(v_j(x) - c_e\big) \tau \cdot \nabla J\left(\frac{x}{\varepsilon}\right) \right|. \tag{4.11}$$

Fix j. We first recall that there exists a one-to-one and onto mapping $K : \overline{T} \to T_j$ from the reference element $\overline{T}$ onto T_j satisfying $\|\nabla K\|_{L^\infty} \leq CH$ and $\|\nabla K^{-1}\|_{L^\infty} \leq CH^{-1}$. In particular, there exists an edge $\overline{e}$ of $\overline{T}$ such that $K(\overline{e}) = e$. We introduce the functions $u_H(y) = u(K(y))$, $v_H(y) = v_j(K(y)) - c_e$ defined on the reference element, and observe that

$$\int_e u(x) \big(v_j(x) - c_e\big) \tau \cdot \nabla J\left(\frac{x}{\varepsilon}\right) \mathrm{d}x$$
$$\sim H^{d-1} \int_{\overline{e}} u_H(y) v_H(y) \tau \cdot \nabla J\left(\frac{K(y)}{\varepsilon}\right) \mathrm{d}y. \tag{4.12}$$

We now claim that

$$\left| \int_{\overline{e}} u_H(y) v_H(y) \tau \cdot \nabla J\left(\frac{K(y)}{\varepsilon}\right) \mathrm{d}y \right|$$
$$\leq C\sqrt{\frac{\varepsilon}{H}} \|J\|_{C^1(\mathbb{R}^d)} \|u_H\|_{H^{1/2}(\overline{e})} \|v_H\|_{H^{1/2}(\overline{e})}. \tag{4.13}$$

This inequality is obtained by interpolation. Suppose indeed, in the first step, that u_H and v_H belong to $H^1(\overline{e})$. Using that the mapping K is affine on the edges and thus is of constant gradient, we first see that

$$\int_{\overline{e}} u_H(y)v_H(y)\tau \cdot \nabla J\left(\frac{K(y)}{\varepsilon}\right)\mathrm{d}y$$
$$= C\frac{\varepsilon}{H}\int_{\overline{e}} u_H(y)v_H(y)\tau \cdot \nabla\left[J\left(\frac{K(y)}{\varepsilon}\right)\right]\mathrm{d}y. \tag{4.14}$$

By integration by parts, we next observe that

$$\frac{\varepsilon}{H}\int_{\overline{e}} u_H(y)v_H(y)\tau \cdot \nabla\left[J\left(\frac{K(y)}{\varepsilon}\right)\right]\mathrm{d}y$$
$$= \frac{\varepsilon}{H}\int_{\partial\overline{e}} u_H(y)v_H(y)\tau \cdot \nu J\left(\frac{K(y)}{\varepsilon}\right)\mathrm{d}y$$
$$- \frac{\varepsilon}{H}\int_{\overline{e}} J\left(\frac{K(y)}{\varepsilon}\right)\tau \cdot \nabla\big(u_H(y)v_H(y)\big)\mathrm{d}y, \tag{4.15}$$

where ν is the outward normal unit vector to $\partial\overline{e}$ tangent to $\overline{e}$. Collecting (4.14)–(4.15), and using the Cauchy-Schwarz inequality, we obtain that

$$\left|\int_{\overline{e}} u_H(y)v_H(y)\tau \cdot \nabla J\left(\frac{K(y)}{\varepsilon}\right)\mathrm{d}y\right|$$
$$\leq C\frac{\varepsilon}{H}\|J\|_{C^0(\mathbb{R}^d)}\big[\|u_H\|_{L^2(\partial\overline{e})}\|v_H\|_{L^2(\partial\overline{e})} + 2\|u_H\|_{H^1(\overline{e})}\|v_H\|_{H^1(\overline{e})}\big]$$
$$\leq C\frac{\varepsilon}{H}\|J\|_{C^0(\mathbb{R}^d)}\|u_H\|_{H^1(\overline{e})}\|v_H\|_{H^1(\overline{e})}, \tag{4.16}$$

where the last inequality above follows from the trace inequality which is valid with a constant C depending only on $\overline{e}$. On the other hand, for u_H and v_H that only belong to $L^2(\overline{e})$, we obviously have

$$\left|\int_{\overline{e}} u_H(y)v_H(y)\tau \cdot \nabla J\left(\frac{K(y)}{\varepsilon}\right)\mathrm{d}y\right|$$
$$\leq \|\nabla J\|_{C^0(\mathbb{R}^d)}\|u_H\|_{L^2(\overline{e})}\|v_H\|_{L^2(\overline{e})}. \tag{4.17}$$

By interpolation between (4.16)–(4.17) (see [9, Theorem 4.4.1]), we obtain (4.13).

The sequel of the proof is easy. Collecting (4.12)–(4.13), we deduce that

$$\left|\int_e u(x)\big(v_j(x) - c_e\big)\tau \cdot \nabla J\left(\frac{x}{\varepsilon}\right)\mathrm{d}x\right|$$
$$\leq CH^{d-\frac{3}{2}}\sqrt{\varepsilon}\|J\|_{C^1(\mathbb{R}^d)}\|u_H\|_{H^{1/2}(\overline{e})}\|v_H\|_{H^{1/2}(\overline{e})}$$
$$\leq CH^{d-\frac{3}{2}}\sqrt{\varepsilon}\|J\|_{C^1(\mathbb{R}^d)}\|u_H\|_{H^1(\overline{T})}\|\nabla v_H\|_{L^2(\overline{T})}, \tag{4.18}$$

where we have used in the last line the trace inequality (4.1) and Lemma 4.2 for v_H (recall that $\int_{\overline{e}} v_H = 0$, since, on the one hand, $\int_e v_j - c_e = 0$ and, on the other hand, the mapping K is affine on $\overline{e}$, and hence is of constant gradient).

To return to norms on the actual element T_j rather than on the reference element $\overline{T}$, we use the following relations, established in the proof of Lemma 4.4:

$$\|u\|_{L^2(T_j)} \sim H^{\frac{d}{2}} \|u_H\|_{L^2(\overline{T})},$$

$$|u|_{H^1(T_j)} \sim H^{\frac{d}{2}-1} |u_H|_{H^1(\overline{T})},$$

$$|v_j|_{H^1(T_j)} \sim H^{\frac{d}{2}-1} |v_H|_{H^1(\overline{T})}.$$

We then infer from (4.18) that

$$\left| \int_c u(x)\big(v_j(x) - c_e\big)\tau \cdot \nabla J\left(\frac{x}{\varepsilon}\right) \right|$$

$$\le C H^{d-\frac{3}{2}} \sqrt{\varepsilon} \|J\|_{C^1(\mathbb{R}^d)} \big[H^{-\frac{d}{2}} \|u\|_{L^2(T_j)} + H^{-\frac{d}{2}+1} |u|_{H^1(T_j)}\big] H^{-\frac{d}{2}+1} |v_j|_{H^1(T_j)}$$

$$\le C \sqrt{\frac{\varepsilon}{H}} \|J\|_{C^1(\mathbb{R}^d)} [\|u\|_{L^2(T_j)} + H |u|_{H^1(T_j)}] |v_j|_{H^1(T_j)}.$$

Inserting this bound in (4.11) for $j = 1$ and 2 yields the desired bound (4.10). □

5 Proof of the Main Error Estimate

Now that we have reviewed a number of classical ingredients, we are in the position, in this section, to prove our main result, Theorem 3.2.

As announced above, our proof is based on the estimate (4.9) provided by Lemma 4.7. To bound both terms in the right-hand side of (4.9), we will use the following result, the proof of which is postponed until Sect. 5.2.

Lemma 5.1 *Under the same assumptions as those of Theorem* 3.2, *we have that, for any* $v \in W_H$,

$$\left| \sum_{T \in \mathcal{T}_H} \int_{\partial T} v \left(A_{\text{per}}\left(\frac{x}{\varepsilon}\right) \nabla u^\varepsilon \right) \cdot n \right| \le C\left(\sqrt{\varepsilon} + H + \sqrt{\frac{\varepsilon}{H}}\right) \|v\|_E \|\nabla u^\star\|_{C^1(\overline{\Omega})}, \tag{5.1}$$

where the constant C *is independent of* H, ε, f, $u^\star$ *and* v.

Remark 5.2 A more precise estimate is given in the course of the proof (see (5.23)).

5.1 Proof of Theorem 3.2

Momentarily assuming Lemma 5.1, we now prove our main result.

We argue on estimate (4.9) provided by Lemma 4.7. In the right-hand side of (4.9), we first bound the nonconforming error (the second term). Let $v \in V_H$. We use the definition (2.1) of a_H and (2.3) to compute:

$$\begin{aligned} a_H\big(u^\varepsilon - u_H, v\big) &= \sum_{T \in \mathcal{T}_H} \int_T (\nabla v)^T A_{\mathrm{per}}\left(\frac{x}{\varepsilon}\right) \nabla u^\varepsilon - \int_\Omega f v \\ &= \sum_{T \in \mathcal{T}_H} \int_{\partial T} v \left(A_{\mathrm{per}}\left(\frac{x}{\varepsilon}\right) \nabla u^\varepsilon \right) \cdot n \\ &\quad - \sum_{T \in \mathcal{T}_H} \int_T v \operatorname{div}\left(A_{\mathrm{per}}\left(\frac{x}{\varepsilon}\right) \nabla u^\varepsilon \right) - \int_\Omega f v \\ &= \sum_{T \in \mathcal{T}_H} \int_{\partial T} v \left(A_{\mathrm{per}}\left(\frac{x}{\varepsilon}\right) \nabla u^\varepsilon \right) \cdot n, \end{aligned}$$

using (1.1) and the regularity of u^ε. Observing that, by definition, $v \in V_H \subset W_H$, we can use Lemma 5.1 to majorize the above right-hand side. We obtain

$$\sup_{v \in V_H \setminus \{0\}} \frac{|a_H(u^\varepsilon - u_H, v)|}{\|v\|_E} \leq C\left(\sqrt{\varepsilon} + H + \sqrt{\frac{\varepsilon}{H}}\right) \|\nabla u^\star\|_{C^1(\overline{\Omega})}. \tag{5.2}$$

We now turn to the best approximation error (the first term of the right-hand side of (4.9)). As shown at the end of Sect. 2.1, we can decompose $u^\varepsilon \in H^1_0(\Omega) \subset W_H$ as

$$u^\varepsilon = v_H + v^0, \quad v_H \in V_H, \quad v^0 \in W_H^0.$$

We may compute, again using (1.1) and the regularity of u^ε, that

$$\begin{aligned} \|u^\varepsilon - v_H\|_E^2 &= a_H\big(u^\varepsilon - v_H, u^\varepsilon - v_H\big) \\ &= a_H\big(u^\varepsilon - v_H, v^0\big) \quad \text{(by definition of } v^0\text{)} \\ &= a_H\big(u^\varepsilon, v^0\big) \quad \text{(by orthogonality of } V_H \text{ with } W_H^0\text{)} \\ &= \sum_{T \in \mathcal{T}_H} \int_T (\nabla v^0)^T A_{\mathrm{per}}\left(\frac{x}{\varepsilon}\right) \nabla u^\varepsilon \\ &= \sum_{T \in \mathcal{T}_H} \int_{\partial T} v^0 \left(A_{\mathrm{per}}\left(\frac{x}{\varepsilon}\right) \nabla u^\varepsilon \right) \cdot n + \sum_{T \in \mathcal{T}_H} \int_T v^0 f. \end{aligned} \tag{5.3}$$

Since $v^0 \in W_H^0$, we may use (4.3) and bound the second term of the right-hand side of (5.3) as follows:

$$\begin{aligned}\left|\sum_{T\in\mathcal{T}_H}\int_T v^0 f\right| &\leq \sum_{T\in\mathcal{T}_H}\|v^0\|_{L^2(T)}\|f\|_{L^2(T)} \quad \text{(Cauchy Schwarz inequality)}\\ &\leq CH\sum_{T\in\mathcal{T}_H}\|\nabla v^0\|_{L^2(T)}\|f\|_{L^2(T)}\\ &\leq CH\|v^0\|_E\|f\|_{L^2(\Omega)}, \qquad (5.4)\end{aligned}$$

where we have used in the last line the Cauchy Schwarz inequality and an equivalence of norms. The first term of the right-hand side of (5.3) is bounded by using Lemma 5.1 (since $v^0 \in W_H^0 \subset W_H$), which yields

$$\left|\sum_{T\in\mathcal{T}_H}\int_{\partial T} v^0\left(A_{\text{per}}\left(\frac{x}{\varepsilon}\right)\nabla u^\varepsilon\right)\cdot n\right| \leq C\left(\sqrt{\varepsilon}+H+\sqrt{\frac{\varepsilon}{H}}\right)\|v^0\|_E\|\nabla u^\star\|_{C^1(\overline{\Omega})}. \qquad (5.5)$$

Inserting (5.4)–(5.5) in the right-hand side of (5.3), we deduce that

$$\|u^\varepsilon - v_H\|_E^2 \leq CH\|v^0\|_E\|f\|_{L^2(\Omega)} + C\left(\sqrt{\varepsilon}+H+\sqrt{\frac{\varepsilon}{H}}\right)\|v^0\|_E\|\nabla u^\star\|_{C^1(\overline{\Omega})}.$$

Since $v^0 = u^\varepsilon - v_H$, we may factor out $\|v^0\|_E$ and obtain

$$\|u^\varepsilon - v_H\|_E \leq CH\|f\|_{L^2(\Omega)} + C\left(\sqrt{\varepsilon}+H+\sqrt{\frac{\varepsilon}{H}}\right)\|\nabla u^\star\|_{C^1(\overline{\Omega})}.$$

By the definition of the infimum, we of course have $\inf_{v\in V_H}\|u^\varepsilon - v\|_E \leq \|u^\varepsilon - v_H\|_E$, and thus

$$\inf_{v\in V_H}\|u^\varepsilon - v\|_E \leq CH\|f\|_{L^2(\Omega)} + C\left(\sqrt{\varepsilon}+H+\sqrt{\frac{\varepsilon}{H}}\right)\|\nabla u^\star\|_{C^1(\overline{\Omega})}. \qquad (5.6)$$

Inserting (5.2) and (5.6) in the right-hand side of (4.9), we obtain the desired bound (3.7). This concludes the proof of Theorem 3.2.

5.2 *Proof of Lemma 5.1*

We now establish Lemma 5.1, actually the key step of the proof of Theorem 3.2.

Let $v \in W_H$. Using (1.1) and (3.2), and inserting in the term we are estimating the approximation $u^{\varepsilon,1}$ defined by (3.4) of the exact solution u^ε, we write

$$\begin{aligned}
&\sum_{T\in\mathcal{T}_H}\int_{\partial T} v\left(A_{\text{per}}\left(\frac{x}{\varepsilon}\right)\nabla u^\varepsilon\right)\cdot n \\
&= -\sum_{T\in\mathcal{T}_H}\int_T vf + \sum_{T\in\mathcal{T}_H}\int_T (\nabla v)^{\mathrm{T}} A_{\text{per}}\left(\frac{x}{\varepsilon}\right)\nabla u^\varepsilon \\
&= \sum_{T\in\mathcal{T}_H}\int_T v\,\mathrm{div}(A^\star_{\text{per}}\nabla u^\star) + \sum_{T\in\mathcal{T}_H}\int_T (\nabla v)^{\mathrm{T}} A_{\text{per}}\left(\frac{x}{\varepsilon}\right)\nabla(u^\varepsilon - u^{\varepsilon,1}) \\
&\quad + \sum_{T\in\mathcal{T}_H}\int_T (\nabla v)^{\mathrm{T}} A_{\text{per}}\left(\frac{x}{\varepsilon}\right)\nabla u^{\varepsilon,1} \\
&= \sum_{T\in\mathcal{T}_H}\int_{\partial T} v(A^\star_{\text{per}}\nabla u^\star)\cdot n + \sum_{T\in\mathcal{T}_H}\int_T (\nabla v)^{\mathrm{T}} A_{\text{per}}\left(\frac{x}{\varepsilon}\right)\nabla(u^\varepsilon - u^{\varepsilon,1}) \\
&\quad + \sum_{T\in\mathcal{T}_H}\int_T (\nabla v)^{\mathrm{T}}\left(A_{\text{per}}\left(\frac{x}{\varepsilon}\right)\nabla u^{\varepsilon,1} - A^\star_{\text{per}}\nabla u^\star\right) \\
&= \mathrm{A} + \mathrm{B} + \mathrm{C}.
\end{aligned} \tag{5.7}$$

We now successively bound the three terms A, B and C in the right-hand side of (5.7). Loosely speaking:

(1) The first term A is macroscopic in nature and would be present for the analysis of a classical Crouzeix-Raviart type method. It will eventually contribute for $O(H)$ to the overall estimate (5.1) (and thus to (3.7)).

(2) The second term B is independent from the discretization: it is an "infinite dimensional" term, the size of which, namely $O(\sqrt{\varepsilon})$, is entirely controlled by the quality of approximation of u^ε by $u^{\varepsilon,1}$. It is the term for which we specifically need to put ourselves in the periodic setting.

(3) The third term C would likewise go to zero if the size of the mesh were much larger than the small coefficient ε; it will contribute for the $O(\sqrt{\frac{\varepsilon}{H}})$ term in the estimate (5.1).

Step 1 Bound on the first term of (5.7) We first note that

$$\sum_{T\in\mathcal{T}_H}\int_{\partial T} v(A^\star_{\text{per}}\nabla u^\star)\cdot n = \sum_{e\in\mathcal{E}_H}\int_e [\![v]\!](A^\star_{\text{per}}\nabla u^\star)\cdot n.$$

We now use arguments that are standard in the context of Crouzeix-Raviart finite elements (see [10, p. 281]). Introducing, for each edge e, the constant $c_e =$

$|e|^{-1} \int_e (A^\star_{\rm per} \nabla u^\star) \cdot n$, and using $\int_e [\![v]\!] = 0$ with $v \in W_H$, we write

$$\begin{aligned}
&\left| \sum_{T \in \mathcal{T}_H} \int_{\partial T} v(A^\star_{\rm per} \nabla u^\star) \cdot n \right| \\
&\quad = \left| \sum_{e \in \mathcal{E}_H} \int_e [\![v]\!] (A^\star_{\rm per} \nabla u^\star) \cdot n \right| \\
&\quad \leq \sum_{e \in \mathcal{E}_H} \left| \int_e [\![v]\!] \left((A^\star_{\rm per} \nabla u^\star) \cdot n - c_e \right) \right| \\
&\quad \leq \sum_{e \in \mathcal{E}_H} \| [\![v]\!] \|_{L^2(e)} \| (A^\star_{\rm per} \nabla u^\star) \cdot n - c_e \|_{L^2(e)} \\
&\quad \leq \left[\sum_{e \in \mathcal{E}_H} \| [\![v]\!] \|^2_{L^2(e)} \right]^{\frac{1}{2}} \left[\sum_{e \in \mathcal{E}_H} \| (A^\star_{\rm per} \nabla u^\star) \cdot n - c_e \|^2_{L^2(e)} \right]^{\frac{1}{2}},
\end{aligned}$$

successively using the continuous and discrete Cauchy-Schwarz inequalities in the last two lines. We now use (4.5) and (4.7) to respectively estimate the two factors in the above right-hand side. Doing so, we obtain

$$\begin{aligned}
&\left| \sum_{T \in \mathcal{T}_H} \int_{\partial T} v(A^\star_{\rm per} \nabla u^\star) \cdot n \right| \\
&\quad \leq C \left[\sum_{e \in \mathcal{E}_H} H \sum_{T \in T_e} \| \nabla v \|^2_{L^2(T)} \right]^{\frac{1}{2}} \left[\sum_{\substack{e \in \mathcal{E}_H \\ \text{choose one } T \in T_e}} H \| \nabla^2 u^\star \|^2_{L^2(T)} \right]^{\frac{1}{2}}.
\end{aligned}$$

We hence have that

$$\begin{aligned}
&\left| \sum_{T \in \mathcal{T}_H} \int_{\partial T} v(A^\star_{\rm per} \nabla u^\star) \cdot n \right| \\
&\quad \leq C \left[H \sum_{T \in \mathcal{T}_H} \| \nabla v \|^2_{L^2(T)} \right]^{\frac{1}{2}} \left[\sum_{T \in \mathcal{T}_H} H \| \nabla^2 u^\star \|^2_{L^2(T)} \right]^{\frac{1}{2}} \\
&\quad \leq C H \| v \|_E \| \nabla^2 u^\star \|_{L^2(\Omega)}. \qquad (5.8)
\end{aligned}$$

Step 2 Bound on the second term of (5.7) We note that

$$\begin{aligned}
&\left| \sum_{T \in \mathcal{T}_H} \int_T (\nabla v)^{\rm T} A_{\rm per}\left(\frac{x}{\varepsilon}\right) \nabla \left(u^\varepsilon - u^{\varepsilon,1}\right) \right| \\
&\quad \leq \| A_{\rm per} \|_{L^\infty} \sum_{T \in \mathcal{T}_H} \| \nabla v \|_{L^2(T)} \| \nabla \left(u^\varepsilon - u^{\varepsilon,1}\right) \|_{L^2(T)}
\end{aligned}$$

$$\leq C\|v\|_E \left\| \nabla\left(u^\varepsilon - u^{\varepsilon,1}\right) \right\|_{L^2(\Omega)}$$

$$\leq C\sqrt{\varepsilon}\|v\|_E \|\nabla u^\star\|_{W^{1,\infty}(\Omega)}, \tag{5.9}$$

eventually using (3.5).

Step 3 Bound on the third term of (5.7) To start with, we differentiate $u^{\varepsilon,1}$ defined by (3.4):

$$\nabla u^{\varepsilon,1}(x) = \sum_{i=1}^d \partial_i u^\star(x)\left(e_i + \nabla w_{e_i}\left(\frac{x}{\varepsilon}\right)\right) + \varepsilon \sum_{i=1}^d w_{e_i}\left(\frac{x}{\varepsilon}\right)\partial_i \nabla u^\star(x).$$

The third term of (5.7) thus writes

$$\sum_{T\in\mathcal{T}_H} \int_T (\nabla v)^{\mathrm{T}}\left(A_{\mathrm{per}}\left(\frac{x}{\varepsilon}\right)\nabla u^{\varepsilon,1} - A^\star_{\mathrm{per}}\nabla u^\star\right)$$

$$= \varepsilon \sum_{i=1}^d \sum_{T\in\mathcal{T}_H} \int_T (\nabla v)^{\mathrm{T}} A_{\mathrm{per}}\left(\frac{x}{\varepsilon}\right)\partial_i \nabla u^\star(x) w_{e_i}\left(\frac{x}{\varepsilon}\right)$$

$$+ \sum_{T\in\mathcal{T}_H} \sum_{i=1}^d \int_T (\nabla v)^{\mathrm{T}} \partial_i u^\star(x) G_i\left(\frac{x}{\varepsilon}\right), \tag{5.10}$$

where we have introduced the vector fields

$$G_i(x) = A_{\mathrm{per}}(x)\left(e_i + \nabla w_{e_i}(x)\right) - A^\star_{\mathrm{per}} e_i, \quad 1 \leq i \leq d,$$

which are all $\mathbb{Z}^d$ periodic, divergence free and of zero mean. In addition, in view of the assumption (3.6), which implies (3.8), we see that

$$G_i \text{ is Hölder continuous.} \tag{5.11}$$

We now successively bound the two terms of the right-hand side of (5.10). The first term is quite straightforward to bound. Using Cauchy-Schwarz inequalities and that $w_p \in L^\infty$ (see (3.8)), we simply observe that

$$\left| \varepsilon \sum_{i=1}^d \sum_{T\in\mathcal{T}_H} \int_T (\nabla v)^{\mathrm{T}} A_{\mathrm{per}}\left(\frac{x}{\varepsilon}\right)\partial_i \nabla u^\star(x) w_{e_i}\left(\frac{x}{\varepsilon}\right) \right|$$

$$\leq d\varepsilon \|A_{\mathrm{per}}\|_{L^\infty} \max_i \|w_{e_i}\|_{L^\infty} \sum_{T\in\mathcal{T}_H} \|\nabla v\|_{L^2(T)} \|\nabla^2 u^\star\|_{L^2(T)}$$

$$\leq C\varepsilon \|v\|_E \|\nabla^2 u^\star\|_{L^2(\Omega)}. \tag{5.12}$$

The rest of the proof is actually devoted to bounding the second term of the right-hand side of (5.10), a task that requires several estimations. We first use a classical argument already exposed in [24, p. 27]. The vector field G_i being $\mathbb{Z}^d$ periodic, divergence free and of zero mean, there exists (see [24, p. 6]) a skew symmetric matrix $J^i \in \mathbb{R}^{d\times d}$, such that

$$\forall 1 \le \alpha \le d, \quad [G_i]_\alpha = \sum_{\beta=1}^{d} \frac{\partial [J^i]_{\beta\alpha}}{\partial x_\beta} \tag{5.13}$$

and

$$J^i \in \left(H^1_{\mathrm{loc}}(\mathbb{R}^d)\right)^{d\times d}, \quad J^i \text{ is } \mathbb{Z}^d\text{-periodic}, \quad \int_{(0,1)^d} J^i = 0.$$

In the two-dimensional setting, an explicit expression can be written. We indeed have

$$J^i(x_1,x_2) = \begin{pmatrix} 0 & -\tau^i(x_1,x_2) \\ \tau^i(x_1,x_2) & 0 \end{pmatrix}, \quad x = (x_1,x_2) \in \mathbb{R}^2,$$

with

$$\tau^i(x_1,x_2) = \tau^i(0) + \int_0^1 \left(x_2[G_i]_1(tx) - x_1[G_i]_2(tx)\right)\mathrm{d}t,$$

where $\tau^i(0)$ satisfies $\int_{(0,1)^2} \tau^i = 0$. In view of (5.11), we in particular have that

$$J^i \in \left(C^1(\mathbb{R}^d)\right)^{d\times d}. \tag{5.14}$$

A better regularity (namely $J^i \in (C^{1,\delta}(\mathbb{R}^d))^{d\times d}$ for some $\delta > 0$) actually holds, but we will not need it henceforth.

The same regularity (5.14) can be also proven in any dimension $d \ge 3$, although in a less straightforward manner. Indeed, the components of J^i constructed in [24, p. 7] using the Fourier series can be seen to satisfy the equation

$$-\Delta[J^i]_{\beta\alpha} = \frac{\partial [G_i]_\beta}{\partial x_\alpha} - \frac{\partial [G_i]_\alpha}{\partial x_\beta},$$

complemented with periodic boundary conditions. Hence the function $[J^i]_{\beta\alpha}$, as well as its gradient, is continuous due to the regularity (5.11) and general results on elliptic equations (see [19, Sect. 4.5]).

In view of (5.13), we see that the α-th coordinate of the vector $\partial_i u^\star(\cdot) G_i(\frac{\cdot}{\varepsilon})$ reads

$$\begin{aligned}
\left[\partial_i u^\star(x) G_i\left(\frac{x}{\varepsilon}\right)\right]_\alpha &= \sum_{\beta=1}^{d} \frac{\partial [J^i]_{\beta\alpha}}{\partial x_\beta}\left(\frac{x}{\varepsilon}\right)\partial_i u^\star(x) \\
&= \varepsilon \sum_{\beta=1}^{d} \frac{\partial}{\partial x_\beta}\left([J^i]_{\beta\alpha}\left(\frac{x}{\varepsilon}\right)\partial_i u^\star(x)\right)
\end{aligned}$$

$$-\varepsilon \sum_{\beta=1}^{d} [J^i]_{\beta\alpha}\left(\frac{x}{\varepsilon}\right)\partial_{i\beta} u^\star(x)$$

$$= \varepsilon\big[\widetilde{B}_i^\varepsilon(x)\big]_\alpha - \varepsilon\big[B_i^\varepsilon(x)\big]_\alpha, \tag{5.15}$$

where the vector fields $\widetilde{B}_i^\varepsilon(x) \in \mathbb{R}^d$ and $B_i^\varepsilon(x) \in \mathbb{R}^d$ are defined, for any $1 \le \alpha \le d$, by

$$\big[B_i^\varepsilon(x)\big]_\alpha = \sum_{\beta=1}^{d} [J^i]_{\beta\alpha}\left(\frac{x}{\varepsilon}\right)\partial_{i\beta} u^\star(x) \quad \text{and}$$

$$\big[\widetilde{B}_i^\varepsilon(x)\big]_\alpha = \sum_{\beta=1}^{d} \frac{\partial}{\partial x_\beta}\left([J^i]_{\beta\alpha}\left(\frac{x}{\varepsilon}\right)\partial_i u^\star(x)\right).$$

The vector field $\widetilde{B}_i^\varepsilon$ is divergence free as J^i is a skew symmetric matrix.

The second term of the right-hand side of (5.10) thus reads

$$\sum_{T\in\mathcal{T}_H}\sum_{i=1}^{d}\int_T (\nabla v)^{\mathrm{T}}\partial_i u^\star(x) G_i\left(\frac{x}{\varepsilon}\right)$$

$$= \varepsilon \sum_{T\in\mathcal{T}_H}\sum_{i=1}^{d}\int_T \big(\nabla v(x)\big)^{\mathrm{T}}\big(\widetilde{B}_i^\varepsilon(x) - B_i^\varepsilon(x)\big)$$

$$= \varepsilon \sum_{T\in\mathcal{T}_H}\sum_{i=1}^{d}\int_{\partial T} v(x)\widetilde{B}_i^\varepsilon(x)\cdot n - \varepsilon \sum_{T\in\mathcal{T}_H}\sum_{i=1}^{d}\int_T \big(\nabla v(x)\big)^{\mathrm{T}} B_i^\varepsilon(x), \tag{5.16}$$

successively using (5.15) and an integration by parts of the former term and the divergence-free property of $\widetilde{B}_i^\varepsilon$. An upper bound for the second term can easily be obtained, given that $J^i \in (L^\infty(\mathbb{R}^d))^{d\times d}$ (see (5.14)):

$$\left|\varepsilon \sum_{T\in\mathcal{T}_H}\sum_{i=1}^{d}\int_T \big(\nabla v(x)\big)^{\mathrm{T}} B_i^\varepsilon(x)\right| = \left|\varepsilon \sum_{T\in\mathcal{T}_H}\sum_{i,\alpha,\beta=1}^{d}\int_T \partial_\alpha v(x)[J^i]_{\beta\alpha}\left(\frac{x}{\varepsilon}\right)\partial_{i\beta} u^\star(x)\right|$$

$$\le d^3\varepsilon \max_i \|J^i\|_{L^\infty} \sum_{T\in\mathcal{T}_H} \|\nabla v\|_{L^2(T)}\|\nabla^2 u^\star\|_{L^2(T)}$$

$$\le C\varepsilon\|v\|_E\|\nabla^2 u^\star\|_{L^2(\Omega)}. \tag{5.17}$$

We are now left with bounding the first term of the right-hand side of (5.16), which reads

$$\begin{aligned}
&\varepsilon \sum_{T\in\mathcal{T}_H} \sum_{i=1}^{d} \int_{\partial T} v(x) \widetilde{B}_i^\varepsilon(x)\cdot n \\
&= \varepsilon \sum_{e\in\mathcal{E}_H} \sum_{i=1}^{d} \int_e [\![v(x)]\!]\, \widetilde{B}_i^\varepsilon(x)\cdot n \\
&= \varepsilon \sum_{e\in\mathcal{E}_H} \sum_{i,\alpha,\beta=1}^{d} \int_e [\![v(x)]\!]\, n_\alpha \frac{\partial}{\partial x_\beta}\left([J^i]_{\beta\alpha}\left(\frac{x}{\varepsilon}\right) \partial_i u^\star(x)\right) \\
&= \sum_{e\in\mathcal{E}_H} \sum_{i,\alpha,\beta=1}^{d} \int_e [\![v(x)]\!]\, n_\alpha \frac{\partial [J^i]_{\beta\alpha}}{\partial x_\beta}\left(\frac{x}{\varepsilon}\right) \partial_i u^\star(x) \\
&\quad + \varepsilon \sum_{e\in\mathcal{E}_H} \sum_{i,\alpha,\beta=1}^{d} \int_e [\![v(x)]\!]\, n_\alpha [J^i]_{\beta\alpha}\left(\frac{x}{\varepsilon}\right) \partial_{i\beta} u^\star(x). \qquad (5.18)
\end{aligned}$$

Our final task is to successively bound the two terms of the right-hand side of (5.18).

We begin with the first term. Considering an edge e, we recast the contribution of that edge to the first term of the right-hand side of (5.18) as follows, using the skew symmetry of J:

$$\begin{aligned}
&\sum_{i,\alpha,\beta=1}^{d} \int_e [\![v(x)]\!]\, n_\alpha \frac{\partial [J^i]_{\beta\alpha}}{\partial x_\beta}\left(\frac{x}{\varepsilon}\right) \partial_i u^\star(x) \\
&= \sum_{\substack{i,\alpha,\beta=1\\ \beta>\alpha}}^{d} \int_e [\![v(x)]\!] \left(n_\alpha \frac{\partial [J^i]_{\beta\alpha}}{\partial x_\beta} - n_\beta \frac{\partial [J^i]_{\beta\alpha}}{\partial x_\alpha}\right)\left(\frac{x}{\varepsilon}\right) \partial_i u^\star(x) \\
&= \sum_{\substack{i,\alpha,\beta=1\\ \beta>\alpha}}^{d} \int_e [\![v(x)]\!] \left(\tau_{\alpha\beta}\cdot \nabla [J^i]_{\beta\alpha}\right)\left(\frac{x}{\varepsilon}\right) \partial_i u^\star(x), \qquad (5.19)
\end{aligned}$$

where $\tau_{\alpha\beta}\in\mathbb{R}^d$ is the vector with α-th component set to $-n_\beta$, β-th component set to n_α, and all other components set to 0. Obviously, $\tau_{\alpha\beta}$ is parallel to e. We can thus use Lemma 4.8, and infer from (5.19) that

$$\begin{aligned}
&\left| \sum_{i,\alpha,\beta=1}^{d} \int_e [\![v(x)]\!]\, n_\alpha \frac{\partial [J^i]_{\beta\alpha}}{\partial x_\beta}\left(\frac{x}{\varepsilon}\right) \partial_i u^\star(x) \right| \\
&\leq C\sqrt{\frac{\varepsilon}{H}} \sum_{i,\alpha,\beta=1}^{d} \|[J^i]_{\beta\alpha}\|_{C^1(\mathbb{R}^d)} \sum_{T\in\mathcal{T}_e} |v|_{H^1(T)} \left(\|\partial_i u^\star\|_{L^2(T)} + H |\partial_i u^\star|_{H^1(T)}\right).
\end{aligned}$$

Using the regularity (5.14) of J^i, we deduce that the first term of the right-hand side of (5.18) satisfies

$$\begin{aligned}
&\left| \sum_{e\in\mathcal{E}_H} \sum_{i,\alpha,\beta=1}^{d} \int_e [\![v(x)]\!] n_\alpha \frac{\partial [J^i]_{\beta\alpha}}{\partial x_\beta}\left(\frac{x}{\varepsilon}\right) \partial_i u^\star(x) \right| \\
&\leq C\sqrt{\frac{\varepsilon}{H}} \left[\sum_{e\in\mathcal{E}_H} \sum_{T\in T_e} \|\nabla v\|^2_{L^2(T)} \right]^{\frac{1}{2}} \\
&\quad \times \left[\sum_{e\in\mathcal{E}_H} \sum_{T\in T_e} \|\nabla u^\star\|^2_{L^2(T)} + H^2 \|\nabla^2 u^\star\|^2_{L^2(T)} \right]^{\frac{1}{2}} \\
&\leq C\sqrt{\frac{\varepsilon}{H}} \|v\|_E \|\nabla u^\star\|_{L^2(\Omega)} + C\sqrt{\varepsilon H} \|v\|_E \|\nabla^2 u^\star\|_{L^2(\Omega)}.
\end{aligned} \tag{5.20}$$

We next turn to the second term of the right-hand side of (5.18), which satisfies

$$\begin{aligned}
&\left| \varepsilon \sum_{e\in\mathcal{E}_H} \sum_{i,\alpha,\beta=1}^{d} \int_e [\![v(x)]\!] n_\alpha [J^i]_{\beta\alpha}\left(\frac{x}{\varepsilon}\right) \partial_{i\beta} u^\star(x) \right| \\
&\leq d^3 \varepsilon \left(\max_i \|J^i\|_{C^0(\mathbb{R}^d)} \right) \|\nabla^2 u^\star\|_{C^0(\overline{\Omega})} \sum_{e\in\mathcal{E}_H} \|[\![v]\!]\|_{L^1(e)} \\
&\leq C\varepsilon \|\nabla^2 u^\star\|_{C^0(\overline{\Omega})} \left[\sum_{e\in\mathcal{E}_H} \|[\![v]\!]\|^2_{L^2(e)} \right]^{\frac{1}{2}} \left[\sum_{e\in\mathcal{E}_H} \|1\|^2_{L^2(e)} \right]^{\frac{1}{2}} \\
&\leq C\varepsilon \|\nabla^2 u^\star\|_{C^0(\overline{\Omega})} \left[\sum_{e\in\mathcal{E}_H} H \sum_{T\in T_e} \|\nabla v\|^2_{L^2(T)} \right]^{\frac{1}{2}} \\
&\quad \times \left[\sum_{\substack{e\in\mathcal{E}_H \\ \text{choose one } T\in T_e}} H^{-1} \|1\|^2_{L^2(T)} \right]^{\frac{1}{2}} \\
&\leq C\varepsilon \|\nabla^2 u^\star\|_{C^0(\overline{\Omega})} \|v\|_E |\Omega|^{\frac{1}{2}},
\end{aligned} \tag{5.21}$$

where we have used (4.7) of Corollary 4.6 and (4.4) of Lemma 4.5.

Collecting the estimates (5.10), (5.12), (5.16)–(5.18) and (5.20)–(5.21), we bound the third term of (5.7):

$$\begin{aligned}
&\left| \sum_{T\in\mathcal{T}_H} \int_T (\nabla v)^{\mathrm{T}} \left(A_{\mathrm{per}}\left(\frac{x}{\varepsilon}\right) \nabla u^{\varepsilon,1} - A^\star_{\mathrm{per}} \nabla u^\star \right) \right| \\
&\leq C\sqrt{\varepsilon H} \|v\|_E \|\nabla^2 u^\star\|_{L^2(\Omega)} + C\sqrt{\frac{\varepsilon}{H}} \|v\|_E \|\nabla u^\star\|_{L^2(\Omega)} \\
&\quad + C\varepsilon \|\nabla^2 u^\star\|_{C^0(\overline{\Omega})} \|v\|_E,
\end{aligned} \tag{5.22}$$

where C is independent of ε, H, v and $u^\star$ (but depends on Ω).

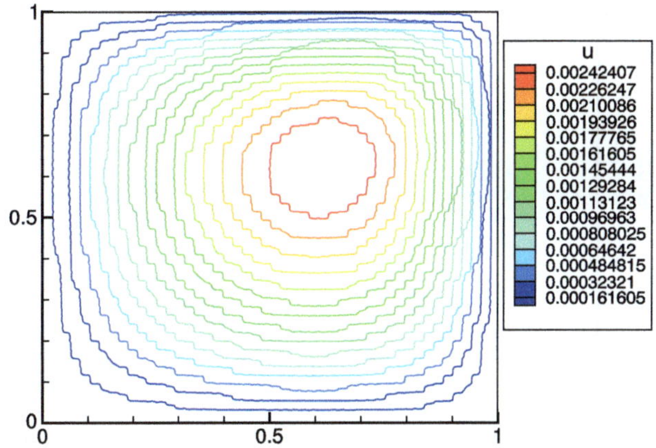

Fig. 1 Reference solution for (1.1) with the choice (6.1)

Step 4 Conclusion Inserting (5.8)–(5.9) and (5.22) in (5.7), we obtain

$$\begin{aligned}&\left|\sum_{T\in\mathcal{T}_H}\int_{\partial T} v\left(A_{\text{per}}\left(\frac{x}{\varepsilon}\right)\nabla u^{\varepsilon}\right)\cdot n\right|\\ &\quad\le C\sqrt{\varepsilon}\|v\|_E\left(\|\nabla u^{\star}\|_{W^{1,\infty}(\Omega)}+\sqrt{\varepsilon}\|\nabla^2 u^{\star}\|_{C^0(\overline{\Omega})}\right)\\ &\qquad + C(H+\sqrt{\varepsilon H})\|v\|_E\|\nabla^2 u^{\star}\|_{L^2(\Omega)}+C\sqrt{\frac{\varepsilon}{H}}\|v\|_E\|\nabla u^{\star}\|_{L^2(\Omega)},\end{aligned} \tag{5.23}$$

which yields the desired bound (5.1). This concludes the proof of Lemma 5.1.

6 Numerical Illustrations

For our numerical tests, we consider (1.1) on the domain $\Omega = (0,1)^2$, with the right-hand side $f(x,y) = \sin(x)\sin(y)$.

First Test-Case We first choose the highly oscillatory matrix

$$A_{\varepsilon}(x,y) = a_{\varepsilon}(x,y)\mathrm{Id}_2, \quad a_{\varepsilon}(x,y) = 1 + 100\cos^2(150x)\sin^2(150y) \tag{6.1}$$

in (1.1). This matrix coefficient is periodic, with period $\varepsilon = \frac{\pi}{150} \approx 0.02$. The reference solution u^{ε} (computed on a fine mesh 1024×1024 of Ω) is shown in Fig. 1.

We show in Fig. 2 the relative errors between the fine scale solution u^{ε} and its approximation provided by various MsFEM type approaches, as a function of the coarse mesh size H.

Our approach is systematically more accurate than the standard (meaning, without the oversampling technique) MsFEM approach. In addition, we see that, for large H, our approach yields an error smaller than or comparable to the best other methods. Likewise, when H is small (but not sufficiently small for the standard

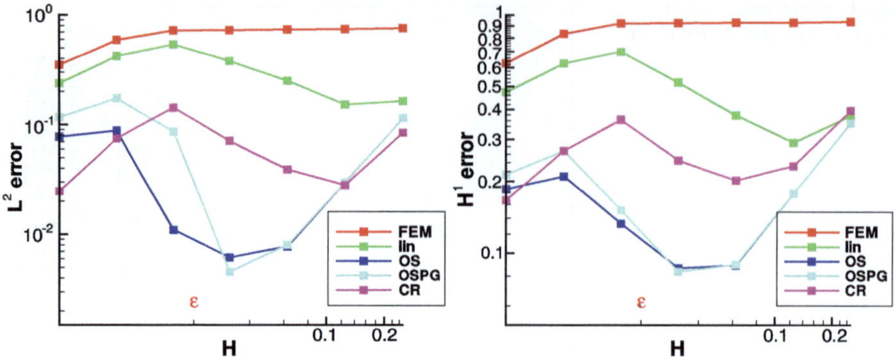

Fig. 2 Test-case (6.1): relative errors (in L^2 (*left*) and H^1-broken (*right*) norms) with various approaches: FEM—the standard Q1 finite elements, lin—MsFEM with linear boundary conditions, OS—MsFEM with oversampling, OSPG—Petrov-Galerkin MsFEM with oversampling, CR—the MsFEM Crouzeix-Raviart approach we propose

FEM approach to be accurate), our approach is also more accurate than the other approaches. For intermediate values of H, our approach is however less accurate than approaches using oversampling (for which we used an oversampling ratio equal to 2). Note that this will no longer be the case for the problem on a perforated domain considered in [25]. Note also that our approach is slightly less expensive than the approaches using oversampling (in terms of computations of the highly oscillatory basis functions) and, much more importantly, has no adjustable parameter.

A comparison with the MsFEM-O variant (described in Remark 3.4) has also been performed but is not included in the figures below. On the particular case considered in this article, we have observed that this approach seems to perform very well. However, it is not clear, in general, whether this approach yields systematically more accurate results than the other MsFEM variants. A more comprehensive assessment of this variant will be performed for the case of perforated domains in [25].

Higher Contrast We now consider the cases

$$A_\varepsilon(x,y) = a_\varepsilon(x,y)\mathrm{Id}_2, \quad a_\varepsilon(x,y) = 1 + 10^3 \cos^2(150x)\sin^2(150y) \tag{6.2}$$

and

$$A_\varepsilon(x,y) = a_\varepsilon(x,y)\mathrm{Id}_2, \quad a_\varepsilon(x,y) = 1 + 10^4 \cos^2(150x)\sin^2(150y) \tag{6.3}$$

in (1.1). In comparison with (6.1), we have increased the contrast by a factor 10 or 100, respectively. Results are shown in Fig. 3, top and bottom rows respectively.

We see that the relative quality of the different approaches is not sensitive to the contrast (at least when the latter does not exceed 10^3). Of course, each method provides an approximation of u^ε that is less accurate than in the case (6.1). However, all methods seem to equally suffer from a higher contrast.

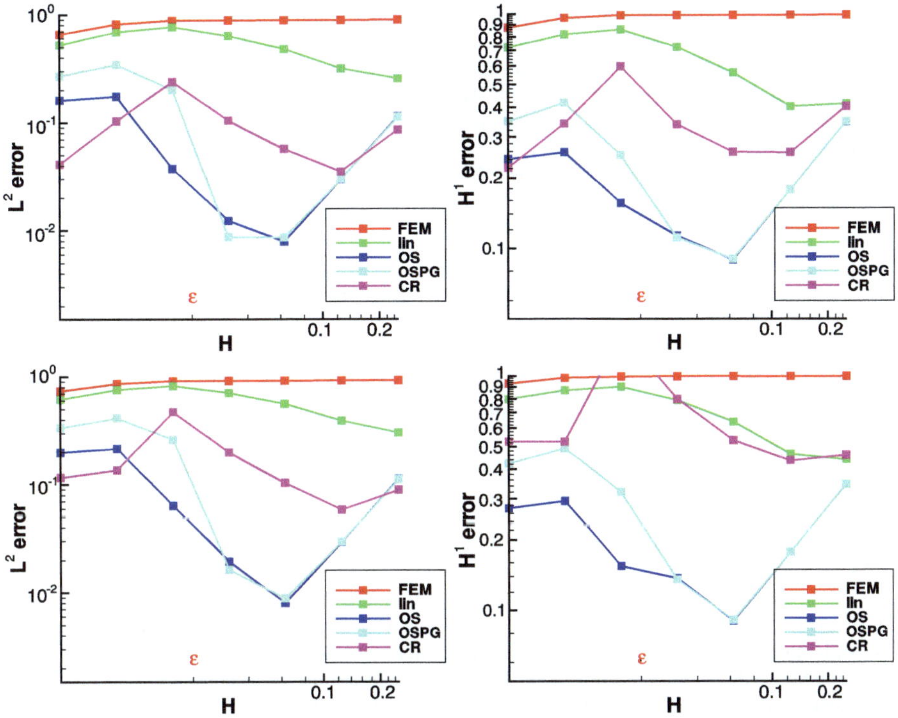

Fig. 3 Test-cases (6.2) (*top row*) and (6.3) (*bottom row*) for higher contrasts: relative errors (in L^2 (*left*) and H^1-broken (*right*) norms) with various approaches: FEM—the standard Q1 finite elements, lin—MsFEM with linear boundary conditions, OS—MsFEM with oversampling, OSPG—Petrov-Galerkin MsFEM with oversampling, CR—the MsFEM Crouzeix-Raviart approach we propose

Acknowledgements The third author acknowledges the hospitality of INRIA. We thank William Minvielle for his remarks on a preliminary version of this article.

References

1. Aarnes, J.: On the use of a mixed multiscale finite element method for greater flexibility and increased speed or improved accuracy in reservoir simulation. Multiscale Model. Simul. **2**(3), 421–439 (2004)
2. Aarnes, J., Heimsund, B.O.: Multiscale discontinuous Galerkin methods for elliptic problems with multiple scales. In: Engquist, B., Lötstedt, P., Runborg, O. (eds.) Multiscale Methods in Science and Engineering. Lecture Notes in Computational Science and Engineering, vol. 44, pp. 1–20. Springer, Berlin (2005)
3. Abdulle, A.: Multiscale method based on discontinuous Galerkin methods for homogenization problems. C. R. Math. Acad. Sci. Paris **346**(1–2), 97–102 (2008)
4. Abdulle, A.: Discontinuous Galerkin finite element heterogeneous multiscale method for elliptic problems with multiple scales. Math. Comput. **81**(278), 687–713 (2012)

5. Arbogast, T.: Implementation of a locally conservative numerical subgrid upscaling scheme for two-phase Darcy flow. Comput. Geosci. **6**(3–4), 453–481 (2002)
6. Arbogast, T.: Mixed multiscale methods for heterogeneous elliptic problems. In: Graham, I.G., Hou, T.Y., Lakkis, O., Scheichl, R. (eds.) Numerical Analysis of Multiscale Problems. Lecture Notes in Computational Science and Engineering, vol. 83, pp. 243–283. Springer, Berlin (2011)
7. Arbogast, T., Boyd, K.J.: Subgrid upscaling and mixed multiscale finite elements. SIAM J. Numer. Anal. **44**(3), 1150–1171 (2006)
8. Bensoussan, A., Lions, J.L., Papanicolaou, G.: Asymptotic analysis for periodic structures. Studies in Mathematics and Its Applications, vol. 5. North-Holland, Amsterdam (1978)
9. Bergh, J., Löfström, J.: Interpolation Spaces. An Introduction. Grundlehren der Mathematischen Wissenschaften, vol. 223. Springer, Berlin (1976)
10. Brenner, S.C., Scott, L.R.: The Mathematical Theory of Finite Element Methods, 3rd edn. Springer, New York (2008)
11. Chen, Z., Cui, M., Savchuk, T.Y.: The multiscale finite element method with nonconforming elements for elliptic homogenization problems. Multiscale Model. Simul. **7**(2), 517–538 (2008)
12. Chen, Z., Hou, T.Y.: A mixed multiscale finite element method for elliptic problems with oscillating coefficients. Math. Comput. **72**(242), 541–576 (2003)
13. Cioranescu, D., Donato, P.: An Introduction to Homogenization. Oxford Lecture Series in Mathematics and Its Applications, vol. 17. Clarendon, Oxford (1999)
14. Crouzeix, M., Raviart, P.A.: Conforming and nonconforming finite element methods for solving the stationary Stokes equations. I. RAIRO. Anal. Numér. **7**(3), 33–75 (1973)
15. Efendiev, Y., Hou, T.Y.: Multiscale Finite Element Method: Theory and Applications. Surveys and Tutorials in the Applied Mathematical Sciences, vol. 4. Springer, New York (2009)
16. Efendiev, Y.R., Hou, T.Y., Wu, X.H.: Convergence of a nonconforming multiscale finite element method. SIAM J. Numer. Anal. **37**(3), 888–910 (2000)
17. Engquist, B., Souganidis, P.: Asymptotic and Numerical Homogenization. Acta Numerica, vol. 17. Cambridge University Press, Cambridge (2008)
18. Ern, A., Guermond, J.L.: Theory and Practice of Finite Elements. Applied Mathematical Sciences, vol. 159. Springer, Berlin (2004)
19. Gilbarg, D., Trudinger, N.S.: Elliptic Partial Differential Equations of Second Order. Classics in Mathematics. Springer, Berlin (2001)
20. Gloria, A.: An analytical framework for numerical homogenization. Part II: Windowing and oversampling. Multiscale Model. Simul. **7**(1), 274–293 (2008)
21. Hou, T.Y., Wu, X.H.: A multiscale finite element method for elliptic problems in composite materials and porous media. J. Comput. Phys. **134**(1), 169–189 (1997)
22. Hou, T.Y., Wu, X.H., Cai, Z.: Convergence of a multiscale finite element method for elliptic problems with rapidly oscillating coefficients. Math. Comput. **68**(227), 913–943 (1999)
23. Hou, T.Y., Wu, X.H., Zhang, Y.: Removing the cell resonance error in the multiscale finite element method via a Petrov-Galerkin formulation. Commun. Math. Sci. **2**(2), 185–205 (2004)
24. Jikov, V.V., Kozlov, S.M., Oleinik, O.A.: Homogenization of Differential Operators and Integral Functionals. Springer, Berlin (1994)
25. Le Bris, C., Legoll, F., Lozinski, A.: A MsFEM type approach for perforated domains. (2013, in preparation)
26. Le Bris, C., Legoll, F., Thomines, F.: Multiscale Finite Element approach for "weakly" random problems and related issues. arXiv:1111.1524 [math.NA]
27. Malqvist, A., Peterseim, D.: Localization of elliptic multiscale problems. arXiv:1110.0692 [math.NA]
28. Owhadi, H., Zhang, L.: Localized bases for finite dimensional homogenization approximations with non-separated scales and high-contrast. Multiscale Model. Simul. **9**, 1373–1398 (2011)
29. Rannacher, R., Turek, S.: Simple nonconforming quadrilateral Stokes element. Numer. Methods Partial Differ. Equ. **8**, 97–111 (1982)

Exact Synchronization for a Coupled System of Wave Equations with Dirichlet Boundary Controls

Tatsien Li and Bopeng Rao

Abstract In this paper, the exact synchronization for a coupled system of wave equations with Dirichlet boundary controls and some related concepts are introduced. By means of the exact null controllability of a reduced coupled system, under certain conditions of compatibility, the exact synchronization, the exact synchronization by groups, and the exact null controllability and synchronization by groups are all realized by suitable boundary controls.

Keywords Exact null controllability · Exact synchronization · Exact synchronization by groups

Mathematics Subject Classification 35B37 · 93B05 · 93B07

1 Introduction

Synchronization is a widespread natural phenomenon. Thousands of fireflies may twinkle at the same time; audiences in the theater can applaud with a rhythmic beat; pacemaker cells of the heart function simultaneously; and field crickets give out a unanimous cry. All these are phenomena of synchronization.

In principle, synchronization happens when different individuals possess likeness in nature, that is, they conform essentially to the same governing equation, and meanwhile, the individuals should bear a certain coupled relation.

The phenomenon of synchronization was first observed by Huygens [4]. The theoretical research on synchronization phenomena dates back to Fujisaka and Ya-

T. Li (✉)
Nonlinear Mathematical Modeling and Methods Laboratory, Shanghai Key Laboratory for Contemporary Applied Mathematic, School of Mathematical Sciences, Fudan University, Shanghai 200433, China
e-mail: dqli@fudan.edu.cn

B. Rao
Institut de Recherche Mathématique Avancée, Université de Strasbourg, 67084 Strasbourg, France
e-mail: bopeng.rao@math.unistra.fr

P.G. Ciarlet et al. (eds.), *Partial Differential Equations: Theory, Control and Approximation*, DOI 10.1007/978-3-642-41401-5_12,

mada's study of synchronization for coupled equations in 1983 (see [2]). The previous studies focused on systems described by ODEs, such as

$$\frac{\mathrm{d}X_i}{\mathrm{d}t} = f(X_i, t) + \sum_{j=1}^{N} A_{ij} X_j, \quad i = 1, \ldots, N, \tag{1.1}$$

where X_i $(i = 1, \ldots, N)$ are n-dimensional vectors, A_{ij} $(i, j = 1, \ldots, N)$ are $n \times n$ matrices, and $f(X, t)$ is an n-dimensional vector function independent of n. The right-hand side of (1.1) shows that every X_i $(i = 1, \ldots, N)$ possesses two basic features, that is, satisfying a fundamental governing equation and bearing a coupled relation among one another.

In this paper, we will consider the synchronization of the following hyperbolic system:

$$\begin{cases} \frac{\partial^2 U}{\partial t^2} - \Delta U + AU = 0 & \text{in } \Omega, \\ U = 0 & \text{on } \Gamma_0, \\ U = H & \text{on } \Gamma_1, \\ t = 0: U = U_0, & \frac{\partial U}{\partial t} = U_1, \end{cases} \tag{1.2}$$

where $U = (u^{(1)}, \ldots, u^{(N)})^{\mathrm{T}}$ is the state variable, $A \in \mathbb{M}^N(\mathbb{R})$ is the coupling matrix, and $H = (h^{(1)}, \ldots, h^{(N)})^{\mathrm{T}}$ is the boundary control. Different from the ODE situation, the coupling of PDE systems can be fulfilled by coupling of the equations or (and) the boundary conditions. Our goal is to synchronize the state variable U through boundary control H. Roughly speaking, the problem is to find a $T > 0$, and through boundary control on $[0, T]$, we have that from time $t = T$ on, the system states tend to be the same. That is to say, we hope to achieve the synchronization of the system state not only at the moment $t = T$ under the action of boundary controls on $[0, T]$, but also when $t \geq T$ withdrawing all the controls. This is forever (instead of short-lived) synchronization, as is desired in many actual applications. Obviously, if the system has the exact boundary null controllability, it must have the exact synchronization, but this is a trivial situation that should be excluded beforehand.

The exact synchronization is linked with the exact null controllability. In fact, let $W = (w^{(1)}, \ldots, w^{(N-1)})^{\mathrm{T}}$ with $w^{(i)} = u^{(i+1)} - u^{(i)}$ $(i = 1, \ldots, N-1)$. Then under some conditions of compatibility on the coupling matrix A, the new state W satisfies a reduced system of $N - 1$ equations as follows:

$$\begin{cases} \frac{\partial^2 W}{\partial t^2} - \Delta W + \overline{A} W = 0 & \text{in } \Omega, \\ W = 0 & \text{on } \Gamma_0, \\ W = \overline{H} & \text{on } \Gamma_1, \\ t = 0: W = W_0, & \frac{\partial W}{\partial t} = W_1, \end{cases} \tag{1.3}$$

where $\overline{A}$ is a matrix of order $N-1$. Under such conditions of compatibility, the exact synchronization of system (1.2) of N equations is equivalent to the exact null controllability of the reduced system (1.3) of $N-1$ equations. Our study will be based on two key points. We will first establish the exact null controllability of system (1.3) via the boundary control $\overline{H}$ of $N-1$ components. We next find some conditions of compatibility on the coupling matrix A to guarantee the reduction of system (1.2) to system (1.3).

There are many works on the exact controllability of hyperbolic systems by means of boundary controls. Generally speaking, one needs N boundary controls for the exact controllability of a system of N wave equations. In the case of less controls, we can not realize the exact controllability in general (see [8]). However, for smooth initial data, the exact controllability of two linear wave equations was proved by means of only one boundary control (see [1, 11]). Li and Rao [9] introduced the asymptotic controllability and established the equivalence between the asymptotic controllability of the original system and the weak observability of the dual system. Moreover, in [12], the optimal polynomial decay rate of energy of distributed systems with less boundary damping terms was studied by means of Riesz basis approach.

The exact synchronization is another way to weaken the notion of exact null controllability. In fact, instead of bringing all the states of system to zero, we only need to steer the states of the system to the same, which is unknown a priori. In terms of degree of freedom, we will use $N-1$ boundary controls to realize the exact synchronization for a system of N equations.

Now we briefly outline the contents of the paper. In Sect. 2, using a recent result on the observability of compactly perturbed systems of Mehrenberger [13], we establish the exact null controllability for (1.3). In Sect. 3, we consider the exact synchronization for the coupled system (1.2). We first give necessary conditions of compatibility on the coupling matrix A for the exact synchronization. We next prove that under these conditions of compatibility, the system (1.2) of N equations can be exactly synchronized by means of $N-1$ boundary controls. In Sect. 4, we generalize the notion of synchronization to the exact synchronization by groups. Section 5 is devoted to a mixed problem of synchronization and controllability. In Sect. 6, we study the behaviors of the final synchronizable state for a system of two wave equations.

2 Exact Controllability for a Coupled System of Wave Equations

Let $\Omega \subset \mathbb{R}^n$ be a bounded open set with smooth boundary Γ of class C^2. Let $\Gamma = \Gamma_1 \cup \Gamma_0$ be a partition of Γ, such that $\overline{\Gamma}_1 \cap \overline{\Gamma}_0 = \emptyset$. Furthermore, we assume that there exists an $x_0 \in \mathbb{R}^n$, such that, by setting $m = x - x_0$, we have

$$(m, \nu) > 0, \quad \forall x \in \Gamma_1, \qquad (m, \nu) \leq 0, \quad \forall x \in \Gamma_0, \tag{2.1}$$

where ν is the unit outward normal vector, and $(\cdot, \cdot)$ denotes the inner product in $\mathbb{R}^n$.

Let

$$W=\big(w^{(1)},\dots,w^{(M)}\big)^{\mathrm{T}},\qquad \overline{H}=\big(\overline{h}^{(1)},\dots,\overline{h}^{(M)}\big)^{\mathrm{T}},\qquad \overline{A}\in\mathbb{M}^{M}(\mathbb{R}).$$

Consider the following mixed problem for a coupled system of wave equations:

$$\frac{\partial^2 W}{\partial t^2}-\Delta W+\overline{A}W=0\quad \text{in } \Omega, \tag{2.2}$$

$$W=0\quad \text{on } \Gamma_0, \tag{2.3}$$

$$W=\overline{H}\quad \text{on } \Gamma_1, \tag{2.4}$$

$$t=0: W=W_0,\qquad \frac{\partial W}{\partial t}=W_1. \tag{2.5}$$

If the coupling matrix $\overline{A}$ is symmetric and positively definite, the exact controllability of (2.2)–(2.5) follows easily from the classical results (see [4, 10]). In this section, we will establish the exact controllability for any coupling matrix $\overline{A}$. We first establish the observability of the corresponding adjoint problem, and then the exact controllability follows from the standard HUM method of Lions.

Now let

$$\Phi=\big(\phi^{(1)},\dots,\phi^{(M)}\big)^{\mathrm{T}}.$$

Consider the corresponding adjoint problem as follows:

$$\frac{\partial^2 \Phi}{\partial t^2}-\Delta \Phi+\overline{A}^{\mathrm{T}}\Phi=0\quad \text{in } \Omega, \tag{2.6}$$

$$\Phi=0\quad \text{on } \Gamma, \tag{2.7}$$

$$t=0: \Phi=\Phi_0,\qquad \frac{\partial \Phi}{\partial t}=\Phi_1. \tag{2.8}$$

It is well-known that the above problem is well-posed in the space $\mathcal{V}\times\mathcal{H}$:

$$\mathcal{V}=\big(H_0^1(\Omega)\big)^{M},\qquad \mathcal{H}=\big(L^2(\Omega)\big)^{M}. \tag{2.9}$$

Moreover, we have the following direct and inverse inequalities.

Theorem 2.1 *Let $T>0$ be suitably large. Then there exist positive constants c and C, such that for any given initial data $(\Phi_0,\Phi_1)\in\mathcal{V}\times\mathcal{H}$, the solution Φ to* (2.6)–(2.8) *satisfies the following inequalities*:

$$c\int_0^T\int_{\Gamma_1}\left|\frac{\partial\Phi}{\partial\nu}\right|^2 d\Gamma dt\le \|\Phi_0\|_{\mathcal{V}}^2+\|\Phi_1\|_{\mathcal{H}}^2\le C\int_0^T\int_{\Gamma_1}\left|\frac{\partial\Phi}{\partial\nu}\right|^2 d\Gamma dt. \tag{2.10}$$

Before proving Theorem 2.1, we first give a uniqueness result.

Lemma 2.1 *Let B be a square matrix of order M, and $\Phi \in H^2(\Omega)$ be a solution to the following system:*

$$\Delta\Phi = B\Phi \quad \text{in } \Omega. \tag{2.11}$$

Assume furthermore that

$$\Phi = 0, \quad \frac{\partial\Phi}{\partial\nu} = 0 \quad \text{on } \Gamma_1. \tag{2.12}$$

Then we have $\Phi \equiv 0$.

Proof Let

$$\widetilde{B} = PBP^{-1} = \begin{pmatrix} \widetilde{b}_{11} & 0 & \dots & 0 \\ \widetilde{b}_{21} & \widetilde{b}_{22} & \dots & 0 \\ & & \dots & \\ \widetilde{b}_{M1} & \widetilde{b}_{M2} & \dots & \widetilde{b}_{MM} \end{pmatrix}, \qquad \widetilde{\Phi} = P\Phi,$$

where $\widetilde{B}$ is a lower triangular matrix of complex entries. Then (2.11)–(2.12) can be reduced to

$$\begin{cases} \Delta\widetilde{\phi}^{(k)} = \sum_{p=1}^{k} \widetilde{b}_{kp}\widetilde{\phi}^{(p)} & \text{in } \Omega, \\ \widetilde{\phi}^{(k)} = 0, \quad \frac{\partial\widetilde{\phi}^{(k)}}{\partial\nu} = 0 & \text{on } \Gamma_1 \end{cases} \tag{2.13}$$

for $k = 1, \dots, M$. In particular for $k = 1$, we have

$$\begin{cases} \Delta\widetilde{\phi}^{(1)} = \widetilde{b}_{11}\widetilde{\phi}^{(1)} & \text{in } \Omega, \\ \widetilde{\phi}^{(1)} = 0, \quad \frac{\partial\widetilde{\phi}^{(1)}}{\partial\nu} = 0 & \text{on } \Gamma_1. \end{cases}$$

Thanks to Carleman's uniqueness result (see [3]), we get

$$\widetilde{\phi}^{(1)} \equiv 0. \tag{2.14}$$

Inserting (2.14) into the second set of (2.13) leads to

$$\begin{cases} \Delta\widetilde{\phi}^{(2)} = \widetilde{b}_{22}\widetilde{\phi}^{(2)} & \text{in } \Omega, \\ \widetilde{\phi}^{(2)} = 0, \quad \frac{\partial\widetilde{\phi}^{(2)}}{\partial\nu} = 0 & \text{on } \Gamma_1, \end{cases}$$

and we can repeat the same procedure. Thus, by a simple induction, we get successively that

$$\widetilde{\phi}^{(k)} \equiv 0, \quad k = 1, \dots, M.$$

This yields that

$$\widetilde{\Phi} \equiv 0 \Rightarrow \Phi \equiv 0.$$

The proof is complete. □

Proof of Theorem 2.1 We rewrite (2.6)–(2.8) as

$$\begin{pmatrix} \Phi \\ \Phi' \end{pmatrix}' = \begin{pmatrix} 0 & I \\ \Delta & 0 \end{pmatrix}\begin{pmatrix} \Phi \\ \Phi' \end{pmatrix} + \begin{pmatrix} 0 & 0 \\ -\overline{A}^{\mathrm{T}} & 0 \end{pmatrix}\begin{pmatrix} \Phi \\ \Phi' \end{pmatrix} = \mathcal{A}\begin{pmatrix} \Phi \\ \Phi' \end{pmatrix} + \mathcal{B}\begin{pmatrix} \Phi \\ \Phi' \end{pmatrix}, \tag{2.15}$$

where $I = I_M$ is the unit matrix of order M. It is easy to see that $\mathcal{A}$ is a skew-adjoint operator with compact resolvent in $\mathcal{V} \times \mathcal{H}$, and $\mathcal{B}$ is a compact operator in $\mathcal{V} \times \mathcal{H}$. Therefore, they generate respectively C^0 groups in the energy space $\mathcal{V} \times \mathcal{H}$.

Following a recent perturbation result of Mehrenberger [13], in order to prove (2.10) for a system of this kind, it is sufficient to check the following assertions:

(i) The direct and inverse inequalities

$$c\int_0^{\mathrm{T}}\int_{\Gamma_1}\left|\frac{\partial\widetilde{\Phi}}{\partial\nu}\right|^2 \mathrm{d}\Gamma\mathrm{d}t \le \|\Phi_0\|^2_{\mathcal{V}} + \|\Phi_1\|^2_{\mathcal{H}} \le C\int_0^{\mathrm{T}}\int_{\Gamma_1}\left|\frac{\partial\widetilde{\Phi}}{\partial\nu}\right|^2 \mathrm{d}\Gamma\mathrm{d}t \tag{2.16}$$

hold for the solution $\widetilde{\Phi} = S_{\mathcal{A}}(t)(\Phi_0, \Phi_1)$ to the decoupled problem (2.6)–(2.8) with $\overline{A} = 0$.

(ii) The system of root vectors of $\mathcal{A} + \mathcal{B}$ forms a Riesz basis of subspaces in $\mathcal{V} \times \mathcal{H}$. That is to say, there exists a family of subspaces $\mathcal{V}_i \times \mathcal{H}_i$ $(i \ge 1)$ composed of root vectors of $\mathcal{A} + \mathcal{B}$, such that for all $x \in \mathcal{V} \times \mathcal{H}$, there exists a unique $x_i \in \mathcal{V}_i \times \mathcal{H}_i$ $(i \ge 1)$, such that

$$x = \sum_{i=1}^{+\infty} x_i, \qquad c_1\|x\|^2 \le \sum_{i=1}^{+\infty}\|x_i\|^2 \le c_2\|x\|^2,$$

where c_1, c_2 are positive constants.

(iii) If $(\Phi, \Psi) \in \mathcal{V} \times \mathcal{H}$ and $\lambda \in \mathbb{C}$, such that

$$(\mathcal{A} + \mathcal{B})(\Phi, \Psi) = \lambda(\Phi, \Psi) \quad \text{and} \quad \frac{\partial\Phi}{\partial\nu} = 0 \quad \text{on } \Gamma_1,$$

then $(\Phi, \Psi) = 0$.

For simplification of notation, we will still denote by $\mathcal{V} \times \mathcal{H}$ the complex Hilbert space corresponding to $\mathcal{V} \times \mathcal{H}$.

Since the assertion (i) is well-known (see [9]), we only have to verify (ii) and (iii).

Verification of (ii). Let $\mu_i^2 > 0$ be an eigenvalue corresponding to an eigenvector e_i of $-\Delta$ with homogeneous Dirichlet boundary condition:

$$\begin{cases} -\Delta e_i = \mu_i^2 e_i & \text{in } \Omega, \\ e_i = 0 & \text{on } \Gamma. \end{cases}$$

Let

$$\mathcal{H}_i \times \mathcal{V}_i = \left\{(\alpha e_i, \beta e_i) : \alpha, \beta \in \mathbb{C}^M\right\}.$$

Obviously, the subspaces $\mathcal{H}_i \times \mathcal{V}_i$ $(i = 1, 2, \ldots)$ are mutually orthogonal, and

$$\mathcal{H} \times \mathcal{V} = \bigoplus_{i \geq 1} \mathcal{H}_i \times \mathcal{V}_i. \tag{2.17}$$

In particular, for any given $x \in H \times \mathcal{V}$ $(i \geq 1)$, there exists an $x_i \in H_i \times \mathcal{V}_i$ $(i \geq 1)$, such that

$$x = \sum_{i=1}^{+\infty} x_i, \qquad \|x\|^2 = \sum_{i=1}^{+\infty} \|x_i\|^2. \tag{2.18}$$

On the other hand, $\mathcal{H}_i \times \mathcal{V}_i$ is an invariant subspace of $\mathcal{A} + \mathcal{B}$ and of finite dimension $2M$. Then, the restriction of $\mathcal{A} + \mathcal{B}$ in the subspace $\mathcal{H}_i \times \mathcal{V}_i$ is a linear bounded operator, and therefore, its root vectors constitute a basis in the finite dimensional complex space $\mathcal{H}_i \times \mathcal{V}_i$. This together with (2.17)–(2.18) implies that the system of root vectors of $\mathcal{A} + \mathcal{B}$ forms a Riesz basis of subspaces in $\mathcal{H} \times \mathcal{V}$.

Verification of (iii). Let $(\Phi, \Psi) \in \mathcal{V} \times \mathcal{H}$ and $\lambda \in \mathbb{C}$, such that

$$(\mathcal{A} + \mathcal{B})(\Phi, \Psi) = \lambda(\Phi, \Psi) \quad \text{and} \quad \frac{\partial \Phi}{\partial \nu} = 0 \quad \text{on } \Gamma_1.$$

Then we have

$$\Psi = \lambda \Phi, \quad \Delta \Phi - \overline{A}^{\mathrm{T}} \Phi = \lambda \Psi,$$

namely,

$$\begin{cases} \Delta \Phi = (\lambda^2 I + \overline{A}^{\mathrm{T}}) \Phi & \text{in } \Omega, \\ \Phi = 0 & \text{on } \Gamma. \end{cases} \tag{2.19}$$

It follows from the classic elliptic theory that $\Phi \in H^2(\Omega)$. Moreover, we have

$$\frac{\partial \Phi}{\partial \nu} = 0 \quad \text{on } \Gamma. \tag{2.20}$$

Then, applying Lemma 2.1 to (2.19)–(2.20), we get $\Phi = 0$, then $\Psi = 0$. The proof is then complete. □

By a standard application of the HUM method, from Theorem 2.1 we get the following result.

Theorem 2.2 *There exists a positive constant $T > 0$, such that for any given initial data*

$$W_0 \in \big(L^2(\Omega)\big)^M, \qquad W_1 \in \big(H^{-1}(\Omega)\big)^M, \tag{2.21}$$

there exist boundary control functions

$$\overline{H} \in \big(L^2(0, T; L^2(\Gamma_1))\big)^M, \tag{2.22}$$

such that (2.2)–(2.5) *admits a unique weak solution*

$$W \in \big(C^0([0, T]; L^2(\Omega))\big)^M, \qquad \frac{\partial W}{\partial t} \in \big(C^0([0, T]; H^{-1}(\Omega))\big)^M, \tag{2.23}$$

satisfying the null final condition

$$t = T : W = 0, \quad \frac{\partial W}{\partial t} = 0. \tag{2.24}$$

Remark 2.1 Note that we do not need any assumption on the coupling matrix $\overline{A}$ in Theorem 2.2.

Remark 2.2 The same result on the controllability for a coupled system of 1-dimensional wave equations in the framework of classical solutions can be found in [7, 14].

3 Exact Synchronization for a Coupled System of Wave Equations

Let

$$U = \big(u^{(1)}, \ldots, u^{(N)}\big)^{\mathrm{T}}, \quad A \in \mathbb{M}^N(\mathbb{R}).$$

Consider the following coupled system of wave equations with Dirichlet boundary controls:

$$\frac{\partial^2 U}{\partial t^2} - \Delta U + AU = 0 \quad \text{in } \Omega, \tag{3.1}$$

$$U = 0 \quad \text{on } \Gamma_0, \tag{3.2}$$

$$U = H \quad \text{on } \Gamma_1, \tag{3.3}$$

$$t = 0 : U = U_0, \quad \frac{\partial U}{\partial t} = U_1. \tag{3.4}$$

According to the result given in the previous section, we have the exact null controllability of the problem (3.1)–(3.4) by means of N boundary controls. If the number of boundary controls is less than N, generally speaking, we can not realize the exact controllability (see [7], for more general discussion).

Definition 3.1 Problem (3.1)–(3.4) is exactly synchronizable at the moment $T > 0$, if for any given initial data $U_0 \in (L^2(\Omega))^N$ and $U_1 \in (H^{-1}(\Omega))^N$, there exist suitable boundary controls given by a part of $H \in (L^2(0, +\infty; L^2(\Gamma_1)))^N$, such that the solution $U = U(t, x)$ to (3.1)–(3.4) satisfies the following final condition:

$$t \geq T : u^{(1)} \equiv u^{(2)} \equiv \cdots \equiv u^{(N)} := u, \tag{3.5}$$

where $u = u(t, x)$ is called the synchronizable state.

Remark 3.1 If problem (3.1)–(3.4) is exactly null controllable, then we have certainly the exact synchronization. This trivial situation should be excluded. Therefore, in Definition 3.1, we should restrict ourselves to the case that the number of the boundary controls is less than N, so that (3.1)–(3.4) can be assumed to be not exactly null controllable.

Theorem 3.1 *Assume that* (3.1)–(3.4) *is exactly synchronizable, but not exactly null controllable. Then the coupling matrix $A = (a_{ij})$ should satisfy the following conditions of compatibility*:

$$\sum_{p=1}^{N} a_{kp} := \widetilde{a}, \quad k = 1, \ldots, N, \tag{3.6}$$

where $\widetilde{a}$ is a constant independent of $k = 1, \ldots, N$.

Proof By synchronization, there exists a $T > 0$ and a scalar function u, such that

$$u^{(k)}(t, x) \equiv u(t, x), \quad t \geq T, \ k = 1, 2, \ldots, N.$$

Then for $t \geq T$, we have

$$\frac{\partial^2 u}{\partial t^2} - \Delta u + \left(\sum_{p=1}^{N} a_{kp} \right) u = 0 \quad \text{in } \Omega, \ k = 1, 2, \ldots, N.$$

In particular, we have

$$t \geq T : \quad \left(\sum_{p=1}^{N} a_{kp} \right) u = \left(\sum_{p=1}^{N} a_{lp} \right) u \quad \text{in } \Omega, k, l = 1, \ldots, N.$$

By the non-exact null controllability, there exists at least an initial datum (U_0, U_1) for which the corresponding solution U, or equivalently u, does not identically vanish for $t \geq T$, whatever boundary controls H are chosen. This yields the conditions of compatibility (3.6). The proof is completed. □

Theorem 3.2 *Assume that the conditions of compatibility* (3.6) *hold. Then the problem* (3.1)–(3.4) *is exactly synchronizable by means of some boundary controls* H *with compact support on* $[0, T]$ *and* $h^{(1)} \equiv 0$.

Proof Let

$$w^{(i)} = u^{(i+1)} - u^{(i)}, \quad i = 1, \ldots, N-1. \tag{3.7}$$

We will transform the problem (3.1)–(3.4) to a reduced problem on the variable $W = (w^{(1)}, \ldots, w^{(N-1)})^{\mathrm{T}}$. By (3.1), we get

$$\frac{\partial^2 w^{(i)}}{\partial t^2} - \Delta w^{(i)} + \sum_{p=1}^{N} (a_{i+1,p} - a_{ip}) u^{(p)} = 0, \quad i = 1, \ldots, N-1. \tag{3.8}$$

Noting (3.7), we have

$$u^{(i)} = \sum_{j=1}^{i-1} w^{(j)} + u^{(1)}, \quad i = 1, \ldots, N.$$

Then a direct computation gives

$$\begin{aligned}
&\sum_{p=1}^{N} (a_{i+1,p} - a_{ip}) u^{(p)} \\
&\quad = \sum_{p=1}^{N} (a_{i+1,p} - a_{ip}) \left(\sum_{j=1}^{p-1} w^{(j)} + u^{(1)} \right) \\
&\quad = \sum_{p=1}^{N} (a_{i+1,p} - a_{ip}) \sum_{j=1}^{p-1} w^{(j)} + \sum_{p=1}^{N} (a_{i+1,p} - a_{ip}) u^{(1)}.
\end{aligned}$$

Because of (3.6), the last term vanishes, and then it follows from (3.8) that

$$\frac{\partial^2 w^{(i)}}{\partial t^2} - \Delta w^{(i)} + \sum_{j=1}^{N-1} \overline{a}_{ij} w^{(j)} = 0, \quad i = 1, \ldots, N-1, \tag{3.9}$$

where

$$\overline{a}_{ij} = \sum_{p=j+1}^{N} (a_{i+1,p} - a_{ip}) = \sum_{p=1}^{j} (a_{ip} - a_{i+1,p}), \quad i, j = 1, \ldots, N-1. \tag{3.10}$$

Correspondingly, for the variable W, we set the new initial data as

$$w_0^{(i)} = u_0^{(i+1)} - u_0^{(i)}, \qquad w_1^{(i)} = u_1^{(i+1)} - u_1^{(i)}, \quad i = 1, \ldots, N-1, \tag{3.11}$$

and the new boundary controls as

$$\overline{h}^{(i)} = h^{(i+1)} - h^{(i)}, \quad i = 1, \ldots, N-1. \tag{3.12}$$

Noting (3.9)–(3.12), the new variable W satisfies the reduced problem (2.2)–(2.5) (in which $M = N - 1$). Then, by Theorem 2.2, there exist boundary controls $\overline{H} \in L^2(0, T; L^2(\Gamma_1))^{N-1}$, such that the corresponding solution $W = W(t, x)$ to the reduced problem (2.2)–(2.5) satisfies the null final condition. Moreover, taking $\overline{H} \equiv 0$ for $t > T$, it is easy to see that

$$t \geq T : w^{(i)} \equiv 0, \quad i = 1, \ldots, N-1. \tag{3.13}$$

In order to determine $h^{(i)}$ $(i = 1, \ldots, N)$ from (3.12), setting $h^{(1)} \equiv 0$, we get

$$h^{(i+1)} = \overline{h}^{(i)} + h^{(i)} = \sum_{j=1}^{i} \overline{h}^{(j)}, \quad i = 1, \ldots, N-1, \tag{3.14}$$

which leads to $H \equiv 0$ for $t \geq T$. Once the controls $h^{(i)}$ $(i = 1, \ldots, N)$ are chosen, we solve the original problem (3.1)–(3.4) to get a solution $U = U(t, x)$. Clearly, the exact synchronization condition (3.5) holds for the solution U. Moreover, from the expression (3.14), we see that $h^{(1)} \equiv 0$ and H are with compact support on $[0, T]$, since $\overline{H}$ are with compact support on $[0, T]$. The proof is complete. □

Remark 3.2 In Definition 3.1, the synchronization condition (3.5) should be required for all $t \geq T$. In fact, assuming that (3.5) is realized only at some moment $T > 0$, if we set hereafter $H \equiv 0$ for $t > T$, then the corresponding solution does not satisfy automatically the synchronization condition (3.5) for $t > T$. This is different from the exact null controllability, where the solution vanishes with $H \equiv 0$ for $t \geq T$. To illustrate it, let us consider the following system:

$$\begin{cases} \dfrac{\partial^2 u}{\partial t^2} - \Delta u = 0 & \text{in } \Omega, \\ \dfrac{\partial^2 v}{\partial t^2} - \Delta v = u & \text{in } \Omega, \\ u = 0 & \text{on } \Gamma, \\ v = h & \text{on } \Gamma. \end{cases} \tag{3.15}$$

Since the first equation is separated from the second one, for any given initial data (u_0, u_1), we can first find a solution u. Once u is determined, we look for a boundary control h, such that the solution v to the second equation satisfies the final synchronization conditions

$$t = T : v = u, \quad \frac{\partial v}{\partial t} = \frac{\partial u}{\partial t}. \tag{3.16}$$

If we set $h \equiv 0$ for $t > T$, generally speaking, we can not get $v \equiv u$ for $t \geq T$. So, in order to keep the synchronization for $t \geq T$, we have to maintain the boundary control h in action for $t \geq T$. However, for the sake of applications, it is more interesting to get the exact synchronization by some boundary controls with compact support. Theorems 3.1 and 3.2 guarantee that this can be realized if the coupling matrix A satisfies the conditions of compatibility (3.6).

Remark 3.3 In the reduction of the problem (3.1)–(3.4), we have taken $w^{(i)} = u^{(i+1)} - u^{(i)}$ with $w_0^{(i)} = u_0^{(i+1)} - u_0^{(i)}$ and $w_1^{(i)} = u_1^{(i+1)} - u_1^{(i)}$ for $i = 1, \ldots, N-1$, but it is only a possible choice for proving Theorem 3.2. Since the boundary controls $\overline{h}^{(i)}$ $(i = 1, \ldots, N-1)$ for the exact controllability of the reduced problem depend on the initial data $(w_0^{(i)}, w_1^{(i)})$ $(i = 1, \ldots, N-1)$, we should find a suitable permutation σ of $\{1, 2, \ldots, N\}$, such that, setting $w^{(i)} = u^{\sigma(i+1)} - u^{\sigma(i)}$ for $i = 1, \ldots, N-1$, the corresponding initial data $(w_0^{(i)}, w_1^{(i)})$ $(i = 1, \ldots, N-1)$ have the smallest energy. On the other hand, in the resolution of (3.12), we have chosen $h^{(1)} \equiv 0$ as a possible choice. A good strategy consists in finding some i_0, such that, by setting $h^{(i_0)} \equiv 0$, the final state u has the smallest energy. These problems would be very interesting.

Theorem 3.3 *Assume that the conditions of compatibility* (3.6) *hold. Then the set of the values* (u, u_t) *at the moment* $t = T$ *of the synchronizable state* $u = (t, x)$ *is actually the whole space* $L^2(\Omega) \times H^{-1}(\Omega)$ *as the initial data* U_0 *and* U_1 *vary in the space* $(L^2(\Omega))^N \times (H^{-1}(\Omega))^N$.

Proof For $t \geq T$, the synchronizable state $u = u(t, x)$ defined by (3.5) satisfies the following wave equation with homogenous Dirichlet boundary condition:

$$\begin{cases} \frac{\partial^2 u}{\partial t^2} - \Delta u + \widetilde{a} u = 0 & \text{in } \Omega, \\ u = 0 & \text{on } \Gamma, \end{cases} \tag{3.17}$$

where $\widetilde{a}$ is given by (3.6). Hence, the evolution of the synchronizable state $u = u(t, x)$ with respect to t is completely determined by the values of (u, u_t) at the moment $t = T$:

$$t = T : u = \widehat{u}_0, \quad u_t = \widehat{u}_1. \tag{3.18}$$

Now for any given $(\widehat{u}_0, \widehat{u}_1) \in L^2(\Omega) \times H^{-1}(\Omega)$, by solving the backward problem (3.17)–(3.18) on the time interval $[0, T]$, we get the corresponding solution $u = u(t, x)$ with its value (u, u_t) at $t = 0$,

$$t = 0 : u = u_0, \quad u_t = u_1. \tag{3.19}$$

Then, under the conditions of compatibility (3.6), the function

$$U(t, x) = (u, u \dots, u)^{\mathrm{T}}(t, x) \tag{3.20}$$

is the solution to (3.1)–(3.3) with the null control $H \equiv 0$ and the initial condition

$$t = 0 : U = U_0 = (u_0, u_0 \dots, u_0)^{\mathrm{T}}, \quad U_t = U_1 = (u_1, u_1 \dots, u_1)^{\mathrm{T}}. \tag{3.21}$$

Therefore, from the initial condition (3.21), by solving (3.1)–(3.3) with null boundary controls, we can reach any given synchronizable state $(\widehat{u}_0, \widehat{u}_1)$ at the moment $t = T$. This fact shows that any given state $(\widehat{u}_0, \widehat{u}_1) \in L^2(\Omega) \times H^{-1}(\Omega)$ can be expected to be a synchronizable state. Consequently, the set of the values $(\widehat{u}_0, \widehat{u}_1)$ of the synchronizable state $u = (t, x)$ is actually the whole space $L^2(\Omega) \times H^{-1}(\Omega)$ as the initial data U_0 and U_1 vary in the space $(L^2(\Omega))^N \times (H^{-1}(\Omega))^N$. The proof is complete. □

Definition 3.2 Problem (3.1)–(3.4) is exactly anti-synchronizable at the moment $T > 0$, if for any given initial data $U_0 \in (L^2(\Omega))^N$ and $U_1 \in (H^{-1}(\Omega))^N$, there exist suitable boundary controls given by a part of $H \in (L^2(0, +\infty; L^2(\Gamma_1)))^N$, such that the solution $U = U(t, x)$ to (3.1)–(3.4) satisfies the final condition

$$t \geq T : u^{(1)} \equiv \cdots \equiv u^{(m)} \equiv -u^{(m+1)} \equiv \cdots \equiv -u^{(N)}. \tag{3.22}$$

Theorem 3.4 *Assume that* (3.1)–(3.4) *is exactly anti-synchronizable, but not exactly null controllable. Then the coupling matrix* $A = (a_{ij})$ *should satisfy the following conditions of compatibility*:

$$\begin{cases} \sum_{p=1}^{m} a_{kp} - \sum_{p=m+1}^{N} a_{kp} = \widetilde{a}, & k = 1, \dots, m, \\ \sum_{p=1}^{m} a_{kp} - \sum_{p=m+1}^{N} a_{kp} = -\widetilde{a}, & k = m+1, \dots, N, \end{cases} \tag{3.23}$$

where $\widetilde{a}$ *is a constant independent of* $k = 1, \dots, N$.

Inversely, assume that the conditions of compatibility (3.23) *hold, and then* (3.1)–(3.4) *is exactly anti-synchronizable by means of some boundary controls* H *with compact support and* $h^{(1)} \equiv 0$.

Proof Let us define

$$\widehat{u}^{(i)} = \begin{cases} u^{(i)}, & 1 \leq i \leq m, \\ -u^{(i)}, & m+1 \leq i \leq N \end{cases}$$

and

$$\widehat{a}_{ij} = \begin{cases} a_{ij}, & 1 \le i, j \le m \text{ or } m+1 \le i, j \le N, \\ -a_{ij}, & 1 \le i \le m,\ m+1 \le j \le N \text{ or } m+1 \le i \le N,\ 1 \le j \le m. \end{cases}$$

Then $\widehat{U} = (\widehat{u}^{(1)}, \ldots, \widehat{u}^{(N)})^{\mathrm{T}}$ satisfies (3.1)–(3.4) with the coupling matrix $\widehat{A} = (\widehat{a}_{ij})$ instead of A. By Theorems 3.1 and 3.2, we obtain that

$$\sum_{p=1}^{m} \widehat{a}_{kp} + \sum_{p=m+1}^{N} \widehat{a}_{kp} = \widetilde{a}, \quad k = 1, 2, \ldots, N \tag{3.24}$$

are necessary and sufficient for the exact synchronization of $\widehat{U}$ by means of $N-1$ boundary controls with compact support. Using the definition of the coefficients $\widehat{a}_{ij}$, we see that (3.24) is precisely (3.23). The proof is complete. □

4 Exact Synchronization by Groups

In this section, we will study the exact synchronization by groups. Roughly speaking, let us rearrange the components of U, for example, in two groups, and we look for some boundary controls H, such that $(u^{(1)}, \ldots, u^{(m)})$ and $(u^{(m+1)}, \ldots, u^{(N)})$ are independently synchronized.

Definition 4.1 Problem (3.1)–(3.4) is exactly synchronizable by 2-groups at the moment $T > 0$, if for any given initial data $U_0 \in (L^2(\Omega))^N$ and $U_1 \in (H^{-1}(\Omega))^N$, there exist suitable boundary controls given by a part of $H \in (L^2(0, \infty; L^2(\Gamma_1)))^N$, such that the solution $U = U(t, x)$ to (3.1)–(3.4) satisfies the final condition

$$t \ge T: \begin{cases} u^{(1)} \equiv \cdots \equiv u^{(m)} := u, \\ u^{(m+1)} \equiv \cdots \equiv u^{(N)} := v, \end{cases} \tag{4.1}$$

and $\widetilde{U} = (u, v)^{\mathrm{T}}$ is called to be the synchronizable state by 2-groups.

Our object is to realize the exact synchronization by 2-groups by means of $N-2$ boundary controls. Of course, generally speaking, we can divide the components of U into p groups, and consider the exact synchronization by p-groups. Here we focus our attention only on two groups, but the results obtained in this section can be easily extended to the general case. On the other hand, it is clear that any given exactly synchronizable system is exactly synchronizable by 2-groups. In what follows, we study only the case that the problem is independently synchronizable by 2-groups, and thus the linear independence of components of the synchronizable state $(u, v)^{\mathrm{T}}$ excludes the exact synchronization of (3.1)–(3.4).

Theorem 4.1 *Assume that* (3.1)–(3.4) *is exactly synchronizable by* 2*-groups. Furthermore, assume that at least for some initial data* U_0 *and* U_1*, the synchronizable states* u *and* v *are linearly independent. Then the coupling matrix* $A = (a_{ij})$ *should satisfy the following conditions of compatibility*:

$$\sum_{p=1}^{m} a_{kp} = \sum_{p=1}^{m} a_{lp}, \qquad \sum_{p=m+1}^{N} a_{kp} = \sum_{p=m+1}^{N} a_{lp} \tag{4.2}$$

for $k, l = 1, \ldots, m$ *and* $k, l = m+1, \ldots, N$*, respectively.*

Proof Since (3.1)–(3.4) is exactly synchronizable by 2-groups, for any given initial data U_0 and U_1, there exists a boundary control H, such that (4.1) holds. It follows that for $t \geq T$, we have

$$\begin{cases} \frac{\partial^2 u}{\partial t^2} - \Delta u + (\sum_{p=1}^{m} a_{kp})u + (\sum_{p=m+1}^{N} a_{kp})v = 0 & \text{in } \Omega,\ k = 1, \ldots, m, \\ \frac{\partial^2 v}{\partial t^2} - \Delta v + (\sum_{p=1}^{m} a_{kp})u + (\sum_{p=m+1}^{N} a_{kp})v = 0 & \text{in } \Omega,\ k = m+1, \ldots, N. \end{cases}$$

Therefore, we have

$$t \geq T: \left(\sum_{p=1}^{m} a_{kp} - \sum_{p=1}^{m} a_{lp}\right)u + \left(\sum_{p=m+1}^{N} a_{kp} - \sum_{p=m+1}^{N} a_{lp}\right)v = 0 \quad \text{in } \Omega \tag{4.3}$$

for $k, l = 1, \ldots, m$ and $k, l = m+1, \ldots, N$, respectively. Since at least for some initial data U_0 and U_1, the synchronizable states u and v are linearly independent, (4.2) follows directly from (4.3). □

Theorem 4.2 *Assume that the conditions of compatibility* (4.2) *hold. Then* (3.1)–(3.4) *is exactly synchronizable by* 2*-groups by means of some boundary controls* H *with compact support and* $h^{(1)} \equiv h^{(m+1)} \equiv 0$.

Proof Let

$$\begin{cases} w^{(j)} = u^{(j+1)} - u^{(j)}, & j = 1, \ldots, m-1, \\ w^{(j)} = u^{(j+2)} - u^{(j+1)}, & j = m, \ldots, N-2. \end{cases} \tag{4.4}$$

Then we have

$$\begin{cases} u^{(j)} = \sum_{s=1}^{j-1} w^{(s)} + u^{(1)}, & j = 1, \ldots, m, \\ u^{(j)} = \sum_{s=m}^{j-2} w^{(s)} + u^{(m+1)}, & j = m+1, \ldots, N. \end{cases} \tag{4.5}$$

By (3.1), it is easy to see that for $1 \leq i \leq N-2$, we have

$$\frac{\partial^2 w^{(i)}}{\partial t^2} - \Delta w^{(i)} + \sum_{p=1}^{N} (a_{i+1,p} - a_{ip}) u^{(p)} = 0.$$

By a direct computation, noting (4.2), we have

$$\begin{aligned}
&\sum_{p=1}^{N}(a_{i+1,p}-a_{ip})u^{(p)}\\
&=\sum_{p=1}^{m}(a_{i+1,p}-a_{ip})u^{(p)}+\sum_{p=m+1}^{N}(a_{i+1,p}-a_{ip})u^{(p)}\\
&=\sum_{p=1}^{m}(a_{i+1,p}-a_{ip})\left(\sum_{s=1}^{p-1}w^{(s)}+u^{(1)}\right)\\
&\quad+\sum_{p=m+1}^{N}(a_{i+1,p}-a_{ip})\left(\sum_{s=m}^{p-2}w^{(s)}+u^{(m+1)}\right)\\
&=\sum_{p=1}^{m}(a_{i+1,p}-a_{ip})\sum_{s=1}^{p-1}w^{(s)}+\sum_{p=m+1}^{N}(a_{i+1,p}-a_{ip})\sum_{s=m}^{p-2}w^{(s)}\\
&\quad+\sum_{p=1}^{m}(a_{i+1,p}-a_{ip})u^{(1)}+\sum_{p=m+1}^{N}(a_{i+1,p}-a_{ip})u^{(m+1)}\\
&=\sum_{s=1}^{m-1}\sum_{p=s+1}^{m}(a_{i+1,p}-a_{ip})w^{(s)}+\sum_{s=m}^{N-2}\sum_{p=s+2}^{N}(a_{i+1,p}-a_{ip})w^{(s)}.
\end{aligned}$$

Then

$$\frac{\partial^2 w^{(i)}}{\partial t^2}-\Delta w^{(i)}+\sum_{s=1}^{N-2}\overline{a}_{is}w^{(s)}=0,\quad 1\le i\le N-2, \tag{4.6}$$

where

$$\overline{a}_{is}=\begin{cases}\sum_{p=s+1}^{m}(a_{i+1,p}-a_{ip}), & 1\le i\le N-2,\ 1\le s\le m-1,\\ \sum_{p=s+2}^{N}(a_{i+1,p}-a_{ip}), & 1\le i\le N-2,\ m\le s\le N-2.\end{cases} \tag{4.7}$$

Corresponding to (4.4), for the variable $W=(w^{(1)},\dots,w^{(N-2)})^{\mathrm{T}}$, we put

$$\overline{h}^{(j)}=\begin{cases}h^{(j+1)}-h^{(j)}, & j=1,\dots,m-1,\\ h^{(j+2)}-h^{(j+1)}, & j=m,\dots,N-2,\end{cases} \tag{4.8}$$

and set the new initial data as follows:

$$w_0^{(j)}=\begin{cases}w_0^{(j+1)}-w_0^{(j)}, & j=1,\dots,m-1,\\ w_0^{(j+2)}-w_0^{(j+1)}, & j=m,\dots,N-2,\end{cases} \tag{4.9}$$

$$w_1^{(j)} = \begin{cases} w_1^{(j+1)} - w_1^{(j)}, & j = 1, \dots, m-1, \\ w_1^{(j+2)} - w_1^{(j+1)}, & j = m, \dots, N-2. \end{cases} \tag{4.10}$$

Noting (4.6)–(4.10), we get again a reduced problem (2.2)–(2.5) on the new variable W with $N-2$ components. By Theorem 2.2 (in which $M = N-2$), there exist boundary controls $\overline{H} \in (L^2(0, T; L^2(\Gamma_1)))^{N-2}$, such that the solution $W = W(t, x)$ to the reduced problem (2.2)–(2.5) satisfies the null final condition. Moreover, taking $\overline{H} \equiv 0$ for $t > T$, we have

$$t \geq T : w^{(i)}(t, x) \equiv 0, \quad i = 1, \dots, N-2. \tag{4.11}$$

In order to determine $h^{(i)}$ $(i = 1, \dots, N)$ from (4.8), setting

$$h^{(1)} \equiv h^{(m+1)} \equiv 0, \tag{4.12}$$

we get

$$\begin{cases} h^{(i+1)} = \sum_{j=1}^{i} \overline{h}^{(j)}, & i = 1, \dots, m-1, \\ h^{(i+1)} = \sum_{j=m}^{i-1} \overline{h}^{(j)}, & i = m+1, \dots, N-1, \end{cases} \tag{4.13}$$

which leads to $H \equiv 0$ for $t > T$. Once the controls $h^{(i)}$ $(i = 1, \dots, N)$ are chosen, we solve the original problem (3.1)–(3.4) to get a solution $U = U(t, x)$, which clearly satisfies the final condition (4.1). Thus the proof is complete. □

Theorem 4.3 *Assume that the conditions of compatibility* (4.2) *hold. Then the set of the values* (u, v, u_t, v_t) *of the synchronizable state* $(u, v) = (u(t, x), v(t, x))$ *of* (3.1)–(3.4) *is actually the whole space* $(L^2(\Omega))^2 \times (H^{-1}(\Omega))^2$ *as the initial data* U_0 *and* U_1 *vary in the space* $(L^2(\Omega))^N \times (H^{-1}(\Omega))^N$. *In particular, there exist initial data* (U_0, U_1) *and boundary controls* H *with compact support and* $h^{(1)} \equiv h^{(m+1)} \equiv 0$, *such that the synchronizable states by* 2*-groups* u *and* v *of the problem* (3.1)–(3.4) *are linearly independent.*

Proof Let $\widetilde{A} = (\widetilde{a}_{ij})$ be the 2×2 matrix with the entries

$$\begin{cases} \widetilde{a}_{11} = \sum_{p=1}^{m} a_{kp}, & \widetilde{a}_{12} = \sum_{p=m+1}^{N} a_{kp}, & k = 1, \dots, m, \\ \widetilde{a}_{21} = \sum_{p=1}^{m} a_{kp}, & \widetilde{a}_{22} = \sum_{p=m+1}^{N} a_{kp}, & k = m+1, \dots, N. \end{cases} \tag{4.14}$$

For $t \geq T$, the synchronizable state by 2-groups $\widetilde{U} = (u, v)^{\mathrm{T}}$ defined by (4.1) satisfies the following coupled system of wave equations:

$$\frac{\partial^2}{\partial t^2} \widetilde{U} - \Delta \widetilde{U} + \widetilde{A} \widetilde{U} = 0 \quad \text{in } \Omega \tag{4.15}$$

with the homogeneous boundary condition

$$\widetilde{U} = 0 \quad \text{on } \Gamma. \tag{4.16}$$

Thus, the evolution of $\widetilde{U} = \widetilde{U}(t, x)$ with respect to t is completely determined by the value of $(\widetilde{U}, \widetilde{U}_t)$ at the time $t = T$. In a way similar to that of Theorem 3.3, we get that the set of values $(\widetilde{U}, \widetilde{U}_t)$ at the moment $t = T$ of the synchronizable state by 2-groups $\widetilde{U} = (u, v)^{\mathrm{T}}$ is actually the whole space $(L^2(\Omega))^2 \times (H^{-1}(\Omega))^2$. The proof is complete. □

Remark 4.1 Under the condition that, at least for some initial data U_0 and U_1, the synchronizable states u and v are linearly independent, we have shown that the conditions of compatibility (4.2) are necessary and sufficient for the synchronization by 2-groups of the problem (3.1)–(3.4). These conditions are still sufficient for the synchronization by 2-groups of the problem (3.1)–(3.4) without the linear independence of u and v. But we do not know if they are also necessary in that case. This seems to be an open problem to our knowledge.

5 Exact Null Controllability and Synchronization by Groups

In general, a system of N wave equations is not exactly controllable by means of less N boundary controls (see [7, 10]). By Theorem 3.2, (3.1)–(3.4) is exactly synchronizable by means of $N - 1$ boundary controls. These results are quite logical from the viewpoint of the degree of freedom system and the number of controls. It suggests us to consider the partial controllability for a system of N equations by means of less boundary controls. Since this is still an open problem in the general situation, we would like to weaken our request by asking if it is possible or not, based on the idea of synchronization, to realize the exact null controllability of $N - 2$ components of the solution to a system of N equations by means of $N - 1$ boundary controls. This is the goal of this section, in which we will discuss this problem in a more general situation.

Definition 5.1 Problem (3.1)–(3.4) is exactly null controllable and synchronizable by 2-groups at the moment $T > 0$, if for any given initial data $U_0 \in (L^2(\Omega))^N$ and $U_1 \in (H^{-1}(\Omega))^N$, there exist suitable boundary controls given by a part of $H \in (L^2(0, +\infty; L^2(\Gamma_1)))^N$, such that the solution $U = U(t, x)$ to (3.1)–(3.4) satisfies the final condition

$$t \geq T : u^{(1)} \equiv \cdots \equiv u^{(m)} \equiv 0, \quad u^{(m+1)} \equiv \cdots \equiv u^{(N)} := u, \tag{5.1}$$

and $u = u(t, x)$ is called the partially synchronizable state.

Theorem 5.1 *Assume that the problem* (3.1)–(3.4) *is exactly null controllable and synchronizable by* 2*-groups, but not exactly null controllable. Then the coupling matrix* $A = (a_{ij})$ *should satisfy the following conditions of compatibility*:

$$\begin{cases} \sum_{p=m+1}^{N} a_{kp} = 0, & k = 1, \dots, m, \\ \sum_{p=m+1}^{N} a_{kp} = \sum_{p=m+1}^{N} a_{lp}, & k, l = m+1, \dots, N. \end{cases} \tag{5.2}$$

Proof By the exact null controllability and synchronization by 2-groups, there exist a $T > 0$ and a scalar function u, such that

$$t \geq T : \begin{cases} u^{(k)}(t,x) \equiv 0, & k = 1, \ldots, m, \\ u^{(k)}(t,x) \equiv u(t,x), & k = m+1, \ldots, N. \end{cases}$$

Then for $t \geq T$, we have

$$\begin{cases} (\sum_{p=m+1}^{N} a_{kp})u = 0 & \text{in } \Omega,\ k = 1, \ldots, m, \\ \frac{\partial^2 u}{\partial t^2} - \Delta u + (\sum_{p=m+1}^{N} a_{kp})u = 0 & \text{in } \Omega,\ k = m+1, \ldots, N. \end{cases}$$

Since the problem is not exactly null controllable, we may assume that $u \not\equiv 0$ and this yields the conditions of compatibility (5.2). □

Theorem 5.2 *Assume that the conditions of compatibility* (5.2) *hold. Then the problem* (3.1)–(3.4) *is exactly null controllable and synchronizable by 2-groups by means of some boundary controls H with compact support on* $[0, T]$ *and* $h^{(m+1)} \equiv 0$.

Proof Let

$$\begin{cases} w^{(j)} = u^{(j)}, & j = 1, \ldots, m, \\ w^{(j)} = u^{(j+1)} - u^{(j)}, & j = m+1, \ldots, N-1. \end{cases} \tag{5.3}$$

We have

$$u^{(j)} = \sum_{s=m+1}^{j-1} w^{(s)} + u^{(m+1)}, \quad j = m+1, \ldots, N. \tag{5.4}$$

Then the first m equations of (3.1) become

$$\frac{\partial^2 w^{(i)}}{\partial t^2} - \Delta w^{(i)} + \sum_{p=1}^{m} a_{ip} w^{(p)} + \sum_{p=m+1}^{N} a_{ip} u^{(p)} = 0, \quad i = 1, \ldots, m.$$

Using (5.4) and the first condition in (5.2), we have

$$\begin{aligned} \sum_{p=m+1}^{N} a_{ip} u^{(p)} &= \sum_{p=m+1}^{N} \sum_{s=m+1}^{p-1} a_{ip} w^{(s)} + \left(\sum_{p=m+1}^{N} a_{ip} \right) u^{(m+1)} \\ &= \sum_{p=m+1}^{N-1} \sum_{s=p+1}^{N} a_{is} w^{(p)}. \end{aligned}$$

Then

$$\frac{\partial^2 w^{(i)}}{\partial t^2} - \Delta w^{(i)} + \sum_{p=1}^{N-1} \overline{a}_{ip} w^{(p)} = 0, \quad 1 \le i \le m, \tag{5.5}$$

where

$$\overline{a}_{ip} = \begin{cases} a_{ip}, & 1 \le i \le m, \ 1 \le p \le m, \\ \sum_{s=p+1}^{N} a_{is}, & 1 \le i \le m, \ m+1 \le p \le N-1. \end{cases} \tag{5.6}$$

Next, by (3.1) and noting the first part of (5.3), for $m+1 \le i \le N-1$, we have

$$\frac{\partial^2 w^{(i)}}{\partial t^2} - \Delta w^{(i)} + \sum_{p=1}^{m} (a_{i+1,p} - a_{ip}) w^{(p)} + \sum_{p=m+1}^{N} (a_{i+1,p} - a_{ip}) u^{(p)} = 0.$$

By a direct computation, noting the second condition in (5.2), we have

$$\begin{aligned} &\sum_{p=m+1}^{N} (a_{i+1,p} - a_{ip}) u^{(p)} \\ &= \sum_{p=m+1}^{N} (a_{i+1,p} - a_{ip}) \left(\sum_{s=m+1}^{p-1} w^{(s)} + u^{(m+1)} \right) \\ &= \sum_{p=m+1}^{N} (a_{i+1,p} - a_{ip}) \sum_{s=m+1}^{p-1} w^{(s)} + \sum_{p=m+1}^{N} (a_{i+1,p} - a_{ip}) u^{(m+1)} \\ &= \sum_{s=m+1}^{N-1} \sum_{p=s+1}^{N} (a_{i+1,p} - a_{ip}) w^{(s)}. \end{aligned}$$

Then

$$\frac{\partial^2 w^{(i)}}{\partial t^2} - \Delta w^{(i)} + \sum_{p=1}^{N-1} \overline{a}_{ip} w^{(p)} = 0, \quad m+1 \le i \le N-1, \tag{5.7}$$

where

$$\overline{a}_{ip} = \begin{cases} a_{i+1,p} - a_{ip}, & m+1 \le i \le N-1, \ 1 \le p \le m, \\ \sum_{s=p+1}^{N} (a_{i+1,s} - a_{is}), & m+1 \le i \le N-1, \ m+1 \le p \le N-1. \end{cases} \tag{5.8}$$

Corresponding to (5.3), for the variable $W = (w^{(1)}, \dots, w^{(N-1)})^{\mathrm{T}}$, we put

$$\overline{h}^{(i)} = \begin{cases} h^{(i)}, & 1 \le i \le m, \\ h^{(i+1)} - h^{(i)}, & m+1 \le i \le N-1, \end{cases} \tag{5.9}$$

and set the new initial data as follows:

$$w_0^{(i)} = \begin{cases} u_0^{(i)}, & 1 \le i \le m, \\ u_0^{(i+1)} - u_0^{(i)}, & m+1 \le i \le N-1, \end{cases} \tag{5.10}$$

$$w_1^{(i)} = \begin{cases} u_1^{(i)}, & 1 \le i \le m, \\ u_1^{(i+1)} - u_1^{(i)}, & m+1 \le i \le N-1. \end{cases} \tag{5.11}$$

Noting (5.5), (5.7) and (5.10)–(5.11), we get again a reduced problem (2.2)–(2.5) on the new variable W with $N-1$ components. By Theorem 2.2 (in which $M = N-1$), there exist controls $\overline{H} \in (L^2(0, T; L^2(\Gamma_1)))^{N-1}$, such that the solution $W = W(t, x)$ to the reduced problem (2.2)–(2.5) satisfies the null final condition. Moreover, taking $\overline{H} \equiv 0$ for $t > T$, we have

$$t \ge T : w^{(i)}(t, x) \equiv 0, \quad i = 1, \dots, N-1. \tag{5.12}$$

In order to determine $h^{(i)}$ $(i = 1, \dots, N)$ from (5.9), setting $h^{(m+1)} \equiv 0$, we get

$$\begin{cases} h^{(i)} = \overline{h}^{(i)}, & i = 1, \dots, m, \\ h^{(i+1)} = \sum_{j=m+1}^{i} \overline{h}^{(j)}, & i = m+1, \dots, N-1, \end{cases} \tag{5.13}$$

which leads to $H \equiv 0$ for $t > T$. Once the controls $h^{(i)}$ $(i = 1, \dots, N)$ are chosen, we solve the original problem (3.1)–(3.4) to get a solution $U = U(t, x)$, which clearly satisfies the final condition (5.1). Then the proof is complete. □

Remark 5.1 Let

$$\sum_{p=m+1}^{N} a_{kp} = \widetilde{a}, \quad k = m+1, \dots, N, \tag{5.14}$$

where $\widetilde{a}$ is a constant independent of $k = m+1, \dots, N$ (see (5.2)). For $t \ge T$, the partially synchronizable state $u = u(t, x)$ satisfies the following wave equation:

$$\frac{\partial^2 u}{\partial t^2} - \Delta u + \widetilde{a} u = 0 \quad \text{in } \Omega \tag{5.15}$$

with the homogeneous boundary condition

$$u = 0 \quad \text{on } \Gamma. \tag{5.16}$$

Hence the evolution of $u = u(t,x)$ with respect to t can be completely determined by its initial values (u, u_t) at the moment $t = T$. Moreover, in a way similar to that of Theorem 3.3, the set of values (u, u_t) at the moment $t = T$ of the partially synchronizable state u is actually the whole space $L^2(\Omega) \times H^{-1}(\Omega)$ as the initial values U_0 and U_1 vary in the space $(L^2(\Omega))^N \times (H^{-1}(\Omega))^N$.

Remark 5.2 Taking $m = N-2$ in Theorem 5.2, under the corresponding conditions of compatibility, we can use $N-1$ (instead of $N-2$!) boundary controls to realize the exact null controllability for $N-2$ state variables in U.

6 Approximation of the Final State for a System of Vibrating Strings

Once the exact synchronization is realized at the moment T, the final state $u = u(t,x)$ for $t \geq T$ will be governed by a corresponding wave equation with homogeneous Dirichlet boundary condition (see also (3.17))

$$\begin{cases} \frac{\partial^2 u}{\partial t^2} - \Delta u + \widetilde{a} u = 0 & \text{in } \Omega, \\ u = 0 & \text{on } \Gamma, \end{cases} \tag{6.1}$$

where the constant $\widetilde{a}$ is given by (3.6). However, for lack of the value of the final state at the moment T, we do not know how the final state u will evolve henceforth. The goal of this section is to give an approximation of the final state $u = u(t,x)$ for $t \geq T$ for a coupled system of vibrating strings with a perturbation of a synchronizable state as the initial data.

Let $0 < a < 1$. Consider the following 1-dimensional problem:

$$\begin{cases} u'' - u_{xx} - av = 0, & 0 < x < \pi, \\ v'' - v_{xx} - au = 0, & 0 < x < \pi, \\ u(t,0) = u(t,\pi) = 0, & \\ v(t,0) = 0, & v(t,\pi) = h(t), \\ t = 0: (u,v) = (u_0, v_0), & (u_t, v_t) = (u_1, v_1). \end{cases} \tag{6.2}$$

Using the spectral analysis as in [8], we can prove that (6.2) is asymptotically controllable, but not exactly controllable. By Theorem 3.2, this problem is exactly synchronizable by means of boundary control h. More precisely, by setting

$$y = v - u, \tag{6.3}$$

for $T > 2\pi$, there exists a boundary control $h \in L^2(0,T)$, which realizes the exact null controllability at the moment T for the following reduced problem:

$$\begin{cases} y'' - y_{xx} + ay = 0, & 0 < x < \pi, \\ y(t,0) = 0, & y(t,\pi) = h(t), \\ t = 0: y = v_0 - u_0, & y' = v_1 - u_1. \end{cases} \tag{6.4}$$

Moreover, by the HUM method (see [9]) or the moment method (see [6]), there exists a positive constant $C > 0$, such that

$$\|h\|_{L^2(0,T)} \le C\|(v_0 - u_0, v_1 - u_1)\|_{L^2(0,\pi)\times H^{-1}(0,\pi)}. \tag{6.5}$$

In what follows, we will give an expression of the final state $u = v$ of the problem (6.2) for $t \ge T$. To this end, setting

$$w = u + v, \qquad w_0 = u_0 + v_0, \qquad w_1 = u_1 + v_1, \tag{6.6}$$

we consider the following anti-synchronization problem:

$$\begin{cases} w'' - w_{xx} - aw = 0, & 0 < x < \pi, \\ w(t,0) = 0, & w(t,\pi) = h(t), \\ t = 0: w = w_0, & w' = w_1. \end{cases} \tag{6.7}$$

Assume that $w_0 \in L^2(0,\pi)$ and $w_1 \in H^{-1}(0,\pi)$, whose coefficients a_j^0 and b_j^0 $(j \ge 1)$ on the orthonormal basis $\left(\sqrt{\frac{2}{\pi}}\sin(jx)\right)_{j\ge1}$ in $L^2(0,\pi)$ and $\left(\sqrt{\frac{2}{\pi}}j \times \sin(jx)\right)_{j\ge1}$ in $H^{-1}(0,\pi)$ are respectively given by

$$a_j^0 = \sqrt{\frac{2}{\pi}}\int_0^\pi w_0(x)\sin(jx)\mathrm{d}x, \qquad jb_j^0 = \sqrt{\frac{2}{\pi}}\int_0^\pi w_1(x)\sin(jx)\mathrm{d}x. \tag{6.8}$$

Correspondingly, for any given $t \ge T$, the coefficients $a_j(t)$ and $b_j(t)$ $(j \ge 1)$ of the final state $u(t,x)$ and $u'(t,x)$ on the orthonormal basis $\left(\sqrt{\frac{2}{\pi}}\sin(jx)\right)_{j\ge1}$ in $L^2(0,\pi)$ and $\left(\sqrt{\frac{2}{\pi}}j\sin(jx)\right)_{j\ge1}$ in $H^{-1}(0,\pi)$ are respectively given by

$$a_j(t) = \sqrt{\frac{2}{\pi}}\int_0^\pi u(t,x)\sin(jx)\mathrm{d}x, \qquad jb_j(t) = \sqrt{\frac{2}{\pi}}\int_0^\pi u'(t,x)\sin(jx)\mathrm{d}x. \tag{6.9}$$

Now let $\mu_j = \sqrt{j^2 - a}$ $(j \ge 1)$. Multiplying the equation in (6.7) by $\sin(\mu_j s) \times \sin(jx)$ and integrating with respect to s and x on $[0,t] \times [0,\pi]$, by integration by parts, we get

$$\left[\int_0^\pi w'(s,x)\sin(\mu_j s)\sin(jx)\mathrm{d}x\right]_0^t - \left[\mu_j\int_0^\pi w(s,x)\cos(\mu_j s)\sin(jx)\mathrm{d}x\right]_0^t$$

$$-\left[\int_0^t w_x(s,x)\sin(\mu_j s)\sin(jx)\mathrm{d}s\right]_0^\pi + \left[j\int_0^t w(s,x)\sin(\mu_j s)\cos(jx)\mathrm{d}s\right]_0^\pi$$

$$+\left(-\mu_j^2+j^2-a\right)\int_0^\pi\int_0^t w(s,x)\sin(\mu_j s)\sin(jx)\mathrm{d}s\mathrm{d}x = 0.$$

It follows that

$$\sin(\mu_j t)\int_0^\pi w'(t,x)\sin(jx)\mathrm{d}x - \mu_j\cos(\mu_j t)\int_0^\pi w(t,x)\sin(jx)\mathrm{d}x$$

$$+\mu_j\int_0^\pi w_0(x)\sin(jx)\mathrm{d}x + (-1)^j j\int_0^t h(s)\sin(\mu_j s)\mathrm{d}s = 0. \tag{6.10}$$

Noting that $w(t,x)=2u(t,x)$ and $w'(t,x)=2u'(t,x)$ for $t\geq T$, by (6.9), we have

$$\int_0^\pi w(t,x)\sin(jx)\mathrm{d}x = 2\int_0^\pi u(t,x)\sin(jx)\mathrm{d}x = \sqrt{2\pi}a_j(t), \tag{6.11}$$

$$\int_0^\pi w'(t,x)\sin(jx)\mathrm{d}x = 2\int_0^\pi u'(t,x)\sin(jx)\mathrm{d}x = \sqrt{2\pi}jb_j(t). \tag{6.12}$$

Inserting (6.12)–(6.11) into (6.10) and noting (6.8), we get

$$a_j(t)\mu_j\cos(\mu_j t) - jb_j(t)\sin(\mu_j t) = \frac{1}{2}\mu_j a_j^0 + (-1)^j j\sqrt{\frac{1}{2\pi}}\int_0^t h(s)\sin(\mu_j s)\mathrm{d}s. \tag{6.13}$$

Similarly, multiplying the equation in (6.7) by $\cos(\mu_j s)\sin(jx)$ and integrating with respect to s and x on $[0,t]\times[0,\pi]$, by integration by parts, we get

$$\left[\int_0^\pi w'(s,x)\cos(\mu_j s)\sin(jx)\mathrm{d}x\right]_0^t + \left[\mu_j\int_0^\pi w(s,x)\sin(\mu_j s)\sin(jx)\mathrm{d}x\right]_0^t$$

$$-\left[\int_0^t w_x(s,x)\cos(\mu_j s)\sin(jx)\mathrm{d}s\right]_0^\pi + \left[j\int_0^t w(s,x)\cos(\mu_j s)\cos(jx)\mathrm{d}s\right]_0^\pi$$

$$+\left(-\mu_j^2+j^2-a\right)\int_0^\pi\int_0^t w(s,x)\cos(\mu_j s)\sin(jx)\mathrm{d}s\mathrm{d}x = 0.$$

It follows that

$$\cos(\mu_j t)\int_0^\pi w'(t,x)\sin(jx)\mathrm{d}x + \mu_j\sin(\mu_j t)\int_0^\pi w(t,x)\sin(jx)\mathrm{d}x$$

$$-\int_0^\pi w_1(x)\sin(jx)\mathrm{d}x + (-1)^j j\int_0^t h(s)\cos(\mu_j s)\mathrm{d}s = 0. \tag{6.14}$$

Then noting (6.12)–(6.11) and (6.8), we get

$$a_j(t)\mu_j\sin(\mu_j t) + jb_j(t)\cos(\mu_j t) = \frac{1}{2}jb_j^0 - (-1)^j j\sqrt{\frac{1}{2\pi}}\int_0^t h(s)\cos(\mu_j s)\mathrm{d}s. \tag{6.15}$$

It follows from (6.13) and (6.15) that

$$a_j(t) = \frac{\mu_j a_j^0\cos(\mu_j t) + jb_j^0\sin(\mu_j t)}{2\mu_j} + (-1)^j\frac{j}{\mu_j}\sqrt{\frac{1}{2\pi}}\int_0^t h(s)\sin\big[\mu_j(s-t)\big]\mathrm{d}s, \tag{6.16}$$

$$b_j(t) = \frac{-\mu_j a_j^0\sin(\mu_j t) + jb_j^0\cos(\mu_j t)}{2j} - (-1)^j\sqrt{\frac{1}{2\pi}}\int_0^t h(s)\cos\big[\mu_j(s-t)\big]\mathrm{d}s. \tag{6.17}$$

Now assume that (v_0, v_1) is a small perturbation of (u_0, u_1), so that by (6.5), the optimal control h is small in $L^2(0, T)$. Then

$$\begin{cases} \widetilde{a}_j(t) = \frac{\mu_j a_j^0\cos(\mu_j t)+jb_j^0\sin(\mu_j t)}{2\mu_j}, \\ \widetilde{b}_j(t) = \frac{-\mu_j a_j^0\sin(\mu_j t)+jb_j^0\cos(\mu_j t)}{2j} \end{cases} \tag{6.18}$$

provide an approximation of the coefficients $a_j(t)$ and $b_j(t)$, respectively. Indeed, let

$$\widetilde{u}(t,x) = \sum_{j=1}^{+\infty}\widetilde{a}_j(t)\sin(jx), \qquad \widetilde{u}'(t,x) = \sum_{j=1}^{+\infty} j\widetilde{b}_j(t)\sin(jx). \tag{6.19}$$

$(\widetilde{u}, \widetilde{u}')$ would be a good approximation of the final state (u, u') for $t \geq T$. In fact, we have the following result.

Theorem 6.1 *Let $T > 2\pi$. Assume that*

$$(u_0,\ u_1) \in L^2(0,\pi)\times H^{-1}(0,\pi), \qquad (v_0,\ v_1)\in L^2(0,\pi)\times H^{-1}(0,\pi). \tag{6.20}$$

Then for all $t \geq T$, we have

$$\begin{aligned} &\big\|\big(u(t,\cdot) - \widetilde{u}(t,\cdot),\ u'(t,\cdot) - \widetilde{u}'(t,\cdot)\big)\big\|_{L^2(0,\pi)\times H^{-1}(0,\pi)} \\ &\quad\leq C\|(v_0 - u_0, v_1 - u_1)\|_{L^2(0,\pi)\times H^{-1}(0,\pi)}. \end{aligned} \tag{6.21}$$

Proof Define

$$s_j(t) = \int_0^t h(s)\sin[\mu_j(s-t)]ds, \qquad c_j(t) = \int_0^t h(s)\cos[\mu_j(s-t)]ds. \quad (6.22)$$

Noting that the reals $\{\mu_j\}_{j\geq 1}$ are distinct, and for all $j \geq 1$, we have the following gap condition:

$$\mu_{j+1} - \mu_j = \frac{2j+1}{\sqrt{(j+1)^2 - a} + \sqrt{j^2 - a}} \geq \frac{2j+1}{2\sqrt{(j+1)^2 - a}} \geq \frac{3}{2\sqrt{4-a}} > 0, \quad (6.23)$$

where the last inequality is due to the growth of the function $x \to \frac{2x+1}{2\sqrt{(x+1)^2 - a}}$ $(x \geq 0)$. Then for any fixed $t \geq T$, the system $\{\sin[\mu_j(s-t)], \cos[\mu_j(s-t)]\}_{j\in\mathbb{N}}$ is a Riesz sequence in $L^2(0,T)$. Consequently, there exists a positive constant $C > 0$ independent of t, such that the following Bessel's inequality holds for all $t \geq T$ (see [5, 15]):

$$\sum_{j=1}^{+\infty}\big(|s_j(t)|^2 + |c_j(t)|^2\big) \leq C\|h\|^2_{L^2(0,T)} \leq C\|(v_0 - u_0, v_1 - u_1)\|^2_{L^2(0,\pi)\times H^{-1}(0,\pi)}, \quad (6.24)$$

where the last inequality is due to (6.5). Then, it follows from (6.16)–(6.17) that

$$\sum_{j=1}^{+\infty}\big(|a_j(t) - \widetilde{a}_j(t)|^2 + |b_j(t) - \widetilde{b}_j(t)|^2\big)$$

$$\leq C\sum_{j=1}^{+\infty}\big(|s_j(t)|^2 + |c_j(t)|^2\big) \leq C\|(v_0 - u_0, v_1 - u_1)\|^2_{L^2(0,\pi)\times H^{-1}(0,\pi)}, \quad (6.25)$$

which yields (6.21). The proof is complete. □

Remark 6.1 Let (v_0, v_1) be a perturbation of (u_0, u_1). Then the norm of their difference $\|(v_0 - u_0, v_1 - u_1)\|_{L^2(0,\pi)\times H^{-1}(0,\pi)}$ is a small quantity, so that (6.21) shows that $(\widetilde{u}, \widetilde{u}')$ is indeed a good approximation of the final state (u, u'). Furthermore, noting that $\frac{j}{\mu_j} \sim 1$ for j large enough, we have

$$|\widetilde{a}_j(t)|^2 + |\widetilde{b}_j(t)|^2 \sim \frac{1}{4}\big(|a_j^0|^2 + |b_j^0|^2\big). \quad (6.26)$$

Noting that a_j^0 $(j \geq 1)$ are the coefficients of $w_0 = u_0 + v_0$ in $L^2(0,\pi)$ and b_j^0 $(j \geq 1)$ are the coefficients of $w_1 = u_1 + v_1$ in $H^{-1}(0,\pi)$, (6.26) shows that the approximate final state $(\widetilde{u}, \widetilde{u}')$ has the same norm as that of the average of the initial data for high frequencies.

References

1. Alabau-Boussouira, F.: A two-level energy method for indirect boundary observability and controllability of weakly coupled hyperbolic systems. SIAM J. Control Optim. **42**, 871–906 (2003)
2. Fujisaka, H., Yamada, T.: Stability theory of synchronized motion in coupled-oscillator systems. Prog. Theor. Phys. **69**, 32–47 (1983)
3. Garofalo, N., Lin, F.: Unique continuation for elliptic operators: a geometric-variational approach. Commun. Pure Appl. Math. **40**, 347–366 (1987)
4. Huygens, C., Horologium Oscillatorium Sive de Motu Pendulorum ad Horologia Aptato Demonstrationes Geometricae, Apud F. Muguet, Parisiis (1673)
5. Komornik, V., Loreti, P.: Fourier Series in Control Theory. Springer, New York (2005)
6. Krabs, W.: On Moment Theory and Controllability of One-dimensional Vibrating Systems and Heating Processes. Lecture Notes in Control and Information Sciences, vol. 173. Springer, Berlin (1992)
7. Yu, L.: Exact boundary controllability for a kind of second-order quasilinear hyperbolic systems and its applications. Math. Methods Appl. Sci. **33**, 273–286 (2010)
8. Li, T.T., Rao, B.P.: Strong (weak) exact controllability and strong (weak) exact observability for quasilinear hyperbolic systems. Chin. Ann. Math. **31B**(5), 723–742 (2010)
9. Li, T.T., Rao, B.P.: Asymptotic controllability for linear hyperbolic systems. Asymptot. Anal. **72**, 169–187 (2011)
10. Lions, J.L., Exacte, C.: Perturbations et Stabilisation de Systèms Distribués vol. 1. Masson, Paris (1988)
11. Liu, Z., Rao, B.P.: A spectral approach to the indirect boundary control of a system of weakly coupled wave equations. Discrete Contin. Dyn. Syst. **23**, 399–414 (2009)
12. Loreti, P., Rao, B.P.: Optimal energy decay rate for partially damped systems by spectral compensation. SIAM J. Control Optim. **45**, 1612–1632 (2006)
13. Mehrenberger, M.: Observability of coupled systems. Acta Math. Hung. **103**, 321–348 (2004)
14. Wang, K.: Exact boundary controllability for a kind of second-order quasilinear hyperbolic systems. Chin. Ann. Math. **32B**(6), 803–822 (2011)
15. Young, R.: An Introduction to Nonharmonic Fourier Series. Academic Press, New York (1980)

Mixing Monte-Carlo and Partial Differential Equations for Pricing Options

Tobias Lipp, Grégoire Loeper, and Olivier Pironneau

Abstract There is a need for very fast option pricers when the financial objects are modeled by complex systems of stochastic differential equations. Here the authors investigate option pricers based on mixed Monte-Carlo partial differential solvers for stochastic volatility models such as Heston's. It is found that orders of magnitude in speed are gained on full Monte-Carlo algorithms by solving all equations but one by a Monte-Carlo method, and pricing the underlying asset by a partial differential equation with random coefficients, derived by Itô calculus. This strategy is investigated for vanilla options, barrier options and American options with stochastic volatilities and jumps optionally.

Keywords Monte-Carlo · Partial differential equations · Heston model · Financial mathematics · Option pricing

Mathematics Subject Classification 91B28 · 65L60 · 82B31

1 Introduction

Since the pioneering work has been achieved by Phelim Boyle [6], Monte-Carlo (or MC for short) methods introduced and shaped financial mathematics as barely any other method can compare. They are often appreciated for their flexibility and applicability in high dimensions, although they bear as well a number of drawbacks: error terms are probabilistic and a high level of accuracy can be computationally burdensome to achieve. In low dimensions, deterministic methods as quadrature and

T. Lipp (✉) · O. Pironneau
LJLL-UPMC, Boite 187, Place Jussieu, 75252 Paris cedex 5, France
e-mail: tobias.lipp@web.de

O. Pironneau
e-mail: Olivier.Pironneau@upmc.fr

G. Loeper
BNP-Paribas, 20 Boulevard des Italiens, 75009 Paris, France
e-mail: gregoire.loeper@bnpparibas.com

P.G. Ciarlet et al. (eds.), *Partial Differential Equations: Theory, Control and Approximation*, DOI 10.1007/978-3-642-41401-5_13,

quadrature based methods are strong competitors. They allow deterministic error estimations and give precise results.

We propose several methods for pricing basket options in a Black-Scholes framework. The methods are based on a combination of Monte-Carlo, quadrature and partial differential equations (or PDE for short) methods. The key idea was studied by two of the authors a few years ago in [14], and it tries to uncouple the underlying system of stochastic differential equations (or SDE for short), and then applies the last-mentioned methods appropriately.

In Sect. 2, we begin with a numerical assessment on the use of Monte-Carlo methods to generate boundary conditions for stochastic volatility models, but this is a side remark independent of what follows.

The way of mixing MC and PDE for stochastic volatility models is formulated in Sect. 3. A numerical evaluation of the method is made by using closed form solutions to the PDE. In Sects. 6 and 4, the method is extended to the case of American options and to the case where the underlying asset is modeled with jump-diffusion processes.

In Sect. 5, a method reducing the number of samples is given based on the smooth dependence of the option price on the volatility.

Finally, in Sect. 7, the strategy is extended to multidimensional problems like basket options, and numerical results are also given.

Moreover, some related results can be found in [5, 8, 9, 16–18].

2 Monte-Carlo Algorithm to Generate Boundary Conditions for the PDE

The diffusion process that we have chosen for our examples is the Heston stochastic volatility model (see [12]). Under a risk neutral probability, the risky asset S_t and the volatility σ_t follow the diffusion process

$$\mathrm{d}S_t = S_t\big(r\mathrm{d}t + \sigma_t \mathrm{d}W_t^1\big), \tag{2.1}$$

$$\mathrm{d}v_t = k(\theta - v_t)\mathrm{d}t + \delta\sqrt{v_t}\mathrm{d}W_t^2, \tag{2.2}$$

and the put option price is given by

$$P_t = \mathrm{e}^{-r(T-t)}\mathbb{E}\big[(K - S_T)^+ | S_t, v_t\big], \tag{2.3}$$

where $v_t = \sigma_t^2$, $\mathbb{E}(\mathrm{d}W_t^1 \cdot \mathrm{d}W_t^2) = \rho \mathrm{d}t$, $\mathbb{E}(\cdot)$ is the expectation with respect to the risk neutral measure, and r is the interest rate on a risk less commodity.

The pair (W^1, W^2) is a two-dimensional correlated Brownian motion, with the correlation between the two components being equal to ρ. As it is usually observed in equity option markets, options with low strikes have an implied volatility higher than that of options at the money or with high strikes, and it is known as the smile. This phenomenon can be reproduced in the model by choosing a negative value of ρ.

The time is discretized into N steps of length δt. Denoting by T the maturity of the option, we have $T = N\delta t$. Full Monte-Carlo simulation (see [10]) consists in a time loop starting at S_0, $v_0 = \sigma_0^2$ of

$$v_{i+1} = v_i + k(\theta - v_i)\delta t + \sigma_i\sqrt{\delta t}N_{0,1}^2\delta \quad \text{with } \sigma_i = \sqrt{v_i}, \tag{2.4}$$

$$S_{i+1} = S_i\big(1 + r\delta t + \sigma_i\sqrt{\delta t}\big(N_{0,1}^1\rho + N_{0,1}^2\sqrt{1-\rho^2}\big)\big), \tag{2.5}$$

where $N_{0,1}^j$ $(j = 1, 2)$ are realizations of two independent normal Gaussian variables. Then set $P_0 = \frac{e^{-rT}}{M}\sum(K - S_N^m)^+$, where $\{S_N^m\}_{m=1}^M$ are M realizations of S_N.

The method is slow, and at least 300000 samples are necessary for a precision of 0.1 %. Of course acceleration methods exist (quasi-Monte-Carlo, multi-level Monte-Carlo etc.), but alternatively, we can use the PDE derived by Itô calculus for u below and set $P_0 = u(S_0, v_0, T)$.

If the return to volatility is 0 (i.e., zero risk premium on the volatility (see [1])), then $u(S, y, \tau)$ is given by

$$\begin{aligned}
&\partial_\tau u - \frac{yS^2}{2}\partial_{SS}u - \rho\lambda Sy\partial_{Sy}u - \frac{\lambda^2 y}{2}\partial_{yy}u - rS\partial_S u - k(\theta - y)\partial_y u + ru = 0,\\
&u(S, y, 0) = (K - S)^+.
\end{aligned} \tag{2.6}$$

Now instead of integrating (2.6) on $\mathbb{R}^+ \times \mathbb{R}^+ \times (0, T)$, let us integrate it on $\Omega \times (0, T)$, $\Omega \subset \mathbb{R}^+ \times \mathbb{R}^+$, and add Dirichlet conditions on $\partial\Omega$ computed with MC by solving (2.4)–(2.5).

Notice that this domain reduction does not change the numerical complexity of the problem. Indeed to reach a precision ε with the PDE, one needs at least $O(\varepsilon^{-3})$ operations to compute the option at all points of a grid of size ε with a time step of size ε. Monte-Carlo needs $O(\varepsilon^{-2})$ per point S_0, v_0, and there are $O(\varepsilon^{-1})$ points on the artificial boundary, when the number of discretization points in the full domain is $O(\varepsilon^{-2})$. However, the computation shown in Fig. 1 validates the methodology, and it may be attractive to use it to obtain more precision on a small domain.

3 Monte-Carlo Mixed with a 1-Dimensional PDE

Let us rewrite (2.1) as

$$dS_t = S_t\big[rdt + \sigma_t\sqrt{1-\rho^2}d\widetilde{W}_t^{(1)} + \sigma_t\rho d\widetilde{W}_t^{(2)}\big], \tag{3.1}$$

where $\widetilde{W}_t^1$, $\widetilde{W}_t^2$ are now independent Brownian motions.

(a) The computational domain is $(0, y_{\max}) \times (0, v_{\max})$ with Neumann condition at $v = v_{\max}$.

(b) The computational domain is the half circle shown on (a); the Dirichlet boundary condition on the circle is obtained by a spline approximation (shown at the bottom) of the solution to Heston's model solved by MC on a few points on the circle.

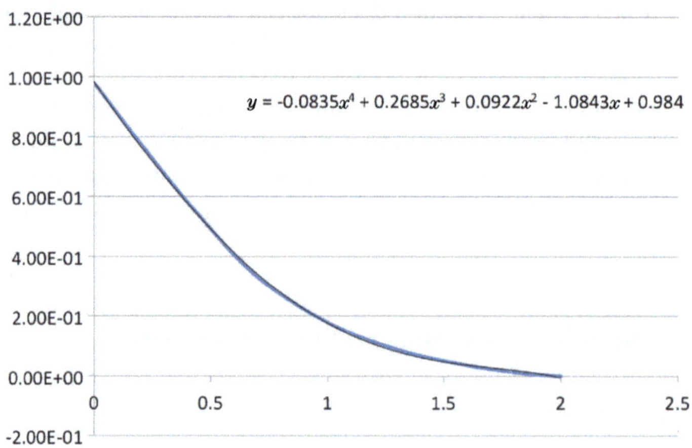

(c) Computation result.

Fig. 1 Put option with Heston's model computed by solving the PDE by implicit Euler + FEM using the public domain package freefem++ (see [11])

Drawing a trajectory of v_t by (2.4), with the same δt and the same discrete trajectory $W^{(2)}_{i+1} = W^{(2)}_i + N^2_{0,1}\sqrt{\delta t}$, we consider

$$\mathrm{d}S_t = S_t\big[\mu_t \mathrm{d}t + \sigma_t\sqrt{1-\rho^2}\mathrm{d}\widetilde{W}^{(1)}_t\big], \tag{3.2}$$

$$\mu_t = r + \rho\sigma_t \frac{W^{(2)}_{i+1} - W^{(2)}_i}{\delta t} - \frac{1}{2}\rho^2\sigma_t^2, \quad t \in [t_i, t_{i+1}[. \tag{3.3}$$

Proposition 3.1 *As $\delta t \to 0$, S_t given by* (2.4) *and* (3.2)–(3.3) *converges to the solution to Heston's model* (2.1)–(2.2). *Moreover, the put $P = \mathrm{e}^{-rT}\mathbb{E}(K - S_T)^+$ is also the expected value of $u(S_0, 0)$, with u given by*

$$\partial_t u + \frac{1}{2}(1-\rho^2)\sigma_t^2 S^2 \partial_{SS} u + S\mu_t \partial_S u - ru = 0, \quad u(S,T) = (K-S)^+ \tag{3.4}$$

with σ_t given by (2.4) *and μ_t given by* (3.3).

Proof By Itô's formula, we have

$$\begin{aligned}
\mathrm{d}\log(S_t) &= \frac{\mathrm{d}S_t}{S_t} + \frac{1}{2}(\log S)''\big(S_t^2\sigma_t^2(1-\rho^2)\mathrm{d}t\big) = \frac{\mathrm{d}S_t}{S_t} - \frac{\sigma_t^2}{2}(1-\rho^2)\mathrm{d}t \\
&= \mu_t \mathrm{d}t + \sqrt{1-\rho^2}\sigma_t \mathrm{d}\widetilde{W}^{(1)}_t - (1-\rho^2)\frac{\sigma_t^2}{2}\mathrm{d}t \\
&\approx r\delta t + \rho\sigma_t \delta W^{(2)}_t - \frac{\rho^2\sigma_t^2}{2}\delta t + \sqrt{1-\rho^2}\sigma_t\delta\widetilde{W}^{(1)}_t - (1-\rho^2)\frac{\sigma_t^2}{2}\delta t \\
&\approx r\mathrm{d}t + \rho\sigma_t \mathrm{d}W^{(2)}_t + \sqrt{1-\rho^2}\sigma_t \mathrm{d}\widetilde{W}^{(1)}_t - \frac{\sigma_t^2}{2}\mathrm{d}t.
\end{aligned} \tag{3.5}$$

Consequently,

$$S_t = S_0 \exp\bigg(\int_0^t \mu_t \mathrm{d}t + \int_0^t \sqrt{1-\rho^2}\sigma_t \mathrm{d}W^{(1)}_t - \int_0^t \frac{1}{2}(1-\rho^2)\sigma_t^2 \mathrm{d}t\bigg). \tag{3.6}$$

□

Proposition 3.2 *If we restrict the MC samples to those that give $0 < \sigma_m \le \sigma_t \le \sigma_M$, for some given σ_m, σ_M, then equations* (2.4) *and* (3.3)–(3.4) *are well-posed.*

Proof Let

$$\Lambda_\tau = \int_{T-\tau}^{T} \mu_\xi \mathrm{d}\xi, \quad y = \frac{S}{K}\mathrm{e}^{\Lambda(\tau)}. \tag{3.7}$$

Then $u(t,S) = v(T-t, \frac{S}{K}\mathrm{e}^{\Lambda(\tau)})$, where v is the solution to

$$\partial_\tau v - \frac{1}{2}(1-\rho^2)\sigma_{T-\tau}^2 y^2 \partial_{yy} v = 0, \quad v(0,y) = (1-y)^+. \tag{3.8}$$

Table 1 Precision versus ρ

ρ	−0.5	0	0.5	0.9
Heston MC	11.135	10.399	9.587	8.960
Heston MC+BS	11.102	10.391	9.718	8.977
Speed-up	42	44	42	42

If $0 < \sigma_m \le \sigma_t \le \sigma_M$ almost surely and for all t, then the solution exists in the sense of Barth et al. [3]. □

Remark 3.1 Note that (3.6) is also

$$\overline{\sigma}^2 = \frac{1-\rho^2}{T}\int_0^T \sigma_t^2 \mathrm{d}t, \qquad m = r - \frac{\overline{\sigma}^2}{2} + \frac{\rho}{T}\sum_i \sigma_{t_i}\left(W_{t_{i+1}}^{(2)} - W_{t_i}^{(2)}\right), \tag{3.9}$$

$$S_T(x) = S_0 \exp(mT + \overline{\sigma}Tx). \tag{3.10}$$

Therefore,

$$\mathbb{E}\big[u(S_0,0)\big] = \mathrm{e}^{-rT}\int_{\mathbb{R}^+}\left(K - S_0 \mathrm{e}^{mT+\overline{\sigma}Tx}\right)^+ \frac{\mathrm{e}^{-\frac{x^2}{2T}}}{\sqrt{2\pi T}}\mathrm{d}x. \tag{3.11}$$

There is a closed form for this integral, namely the Black-Scholes (or BS for short) formula with the interest rate r, the dividend $m + r$ and the volatility $\overline{\sigma}$.

3.1 Numerical Tests

In the simulations, the parameters are $S_0 = 100$, $K = 90$, $r = 0.05$, $\sigma_0 = 0.6$, $\theta = 0.36$, $k = 5$, $\lambda = 0.2$, $T = 0.5$. We compared a full MC solution with M samples to the new algorithm with M' samples for μ_t and σ_t given by (2.4). The Black-Scholes formula is used as indicated in Remark 3.1.

To observe the precision with respect to ρ (see Table 1), we have taken a large number of Monte-Carlo samples, i.e., $M = 3 \times 10^5$ and $M' = 10^4$. Similarly, the number of time steps is 300 with 400 mesh points and $S_{\max} = 600$ (i.e., $\delta S = 1.5$).

To study the precision, we let M and M' vary. Table 2 shows the results for 5 realizations of both algorithms and the corresponding mean value for P_N and variance.

Note that one needs many more samples for pure MC than those for the mixed strategy MC+BS. This variance reduction explains why MC+BS is much faster.

Table 2 Precision study with respect to M and M'. Five realizations of pure MC and MC+PDE for various M' and M

	MC+BS: $M'=$			MC: $M=$		
	100	1000	10000	3000	30000	300000
P^1	10.475	11.129	11.100	11.564	11.481	11.169
P^2	10.436	11.377	11.120	11.6978	11.409	11.249
P^3	11.025	11.528	11.113	11.734	11.383	11.143
P^4	11.205	11.002	11.113	11.565	11.482	11.169
P^5	11.527	11.360	11.150	11.085	11.519	11.208
$P=\frac{1}{5}\sum P^i$	10.934	11.279	11.119	11.529	11.454	11.187
$\sqrt{\frac{1}{5}\sum(P^i-P)^2}$	0.422	0.188	0.0168	0.232	0.0507	0.0370

4 Lévy Processes

Consider Bates model (see [4]), i.e., an asset modeled with stochastic volatility and a jump process,

$$\mathrm{d}v_t = k(\theta - v_t)\mathrm{d}t + \xi\sqrt{v_t}\mathrm{d}W_t^{(2)}, \quad \sigma_t = \sqrt{v_t}, \tag{4.1}$$

$$\mathrm{d}X_t = \left(r - \frac{\sigma_t^2}{2}\right)\mathrm{d}t + \sigma_t\left(\sqrt{1-\rho^2}\mathrm{d}\widetilde{W}_t^{(1)} + \rho\mathrm{d}\widetilde{W}_t^{(2)}\right) + \eta\mathrm{d}N_t, \tag{4.2}$$

where $X_t = \ln S_t$ and N_t is a Poisson process. As before, this is

$$\mathrm{d}X_t = \widetilde{\mu}_t\mathrm{d}t + \sigma_t\sqrt{1-\rho^2}\,\mathrm{d}\widetilde{W}_t^{(1)} + \eta\mathrm{d}N_t, \tag{4.3}$$

$$\widetilde{\mu}_t = r - \frac{\sigma_t^2}{2} + \rho\sigma_t\frac{\delta W^{(2)}}{\delta t}. \tag{4.4}$$

By Itô, a put on S_t with $u(T) = (K - \mathrm{e}^x)^+$ satisfies

$$\begin{aligned} &\partial_t u - ru + \frac{1}{2}(1-\rho^2)\sigma_t^2\partial_{xx}u + \widetilde{\mu}_t\partial_x u \\ &\quad = -\int_{\mathbb{R}}\left[\left(u(x+z) - u(x)\right)J(z) - \partial_x u(x)(\mathrm{e}^z - 1)J(z)\right]\mathrm{d}z. \end{aligned} \tag{4.5}$$

Let us apply a change of variables $\tau = T - t$, $y = x - \int_{T-\tau}^T \overline{\mu}_t\mathrm{d}t$ with $\overline{\mu}_t = \widetilde{\mu}_t - \int_{\mathbb{R}}(\mathrm{e}^z - 1)J(z)\mathrm{d}z$, and use

$$v(y,\tau) = \mathrm{e}^{(r+\int_{\mathbb{R}}J(z)\mathrm{d}z)\tau}u\left(y + \int_{T-\tau}^T \overline{\mu}_t\mathrm{d}t, T-\tau\right). \tag{4.6}$$

Table 3 9 realizations of $(\frac{1}{T}\int_0^T \sigma_t^2 \mathrm{d}t)^{\frac{1}{2}}$ for $M' = 100$ and 500

M′	T1	T2	T3	T4	T5	T6	T7	T8	T9	Mean
100	0.3470	0.3482	0.3496	0.3484	0.3474	0.3548	0.3492	0.3492	0.3502	0.3493 ± 0.002
500	0.3490	0.3481	0.3488	0.3493	0.3502	0.3501	0.3501	0.3489	0.3488	0.3493 ± 0.0007

Proposition 4.1

$$\partial_\tau v - \frac{1}{2}(1-\rho^2)\sigma_{T-\tau}^2 \partial_{yy} v - \int_{\mathbb{R}} v(y+z)J(z)\mathrm{d}z = 0, \quad v(y,0) = \left(K - \mathrm{e}^y\right)^+. \tag{4.7}$$

Proof Let $\overline{r} = r + \int_{\mathbb{R}} J(z)\mathrm{d}z$. Then

$$\begin{aligned} \partial_\tau v &= \mathrm{e}^{\overline{r}\tau}\left[-\left(r + \int_{\mathbb{R}} J(z)\mathrm{d}z\right)u + \overline{\mu}_{T-\tau}\partial_x u - \partial_t u\right], \\ \partial_y v &= \mathrm{e}^{\overline{r}\tau}\partial_x u, \quad \partial_{yy} v = \mathrm{e}^{\overline{r}\tau}\partial_{xx} u. \end{aligned} \tag{4.8}$$

Therefore,

$$\begin{aligned} &\mathrm{e}^{-\overline{r}\tau}\left[\partial_\tau v - \frac{1}{2}(1-\rho^2)\sigma_\tau^2 \partial_{yy} v - \int_{\mathbb{R}} v(y+z)J(z)\mathrm{d}z\right] \\ &= \left(r + \int_{\mathbb{R}} J(z)\mathrm{d}z\right)u + \overline{\mu}_t \partial_x u - \partial_t u - \left(1-\rho^2\right)\frac{\sigma_t^2}{2}\partial_{xx} u - \int_{\mathbb{R}} u(x+z)J(z)\mathrm{d}z, \end{aligned}$$

which is zero by (4.5). □

Remark 4.1 Once more, we notice that the PDE depends on time integrals of $\widetilde{\mu}_t$ and σ_t, and integrals damp the randomness and make the partial integro-differential equation (or PIDE for short) (4.7) easier to solve. Table 3 displays 9 realizations of $\sqrt{\frac{1}{T}\int_0^T \sigma_t^2 \mathrm{d}t}$ for $M' = 100$ and 500.

Remark 4.2 Let $\overline{f}_\tau = \frac{1}{\tau}\int_{T-\tau}^T f(t)\mathrm{d}t$. From (4.6), we see that the option price is recovered by

$$\begin{aligned} u(S,t) = \mathrm{e}^{-(r+\int_{\mathbb{R}} J(z)\mathrm{d}z)(T-t)} v\Bigg(\ln S - \Bigg(r - \frac{\overline{\sigma_t^2}|_t}{2} - \int_{\mathbb{R}} (\mathrm{e}^z - 1)J(z)\mathrm{d}z \\ + \rho\sigma_t \overline{\frac{\delta W^{(2)}}{\delta t}}\bigg|_t\Bigg)(T-t), T-t\Bigg), \end{aligned}$$

where v is the solution to (4.7). For a European put option, with the standard diffusion-Lévy process model and the dividend q, the formula is

$$u(S,t)=\mathrm{e}^{-(r+\int_{\mathbb{R}}J(z)\mathrm{d}z)(T-t)}v\bigg(\ln S-\bigg(r-q-\frac{\sigma^2}{2}$$
$$-\int_{\mathbb{R}}(\mathrm{e}^z-1)J(z)\mathrm{d}z\bigg)(T-t),T-t\bigg), \tag{4.9}$$
$$\partial_\tau v-\frac{1}{2}\sigma^2\partial_{yy}v-\int_{\mathbb{R}}v(y+z)J(z)\mathrm{d}z=0,\quad v(y,0)=\big(K-\mathrm{e}^y\big)^+.$$

It means that any solver for the European put option, with the standard diffusion-Lévy process model and the dividend q, can be used provided that the following modifications are made:

(1) In the solver, change σ^2 into $(1-\rho^2)\overline{\sigma_t^2}|_t$.
(2) Change q into $q+\rho^2\overline{\sigma_t^2}|_t-\rho\overline{\sigma_t\frac{\delta W^{(2)}}{\delta t}}|_t$.

4.1 The Numerical Solution to the PIDE by the Spectral Method

Let the Fourier transform operators be

$$\mathbb{F}(u)=\int_{\mathbb{R}}\mathrm{e}^{-\mathrm{i}\omega x}u(x)\mathrm{d}x\quad\text{and}\quad\mathbb{F}^{-1}(\widehat{u})=\frac{1}{2\pi}\int_{\mathbb{R}}\mathrm{e}^{\mathrm{i}\omega x}\widehat{u}(\omega)\mathrm{d}\omega. \tag{4.10}$$

Applying the operator $\mathbb{F}$ to the PIDE (4.7) for a call option gives

$$\partial_\tau\widehat{v}-\Psi\widehat{v}=0\quad\text{in }\mathbb{R},\quad\widehat{v}(\omega,0)=\mathbb{F}\big(\mathrm{e}^x-K\big)^+, \tag{4.11}$$

where Ψ is

$$\Psi(\omega)=-\big(1-\rho^2\big)\frac{\sigma_t^2}{2}\omega^2-\varphi(\omega),\quad\varphi(\omega)=\int_{\mathbb{R}}\mathrm{e}^{\mathrm{i}\omega y}J(y)\mathrm{d}y. \tag{4.12}$$

So, with m indicating a realization, the solution is

$$u\bigg(x-\int_{T-\tau}^{T}\overline{\mu}_t\mathrm{d}t\bigg)$$
$$=\frac{1}{M'}\sum_m\mathrm{e}^{-rT}\big(K-\mathbb{F}^{-1}[\{\mathbb{F}v^0\}(\omega)\mathrm{e}^{-\varphi(w)\tau-\omega^2\frac{1-\rho^2}{2}\int_{T-\tau}^{T}\sigma_t^{m2}\mathrm{d}t}]\big)^+ \tag{4.13}$$

with $\widetilde{\mu}_t$ given by (3.3) and $\overline{\mu}_t=\widetilde{\mu}_t+\int_{\mathbb{R}}(\mathrm{e}^z-1)J(z)\mathrm{d}z$.

Remark 4.3 The Car-Madan trick in [7] must be used, and v^0 must be replaced by $e^{-\eta S}(S-K)^+$, which has a Fourier transform, in the case of a call option. Then in (4.13) $\mathbb{F}^{-1}\widehat{\chi}$ must be changed into

$$\frac{K^\eta}{\pi}\int_0^\infty \Re\big(e^{-i\omega S}\widehat{\chi}(\xi+i\eta)\big)d\xi.$$

Remark 4.4 As an alternative to the fast Fourier transform (or FFT for short) methods, following Lewis [13], for a call option, when $\Im\omega > 1$,

$$\mathbb{F}v^0 = \mathbb{F}\big(e^y - K\big)^+ = -\frac{e^{\ln K(i\omega+1)}}{\omega^2 - i\omega}. \tag{4.14}$$

Using such extended calculus in the complex plane, Lewis obtained for the call option,

$$u(S,T) = S - \frac{\sqrt{KS}}{\pi}\int_0^\infty \Re\left[e^{iuk}\phi_T\left(u-\frac{i}{2}\right)\right]\frac{du}{u^2+\frac{1}{4}} \tag{4.15}$$

with $k = \ln\frac{S}{K}$, where ϕ_t is the characteristic function of the process, which, in the case of (4.7) with Merton Kernel (see [15])

$$J(x) = \lambda\frac{e^{-\frac{(x-\mu)^2}{\delta^2}}}{\sqrt{2\pi\delta^2}},$$

is

$$\phi_T(u) = \exp\left(iuwT - \frac{1}{2}u^2\Sigma^2T + T\lambda\big(e^{-\frac{\delta^2u^2}{2}+i\mu u} - 1\big)\right)$$

with $\Sigma^2 = \frac{1}{T}\int_0^T\sigma_\tau^2 d\tau$ and $w = \frac{1}{2}\Sigma^2 - \lambda(e^{\frac{\delta^2}{2}+\mu} - 1)$. The method has been tested with the following parameters:

$$\begin{aligned}&T=1,\quad \mu=-0.5,\quad \lambda=0.1,\quad \delta=0.4,\quad K=1,\quad r=0.03,\quad \sigma_0=0.4,\\ &\theta=0.4,\quad \kappa=2,\quad \rho=-0.5,\quad \xi=0.25,\quad M'=10000,\quad \delta t=0.001.\end{aligned} \tag{4.16}$$

Results for a put are reported in Fig. 2. The method is not precise out of the money, i.e., $S > K$. The central processing unit (or CPU for short) is 0.8″ per point on the curve.

4.2 Numerical Results

The method has been tested numerically. The coefficients for the Heston+Merton-Lévy are $T = 1$, $r = 0$, $\xi = 0.3$, $v_0 = 0.1$, $\theta = 0.1$, $k = 2$, $\lambda = 0.3$, $\rho = 0.5$. This gives an average volatility 0.27. For the Heston and the pure Black-Scholes for comparison, $T = 1$, $r = 0$, $\sigma = 0.3$, $\lambda = 5$, $m = -0.01$, $v = 0.01$.

The results are shown in Fig. 3.

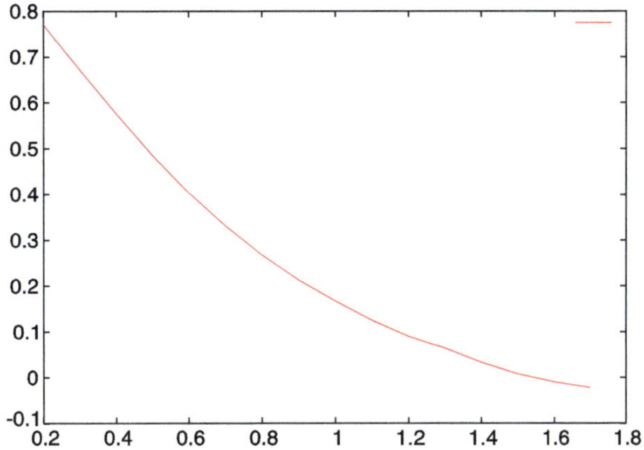

Fig. 2 Put calculated with Bates' model by mixing MC with Lewis' formula (see (4.15))

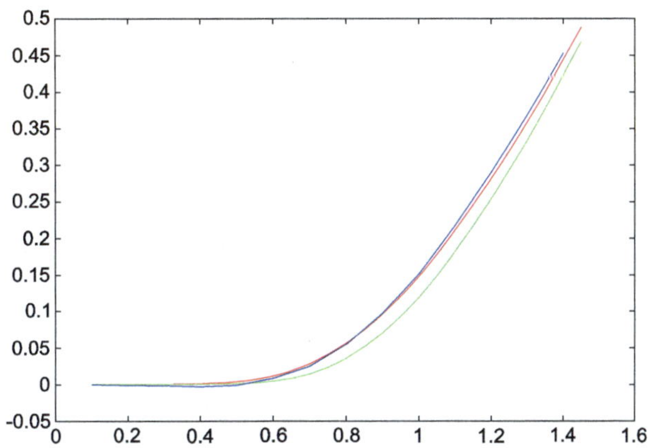

Fig. 3 Call calculated by a Heston+Merton-Lévy by mixed MC-Fourier (see the blue curve), and compared with the solution to the 2-dimensional PIDE Black-Scholes+Lévy (see *the red curve*), and a pure Black-Scholes (see *the green curve*)

5 Conditional Expectation with Spot and Volatility

If the full surface $\sigma_0, S_0 \to u(\sigma_0, S_0, 0)$ is required, MC+PDE becomes prohibitively expensive, much like MC is too expensive if $S_0 \to u(S_0, 0)$ is required for all S.

However, notice that after some time t_1 the stochastic differential equation (or SDE for short) for σ_t will generate a large number of sample values σ_1. Let us take advantage of this to compute $u(\sigma_1, S_1, t_1)$.

5.1 Polynomial Fits

Let $\tau = T - t_1$ for some fixed t_1.

Instead of gathering all $u(\cdot, \tau)$ corresponding to the samples σ_τ^m with the same initial value σ_0 at $t = 0$, we focus on the time interval (t_1, T), consider that σ_t^m is a stochastic volatility initiated by $\sigma_{t_1}^{(m)}$, then search for the best polynomial fit in terms of σ for u, i.e., a projection on the basis $\phi_k(\sigma)$ of $\mathbb{R}$, and solve

$$\min_\alpha J(\alpha) := \frac{1}{M}\sum_m \frac{1}{L}\int_0^L \left\| \sum_k \alpha_k(S)\phi^k\left(\sigma_\tau^{(m)}\right) - u^{(m)}(S,\tau)\right\|^2 \mathrm{d}S.$$

It leads to solving, for each $S_i = i\delta S$,

$$\left(\frac{1}{M}\sum_m \phi_k\left(\sigma_\tau^{(m)}\right)\phi_l\left(\sigma_\tau^{(m)}\right)\right)\alpha_k^i = \frac{1}{M}\sum_m u^{(m)}(S_i,\tau)\phi_l\left(\sigma_\tau^{(m)}\right). \tag{5.1}$$

5.2 Piecewise Constant Approximation on Intervals

We begin with a local basis of polynomials, namely, $\phi_k(\sigma) = 1$ if $\sigma \in (\sigma_k, \sigma_{k+1})$ and $\phi_k(\sigma) = 0$ otherwise.

Algorithm 5.1

(1) Choose $\sigma_m, \sigma_M, \delta\sigma, \sigma_0$.
(2) Initialize an array $n[j] = 0$, $j = 0, \dots, J := \frac{\sigma_M - \sigma_m}{\delta\sigma}$.
(3) Compute M realizations $\{\sigma_{t_i}^{(m)}\}$ by MC on the volatility equation.
(4) For each realization, compute $u(\cdot,\tau)$ by solving the PDE.
(5) Set $j = \frac{\sigma_\tau^{(m)} - \sigma_m}{\delta\sigma}$ and $n[j] += 1$, and store $u(\cdot,\tau)$ in $w(\cdot)[j]$.
(6) The answer is $u(\sigma; S,\tau) = \frac{w(S)[j]}{n[j]}$ with $j = \frac{\sigma - \sigma_m}{\delta\sigma}$.

5.3 Polynomial Projection

Now we choose $\phi_k(\sigma) = \sigma^k$.

Algorithm 5.2

(1) Choose $\sigma_m, \sigma_M, \delta\sigma, \sigma_0$.
(2) Set $A[\cdot][\cdot] = 0$, $b[\cdot][\cdot] = 0$.
(3) Compute M realizations $\{\sigma_{t_i}^{(m)}\}$ by MC on the volatility equation and for each realization.
   (i) Compute $u(\cdot,\tau)$ by solving the PDE.
   (ii) Do $A[j][k] += \frac{1}{M}\sum_m (\sigma_\tau^{(m)})^{j+k}$, $j,k = 1,\dots,K$.
   (iii) Do $b[i][k] += \frac{1}{M} u(i\delta S,\tau)(\sigma_\tau^{(m)})^k$, $k = 1,\dots,K$.
(4) The answer is found by solving (5.1) for each $i = 1,\dots,N$.

5.4 The Numerical Test

A Vanilla put with the same characteristics as in Sect. 3.1 has been computed by Algorithm 5.2 for a maturity of 3 years. The surface $S_{t_1}, \sigma_{t_1} \to u$ is shown after $t_1 = 1.5$ years in Fig. 4. The implied volatility is also shown.

6 American and Bermudan Options

For American options, we must proceed step by step backward in time as in the dynamic programming for binary trees (see [2]).

Consider M' realizations $[\{\sigma_t^m\}_{t\in(0,T)}]_{m=1}^{M'}$, giving $[\{\mu_t^m\}_{t\in(0,T)}]_{m=1}^{M'}$ by (3.3). At time $t_n = T$, the price of the contract is $(K-S)^+$. At time $t_{n-1} = T - \delta t$, it is given by the maximum of the European contract, knowing S and σ at t_{n-1} and $(K-S)^+$, i.e.,

$$u_{n-1}(S) = \max\left\{\frac{1}{|M_\sigma|}\sum_{m\in M_\sigma} u_{n-1}^m(S), (K-S)^+\right\}, \tag{6.1}$$

where u_{n-1}^m is the solution at t_{n-1} to

$$\begin{aligned} &\partial_t u + \left(1-\rho^2\right)\frac{(S\sigma_t^m)^2}{2}\partial_{SS}u + S\mu_t^m\partial_S u - ru = 0, \quad t \in (t_{n-1}, t_n), \\ &u_n := u(S, t_n), \end{aligned} \tag{6.2}$$

where u_n is known, and M_σ is the set of trajectories which give a volatility equal to σ at time t.

Here we have used the piecewise constant approximation intervals to compute the European premium. Alternatively, one could use any projection method, and the backward algorithm follows the same lines.

As with American options with binary trees, convergence with optimal order will hold only if δt is small enough. M_σ is built as in the previous section.

To prove the concept, we computed a Bermudan contract at $\frac{1}{2}T$ by the above method, using the polynomial basis for the projection. The parameters are the same as above except $K = 100$. The results are displayed in Fig. 5. To obtain the price of the option at time zero, the surface of Fig. 5, i.e., (6.1), must be used as time-boundary conditions for the MC-PDE mixed solver for $t \in (0, \frac{1}{2}T)$, while for Americans, this strategy is applied at every time step, but here it is done once only at $\frac{1}{2}T$.

7 Systems of Dimension Greater than 2

Stochastic volatility models with several SDEs for the volatilities are now in use. However, in order to assess the mixed MC-PDE method, we need to work on a

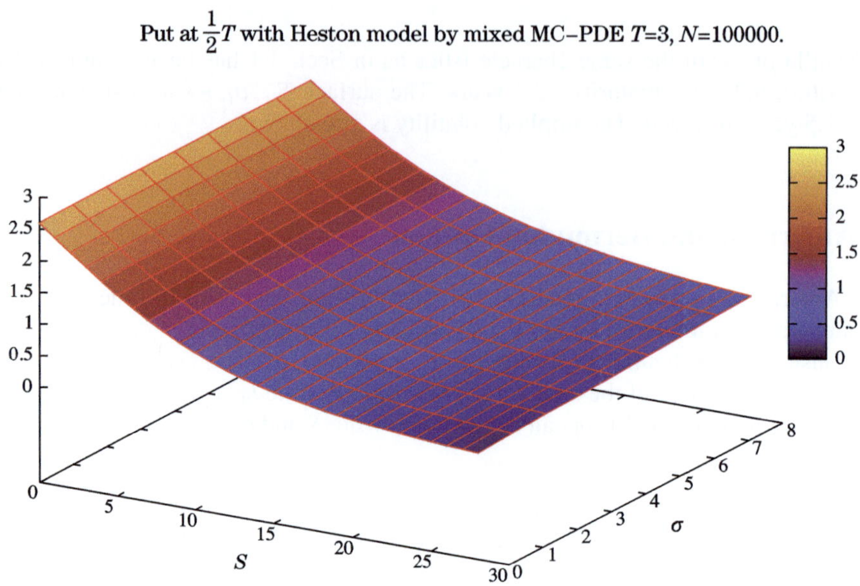

(a) Local volatility of a vanilla put with 3 years maturity after 1.5 years, computed with a Heston model by the mixed MC-PDE algorithm with polynomial projection.

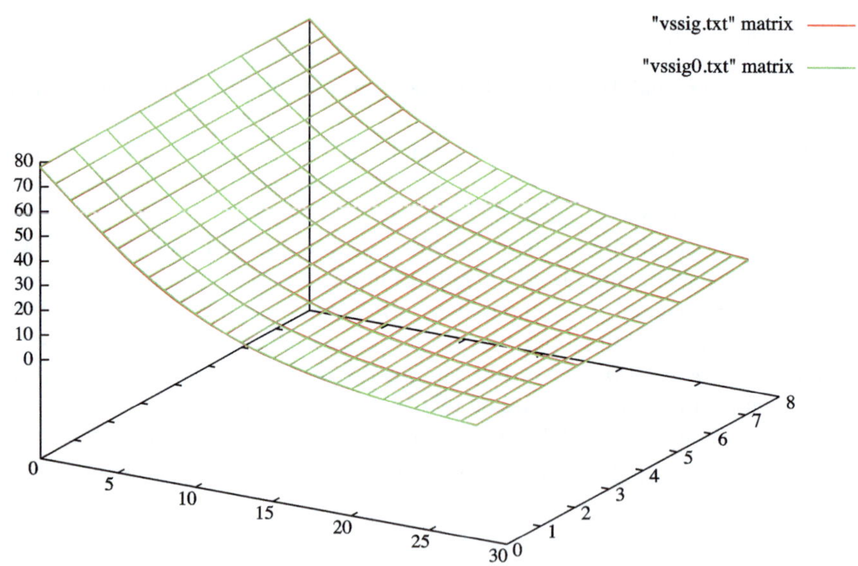

(b) Comparison on the price of the put computed with full MC Heston.

Fig. 4 Both surfaces (**a**) and (**b**) are on top of each other, indistinguishable

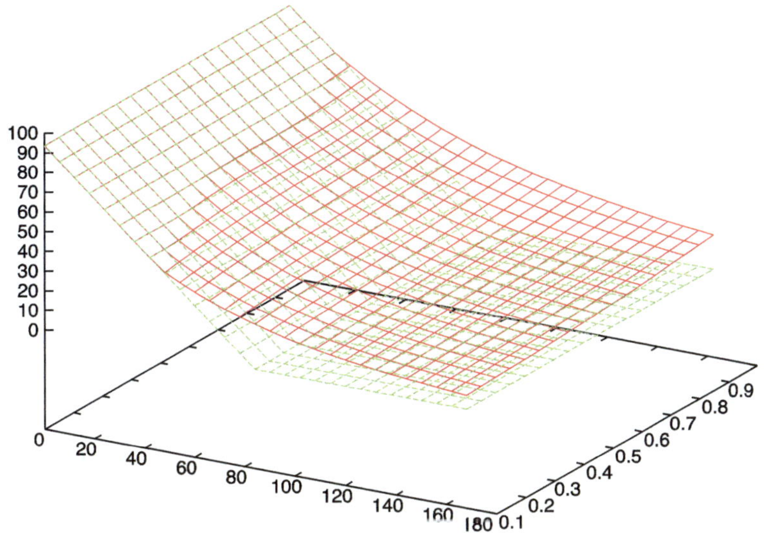

Fig. 5 A Bermuda option at $\frac{1}{2}T$ with Heston's model compared with $(K-S)^+$

systems for which an exact or precise solution is easily available. Therefore, we will investigate basket options instead.

7.1 Problem Formulation

We consider an option P on three assets whose dynamics are determined by the following system of stochastic differential equations:

$$\mathrm{d}S_{i,t} = S_{i,t}(r\mathrm{d}t + \mathrm{d}W_{i,t}), \quad t > 0,\ i = 1, 2, 3 \tag{7.1}$$

with initial conditions $S_{i,t=0} = S_{i,0},\ S_{i,0} \in \mathbb{R}^+$. The parameter r ($r \in \mathbb{R}_{\geq 0}$) is constant, and $W_i := \sum_{j=1}^{3} a_{ij} B_j$ are linear combinations of standard Brownian motions B_j, such that

$$\mathrm{Cov}[W_{i,t}, W_{j,t}] = \rho_{ij}\sigma_i\sigma_j t, \quad t > 0.$$

We further assume that $\Xi := (\rho_{ij}\sigma_i\sigma_j)_{i,j=1}^{3}$ is symmetric positive definite with

$$\rho_{ij} = 1 \quad (i = j) \quad \text{or} \quad \rho_{ij} \in (-1, 1) \quad \text{otherwise.}$$

The coefficients a_{ij} ($a_{ij} \in \mathbb{R}$) have to be chosen, such that

$$\begin{aligned}\mathrm{Cov}[W_{i,t}, W_{j,t}] &= E[W_{i,t} W_{j,t}] \\ &= E\big[(a_{i1}B_{1,t} + a_{i2}B_{2,t} + a_{i3}B_{3,t})(a_{j1}B_{1,t} + a_{j2}B_{2,t} + a_{j3}B_{3,t})\big]\end{aligned}$$

$$= a_{i1}a_{j1}E\left[B_{1,t}^2\right] + a_{i2}a_{j2}E\left[B_{2,t}^2\right] + a_{i3}a_{j3}E\left[B_{3,t}^2\right]$$
$$= (a_{i1}a_{j1} + a_{i2}a_{j2} + a_{i3}a_{j3})t, \quad t > 0,$$

or equivalently,

$$AA^{\mathrm{T}} = \Xi,$$

where $A := (a_{ij})_{i,j=1}^3$. Without loss of generality, we may set the strict upper triangular components of A to zero and find

$$A = \begin{pmatrix} \sigma_1 & 0 & 0 \\ \sigma_2\rho_{21} & \sigma_2\sqrt{1-\rho_{12}^2} & 0 \\ \sigma_3\rho_{31} & \sigma_3\frac{\rho_{32}-\rho_{21}\rho_{31}}{\sqrt{1-\rho_{12}^2}} & \sigma_3\sqrt{1-\rho_{31}^2-\left(\frac{\rho_{32}-\rho_{21}\rho_{31}}{\sqrt{1-\rho_{12}^2}}\right)^2} \end{pmatrix}.$$

The option P has the maturity T ($T \in \mathbb{R}^+$), the strike K ($K \in \mathbb{R}^+$) and the payoff function $\varphi : \mathbb{R}^{+3} \to \mathbb{R}$,

$$\varphi(x) = \left(K - \sum_{i=1}^{3} x_i\right)^+, \quad x = (x_1, x_2, x_3)^{\mathrm{T}} \in \mathbb{R}^{+3}.$$

The Black-Scholes price of P at time 0 is

$$P_0 = \mathrm{e}^{-rT} E^*\left[\left(K - \sum_{i=1}^{3} S_{i,T}\right)^+\right], \tag{7.2}$$

where E^* denotes the expectation with respect to the risk-neutral measure.

7.2 The Uncoupled System

In order to combine different types of methods (Monte-Carlo, quadrature and/or PDE methods), we will uncouple the SDE in (7.1), we start with a change of variable to logarithmic prices. Let $s_{i,t} := \log(S_{i,t})$, $i = 1, 2, 3$, and then Itô's lemma shows that

$$\mathrm{d}s_{i,t} = r_i \mathrm{d}t + \mathrm{d}W_{i,t}, \quad t > 0 \tag{7.3}$$

with initial conditions $s_{i,t=0} = s_{i,0} := \log(S_{i,0})$. The parameters r_i ($i = 1, 2, 3$) have been defined as $r_i = r - \frac{a_{i1}^2}{2} - \frac{a_{i2}^2}{2} - \frac{a_{i3}^2}{2} = r - \frac{\sigma_i^2}{2}$. In the rest of the section, the time index of any object is omitted to simplify the notation.

We note that (7.3) can be written as

$$\begin{pmatrix} ds_1 - r_1 dt \\ ds_2 - r_2 dt \\ ds_3 - r_3 dt \end{pmatrix} = \begin{pmatrix} a_{11} & 0 & 0 \\ a_{21} & a_{22} & 0 \\ a_{31} & a_{32} & a_{33} \end{pmatrix} \begin{pmatrix} dB_1 \\ dB_2 \\ dB_3 \end{pmatrix}.$$

Then, uncoupling reduces to Gaussian elimination. Using the Frobenius matrices

$$F_1 := \begin{pmatrix} 1 & 0 & 0 \\ -\frac{a_{21}}{a_{11}} & 1 & 0 \\ -\frac{a_{31}}{a_{11}} & 0 & 1 \end{pmatrix}, \qquad F_2 := \begin{pmatrix} 1 & 0 & 0 \\ 0 & 1 & 0 \\ 0 & -\frac{a_{32}}{a_{22}} & 1 \end{pmatrix},$$

we write

$$F_2 F_1 (ds + r dt) = \mathrm{Diag}(a_{11}, a_{22}, a_{33}) dB,$$

where $s = (s_1, s_2, s_3)^T$, $r = (r_1, r_2, r_3)^T$ and $B = (B_1, B_2, B_3)^T$. We set $L^{-1} := F_2 F_1$, and define

$$\widetilde{s} := L^{-1} s \quad \text{and} \quad \widetilde{S} := e^{L^{-1} s}.$$

Remark 7.1 (i) The processes $\widetilde{s}_1$, $\widetilde{s}_2$ and $\widetilde{s}_3$ are independent of each other, and are analogous with $\widetilde{S}_1$, $\widetilde{S}_2$ and $\widetilde{S}_3$, respectively.

(ii) Let $\widetilde{r} := L^{-1} r$. Then

$$d\widetilde{s} = \widetilde{r} dt + \mathrm{Diag}(a_{11}, a_{22}, a_{33}) dB.$$

(iii) The coupled system expressed in terms of the uncoupled system is $s = L\widetilde{s}$.

(iv) In the next section, we will make use of the triangular structure of $L = (L_{ij})_{i,j=1}^3$ and $L^{-1} = ((L^{-1})_{ij})_{i,j=1}^3$,

$$L = \begin{pmatrix} 1 & 0 & 0 \\ \frac{a_{21}}{a_{11}} & 1 & 0 \\ \frac{a_{31}}{a_{11}} & \frac{a_{32}}{a_{22}} & 1 \end{pmatrix} \quad \text{and} \quad L^{-1} = \begin{pmatrix} 1 & 0 & 0 \\ -\frac{a_{21}}{a_{11}} & 1 & 0 \\ \frac{a_{21}a_{32}}{a_{11}a_{22}} - \frac{a_{31}}{a_{11}} & -\frac{a_{32}}{a_{22}} & 1 \end{pmatrix}.$$

(v) The notation has been symbolic and the derivation heuristic.

7.3 Mixed Methods

We describe nine combinations of Monte-Carlo, quadrature (or QUAD for short) and/or PDE methods.

Convention If Z is a stochastic process, we denote by Z^m a realization of the process. Let M' stand for a fixed number of Monte-Carlo samples.

Basic Methods (i) MC3 method

Simulate M' trajectories of (S_1, S_2, S_3). An approximation of the option price P_0 is

$$P_0^a := \mathrm{e}^{-rT}\frac{1}{M'}\sum_{m=1}^{M'}\varphi\big(S_{1,T}^m, S_{2,T}^m, S_{3,T}^m\big).$$

(ii) QUAD3 method

In order to use a quadrature formula, we replace the risk neutral measure in

$$P_0 = \mathrm{e}^{-rT}E^*\big[\big(K - \mathrm{e}^{(L\widetilde{s}_T)_1} - \mathrm{e}^{(L\widetilde{s}_T)_2} - \mathrm{e}^{(L\widetilde{s}_T)_3}\big)^+\big]$$

by the Lebesgue-measure. Note

$$\widetilde{s}_{i,t} \sim N\big(\mu_{i,t}, a_{ii}^2 t\big), \quad 1 \le i \le 3,$$

where $\mu_{i,t} = \widetilde{s}_{i,0} + \widetilde{r}_i t$. Let $f_{i,t}$ be the density of $\widetilde{s}_{i,t}$, i.e.,

$$f_{i,t}(x_i) = \frac{1}{\sqrt{2\pi}a_{ii}\sqrt{t}}\mathrm{e}^{-\frac{1}{2}(\frac{x_i-\mu_{i,t}}{a_{ii}\sqrt{t}})^2}, \quad x_i \in \mathbb{R},\ 1 \le i \le 3.$$

Due to the independence of $\widetilde{s}_{1,t}$, $\widetilde{s}_{2,t}$ and $\widetilde{s}_{3,t}$, the density of

$$\big(K - \mathrm{e}^{(L\widetilde{s}_T)_1} - \mathrm{e}^{(L\widetilde{s}_T)_2} - \mathrm{e}^{(L\widetilde{s}_T)_3}\big)^+$$

is

$$(x_1, x_2, x_3) \mapsto f_{1,T}(x_1)f_{2,T}(x_2)f_{3,T}(x_3), \quad (x_1, x_2, x_3) \in \mathbb{R}^3.$$

The formula for the option price becomes

$$P_0 = \mathrm{e}^{-rT}\int_{\mathbb{R}^3}\big(K - \mathrm{e}^{(Lx)_1} - \mathrm{e}^{(Lx)_2} - \mathrm{e}^{(Lx)_3}\big)^+ f_{1,T}(x_1)f_{2,T}(x_2)f_{3,T}(x_3)\mathrm{d}x.$$

Now, a quadrature formula can be used to compute the integral.

The methods, which are based on a combination of quadrature and some other methods, will be presented for the case, where the trapezoidal rule is used. Next we show how the trapezoidal rule can be used to compute the integral. This allows us to introduce the notation for the description of methods, which are combinations of quadrature and some other methods.

To compute the integral, we truncate the domain of integration to κ standard deviations around the means $\mu_{1,T}$, $\mu_{2,T}$ and $\mu_{3,T}$. Let

$$x_{i,0} = \mu_{i,T} - \kappa a_{ii}^2,$$

$$x_{i,n} = x_{i,0} + n\delta x_i, \quad n = 1, \ldots, N_Q,$$

$1 \le i \le 3$, where $\delta x_i = \frac{2\kappa}{N_Q}$, N_Q.

The option price P_0 is then approximated by

$$P_0^a := \mathrm{e}^{-rT} \sum_{n_1,n_2,n_3=1}^{N} \left(\prod_{i=1}^{3} \chi_{n_i} \delta x_i f_{i,T}(x_{i,n_i}) \right) \left(K - \mathrm{e}^{(Lx_n)_1} - \mathrm{e}^{(Lx_n)_2} - \mathrm{e}^{(Lx_n)_3}\right)^+,$$

where $x_n := (x_{1,n_1}, x_{2,n_2}, x_{3,n_3})^{\mathrm{T}}$ and

$$\chi_n = \begin{cases} 0.5, & \text{if } n = 0 \text{ or } n = N_{\mathrm{Q}}, \\ 1, & \text{otherwise.} \end{cases}$$

(iii) MC2-PDE1 method (combination of two methods)
Note

$$\begin{aligned} P_0 &= \mathrm{e}^{-rT} E^*\left[\left(K - S_{1,T} - S_{2,T} - S_{1,T}^{-2(L^{-1})_{31}} S_{2,T}^{-(L^{-1})_{32}} \widetilde{S}_{3,T}\right)^+\right] \\ &= \mathrm{e}^{-rT} E^*\left[(\overline{K} - \widetilde{\widetilde{S}}_{3,T})^+\right], \end{aligned}$$

where

$$\overline{K} := K - S_{1,T} - S_{2,T},$$

and $\widetilde{\widetilde{S}}_3$ is the solution to the stochastic initial value problem

$$\begin{aligned} \mathrm{d}\widetilde{\widetilde{S}}_{3,t} &= \widetilde{\widetilde{S}}_{3,t}(\widetilde{\widetilde{r}}_3 \mathrm{d}t + a_{33} \mathrm{d}B_{3,t}), \\ \widetilde{\widetilde{S}}_{3,t=0} &= \alpha \widetilde{S}_{3,0} \end{aligned}$$

with parameters $\widetilde{\widetilde{r}}_3 := \widetilde{r}_3 + \frac{a_{33}^2}{2}$ and $\alpha = S_{1,T}^{-2(L^{-1})_{31}} S_{2,T}^{-(L^{-1})_{32}}$.

The method is then as follows. Simulate M' realizations of (S_1, S_2) and set $\overline{K}^m = K - S_{1,T}^m - S_{2,T}^m$ and $\alpha^m = {S_{1,T}^m}^{-2(L^{-1})_{31}} {S_{2,T}^m}^{-(L^{-1})_{32}}$. Compute an approximation of P_0 by

$$P_0^a := \frac{1}{M'} \sum_{m=1}^{M'} u\left(x_3, t; \overline{K}^m\right)\big|_{x_3 = \alpha^m \widetilde{S}_{3,0}, t=T},$$

where u is the solution to the initial value problem for the one-dimensional Black-Scholes PDE with the parametrized (β) initial condition

$$\frac{\partial u}{\partial t} - \frac{(a_{33} x_3)^2}{2} \frac{\partial^2 u}{\partial x_3^2} - \widetilde{\widetilde{r}}_3 x_3 \frac{\partial u}{\partial x_3} + \widetilde{\widetilde{r}}_3 u = 0 \quad \text{in } \Omega \times (0, T), \tag{7.4a}$$

$$u(t = 0) = u_0 \quad \text{in } \Omega, \tag{7.4b}$$

where $\Omega = \mathbb{R}^+$ and

$$u_0(x_3;\beta) := (\beta - x_3)^+, \quad x_3 > 0.$$

(iv) QUAD2-PDE1 method
Note

$$P_0 = \mathrm{e}^{-rT} \int_{\mathbb{R}^2} E^*\big[\big(K - \mathrm{e}^{L_{11}x_1} - \mathrm{e}^{L_{21}x_1 + L_{22}x_2} - \mathrm{e}^{L_{31}x_1 + L_{32}x_2}\mathrm{e}^{L_{33}\widetilde{s}_{3,T}}\big)^+\big] f_{1,T}(x_1) f_{2,T}(x_2)\mathrm{d}x_1\mathrm{d}x_2.$$

The option price P_0 is approximated by

$$P_0^a := \sum_{n_1,n_2=1}^{N_Q} \left(\prod_{i=1}^{2} \chi_{n_i}\delta x_i f_{i,T}(x_{i,n_i})\right) u(x_3, t; \overline{K}_{n_1n_2})|_{x_3=\alpha_{n_1n_2}\widetilde{S}_{3,0}, t=T},$$

where

$$\overline{K}_{n_1n_2} := K - \mathrm{e}^{L_{11}x_{1,n_1}} - \mathrm{e}^{L_{21}x_{1,n_1} + L_{22}x_{2,n_2}},$$
$$\alpha_{n_1n_2} := \mathrm{e}^{L_{31}x_{1,n_1} + L_{32}x_{2,n_2}},$$

and u denotes the solution to (7.4a)–(7.4b).
(v) MC1-PDE2 method
Note

$$P_0 = \mathrm{e}^{-rT} E^*\big[\big(K - S_{1,T} - S_{2,T} - S_{1,T}^{-2(L^{-1})_{31}} S_{2,T}^{-(L^{-1})_{32}} \widetilde{S}_{3,T}\big)^+\big].$$

Simulate M' realizations of $\widetilde{S}_3$. The option price P_0 is then approximated by

$$P_0^a := \frac{1}{M'} \sum_{m=1}^{M'} u\big(x_1, x_2, t; \widetilde{S}_{3,T}^m\big)\big|_{x_1=S_{1,0}, x_2=S_{2,0}, t=T},$$

where u denotes the solution to the initial value problem for the 2-dimensional Black-Scholes PDE with the parameterized (β) initial condition

$$u_0(x_1, x_2, 0; \beta) = \big(K - x_1 - x_2 - x_1^{-2(L^{-1})_{31}} x_2^{-(L^{-1})_{32}} \beta\big)^+, \quad x_1, x_2 > 0.$$

The problem is

$$\frac{\partial u}{\partial t} - \sum_{i,j=1}^{2} x_i x_j \varrho_{ij} \frac{\partial u}{\partial x_i}\frac{\partial u}{\partial x_j} - r\sum_{i=1}^{2} x_i \frac{\partial u}{\partial x_i} + ru = 0 \quad \text{in } \Omega \times (0, T), \tag{7.5a}$$

$$u(t=0) = u_0 \quad \text{in } \Omega, \tag{7.5b}$$

where $\Omega = \mathbb{R}^+ \times \mathbb{R}^+$ and

$$\varrho = (\varrho_{ij})_{i,j=1,\dots,3} = \frac{1}{2}\begin{pmatrix} a_{11}^2 & a_{11}a_{21} \\ a_{11}a_{21} & a_{21}^2 + a_{22}^2 \end{pmatrix}. \tag{7.6}$$

(vi) QUAD1-PDE2 method
Note

$$P_0 = \mathrm{e}^{-rT} \int_{\mathbb{R}} E^*\big[\big(K - S_{1,T} - S_{2,T} - S_{1,T}^{-2(L^{-1})_{31}} S_{2,T}^{-(L^{-1})_{32}} \mathrm{e}^{x_3}\big)^+\big] f_{3,T}(x_3)\mathrm{d}x_3.$$

With the notation above, another approximation of the option price P_0 is

$$\begin{aligned} P_0^a &:= \sum_{n=1}^{N_Q} \delta x_3 f_{3,T}(x_{3,n}) \mathrm{e}^{-rT} E^*\big[\big(K - S_{1,T} - S_{2,T} - S_{1,T}^{2(L^{-1})_{31}} S_{2,T}^{-(L^{-1})_{32}} \mathrm{e}^{x_{3,n}}\big)^+\big] \\ &= \sum_{n=1}^{N_Q} \delta x_3 f_{3,T}(x_{3,n}) u(x_1, x_2, t; x_{3,n})|_{x_1=S_{1,0}, x_2=S_{2,0}, t=T}, \end{aligned}$$

where u is the solution to the initial value problem (7.5a)–(7.5b).
(vii) MC1-QUAD2 method
Reformulating (7.2), we deduce

$$\begin{aligned} P_0 = \mathrm{e}^{-rT} E^* &\int_{\mathbb{R}^2} \big(K - \mathrm{e}^{(Lx)_1} - \mathrm{e}^{(Lx)_2} - \mathrm{e}^{L_{31}x_1 + L_{32}x_2 + \widetilde{s}_{3,T}}\big)^+ \\ &\times f_{1,T}(x_1) f_{2,T}(x_2) \mathrm{d}x_1 \mathrm{d}x_2, \end{aligned}$$

and obtain the following method.
Compute M' realizations of $\widetilde{s}_{3,T}$, and approximate P_0 by

$$\begin{aligned} P_0^a := \mathrm{e}^{-rT} \frac{1}{M'} &\sum_{n_1,n_2=1}^{N_Q} \sum_{m=1}^{M'} \left(\prod_{i=1}^{2} \chi_{n_i} \delta x_i f_{i,T}(x_{i,n_i}) \right) \\ &\cdot \big(K - \mathrm{e}^{x_{1,n_1}} - \mathrm{e}^{L_{21}x_{1,n_1} + x_{2,n_2}} - \mathrm{e}^{L_{31}x_{1,n_1} + L_{32}x_{2,n_2} + \widetilde{s}_{3,T}^m}\big)^+. \end{aligned}$$

(viii) MC2 QUAD1 method
Note

$$P_0 = \mathrm{e}^{-rT} \int_{\mathbb{R}} E^*\big[\big(K - S_{1,T} - S_{2,T} - S_{1,T}^{-2(L^{-1})_{31}} S_{2,T}^{-(L^{-1})_{32}} \mathrm{e}^{x_3}\big)^+\big] f_{3,T}(x_3)\mathrm{d}x_3.$$

The method is as follows. Simulate M' realizations of (S_1, S_2), and compute

$$\begin{aligned} P_0^a := \mathrm{e}^{-rT} \frac{1}{M'} &\sum_{m=1}^{M'} \sum_{n=1}^{N_Q} \chi_n \delta x_3 f_{3,T}(x_{3,n}) \big(K - S_{1,T}^m - S_{2,T}^m \\ &- S_{1,T}^m{}^{-2(L^{-1})_{31}} S_{2,T}^m{}^{-(L^{-1})_{32}} \mathrm{e}^{x_{3,n}}\big)^+. \end{aligned}$$

(ix) MC1-QUAD1-PDE1 method (combination of three methods)
Note

$$P_0 = \int_{\mathbb{R}} f_{2,T}(x_2) e^{-rT} E^*\big[\big(K - e^{\widetilde{s}_{1,T}} - e^{L_{21}\widetilde{s}_{1,T}+x_2} - e^{(-2(L^{-1})_{31}-(L^{-1})_{32}L_{21})\widetilde{s}_{1,T}-(L^{-1})_{32}x_2}\widetilde{S}_{3,T}\big)^+\big] dx_2.$$

Then an approximation to P_0 is

$$P_0^a := \frac{1}{M'} \sum_{m=1}^{M'} \sum_{n=1}^{N_Q} \chi_2 \delta x_2 f_{2,T}(x_{2,n}) u\big(x_3, t; \overline{K}_n^m\big)\big|_{x_3=\alpha_n^m \widetilde{S}_{3,0}, t=T},$$

where

$$\overline{K}_n^m := K - e^{\widetilde{s}_{1,T}^m} - e^{L_{21}\widetilde{s}_{1,T}^m + x_{2,n_2}},$$

$$\alpha_n^m := e^{(-2(L^{-1})_{31}-(L^{-1})_{32}L_{21})\widetilde{s}_{1,T}^m-(L^{-1})_{32}x_{2,n}},$$

and u denotes the solution to (7.4a)–(7.4b).

7.4 Numerical Results

This section provides a documentation of numerical results. We have considered European put options on baskets of three and five assets, and used mixed methods to compute their prices. If the method is stochastic, i.e., if a part of it is Monte-Carlo simulation, then we have run the method with different seed values several times (N_S) and computed mean (m) and standard deviation (s) of the price estimates. If the method is deterministic, we have chosen the discretization parameters, such that the first three digits of P_0^a remained fix, while the discretization parameters have been further refined. Instead of solving the 1-dimensional Black-Scholes PDE, we have used the Black-Scholes formula.

(i) European put on three assets

The problem is to compute the price of a European put option on a basket of three assets in the framework outlined in Sect. 7.1.

We have chosen the parameters as follows: $K = 150$, $T = 1$, $r = 0.05$, $S_0 = (55, 50, 45)$,

$$\rho = \begin{pmatrix} 1 & -0.1 & -0.2 \\ -0.1 & 1 & -0.3 \\ -0.2 & -0.3 & 1 \end{pmatrix}, \qquad \sigma = \begin{pmatrix} 0.3 & 0.2 & 0.25 \end{pmatrix}^{\mathrm{T}}.$$

We have used various (mixed) methods to compute approximations to P_0 (see (7.2)).

Table 4 Pricing a European put option on a basket of three assets, i.e., estimates of the option price at time 0. *Columns 1–3*: the method used to approximate P_0. *Columns 4–6*: the discretization parameters. M' is the number of Monte-Carlo samples, N_Q is the number of quadrature points, N_S is the number of samples used to compute the mean (m) and the standard deviation (s). *Column 9*: the computing time

MC	PDE	QUAD	M'	N_Q	N_S	m	s	CPU
3	–	–	10^7	–	10	3.988	0.002	22.46
3	–	–	25000	–	100	3.994	0.046	0.147
2	1	–	25000	–	100	3.989	0.029	0.162
1	2	–	100	2601	10	3.886	0.195	372.5
–	–	3	–	–	–	3.984	–	0.005
–	1	2	–	–	–	3.987	–	0.005
–	2	1	–	2601	–	4.016	–	42.24
1	–	2	25000	–	100	3.991	0.022	2.723
2	–	1	25000	–	100	3.987	0.032	0.369
1	1	1	25000	–	100	3.990	0.023	0.514

We have used freefem++, and the rest is programmed in C++. The implementation in freefem++ requires a localization and the weak formulation of the Black-Scholes PDE. The triangulation of the computational domain and the discretization of the Black-Scholes PDE by conforming P1 finite elements are done by freefem++.

A reference result for P_0 has been computed by using the Monte-Carlo method with 10^7 samples.

The numerical results are displayed in Table 4. One can see that the computational load for the PDE2 methods (i.e., MC1-PDE2, QUAD1-PDE2) is much larger than that for the other methods. Furthermore, the results seem to be less precise than those in the other cases. The results have been obtained very fast if just quadrature (i.e., QUAD3) or quadrature in combination with the Black-Scholes formula (i.e., QUAD2-PDE1) was used. In these cases, the results seem to be very precise although the discretization has been coarse ($N_Q = 12$). Comparison of the results obtained by the MC3 method with the results obtained by the MC2-PDE1 method shows that the last mentioned seems to be superior. The computing time is about equal, but the standard deviation for MC2-PDE1 is much less than that for MC3.

(ii) European put on five assets

Let P be a European put option on a basket of five assets, with payoff

$$\varphi(x) = \left(K - \sum_{i=1}^{5} x_i\right)^+ .$$

The system of stochastic differential equations, which describes the dynamics of the underlying assets, has the usual form. We have set $K = 250$, $T = 1$, $r = 0.05$,

$$S_0 = (40, 45, 50, 55, 60)^{\mathrm{T}},$$

$$\sigma = (0.3, 0.275, 0.25, 0.225, 0.2)^{\mathrm{T}},$$

Table 5 Pricing a European put option on a basket of five assets, i.e., estimates of the option price at time 0. *Columns 1–3*: the method used to approximate P_0. *Columns 4–6*: the discretization parameters. M' is the number of Monte-Carlo samples, N_Q is the number of quadrature points, N_S is the number of samples used to compute the mean (m) and the standard deviation (s). *Columns 7–9*: the numerical results. *Column 7*: the mean of P_0. *Column 8*: the standard deviation of P_0. *Column 9*: the computing time

MC	PDE	QUAD	M'	N_Q	N_S	m	s	τ
5	–	–	10^7	–	10	1.159	0.001	27.67
5	–	–	25000	–	100	1.161	0.019	0.162
4	–	1	25000	–	100	1.156	0.015	0.174
–	–	5	–	10	–	1.161	–	0.082
–	1	4	–	10	–	1.159	–	0.036
3	1	1	25000	10	100	1.158	0.013	0.442

$$\rho = \begin{pmatrix} 1 & -0.37 & -0.40 & -0.44 & -0.50 \\ -0.37 & 1 & -0.50 & -0.46 & -0.05 \\ -0.40 & -0.50 & 1 & 0.51 & 0.29 \\ -0.44 & -0.46 & 0.51 & 1 & 0.20 \\ -0.50 & -0.05 & 0.29 & 0.20 & 1 \end{pmatrix}.$$

We approximated the price of P at time 0 by various (mixed) methods. The results are displayed in Table 5. One can see that for all tested methods the (mean) price has been close (± 0.003) to the reference price (1.159). Since $N_Q = 10$ turned out to be enough, the computational effort has been very low for QUAD5 and QUAD4-PDE1. In the case, the method is stochastic, and deterministic methods allow to reduce the variance, such as in MC4-QUAD1 and MC4-PDE1-QUAD1.

8 Conclusion

Mixing Monte-Carlo methods with partial differential equations allows the use of closed formula on problems which do not have any otherwise. In these cases, the numerical methods are much faster than full MC or full PDE. The method works also for nonconstant coefficient models with and without jump processes and also for American contracts, although proofs of convergence have not been given here.

For multi-dimensional problems, we tested all possibilities of mixing MC and PDE and also quadrature on semi-analytic formula, and we found that the best is to apply PDE methods to one equation only.

The speed-up technique by polynomial fit has been discussed also, but we plan to elaborate on such ideas in the future particularly in the context of reduced basis, such as POD (proper orthogonal decomposition), ideally suited to the subproblems arising from MC+PDE, because the same PDE has to be solved many times for different time dependent coefficients.

References

1. Achdou, Y., Pironneau, O.: Numerical Methods for Option Pricing. SIAM, Philadelphia (2005)
2. Amin, K., Khanna, A.: Convergence of American option values from discrete- to continuous-time financial models. Math. Finance **4**, 289–304 (1994)
3. Barth, A., Schwab, C., Zollinger, N.: Multi-level Monte Carlo finite element method for elliptic PDEs with stochastic coefficients. Numer. Math. **119**(1), 123–161 (2011)
4. Bates, D.S.: Jumps and stochastic volatility: exchange rate processes implicit Deutsche mark options. Rev. Financ. Stud. **9**(1), 69–107 (1996)
5. Black, F., Scholes, M.: The pricing of options and corporate liabilities. J. Polit. Econ. **81**, 637–659 (1973)
6. Boyle, P.: Options: a Monte Carlo approach. J. Financ. Econ. **4**, 323–338 (1977)
7. Carr, P., Madan, D.: Option valuation using the fast Fourier transform. J. Comput. Finance **2**(4), 61–73 (1999)
8. Dupire, B.: Pricing with a smile. Risk 18–20 (1994)
9. George, P.L., Borouchaki, H.: Delaunay triangulation and meshing. Hermès, Paris (1998). Application to Finite Elements. Translated from the original, Frey, P.J., Canann, S.A. (eds.), French (1997)
10. Glasserman, P.: Monte-Carlo Methods in Financial Engineering, Stochastic Modeling and Applied Probability vol. 53. Springer, New York (2004)
11. Hecht, F., Pironneau, O., Le Yaric, A., et al.:. freefem++ documentation. http://www.freefem.org
12. Heston, S.: A closed form solution for options with stochastic volatility with application to bond and currency options. Rev. Financ. Stud. **6**(2), 327–343 (1993)
13. Lewis, A.: A simple option formula for general jump-diffusion and other exponential Lévy processes (2001). http://www.optioncity.net
14. Loeper, G., Pironneau, O.: A mixed PDE/Monte-Carlo method for stochastic volatility models. C. R. Acad. Sci., Ser. 1 Math. **347**, 559–563 (2009)
15. Merton, R.C.: Option pricing when underlying stock returns are discontinuous. J. Financ. Econ. **3**, 125–144 (1976)
16. Pironneau, O.: Dupire Identities for Complex Options. C. R. Math. Acad. Sci. (2013, to appear)
17. Li, X.S., Demmel, J.W., Gilbert, J.R.: The superLU library. http://crd.lbl.gov/xiaoye/SuperLU
18. Wilmott, P., Howison, S., Dewynne, J.: The Mathematics of Financial Derivatives. Cambridge University Press, Cambridge (1995)

$h - P$ Finite Element Approximation for Full-Potential Electronic Structure Calculations

Yvon Maday

Abstract The (continuous) finite element approximations of different orders for the computation of the solution to electronic structures was proposed in some papers and the performance of these approaches is becoming appreciable and is now well understood. In this publication, the author proposes to extend this discretization for full-potential electronic structure calculations by combining the refinement of the finite element mesh, where the solution is most singular with the increase of the degree of the polynomial approximations in the regions where the solution is mostly regular. This combination of increase of approximation properties, done in an a priori or a posteriori manner, is well-known to generally produce an optimal exponential type convergence rate with respect to the number of degrees of freedom even when the solution is singular. The analysis performed here sustains this property in the case of Hartree-Fock and Kohn-Sham problems.

Keywords Electronic structure calculation · Density functional theory · Hartree-Fock model · Kohn-Sham model · Nonlinear eigenvalue problem · $h - P$ version · Finite element method

Mathematics Subject Classification 65N25 · 65N30 · 65T99 · 35P30 · 35Q40 · 81Q05

1 Introduction

The basic problem in quantum chemistry starts from the postulate of the existence of a time dependent complex function of the coordinates $\boldsymbol{x}$ called the wave function Ψ that contains all possible information about the system we want to consider. The evolution of this wave function depends on its current state through the fol-

Y. Maday (✉)
UPMC University, Paris 06, UMR 7598 LJLL, Paris 75005, France
e-mail: maday@ann.jussieu.fr

Y. Maday
Institut Universitaire de France and Division of Applied Mathematics, Brown University, Providence, RI, USA

P.G. Ciarlet et al. (eds.), *Partial Differential Equations: Theory, Control and Approximation*, DOI 10.1007/978-3-642-41401-5_14,

lowing equation proposed by Schrödinger: It involves a potential-energy function V that takes into account internal or external interactions as, for instance, those of electrostatic nature; for a single particle, it takes the form

$$\mathrm{i}\hbar\frac{\partial\Psi}{\partial t} = \mathcal{H}\Psi \equiv -\frac{\hbar^2}{2m}\nabla_{\boldsymbol{x}}^2\Psi + V\Psi.$$

The understanding of what the wave function represents was provided by Born who postulated, after Schrödinger, that $|\Psi(x,t)|^2\mathrm{d}\boldsymbol{x}$ represents the probability density of finding at time t the particle at position $\boldsymbol{x}$. The wave function Ψ is thus normalized in such a way that the spatial L^2 norm of Ψ is 1. The strength of the concept comes from the fact that it applies to any system, in particular to molecules; the coordinates $\boldsymbol{x}$ are then the positions of each particle (electrons and nuclei) of the system: hence $\boldsymbol{x}$ belongs to $\mathbb{R}^{3(N+M)}$, where N is the number of electrons and M is the number of nuclei. The Schrödinger's equations contains all the physical information on the system it is applied to, it does not involve any empirical parameter except some fundamental constants of physics like the Planck constant, the mass and charge of the electrons and nuclei .... It is thus a fantastic tool to better understand, predict and control the properties of matter from the fundamental background. The very simple Schrödinger equation in appearance is however set in a much too high dimensional framework: $1 + 3(N + M)$, so that it is not tractable for most problems of interest, except that a Quantum Monte Carlo (or QMC for short) approach is used to model and approximate the solutions. These QMC methods allow now to have access to properties other than the energy, including dipole and quadrupole moments, as well as matrix elements between different electronic states. Development and implementation of linear scaling QMC, analytical forces, wave function optimization, and embedding techniques are being pursued (see, e.g., [35, 36]).

For direct methods, though, simplifications need to be proposed to make this much too high dimensional problem accessible to numerical discretizations and simulations. Taking into account the time is quite easy from the principle of the separation of variables in case where the potential V does not depend on time. As it is classical in this approach, the problem becomes time independent and takes the form of an eigenvalue problem:

$$-\frac{\hbar^2}{2m}\nabla_{\boldsymbol{x}}^2\Psi + V\Psi = E\Psi, \tag{1.1}$$

where E has the dimension of an energy.

Through the variation principle, the various solutions to this (linear) eigenproblem, starting by the one associated with the smallest eigenvalue, are associated with an Hamiltonian energy $\langle\Psi \mid \mathcal{H}\Psi\rangle$, and, the ground state energy of the molecule corresponds to the smallest eigenvalue in (1.1). This interpretation through a variation principle does not simplify the matter but leads to tractable simplified models. The first one—known as the Born Oppenheimer approximation (see [5])—allows to separate the behavior of nuclei and electrons taking into account their large difference of masses. By considering the nuclei as fixed (or moving very slowly), the problem focuses on the behavior of the electrons—in the so called electronic structure calculation—and is thus related to the wave function Ψ that depends on N variables in $\mathbb{R}^3$ (the position of the electrons) and is parametrized by the position of the M nuclei in the associated Hamiltonian.

In order to comply with the Pauli principle of exclusion, the electronic wave function has to be antisymmetric with respect to the electron positions. The electronic problem thus consists in the minimization of the Schrödinger's Hamiltonian over all L^2 normalized, antisymetric wave functions. By minimizing instead on a smaller set of functions provides a tractable problem at the price of yielding to a larger ground state energy. This is the matter of the Hartree Fock problem that consists in minimizing the actual Schrödinger's energy over all wave functions that are written as a so called Slater determinant, i.e., a determinant: $\det[\phi_i(\boldsymbol{x}_j)]$, where the one electron orbitals ϕ_i $(i = 1, \ldots, N)$ are unknown functions over $\mathbb{R}^3$. The minimization problem over such Slater determinants leads to a minimization problem involving a new energy.

Let us describe this model associated to a so called closed-shell system with an even number $\mathcal{N} = 2N$ of electrons, the electronic state is described by N orbitals $\Phi = (\phi_1, \ldots, \phi_N)^{\mathrm{T}} \in (H^1(\mathbb{R}^3))^N$ satisfying the orthonormality conditions

$$\int_{\mathbb{R}^3} \phi_i \phi_j \mathrm{d}\boldsymbol{x} = \delta_{ij},$$

and the associated electronic density

$$\rho_\Phi(x) := 2\sum_{i=1}^{N} |\phi_i(x)|^2.$$

The factor 2 in the above expression accounts for the spin. In closed-shell systems, each orbital is indeed occupied by two electrons, one with spin up and one with spin down.

We then introduce the admissible space for molecular orbitals

$$\mathcal{M} = \left\{ \Phi = (\phi_1, \ldots, \phi_N)^{\mathrm{T}} \in \left(H^1_\#(\Gamma)\right)^N \mid \int_\Gamma \phi_i \phi_j \mathrm{d}\boldsymbol{x} = \delta_{ij} \right\}.$$

In the case where the molecular system we consider is in vacuo and consists of M nuclei of charges $(z_1, \ldots, z_M) \in (\mathbb{N} \setminus \{0\})^M$ located at the positions $(R_1, \ldots, R_M) \in (\mathbb{R}^3)^M$ of the physical space, and of N pairs of electrons, the so called Hartree Fock problem reads: Find Φ^0 such that

$$\mathcal{I}_N^{\mathrm{HF}}(V^{\mathrm{nuc}}) \equiv \mathcal{E}^{\mathrm{HF}}(\Phi^0) = \inf\{\mathcal{E}^{\mathrm{HF}}(\Phi),\ \Phi \in \mathcal{M}\} \tag{1.2}$$

with

$$\mathcal{E}^{HF}(\{\phi_i\}) = \sum_{i=1}^{N} \int_{\mathbb{R}^3} |\nabla\phi_i|^2 \mathrm{d}\boldsymbol{x} + \int_{\mathbb{R}^3} V^{\mathrm{nuc}} \rho_\Phi \mathrm{d}\boldsymbol{x} + \frac{1}{2}\int_{\mathbb{R}^3}\int_{\mathbb{R}^3} \frac{\rho_\Phi(\boldsymbol{x})\rho_\Phi(\boldsymbol{x}')}{|\boldsymbol{x}-\boldsymbol{x}'|} \mathrm{d}\boldsymbol{x}\mathrm{d}\boldsymbol{x}'$$

$$- \frac{1}{2}\int_{\mathbb{R}^3}\int_{\mathbb{R}^3} \frac{|\tau_\Phi(\boldsymbol{x},\boldsymbol{x}')|^2}{|\boldsymbol{x}-\boldsymbol{x}'|} \mathrm{d}\boldsymbol{x}\mathrm{d}\boldsymbol{x}', \tag{1.3}$$

$$\tau_\Phi(\boldsymbol{x},\boldsymbol{x}') = 2\sum_{i=1}^{N}\phi_i(\boldsymbol{x})\phi_i(\boldsymbol{x}'), \qquad \rho_\Phi(\boldsymbol{x}) = 2\sum_{i=1}^{N}|\phi_i(\boldsymbol{x})|^2,$$

$$V^{\text{nuc}}(\boldsymbol{x}) = -\sum_{k=1}^{M}\frac{z_k}{|\boldsymbol{x}-R_k|}. \tag{1.4}$$

An alternative formalism, different in nature but that ends up to a similar mathematical problem, is based on the key result of Hohenberg and Kohn [30] that shows that ground state properties of a system is fully described by the electronic density. This led to the density functional theory, with the instrumental approach of Kohn and Sham [31]. Indeed, from [30] the existence of an energy functional of the electronic density was established, this result is weakened however by the lack, even as of today, of knowledge of its proper functional form. It follows from the Hohenberg-Kohn theorem (see [30, 37, 38, 56]), that there exists an exact functional, that is a functional of the electronic density ρ that provides the ground state electronic energy and density of the $\mathcal{N}$-body electronic Schrödinger equation. The work of Kohn and Sham addressed this issue by providing approximations of the energy functional and laid the foundations for the practical application of DFT to materials systems.

The Kohn-Sham approach reduces the many-body problem of interacting electrons into an equivalent problem of non-interacting electrons in an effective mean field that is governed by the electron density. It is formulated in terms of an unknown exchange-correlation term that includes the quantum-mechanical interactions between electrons. Even though this exchange-correlation term is approximated and takes the form of an explicit functional of electron density, these models were shown to predict a wide range of materials properties across various materials systems. The development of increasingly accurate and computationally tractable exchange-correlation functionals is still an active research area in electronic structure calculations.

In the Kohn-Sham model, also described in the closed-shell configuration, the ground state is obtained by solving the minimization problem: Find Φ^0 such that

$$\mathcal{I}_N^{KS}(V) \equiv \mathcal{E}^{\text{KS}}(\Phi^0) = \inf\{\mathcal{E}^{\text{KS}}(\Phi),\ \Phi \in \mathcal{M}\}, \tag{1.5}$$

where the Kohn-Sham energy functional reads

$$\mathcal{E}^{\text{KS}}(\Phi) := \sum_{i=1}^{N}\int_{\mathbb{R}^3}|\nabla\phi_i|^2\mathrm{d}\boldsymbol{x} + \int_{\mathbb{R}^3}V^{\text{nuc}}\rho_\Phi\,\mathrm{d}\boldsymbol{x} + \frac{1}{2}\int_{\mathbb{R}^3}\int_{\mathbb{R}^3}\frac{\rho_\Phi(\boldsymbol{x})\rho_\Phi(\boldsymbol{x}')}{|\boldsymbol{x}-\boldsymbol{x}'|}\mathrm{d}\boldsymbol{x}\,\mathrm{d}\boldsymbol{x}' + E_{\text{xc}}(\rho_\Phi). \tag{1.6}$$

The first term models the kinetic energy of Φ, the second term models the interactions between nuclei and electrons, and the third term models the interaction between electrons. The fourth term, called the exchange-correlation functional actually collects the errors made in the approximations of the kinetic energy and of the interactions between electrons by the first and third terms of the Kohn-Sham functional, respectively, as follows from the Hohenberg-Kohn theorem. The lack of precise knowledge for the Kohn-Sham functional is localized on this exchange-correlation term only. It therefore has to be approximated in practice. The local density approximation (or LDA for short) consists in approximating the exchange-correlation functional by

$$\int_{\mathbb{R}^3} e_{\mathrm{xc}}^{\mathrm{LDA}}\big(\rho(\boldsymbol{x})\big)\mathrm{d}\boldsymbol{x},$$

where $e_{\mathrm{xc}}^{\mathrm{LDA}}(\overline{\rho})$ is an approximation of the exchange-correlation energy per unit volume in a uniform electron gas with density $\overline{\rho}$. The resulting Kohn-Sham LDA model is well understood from a mathematical viewpoint (see [1, 33]). On the other hand, the existence of minimizers for Kohn-Sham models based on more refined approximations of the exchange-correlation functional, such as generalized gradient approximations (see [1]) or exact local exchange potentials (see [12]) in the general case, is still an open problem.

Note that the Kohn-Sham problem can be split-up into two problems of minimization, with one among them being stated as a pure density problem. We first define the set of admissible densities:

$$\mathcal{R}_N = \left\{\rho \geq 0, \sqrt{\rho} \in H^1\big(\mathbb{R}^3\big), \int_{\mathbb{R}^3} \rho \mathrm{d}\boldsymbol{x} = N\right\}, \tag{1.7}$$

then we propose the first problem

$$\begin{aligned} T_{KS}(\rho) = \inf\Bigg\{ &\sum_{i=1}^{N} \int_{\mathbb{R}^3} |\nabla \phi_i|^2 \mathrm{d}\boldsymbol{x},\, \Phi = (\phi_i)_{i=1,\dots,N}, \forall i, j = 1, \dots, N, \\ &\int_{\mathbb{R}^3} \phi_i \phi_j \mathrm{d}\boldsymbol{x} = \delta_{i,j},\, \rho = \rho_{\Phi} \Bigg\} \end{aligned} \tag{1.8}$$

followed by the pure density functional problem

$$\begin{aligned} \mathcal{I}_N^{KS}(V) = \inf\Bigg\{ \mathcal{F}(\rho) = T_{KS}(\rho) + \int_{\mathbb{R}^3} V^{\mathrm{nuc}} \mathrm{d}\boldsymbol{x}\rho + \frac{1}{2}\int_{\mathbb{R}^3}\int_{\mathbb{R}^3} \frac{\rho(\boldsymbol{x})\rho(\boldsymbol{x}')}{|\boldsymbol{x}-\boldsymbol{x}'|}\mathrm{d}\boldsymbol{x}\mathrm{d}\boldsymbol{x}' \\ + E_{\mathrm{xc}}(\rho)\Bigg\}. \end{aligned} \tag{1.9}$$

There are a lot of variations on the frame of the simulation of these equations. First we may be interested in simulating a molecule alone, small or big, the molecule may also have neighbors, and these can be taken into account exactly or in an average manner like for molecules in solvation (see [52]). When there are many molecules, these can be arranged in a periodic array that is exactly periodic or contains some local defects then the simulation will be done on a very large box composed of many cells, one of them containing the defect. In this case, the simulation domain, sometimes referred to as the supercell, is no longer the whole space $\mathbb{R}^3$, as in (1.5); it is the unit cell Γ of some periodic lattice of $\mathbb{R}^3$. In the periodic Kohn-Sham framework, the periodic boundary conditions are imposed to the Kohn-Sham orbitals (Born-von Karman PBC). Imposing PBC at the boundary of the simulation cell is the standard method to compute condensed phase properties with a limited number of atoms in the simulation cell, hence at a moderate computational cost.

Both minimization problems (1.2)–(1.5) lead to the resolution of a nonlinear eigenvalue problem, where the eigensolutions are atomic orbitals, function over $\mathbb{R}^3$, that thus become tractable to numerical simulations. In order to formulate these eigenproblems, we have to introduce the Hamiltonian for the Hartree-Fock or Kohn-Sham energies:

$$\mathcal{H}_{\Phi}^{\mathrm{HF}} = -\frac{1}{2}\Delta + \left(V^{\mathrm{nuc}} + V_{\rho_\Phi}^{\mathrm{Coulomb}} - V_{\tau_\Phi}^{\mathrm{Exchange}}\right) = h + \mathcal{V}_\Phi,$$

where

$$h = -\frac{1}{2}\Delta + V^{\mathrm{nuc}}, \qquad \mathcal{V}_\Phi = V_{\rho_\Phi}^{\mathrm{Coulomb}} - V_{\tau_\Phi}^{\mathrm{Exchange}}, \tag{1.10}$$

where V^{Coulomb} and V^{Exchange} are defined for any ψ by

$$V_\rho^{\mathrm{Coulomb}}\psi(\boldsymbol{x}) = \left(\rho \star \frac{1}{|\boldsymbol{x}|}\right)\psi(\boldsymbol{x}), \qquad V_\tau^{\mathrm{Exchange}}\psi(\boldsymbol{x}) = \int_{\mathbb{R}^3} \frac{\tau(\boldsymbol{x},\boldsymbol{y})}{|\boldsymbol{x}-\boldsymbol{y}|}\psi(\boldsymbol{y})\mathrm{d}\boldsymbol{y},$$
$$\forall \boldsymbol{x} \in \mathbb{R}^3. \tag{1.11}$$

We notice that $\mathcal{E}^{HF\prime}(\Phi^0) = 4\mathcal{H}_{\Phi^0}^{\mathrm{HF}}\Phi^0$ and thus the Euler equations associated with the minimization problem (1.2) read

$$\mathcal{H}_{\Phi^0}^{\mathrm{HF}}\phi_i^0 = \sum_{j=1}^{N} \lambda_{ij}^0 \phi_j^0, \quad \forall 1 \le i \le N, \tag{1.12}$$

where the $N \times N$ matrix $\Lambda_N^0 = (\lambda_{ij}^0)$, which is the Lagrange multiplier of the matrix constraint $\int_\Gamma \phi_i \phi_j \mathrm{d}\boldsymbol{x} = \delta_{ij}$, is symmetric.

In fact, the problem (1.2) has an infinity of minimizers since any unitary transform of the Hartree-Fock orbitals Φ^0 is also a minimizer of the Hartree-Fock energy. This is a consequence of the following invariance property:

$$U\Phi \in \mathcal{M} \quad \text{and} \quad \mathcal{E}^{HF}(U\Phi) = \mathcal{E}^{HF}(\Phi), \quad \forall \Phi \in \mathcal{M}, \ \forall U \in \mathcal{U}(N), \tag{1.13}$$

where $\mathcal{U}(N)$ is the group of the real unitary matrices:

$$\mathcal{U}(N) = \left\{U \in \mathbb{R}^{N\times N} \mid U^T U = 1_N\right\},$$

1_N denoting the identity matrix of rank N. This invariance can be exploited to diagonalize the matrix of the Lagrange multipliers of the orthonormality constraints (see, e.g., [18]), yielding the existence of a minimizer (still denoted by Φ^0), such that

$$\mathcal{H}^{\mathrm{HF}}_{\Phi^0}\phi_i^0 = \epsilon_i^0\phi_i^0 \tag{1.14}$$

for some $\epsilon_1^0 \le \epsilon_2^0 \le \cdots \le \epsilon_N^0$.

Similarly, for the Kohn Sham problem, we introduce the associated Hamiltonian

$$\mathcal{H}^{\mathrm{KS}}_{\Phi} = -\frac{1}{2}\Delta + \left(V^{\mathrm{nuc}} + V^{\mathrm{Coulomb}}_{\rho_\Phi} + \frac{\mathrm{d}e^{\mathrm{LDA}}_{\mathrm{xc}}}{\mathrm{d}\rho}(\rho_\Phi)\right) = h + \mathcal{V}_{\rho_\Phi},$$

where h is the same as above and

$$\mathcal{V}_\rho = V^{\mathrm{Coulomb}}_\rho + \frac{\mathrm{d}e^{\mathrm{LDA}}_{\mathrm{xc}}}{\mathrm{d}\rho}(\rho). \tag{1.15}$$

The same analysis leads to an eigenvalue problem. By using again the invariance through unitary transforms (1.13), that still holds for the Kohn-Sham problem, we get the existence of a minimizer with a set of molecular orbitals still denoted as Φ^0, such that

$$\mathcal{H}^{\mathrm{KS}}_{\Phi^0}\phi_i^0 = \epsilon_i^0\phi_i^0 \tag{1.16}$$

for some $\epsilon_1^0 \le \epsilon_2^0 \le \cdots \le \epsilon_N^0$.

2 About Numerical Methods

2.1 Generalities

For problems set in a periodic framework (analysis of crystals), the approximation by plane waves (Fourier) has traditionally been one of the popular approaches used for solving the Kohn-Sham problem since it allows for an efficient computation of the electrostatic interactions through Fourier transforms. In addition the plane waves (Fourier) approximation is a high order method that is fully deployed if the solutions to be approximated are very regular. Unfortunately, for full potential electronic structure calculations, the nuclear potential is not smeared out and induces singularities in the solutions (atomic orbitals and density) at the level of the nuclei, more precisely cusps in place of the nuclei and rapidly varying wave functions in their vicinity (see [24, 29]). Another drawback of these methods lies in the nonlocality of the basis set that leads to a uniform spatial resolution which can be useless e.g. for materials systems with defects, where higher basis resolution is required in some spatial regions and a coarser resolution suffices elsewhere. In practice of such discretizations, the singular nuclear potential V^{nuc} defined by (1.4) is usually replaced with a smoother potential V^{ion}; this amounts to replacing point nuclei with smeared nuclei. Not surprisingly, the smoother the potential, the faster the convergence of the planewave approximation to the exact solution of (1.2) or (1.5) (see [8]). The nuclear potential V^{nuc} is replaced by a pseudopotential modeling the Coulomb interaction between the valence electrons on the one hand, and the nuclei and the core electrons on the other hand. The pseudopotential consists of two terms: a local component $V_{\rm local}$ (whose associated operator is the multiplication by the function $V_{\rm local}$) and a nonlocal component. As a consequence, the second term in the Kohn-Sham energy functional (1.6) is replaced by

$$\int_\Gamma \rho_\Phi V_{\rm local} \mathrm{d}\boldsymbol{x} + 2 \sum_{i=1}^{N} \langle \phi_i | V_{\rm nl} | \phi_i \rangle.$$

The pseudopotential approximation gives satisfactory results in most cases, but sometimes fails. Note that a mathematical analysis of the pseudopotential approximation is still lacking. Moreover, the core electrons need sometimes be considered since they are responsible for intricate properties. The full-potential/all-electron calculation is thus sometimes necessary. In order to overcome the convergence difficulties of the plane wave approximations, resulting from the cusp singularities one can augment the plane waves bases set as done in the augmented plane wave (or APW for short) method (see [42, 50]), which is among the most accurate methods for performing electronic structure calculations for crystals. We refer to [16] for the numerical analysis of the convergence based on the careful analysis of the properties of the cusp that we shall recall in Sect. 2.2. These APWs provide very good results, at the price however of two remaining drawbacks. The first one is that of the periodic framework that does not fit for single molecules or molecules in solvent. The second one is that the basis come from two different families and the locality of the plane waves (orthonormal basis) is lost.

For efficient computations in the case of all-electron calculations on a large materials system, approximation methods based on Gaussian basis are among the other most classical methods. An example is using the Gaussian package (see [25]). These approaches initially introduced on Hartree-Fock problem, have been developed both for reasons of accuracy and easiness of implementation due to the fact that product of these basis functions arising in nonlinear terms of the potential are easy to evaluate through analytical expressions. The basis functions are centered at each nuclei and are fitted so as to represent well the behavior of the atomic orbital at the level of the cusp and at infinity. There exist a large amount of know how in these methods, that benefit from highly optimized Gaussian basis functions on many molecules. When this expertise does not exist, the approximation properties of the Gaussian expansion are more questionable. We refer e.g. to [9, 10] for the presentation and numerical analysis in this context.

Due to the large nonlinearities encountered in the energies involved in advanced Kohn-Sham models, the complexity of the computations, when it turns to implement the methods, scales as $\mathcal{O}(N^d)$, where N is the number of degrees of freedom, and d can be pretty large ($d \geq 3$). One way is to "squeeze" at most the numerical scheme, performing, at the mathematical level, what computational experts in simulations for electronic structure calculations design when they propose ad-hoc discrete basis (e.g. contracted Gaussian bases sets). The expertise here is based on the mathematical arguments involved in model reduction techniques (the reduced basis approximation), and we refer to [11, 40] for a presentation of these techniques. They are based on adapted (not universal) discretizations and are shown to provide good approximations, but are still in their infancy.

There is thus room for the development of more robust approaches for electronic structure calculations, like for example finite element approximation of low or high order that, in other contexts (fluid mechanics, structure mechanics, wave ...), are of classical use. There has been already quite a lot of experiences in the domain of quantum chemistry even though the relative number of contributions is still small. We refer to [2, 6, 21, 34, 39, 43, 45–47, 51, 53–55, 58, 59] and the references therein for an overview of the contributions in this direction.

In order to be competitive with respect to plane-wave basis or Gaussian type basis, though, the full knowledge and expertise in the finite element machinery has to be requested, indeed, as appears in e.g. [6, 28], the accuracy required for electronic structure calculation involves of the order of 100000 basis functions per atom for $\mathbb{P}_1$ finite elements, which is far too expensive and the use of higher order finite element methods is thus the only viable way. However, the use of high-order finite elements has some consequences on the complexity of the implementation, indeed these require the use of higher-order accurate numerical quadrature rules with larger stencils of points and leads also to increase in the bandwidth of the stiffness matrices that grow cubically with the order of the finite-element, with a mass matrix that, contrarily to what happens for plane wave approximation, is not diagonal. In addition, using high order methods in regions where the solution presents singularities is a waste of resources.

The right question to raise is thus not accuracy with respect to number of degrees of freedom but accuracy with respect to run time. In this respect, the publication [57] analyses in full details on a variety of problems and regularity of solutions, the accuracy achieved by low to high order finite element approximations as a function of the number of degrees of freedom and of the run time. It appears, with respect to this second argument that the use of degrees between 5 and 8 is quite competitive. Of course, the answer depends on the implementation of the discretization method and the exact properties of the solution to be approximated but this indicates a tendency that is confirmed, both by the numerical analysis and by implementation on a large set of other applications.

A recent investigation in the context of orbital-free DFT indicates that the use of higher-order finite elements can significantly improve the computational efficiency of the calculations (see [44, 51]). For instance, a 100 to 1000 fold computational advantage was reported by using a fourth to sixth order finite element in comparison to a linear finite element. This involves a careful implementation of various numerical and computational techniques: (i) an a priori mesh adaption technique to construct a close to optimal finite element discretization of the problem; (ii) an efficient solution strategy for solving the discrete eigenvalue problem by using spectral finite elements in conjunction with Gauss-Lobatto quadrature, and a Chebyshev acceleration technique for computing the occupied eigenspace (see [44]).

As far as we are aware of, the current implementations of the finite element method involve uniform degree of the polynomial approximation. This results in an improved accuracy-per-node ratio that is still polynomial in the number of degrees of freedom. This is actually a bit disappointing since, as is explained in a series of papers by Fournais, Sørensen, Hoffmann-Ostenhof, and Hoffmann-Ostenhof [22–24, 29] the solution is analytic (with exponential convergence to zero at infinity on unbounded domains) at least away from the position of the nuclei, where, if exact singular potential are used, the knowledge on singular behavior of the solution is rather well known (this one being of the shape $e^{-\frac{Zr}{2}}$), which results that the solution is not better than $H^{\frac{5}{2}}$ around the singularities.

In the finite element culture, such behavior—very regular except at some point where the behavior of the pointwise singularity is known—is know to allow for an exponential convergence with respect to the number of degrees of freedom. Indeed, in a series of papers written by Babuška and co-authors [3, 26, 27], a careful analysis is performed that leads to the conclusion that the $h-P$ version of the finite element method allows for an exponential rate of convergence when solving problems with piecewise analytic data. In particular, in the three papers [3, 26, 27], the authors focus on the approximation of the function $(x-\xi)_+^\alpha$ over $(0,1)$ for $\xi \in (0,1)$, this simple function is a prototype of pointwise singular behavior that can be present in the solution of regular problems in geometries with corners or edges, or for problems with non regular coefficients. It is straightforward to check that when

(1) $\alpha > -\frac{1}{2}$ then the function is in $L^2(0,1)$,
(2) $\alpha > \frac{1}{2}$ then the function is in $H^1(0,1)$,
(3) $\alpha > \frac{3}{2}$ then the function is in $H^2(0,1)$,
(4) $\alpha > \frac{5}{2}$ then the function is in $H^3(0,1)$.

We believe that it is interesting to summarize the conclusion of these papers as follows:

(1) For the P version of the FEM (or spectral method (see [13])): if $\xi \in (0,1)$, the convergence of the best fit is of the order of $\frac{C}{P^{\alpha-\frac{1}{2}}}$ (i.e., $\frac{C}{P^r}$ where r is related to the regularity of the function). Note that, if $\xi = 0$ (or $\xi = 1$), the convergence of the best fit is of the order of $\frac{C}{P^{2\alpha-1}}$. This phenomenon is know as the doubling of convergence for singular functions (see, e.g., [4] for more results in this direction).

(2) For the $h-P$ version of the FEM (or spectral element method): The approximation is generally of the order of $h^{\min(\alpha\frac{1}{2}, P+1)}$.

(3) For the $h-P$ version, with a graded mesh, i.e., the size of the mesh diminishes as one gets closer to the singularities and P uniformly increasing: The approximation can be of exponential order with respect to the number of degrees of freedom.

(4) For the optimal $h-P$ version of the finite element method: The approximation can be of exponential order with a better rate (with respect to the above rate) if the degree P that is used in the largest elements increases while the graded mesh is refined in the neighborhood of the singularity. Starting from a uniform mesh, the elements that contain the singularity are recursively refined; starting from the singularity, the degree of the approximation is equal to 1 and linearly increases with the distance to the singularities in the other elements; the error then scales like

$\exp(-cN_h^{\beta})$, where N_h is the number of degrees of freedom of the finite element approximation.

2.2 Regularity Results

The natural question is then: What is the regularity of the density, the solution to the Hartree Fock or Kohn Sham problems?

It is proven (see, e.g., the careful analysis of [22–24]) that the solution to such systems is analytic (with exponential convergence to zero on unbounded domains) at least away from the position of the nuclei, where, if exact singular potential is used there, the solution is not better than $H^{\frac{5}{2}}$ around the singularities (this one being of the shape $e^{-\frac{Zr}{2}}$).

For the same reasons as the doubling of convergence, if the finite element vertices are on the nuclei positions, then there is a doubling of convergence for the P and $h - P$ version of the approximation leading to a convergence rate like P^{-3} for the solution obtained with polynomial degree 4, if the mesh is uniform. The energy is approximated with another doubling of accuracy, i.e., P^{-6}.... This analysis deals only with the polynomial approximation without taking care of any h effect and is consistent with the analysis of the paper [20]. At this level, we want to emphasize on two points for which we refer to [16]. The first one deals with a better understanding of the characteristics of the singularity of the solution of the Hartree Fock problem at the level of nuclei; indeed it appears that locally, when expressed in spherical coordinates around the nuclei, the solution is infinitely differentiable. The second is to indicate that, from this knowledge, it is actually possible to propose combined approximations that allows exponential convergence with respect to the number of degrees of freedom (see also the recent approach in [39]).

In order to be more precise on the regularity of the solutions to such systems, we are going to place ourselves in an adapted framework. We follow [48, 49] to define, over any regular bounded domain Ω that contains (far from the boundary) all nuclei R_j, $j = 1, \ldots, M$, some weighted Sobolev spaces well suited for the numerical analysis of adapted finite element methods (as explained in [17], these weighted Sobolev spaces are well suited to characterize the singular behavior of solutions of general second-order elliptic boundary value problems in polyhedra). First, we define the subdomain Ω_0 which is the complementary in Ω to the union of small enough balls ω_j around each nuclei position R_j, $j = 1, \ldots, M$. In addition, to each nuclei position R_j, $j = 1, \ldots, M$, we associate an exponent β_j, and the following semi-norms for any $m \in \mathbb{N}$:

$$|u|^2_{M^m_{\underline{\beta}}(\Omega)} = |u|_{H^m(\Omega_0)} + \sum_{j=1}^{M} \sum_{\underline{\alpha}=m} \|r_j^{\beta_j+|\underline{\alpha}|} D^{\underline{\alpha}} u\|_{L^2(\omega_i)} \tag{2.1}$$

where r_j denotes the distance to R_j and norm

$$\|u\|^2_{M^m_{\underline{\beta}}(\Omega)} = \sum_{k=0}^{m} |u|^2_{M^k_{\underline{\beta}}(\Omega)}, \tag{2.2}$$

where $D^{\underline{\alpha}}$ denotes the derivative in the local coordinate directions corresponding to the multi-index $\underline{\alpha}$.

The space $M^m_{\underline{\beta}}(\Omega)$ is then the closure of $C_0^\infty(\Omega)$ of all infinitely differentiable functions that vanish on the boundary of Ω.

From the results stated above, we deduce that the solutions of the Hartree-Fock problem ϕ_i^0 for any i $(1 \le i \le N)$ belong to such spaces (they are said asymptotically well behaved) and moreover

$$|\phi_i^0|_{M^m_{\underline{\beta}}(\Omega)} \le C^m m! \tag{2.3}$$

with $\beta_j > -\frac{3}{2}$. In [16], it is indicated that the same type of result can be assumed for the solution to the Kohn-Sham problem, at least for regular enough exchange correlation potential. In what follows we shall assume that the same regularity result holds for those solutions.

3 Galerkin Approximation

3.1 Generalities on the Variational Approximation

Let us consider a family of finite dimensional spaces X_δ, with dimension N_δ. We assume that it is defined through the data of a finite basis set $\{\chi_\mu\}_{1\le\mu\le N_\delta}$. Let us assume that these are subspaces of $H_0^1(\Omega)$ (for the time being this means that the discrete functions should be continuous)

The variational approximations of the Hartree-Fock or Kohn-Sham problems are

$$\mathcal{I}^{HF}_{N,\delta}(V) = \inf\left\{\mathcal{E}^{HF}(\Phi_\delta),\ \Phi_\delta = (\phi_{1,\delta},\ldots,\phi_{N,\delta})^{\mathrm{T}} \in (X_\delta)^N,\right.$$
$$\left.\int_{\mathbb{R}^3} \phi_{i,\delta}\phi_{j,\delta}\mathrm{d}\mathbf{x} = \delta_{ij},\ 1 \le i, j \le N\right\} \tag{3.1}$$

and

$$\mathcal{I}^{KS}_{N,\delta}(V) = \inf\left\{\mathcal{E}^{\mathrm{KS}}(\Phi_\delta),\ \Phi_\delta = (\phi_{1,\delta},\ldots,\phi_{N,\delta})^{\mathrm{T}} \in (X_\delta)^N,\right.$$
$$\left.\int_{\mathbb{R}^3} \phi_{i,\delta}\phi_{j,\delta}\mathrm{d}\mathbf{x} = \delta_{ij},\ 1 \le i, j \le N\right\}. \tag{3.2}$$

The solution to the Galerkin approximation procedure is determined by

$$\phi_{i,\delta} = \sum_{\mu=1}^{N_\delta} C_{\mu i} \chi_\mu .$$

Hence by the determination of the rectangular matrix $C \in \mathcal{M}(N_\delta, N)$ contains the N_δ coefficients of the molecular orbital $\phi_{i,\delta}$ in the basis $\{\chi_\mu\}_{1\le\mu\le N_\delta}$. It is classical in this context to introduce the so called overlap matrix S defined as

$$S_{\mu\nu} = \int_{\mathbb{R}^3} \chi_\mu \chi_\nu \mathrm{d}\boldsymbol{x}, \tag{3.3}$$

so that the constraints $\int_{\mathbb{R}^3} \phi_{i,\delta}\Phi_{j,\delta}\mathrm{d}\boldsymbol{x} = \delta_{ij}$ on the discrete solutions read

$$\delta_{ij} = \sum_{\mu=1}^{N_\delta}\sum_{\nu=1}^{N_\delta} C_{\mu j} S_{\mu\nu} C_{\nu i},$$

or again in matrix form

$$C^* S C = I_N .$$

Similarly,

$$\begin{aligned}
&\sum_{i=1}^{N} \frac{1}{2}\int_{\mathbb{R}^3} |\nabla \phi_{i,\delta}|^2 \mathrm{d}\boldsymbol{x} + \int_{\mathbb{R}^3} \rho_{\Phi_\delta} V^{\mathrm{nuc}} \mathrm{d}\boldsymbol{x} \\
&= \sum_{i=1}^{N} \left(\frac{1}{2}\int_{\mathbb{R}^3} \left| \nabla \sum_{\mu=1}^{N_\delta} C_{\mu i}\chi_\mu \right|^2 \mathrm{d}\boldsymbol{x} + \int_{\mathbb{R}^3} V^{\mathrm{nuc}} \left| \sum_{\mu=1}^{N_\delta} C_{\mu i}\chi_\mu \right|^2 \mathrm{d}\boldsymbol{x} \right) \\
&= \sum_{i=1}^{N}\sum_{\mu=1}^{N_\delta}\sum_{\nu=1}^{N_\delta} h_{\mu\nu} C_{\nu i} C_{\mu i} = \mathrm{Trace}\big(hCC^*\big),
\end{aligned}$$

where $h \in \mathcal{M}_S(N_\delta)$ denotes the matrix of the operator $-\frac{1}{2}\Delta + V^{\mathrm{nuc}}$ in the basis $\{\chi_k\}$:

$$h_{\mu\nu} = \frac{1}{2}\int_{\mathbb{R}^3} \nabla\chi_\mu \cdot \nabla\chi_\nu \mathrm{d}\boldsymbol{x} + \int_{\mathbb{R}^3} V^{\mathrm{nuc}} \chi_\mu \chi_\nu \mathrm{d}\boldsymbol{x}. \tag{3.4}$$

Finally, we can write the Coulomb and exchange terms by introducing first the notations

$$(\mu\nu|\kappa\lambda) = \int_{\mathbb{R}^3}\int_{\mathbb{R}^3} \frac{\chi_\mu(x)\chi_\nu(x)\chi_\kappa(x')\chi_\lambda(x')}{|x-x'|} \mathrm{d}x\mathrm{d}x', \tag{3.5}$$

and, for any matrix X with size $N_\delta \times N_\delta$

$$J(X)_{\mu\nu} = \sum_{\kappa,\lambda=1}^{N_\delta} (\mu\nu \mid \kappa\lambda) X_{\kappa\lambda}, \qquad K(X)_{\mu\nu} = \sum_{\kappa,\lambda=1}^{N_\delta} (\mu\lambda \mid \nu\kappa) X_{\kappa\lambda}.$$

The Coulomb and exchange terms can respectively be expressed as

$$\begin{aligned}\int_{\mathbb{R}^3}\int_{\mathbb{R}^3} \frac{\rho_{\Phi_\delta}(x)\rho_{\Phi_\delta}(x')}{|x-x'|}\mathrm{d}x\mathrm{d}x' &= \sum_{\mu,\nu,\kappa,\lambda=1}^{N_\delta}\sum_{i,j=1}^{N} (\mu\nu|\kappa\lambda) C_{\mu i} C_{\nu i} C_{\kappa j} C_{\lambda j} \\ &= \mathrm{Trace}\big(J\big(CC^*\big)CC^*\big)\end{aligned}$$

and

$$\begin{aligned}\int_{\mathbb{R}^3}\int_{\mathbb{R}^3} \frac{|\tau_{\Phi_\delta}(x,x')|^2}{|x-x'|}\mathrm{d}x\mathrm{d}x' &= \sum_{\mu,\nu,\kappa,\lambda=1}^{N_\delta}\sum_{i,j=1}^{N} (\mu\lambda|\kappa\nu) C_{\mu i} C_{\nu i} C_{\kappa j} C_{\lambda j} \\ &= \mathrm{Trace}\big(K\big(CC^*\big)CC^*\big).\end{aligned}$$

The discrete problem, is thus written equivalently as a minimization problem over the space

$$\mathcal{W}_N = \big\{\mathcal{W}_{N_\delta} \in \mathcal{M}(N_\delta, N),\ C^*SC = 1_N\big\},$$

as

$$\inf\big\{E^{HF}\big(CC^*\big),\ C \in \mathcal{W}_{N_\delta}\big\}, \tag{3.6}$$

where for any $D \in \mathcal{M}_S(N_\delta)$

$$E^{HF}(D) = \mathrm{Trace}(hD) + \frac{1}{2}\mathrm{Trace}\big(J(D)D\big) - \frac{1}{2}\mathrm{Trace}\big(K(D)D\big).$$

The energy can also be written in term of the so-called density matrix

$$D = CC^*$$

leading to the problem

$$\inf\big\{E^{HF}(D),\ D \in \mathcal{P}_N\big\} \tag{3.7}$$

with

$$\mathcal{P}_N = \big\{D \in \mathcal{M}_S(N), DSD = D,\ \mathrm{Trace}(SD) = N\big\}.$$

Similarly the Kohn-Sham problem (3.2) reads

$$\mathcal{I}^{KS}_{N,\delta}(V) = \inf\big\{E^{KS}\big(CC^*\big),\ C \in \mathcal{W}_{N_\delta}\big\} \tag{3.8}$$

with

$$E^{KS}(D) = 2\,\mathrm{Trace}(hD) + 2\,\mathrm{Trace}\big(J(D)D\big) + E_{xc}(D),$$

here $E_{xc}(D)$ denotes the exchange-correlation energy:

$$E_{xc}(D) = \int_{\mathbb{R}^3} \rho(x)\epsilon_{xc}^{\mathrm{LDA}}\big(\rho(x)\big)\mathrm{d}x \quad \text{with } \rho(x) = 2\sum_{i=1}^{N} D_{\mu\nu}\chi_\mu(x)\chi_\nu(x).$$

For general analysis of these discrete problems and the associated approximation results, we refer to [7, 8, 14, 15, 20, 32, 41, 60] and the references therein.

3.2 The Adapted $h - P$ Discrete Spaces

Part 1 Definition of the $h - P$ Discrete Spaces The purpose of this section is to introduce the class of $h - P$ finite elements spaces for the approximation of the minimization problems (1.2) and (1.5) which we want to propose and analyze in this paper. They will be used to get an exponential convergence for the finite element approximation of the solution to the Hartree-Fock and Kohn-Sham problems. We start by truncating the domain $\mathbb{R}^3$ in a regular bounded domain Ω that, for the sake of simplicity, we shall consider to be a ball large enough to contain largely each nuclei, similarly as in papers where this class of approximations was proposed (see [3, 26, 27, 48, 49]).

The first step of the discretization consists in defining an initial triangulation $\mathcal{T}^0$ of Ω composed of hexahedral elements K subject to the following classical assumption:

(1) $\overline{\Omega} = \cup_{K\in\mathcal{T}^0}\overline{K}$,

(2) Each element K is the image of the reference cube $(-1, 1)^3$ under a diffeomorphism, that, in most cases, is an homothetic transformation (except of course on the boundary of Ω, but we shall not carefully analyze the approximation there since the solution is very regular at this level),

(3) The initial triangulation is conforming in the sense that the intersection of two different elements K and K' is either empty, a common vertex, a common edge or a common face,

(4) The initial triangulation is regular and quasi uniform in the sense that there exist two constants $\kappa_1, \kappa_2 > 0$ with $\kappa_2 h \le h_K \le \kappa_1 \rho_K$, where h_K denotes the diameter of K, ρ_K denotes the diameter of the largest circle that can be inscribed into K and $h = \max_K\{h_K\}$.

We assume in addition that the positions of the nuclei $R_1, R_2, \ldots, R_M$ are the vertices of some element in the initial triangulation $\mathcal{T}^0$. Starting from this initial hexahedral triangulation $\mathcal{T}^0$ of Ω, for the $h - P$ procedure, we define a family of so-called σ-geometric triangulations.

First for the triangulation, we refine recursively (by partitioning into 4 hexahedra) each hexahedron that admits a nuclei at the vertex. This partitioning is based

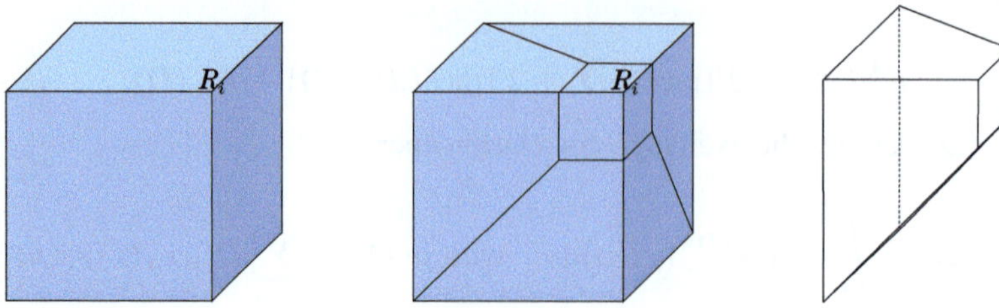

Fig. 1 The partitioning of a hexahedron, that admits a nuclei at R_i as a vertex, in a $(\sigma, 1-\sigma)$-ratio (*the left*: before the refinement, *the middle*: after the refinement, *the right*: an element created in the refined hexahedron that maintains the conformity with adjacent elements)

on a ratio $(\sigma, 1-\sigma)$ (with $0 < \sigma < 1$) of each edge starting from the vertex that coincide at the nucleus, as explained on Fig. 1. Note that the refinement only deals the elements that have a nuclei as a vertex, this refinement preserves the conformity with the non refined elements. The first refinement allows to define the triangulation denoted as $\mathcal{T}^1$. Next, the same process applied to $\mathcal{T}^i$ allows to define $\mathcal{T}^{i+1}$. For any element K of the new triangulations $\mathcal{T}^i$ (that of course are not quasi uniform anymore), h_K still denotes the diameter of K.

Concerning the degree of the approximation that is used over each elements, we start at the level $\mathcal{T}^0$ by using polynomials in Q_1, i.e., tri-affine, over each element. Then each time a triangulation refinement is performed, the new elements, created at this stage, are provided with Q_1 polynomials, while, on the other elements that did not change, the degree of the polynomial is increased by 1 unit in each variable (or by a fixed factor $\sigma > 0$ to be more general). In particular, the partial degree of the polynomial on the elements of a triangulation $\mathcal{T}^i$ that have never been cut is $\leq 1 + i\sigma$ and is uniform in all directions.

The discrete finite element space X^i_{h-P} is composed of all continuous such piecewise polynomials (associated to the triangulation $\mathcal{T}^i$) that vanish on the boundary of Ω. From the analysis in 1D explained in [26], exponential convergence for pointwise singular solutions can be obtained from the combination of the σ-geometric mesh refinement and σ-linear increase of the degree of the polynomials.

Part 2 Analysis of the $h - P$ Approximation The difficulty in this analysis lies in the conjunction of three facts:

(1) The problem is set in three dimensions.
(2) Neither the mesh nor the degree from one element to the other is uniform.
(3) We are interested in approximation results for functions asymptotically well behaved.

For asymptotically well behaved functions, we modify the analysis proposed in [49] that dealt with discontinuous finite element methods. We thus start from the existence of one dimensional operators $\widehat{\pi}_{p,k}$ defined for any integer $k \geq 0$ and any integer $p \geq 2k + 1$: $H^k(-1, 1) \to \mathbb{P}^p(-1, 1)$ such that

$$(\widehat{\pi}_{p,k})^{(j)} u(\pm 1) = u^{(j)}(\pm 1), \quad j = 0, 1, \ldots, k-1. \tag{3.9}$$

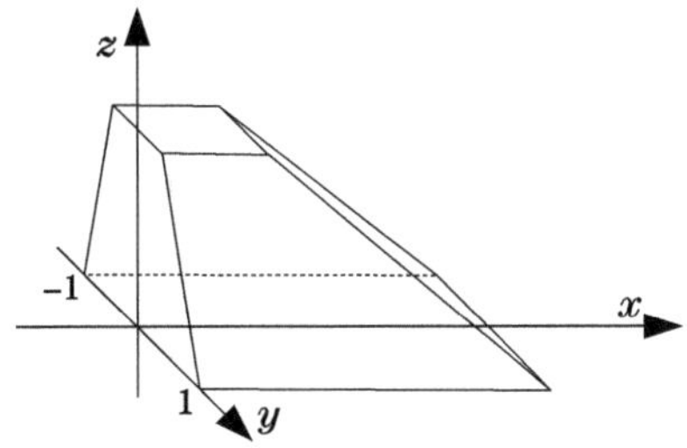

Fig. 2 The reference truncated pyramidal element $\widehat{TP}$

These operators can actually be chosen such that the following proposition holds.

Proposition 3.1 *For every $k \in \mathbb{N}$, there exists a constant $C_k > 0$ such that*

$$\|\widehat{\pi}_{p,k}u\|_{H^k(-1,1)} \leq C_k \|u\|_{H^k(-1,1)}, \quad \forall u \in H^k(-1,1),\ \forall p \geq 2k+1. \tag{3.10}$$

For integers, $p, k \in \mathbb{N}$ with $p > 2k - 1$, $\kappa = p - k + 1$ and for $u \in H^{k+s}(-1,1)$ with $k \leq s \leq \kappa$, there holds the error bound

$$\|(u - \pi_{p,k}u)^{(j)}\|^2_{L^2(\Omega)} \leq \frac{(\kappa - s)!}{(\kappa + s)!} \|u^{(k+s)}\|^2_{L^2(\Omega)} \tag{3.11}$$

for any $j = 0, 1, \ldots, k$.

We then notice that, on the particular mesh of interest $\mathcal{T}^i$ (except for those elements of $\mathcal{T}^i$—that are actually also in $\mathcal{T}^0$—that are close to the boundary and for which there is no problem of regularity nor approximation), there are essentially two reference elements: A perfect cube $\widehat{K} = (-1,1)^3$ and a truncated pyramid $\widehat{TP}$ like the one represented in Fig. 2.

Over such a reference element $\widehat{TP}$, we want to propose a local reference quasi-interpolant and we follow, for this sake, the same construction as in [49] that is dedicated to the cube $\widehat{C}$. It uses the tensorization of the one dimensional operator $\widehat{\pi}_{p,2}$ that leads to the operator over $\widehat{C}$ defined by

$$\widehat{\Pi}^3_{p,2} = \widehat{\pi}^{(x)}_{p,2} \otimes \widehat{\pi}^{(y)}_{p,2} \otimes \widehat{\pi}^{(z)}_{p,2}$$

for which it is proved (see [49, Proposition 5.2])

Theorem 3.1 *For any integer $3 \leq s \leq p$, the operator $\widehat{\Pi}^3_{p,2}$ satisfies*

$$\|u - \widehat{\Pi}^3_{p,2}u\|^2_{H^2_{\text{mix}}(\widehat{C})} \leq \frac{(p-s)!}{(p+s-2)!} \|u\|^2_{H^{s+5}(\widehat{C})} \tag{3.12}$$

and

$$H^2_{\text{mix}}(\widehat{C}) = H^2(-1,1) \otimes H^2(-1,1) \otimes H^2(-1,1).$$

In order to get the same type of result, over $\widehat{TP}$, we modifying the operator $\widehat{\Pi}^3_{p',2}$ defined above over $\widehat{C}$ by using the affine transform from the cube to the truncated pyramid $\widehat{TP}$. This results in an operator with range equal to the set of all polynomials over $\widehat{TP}$ with partial degree $\leq p'$ with respect to x and y but $\leq 3p'$ with respect to the z direction. In order to be an operator of degree $\leq p$, we choose $p' = \frac{p}{3}$, and denote by $\widehat{\Pi}^3_{p,2,TP}$ this operator for which the following theorem holds.

Theorem 3.2 *For any integer* $3 \leq s \leq p$, *the operator* $\widehat{\Pi}^3_{p,2,TP}$ *satisfies*

$$\|u - \widehat{\Pi}^3_{p,2,TP}u\|^2_{H^2_{\text{mix}}(\widehat{TP})} \leq (c_{\widehat{TP}})^s \frac{(p-s)!}{(p+s-2)!}\|u\|^2_{H^{s+5}(\widehat{TP})}, \tag{3.13}$$

where the constant $c_{\widehat{TP}} \geq 1$ *only depends on the shape of* $\widehat{TP}$.

In order to construct a quasi-interpolant in the equivalent DG space to X^i_{h-P} built over $\mathcal{T}^i$, the operator $\widehat{\Pi}^3_{p,2}$ was then used in [49] by scaling it on each element of the mesh $\mathcal{T}^i$ with the appropriate degree to propose a discontinuous approximation of the solution. Here we need to be a little bit more cautious since we want the approximation to be continuous since X^i_{h-P} is a conforming approximation of $H^1_0(\Omega)$.

We nevertheless proceed as in [49]. We first notice that every element $K \in \mathcal{T}^i$, is canonically associated to a reference element $\widehat{K}$ that is either $\widehat{C}$ or $\widehat{TP}$. The mapping that allows to go from K to $\widehat{K}$ is denoted as χ_K and is composed of a rotation, i.e., an homothetic transformation that thus preserve the polynomial degree. From these transformations, we first build from the operators $\widehat{\Pi}^3_{p,2}$ and $\widehat{\Pi}^3_{p,2,TP}$ a totally discontinuous approximation of any $H^1_0(\Omega)$ function u with the appropriate degree as in X^i_{h-P}. We denote by $\widetilde{\Pi}^{DG}_i$ this operator. From the analysis performed in [49] based on the same regularity results as in (2.3), this first nonconforming approximation satisfies (see (5.21), (5.25) and (5.35) in [49]):

$$\begin{aligned}\sum_{K\in\mathcal{T}^i} \frac{p_K^2}{h_K^2}\|u - \widetilde{\Pi}^{DG}_i u\|^2_{L^2(K)} + \sum_{K\in\mathcal{T}^i} \|\nabla(u - \widetilde{\Pi}^{DG}_i u)\|^2_{L^2(K)} \\ \leq C \sum_{K\in\mathcal{T}^i} h_K \|\widehat{u}_{|K} - \widehat{\pi}(\widehat{u}_{|K})\|^2_{H^2_{\text{mix}}},\end{aligned} \tag{3.14}$$

where, for any $K \in \mathcal{T}^i$, we denote by $\widehat{u}_{|K}$ the pull back function associated with $u_{|K}$ through χ_K. The argument of the function $\widehat{u}_{|K}$ is thus points $\mathbf{x} \in \widehat{K}$. It can thus be projected with the appropriate reference operator $\widehat{\pi}$ equal to either $\widehat{\Pi}^3_{p,2}$ or $\widehat{\Pi}^3_{p,2,TP}$.

We have now to make the approximation conforming (continuous) between two adjacent elements. This is done by lifting the discontinuities one after the other starting from the discontinuities at the vertex, then at the sides and then, finally at the faces.

Let us start with the vertices. We consider the set of all elements of $\mathcal{T}^i$ that share a common vertex $\mathbf{a}$ and denote them as $K_{\mathbf{a}}^j$ with $j = 1, \ldots, J$. The non conforming approximation $\widetilde{\Pi}_i^{DG} u$ thus proposes J distinct, but close, values. The rectification first consists in modifying this value so that the new approximation over $K_{\mathbf{a}}^j$ with $j = 2, \ldots, J$ is equal to $[\widetilde{\Pi}_i^{DG} u]_{|K_{\mathbf{a}}^1}(\mathbf{a})$. Assume that for a given $K_{\mathbf{a}}^j$ with $j = 2, \ldots, J$, the associated $\widehat{K}$ is $\widehat{\mathcal{C}}$, and that the associated pull back transformation maps the vertex $\mathbf{a}$ onto $(1, 1, 1)$. The rectification of $\widehat{\pi}(\widehat{u}_{|K_{\mathbf{a}}^j})$ is obtained by adding a quantity

$$\widehat{\text{rect}}_{\mathbf{a},j}(\hat{x}, \hat{y}, \hat{z}) = \varepsilon_{\mathbf{a},j} \mathcal{H}_{p,1}(\hat{x}) \mathcal{H}_{p,1}(\hat{y}) \mathcal{H}_{p,1}(\hat{z}),$$

where $\mathcal{H}_{p,1}$ is the polynomial $\mathcal{H}_{p,1}(\hat{x}) = \alpha(1 - \hat{x}) L_p'(\hat{x})$; here L_p stands for the Legendre polynomial with degree p, and α is such that $\mathcal{H}_{p,1}(1) = 1$.

This modification: $\varepsilon_{\mathbf{a},j}$, is upper bounded by the L^∞ bound between $[\widetilde{\Pi}_i^{DG} u]_{|K_{\mathbf{a}}^1}(\mathbf{a})$ and $[\widetilde{\Pi}_i^{DG} u]_{|K_{\mathbf{a}}^j}(\mathbf{a})$, hence bounded by

$$|\varepsilon_{\mathbf{a},j}| \le c \|\widehat{u}_{|K_{\mathbf{a}}^1} - \widehat{\pi}(\widehat{u}_{|K_{\mathbf{a}}^1})\|_{H^2_{\text{mix}}} + \|\widehat{u}_{|K_{\mathbf{a}}^j} - \widehat{\pi}(\widehat{u}_{|K_{\mathbf{a}}^j})\|_{H^2_{\text{mix}}}. \tag{3.15}$$

From classical considerations we know that

$$\|\mathcal{H}_{p,1}\|_{L^2(-1,1)} \le C\frac{1}{p}, \qquad \|\nabla \mathcal{H}_{p,1}\|_{L^2(-1,1)} \le Cp.$$

Hence

$$\left\|\nabla\left[\varepsilon_{\mathbf{a},j} \mathcal{H}_{p,1}(\hat{x}) \mathcal{H}_{p,1}(\hat{y}) \mathcal{H}_{p,1}(\hat{z})\right]\right\|_{L^2(\widehat{K})} \le C\frac{1}{p}.$$

The associated modification of the approximation $\widetilde{\Pi}_i^{DG} u$ of u, that we denote by $\text{rect}_{\mathbf{a},j}$ over $K_{\mathbf{a}}^j$ is thus upper bounded with an additional factor $c h_K$:

$$\|\nabla[\text{rect}_{\mathbf{a},j}]\|_{L^2(K_{\mathbf{a}}^j)} \le c\frac{h_K}{p_K} \varepsilon_{\mathbf{a},j}.$$

Let us continue on the rectification. We proceed similarly with the edge values, and then the faces values. Let us present the face rectification. We only have two elements K and K', that share a whole common face which we denote by $\mathcal{F}_{K,K'}$. The two approximations (already rectified at each vertex and edge) only differ from the internal values on this face. The difference is thus a function $\varepsilon(x, y)$ (say) that vanishes on the boundary of the face, and that we are going to lift on the element that has the largest degree (say K'). This lifting is again performed thanks to a function $\mathcal{H}_{p,1}(\hat{z})$:

$$\text{rect}_{\mathcal{F}_{K,K'}}(\hat{x}, \hat{y}, \hat{z}) = \varepsilon(\hat{x}, \hat{y}) \mathcal{H}_{p,1}(\hat{z}),$$

the norm of which satisfies

$$\|\nabla[\text{rect}_{\mathcal{F}_{K,K'}}]\|_{L^2(K')} \le c\left[\frac{h_{K'}}{p_{K'}} \|\nabla \varepsilon_{\mathbf{a},j}\|_{L^2(\mathcal{F}_{K,K'})} + \frac{p_{K'}}{h_{K'}} \|\varepsilon_{\mathbf{a},j}\|_{L^2(\mathcal{F}_{K,K'})}\right]. \tag{3.16}$$

By summing up these three type of corrections, we deduce that the new conforming approximation still satisfies:

$$\sum_{K\in\mathcal{T}^i} \|\nabla(u-\Pi_i^{\mathrm{conf}}u)\|_{L^2(K)}^2 \le C\sum_{K\in\mathcal{T}^i} h_K\|\widehat{u_{|K}}-\widehat{\pi}(\widehat{u_{|K}})\|_{H^2_{\mathrm{mix}}}^2. \tag{3.17}$$

We finish up as in [49] where the term on the right-hand side above is bounded, with the regularity (2.3) by upper bounding this contribution by $\exp(-ci)$, where i is the index of the triangulation $\mathcal{T}^i$, hence by $\exp(-c\sqrt[4]{N_i})$ where N_i is the total number of degrees of freedom of X^i_{h-P}.

Let us now introduce the $H^1_0(\Omega)$ orthogonal projection operator $\Pi^i_{H^1,h-P}$ over X^i_{h-P} thus defined as follows:

$$\begin{aligned} &\Pi^i_{H^1,h-P}(\varphi)\in X^i_{h-P},\quad \forall\varphi\in H^1_0(\Omega)\quad\text{and}\\ &\int_\Omega \nabla\big(\varphi-[\Pi^i_{L^2,h-P}(\varphi)]\big)\nabla\psi\,\mathrm{d}\boldsymbol{x}=0,\quad \forall\psi\in X^i_{h-P}. \end{aligned}\tag{3.18}$$

We can state as follows.

Theorem 3.3 *There exists a constant $C_0>0$, such that, for all u that satisfies the regularity assumption* (2.3),

$$\big\|u-\big[\Pi^i_{H^1,h-P}(u)\big]\big\|_{H^1}\le C_0\exp\big(-c\sqrt[4]{N_i}\big). \tag{3.19}$$

4 A Priori Analysis

4.1 The Hartree-Fock Problem

Part 1 Preliminary Analysis Let us first start with the discretization of the Hartree-Fock problem.

Following (3.1), the $h-P$ approximation of the Hartree-Fock problem is

$$\begin{aligned} \mathcal{I}^{HF,i}_{N,h-P}(V)=\inf\Big\{&\mathcal{E}^{HF}(\Phi_{h-P}),\ \Phi_{h-P}=(\phi_{1,h-P},\dots,\phi_{N,h-P})^{\mathrm{T}}\in\big(X^i_{h-P}\big)^N,\\ &\int_\Omega \phi_{i,h-P}\phi_{j,h-P}\mathrm{d}\boldsymbol{x}=\delta_{ij},\ 1\le i,j\le N\Big\}. \end{aligned}\tag{4.1}$$

Remark 4.1 The various integrals appearing in this energy should be—and are—generally computed through numerical quadrature. These affect (sometimes dramatically) the convergence of the discrete ground state to the exact one. We shall not investigate here this effect that is well described in e.g. [7] in a more simple settings.

The lack of uniqueness for the minimization problem, as recalled in (1.13) is a difficulty for the error analysis that requires the understanding of the geometry of the Grassmann manifold $\mathcal{M}$; this was first addressed in [41]. For each $\Phi = (\phi_1, \ldots, \phi_N)^{\mathrm{T}} \in \mathcal{M}$, we denote by

$$T_\Phi \mathcal{M} = \left\{ (\psi_1, \ldots, \psi_N)^{\mathrm{T}} \in \left(H_0^1(\Omega)\right)^N \mid \forall 1 \le i, j \le N, \ \int_\Omega (\phi_i \psi_j + \psi_i \phi_j) \mathrm{d}\mathbf{x} = 0 \right\}$$

the tangent space to $\mathcal{M}$ at Φ, and we also define

$$\Phi^{\perp\perp} = \left\{ \Psi = (\psi_1, \ldots, \psi_N)^{\mathrm{T}} \in \left(H_0^1(\Omega)\right)^N \mid \forall 1 \le i, j \le N, \ \int_\Omega \phi_i \psi_j \mathrm{d}\mathbf{x} = 0 \right\}.$$

Let us recall (see, e.g., [41, Lemma 4]) that

$$T_\Phi \mathcal{M} = \mathcal{A}\Phi \oplus \Phi^{\perp\perp},$$

where $\mathcal{A} = \{A \in \mathbb{R}^{N \times N} \mid A^{\mathrm{T}} = -A\}$ is the space of the $N \times N$ antisymmetric real matrices.

The second order condition associated to the minimization problem (1.2) reads

$$a_{\Phi^0}(W, W) \ge 0, \quad \forall W \in T_{\Phi^0}\mathcal{M},$$

where for all $\Psi = (\psi_1, \ldots, \psi_N)^{\mathrm{T}}$ and $\Upsilon = (\upsilon_1, \ldots, \upsilon_N)^{\mathrm{T}}$ in $(H_0^1(\Omega))^N$,

$$\begin{aligned} a_{\Phi^0}(\Psi, \Upsilon) &= \frac{1}{4}\mathcal{E}^{\mathrm{HF}''}(\Phi^0)(\Psi, \Upsilon) - \sum_{i=1}^N \epsilon_i^0 \int_\Omega \psi_i \upsilon_i \mathrm{d}\mathbf{x} \\ &= \sum_{i=1}^N \langle (\mathcal{H}_{\rho^0}^{\mathrm{KS}} - \epsilon_i^0)\psi_i, \upsilon_i \rangle_{H^{-1}, H_0^1} \mathrm{d}x \\ &\quad + 4 \sum_{i,j=1}^N \int_\Omega \int_\Omega \frac{\phi_i^0(x)\psi_i(x)\phi_j^0(y)\upsilon_j(y)}{|x - y]} \mathrm{d}x\mathrm{d}y \\ &\quad - 2 \sum_{i,j=1}^N \int_\Omega \int_\Omega \frac{\upsilon_i(x)\phi_i^0(y)\phi_j^0(x)\psi_j(y)}{|x - y]} \mathrm{d}x\mathrm{d}y \\ &\quad - 2 \sum_{i,j=1}^N \int_\Omega \int_\Omega \frac{\phi_i^0(x)\upsilon_i(y)\phi_j^0(x)\psi_j(y)}{|x - y]} \mathrm{d}x\mathrm{d}y. \end{aligned} \tag{4.2}$$

It follows from the invariance property (1.13) that

$$a_{\Phi^0}(\Psi, \Psi) = 0 \quad \text{for all } \Psi \in \mathcal{A}\Phi^0.$$

This leads us, as in [41], to make the assumption that a_{Φ^0} is positive definite on $\Phi^{0,\perp\perp}$, so that, as in [41, Proposition 1], it follows that a_{Φ^0} is actually coercive on

$\Phi^{0,\perp\!\!\perp}$ (for the H_0^1 norm). In all what follows, we thus assume that there exists a positive constant c_{Φ^0} such that

$$a_{\Phi^0}(\Psi,\Psi) \geq c_{\Phi^0}\|\Psi\|^2_{H_0^1}, \quad \forall \Psi \in \Phi^{0,\perp\!\!\perp}. \tag{4.3}$$

Remark 4.2 As noticed in [8], in the linear framework, the coercivity condition (4.3) is satisfied if and only if

(i) $\epsilon_1^0,\dots,\epsilon_N^0$ are the lowest N eigenvalues (including multiplicities) of the linear self-adjoint operator $h = -\frac{1}{2}\Delta + V^{\text{nuc}}$,
(ii) there is a gap $c_{\Phi^0} > 0$ between the lowest Nth and $(N+1)$st eigenvalues of h.

The topology of the Grassmann manifold $\mathcal{M}$ quotiented by the equivalence relation through unitary transformations (see e.g. [19]) was analyzed in this context in [41] and in [8]. In particular,

(1) Let $\Phi \in \mathcal{M}$ and $\Psi \in \mathcal{M}$. If $M_{\Psi,\Phi}$ is invertible, then

$$U_{\Psi,\Phi} = M^T_{\Psi,\Phi}(M_{\Psi,\Phi}M^T_{\Psi,\Phi})^{-\frac{1}{2}}$$

is the unique minimizer to the problem $\min_{U\in\mathcal{U}(N)}\|U\Psi-\Phi\|_{(L^2(\Omega))^N}$.

(2) The set

$$\mathcal{M}^\Phi := \Big\{\Psi \in \mathcal{M} \mid \|\Psi-\Phi\|_{L^2} = \min_{U\in\mathcal{U}(N)}\|U\Psi-\Phi\|_{L^2}\Big\},$$

verifies

$$\mathcal{M}^\Phi = \big\{(1_N - M_{W,W})^{\frac{1}{2}}\Phi + W \mid W \in \Phi^{\perp\!\!\perp},\ 0 \leq M_{W,W} \leq 1_N\big\}.$$

Hence, for any $\Phi \in \mathcal{M}$ and any $\Psi \in \mathcal{M}^\Phi$ there exists $W \in \Phi^{\perp\!\!\perp}$ such that

$$\Psi = \Phi + \mathcal{S}(W)\Phi + W, \tag{4.4}$$

where $\mathcal{S}(W) = (1_N - M_{W,W})^{\frac{1}{2}} - 1_N$ is an $N\times N$ symmetric matrix, and the converse also holds. Similarly, at the discrete level, for any $\Phi_{h-P} \in [X^i_{h-P}]^N \cap \mathcal{M}$ and any $\Psi_{h-P} \in V^N_{h-P} \cap \mathcal{M}^{\Phi_{h-P}}$ there exists $W_{h-P} \in [X^i_{h-P}]^N \cap \Phi^{\perp\!\!\perp}_{h-P}$ such that

$$\Psi_{h-P} = \Phi_{h-P} + \mathcal{S}(W_{h-P})\Phi_{h-P} + W_{h-P}, \tag{4.5}$$

where $\mathcal{S}(W_{h-P}) = (1_N - M_{W_{h-P},W_{h-P}})^{\frac{1}{2}} - 1_N$ is an $N\times N$ symmetric matrix, and the converse also holds.

In what follows, we shall compare, as in [8], the error between the solution to (4.1) and Φ^0, the solution to (1.2), with its best approximation in $(X^i_{h-P})^N \cap \mathcal{M}$ (see [8, Lemma 4.3]).

Lemma 4.1 (1) *Let* $\Phi = (\phi_1,\dots,\phi_N)^{\mathrm{T}} \in \mathcal{M}$. *If* $i \in \mathbb{N}$ *is such that*

$$\dim\big(\mathrm{span}\big(\Pi^i_{L^2,h-P}(\phi_1),\dots\Pi^i_{L^2,h-P}(\phi_N)\big)\big) = N,$$

then the unique minimizer of the problem

$$\min_{\Phi_{i,h-P} \in (X^i_{h-P})^N \cap \mathcal{M}} \|\Phi_{i,h-P} - \Phi\|_{[L^2(\Omega)]^N}$$

is

$$\pi^{\mathcal{M}}_{i,h-P}\Phi = (M_{\Pi^i_{L^2,h-P}\Phi, \Pi^i_{L^2,h-P}\Phi})^{-\frac{1}{2}} \Pi^i_{L^2,h-P}\Phi. \tag{4.6}$$

In addition, $\pi^{\mathcal{M}}_{i,h-P}\Phi \in (X^i_{h-P})^N \cap \mathcal{M}^{\Phi}$,

$$\|\pi^{\mathcal{M}}_{i,h-P}\Phi - \Phi\|_{[L^2(\Omega)]^N} \leq \sqrt{2}\|\Pi^i_{L^2,h-P}\Phi - \Phi\|_{[L^2(\Omega)]^N}, \tag{4.7}$$

and for all i *large enough,*

$$\begin{aligned} \|\pi^{\mathcal{M}}_{i,h-P}\Phi - \Phi\|_{[H^1(\Omega)]^N} \leq \|\Phi\|_{[H^1(\Omega)]^N} \|\Pi^i_{L^2,h-P}\Phi - \Phi\|^2_{[L^2(\Omega)]^N} \\ + \|\Pi^i_{L^2,h-P}\Phi - \Phi\|_{[H^1(\Omega)]^N}. \end{aligned} \tag{4.8}$$

(2) *Let* $i \in \mathbb{N}$ *such that* $\dim(X^i_{h-P}) \geq N$ *and* $\Phi_{i,h-P} \in (X^i_{h-P})^N \cap \mathcal{M}$. *Then*

$$\begin{aligned} \left(X^i_{h-P}\right)^N \cap \mathcal{M}^{\Phi_{i,h-P}} = \big\{(1_N - M_{W_{i,h-P}, W_{i,h-P}})^{1/2}\Phi_{i,h-P} + W_{i,h-P} \mid \\ W_{i,h-P} \in \left(X^i_{h-P}\right)^N \cap \Phi^{\perp\perp}_{i,h-P},\ 0 \leq M_{W_{i,h-P}, W_{i,h-P}} \leq 1_N\big\}. \end{aligned} \tag{4.9}$$

The following Lemma 4.2 (see [8]) collects some properties of the function $W \mapsto \mathcal{S}(W)$.

Lemma 4.2 *Let*

$$K = \left\{W \in \left(L^2(\Omega)\right)^N \mid 0 \leq M_{W,W} \leq 1_N\right\},$$

and $\mathcal{S} : K \to \mathbb{R}^{N\times N}_{\mathrm{S}}$ *(the space of the symmetric* $N \times N$ *real matrices) defined by*

$$\mathcal{S}(W) = (1_N - M_{W,W})^{\frac{1}{2}} - 1_N.$$

The function $\mathcal{S}$ *is continuous on* K *and differentiable on the interior* $\overset{\circ}{K}$ *of* K. *In addition,*

$$\|\mathcal{S}(W)\|_F \leq \|W\|^2_{L^2}, \quad \forall W \in K, \tag{4.10}$$

where $\|\cdot\|_{\mathrm{F}}$ *denotes the Frobenius norm. For all* $(W_1, W_2, Z) \in K \times K \times (L^2(\Omega))^N$ *such that* $\|W_1\|_{L^2} \leq \frac{1}{2}$ *and* $\|W_2\|_{L^2} \leq \frac{1}{2}$,

$$\|\mathcal{S}(W_1) - \mathcal{S}(W_2)\|_{\mathrm{F}} \leq 2(\|W_1\|_{L^2} + \|W_2\|_{L^2})\|W_1 - W_2\|_{L^2}, \tag{4.11}$$

$$\left\|\left(\mathcal{S}'(W_1) - \mathcal{S}'(W_2)\right) \cdot Z\right\|_{\mathrm{F}} \leq 4\|W_1 - W_2\|_{L^2}\|Z\|_{L^2}, \tag{4.12}$$

$$\|(\mathcal{S}''(W_1)(Z,Z)\|_{\mathrm{F}} \leq 4\|Z\|_{L^2}^2. \tag{4.13}$$

Now, we recall the following stability results that can be proved following the same lines as in [8].

Lemma 4.3 *There exists $C \geq 0$ such that*

(1) *for all* $(\Upsilon_1, \Upsilon_2, \Upsilon_3) \in ((H_0^1(\Omega))^N)^3$,

$$\begin{aligned}&\left|\left(\mathcal{E}^{\mathrm{HF}''}(\Phi^0 + \Upsilon_1) - \mathcal{E}^{\mathrm{HF}''}(\Phi^0)\right)(\Upsilon_2, \Upsilon_3)\right| \\ &\quad \leq C\left(\|\Upsilon_1\|_{L^2}^{\alpha} + \|\Upsilon_1\|_{H_0^1}^2\right)\|\Upsilon_2\|_{H_0^1}\|\Upsilon_3\|_{H_0^1};\end{aligned}$$

(2) *for all* $(\Upsilon_1, \Upsilon_2, \Upsilon_3) \in ((H^2(\Omega))^N)^3$,

$$\begin{aligned}&\left|\left(\mathcal{E}^{\mathrm{HF}''}(\Phi^0 + \Upsilon_1) - \mathcal{E}^{\mathrm{HF}''}(\Phi^0)\right)(\Upsilon_2, \Upsilon_3)\right| \\ &\quad \leq C\left(\|\Upsilon_1\|_{L^2} + \|\Upsilon_1\|_{L^2}^2\right)\|\Upsilon_2\|_{L^2}\|\Upsilon_3\|_{H^2}.\end{aligned}$$

In addition, for all $(q,r,s) \in \mathbb{R}_+^3$ such that $\frac{3}{2} < q, s > \frac{3}{2}$ and $r \leq \min(q,s)$, and all $0 < M < \infty$, there exists a constant $C \geq 0$ such that

(1) *for all* $(\Upsilon_1, \Upsilon_2, \Upsilon_3) \in (H^q(\Omega))^N \times (H_0^{-r}(\Omega))^N \times (H_0^s(\Omega))^N$ *such that* $\|\Upsilon_1\|_{H^q} \leq M$,

$$\left|\left(\mathcal{E}^{\mathrm{HF}''}(\Phi^0 + \Upsilon_1) - \mathcal{E}^{\mathrm{HF}''}(\Phi^0)\right)(\Upsilon_2, \Upsilon_3)\right| \leq C\|\Upsilon_1\|_{H^q}\|\Upsilon_2\|_{H^{-r}}\|\Upsilon_3\|_{H^s}.$$

Following the same lines as in [8, Lemma 4.7], we deduce from Lemma 4.3 that there exists $C \geq 0$ such that for all $\Psi \in \mathcal{M}$,

$$\mathcal{E}^{\mathrm{HF}}(\Psi) = \mathcal{E}^{\mathrm{HF}}(\Phi^0) + 2a_{\Phi^0}(\Psi - \Phi^0, \Psi - \Phi^0) + R(\Psi - \Phi^0) \tag{4.14}$$

with

$$\left|R(\Psi - \Phi^0)\right| \leq C\left(\|\Psi - \Phi^0\|_{H_0^1}^3 + \|\Psi - \Phi^0\|_{H_0^1}^4\right). \tag{4.15}$$

Part 2 Existence of a Discrete Solution We now use the parametrization of the manifold $X_{h-P}^i \cap \mathcal{M}^{\Phi_{i,h-P}}$ explained in (4.9) to express a given minimizer of the problem (4.1) close to Φ^0 in terms of an element $W_{i,h-P}$ in a neighborhood of 0 expressed as $W_{i,h-P} \in \mathcal{B}_{i,h-P}$, where

$$\mathcal{B}_{i,h-P} := \left\{W^{i,h-P} \in \left[X_{h-P}^i\right]^N \cap \left[\pi_{i,h-P}^{\mathcal{M}}\Phi^0\right]^{\perp\perp} \mid 0 \leq M_{W^{i,h-P},W^{i,h-P}} \leq 1\right\}.$$

Indeed, we can define

$$\mathbf{E}_{i,h-P}\left(W^{i,h-P}\right) = \mathcal{E}^{\mathrm{HF}}\left(\pi_{i,h-P}^{\mathcal{M}}\Phi^0 + \mathcal{S}\left(W^{i,h-P}\right)\pi_{i,h-P}^{\mathcal{M}}\Phi^0 + W^{i,h-P}\right), \tag{4.16}$$

the minimizers of which are in one-to-one correspondence with those of (4.1). Then the same line as in the proof of Lemma 4.8 in [8] leads to the following lemma.

Lemma 4.4 *There exist $r>0$ and i^0 such that for all $i \geq i^0$, the functional $\mathbf{E}_{i,h-P}$ has a unique critical point $W_0^{i,h-P}$ in the ball*

$$\left\{W^{i,h-P} \in \left[X_{h-P}^i\right]^N \cap \left[\pi_{i,h-P}^{\mathcal{M}} \Phi^0\right]^{\perp\perp} \mid \|W^{i,h-P}\|_{H_0^1} \leq r\right\}.$$

Besides, $W_0^{i,h-P}$ is a local minimizer of $\mathbf{E}_{i,h-P}$ over the above ball and we have the estimate

$$\|W_0^{i,h-P}\|_{H_0^1} \leq C\|\pi_{i,h-P}^{\mathcal{M}} \Phi^0 - \Phi^0\|_{H_0^1}. \tag{4.17}$$

Since the solution Φ^0 is asymptotically well behaved, i.e., satisfies (2.3), we obtain from the accuracy offered by the $h-P$ finite elements spaces X_{h-P}^i stated in Theorem 3.3 that there exists a constant C such that

$$\|W_0^{i,h-P}\|_{H_0^1} \leq C\exp\left(-c\sqrt[4]{N_i}\right).$$

Then we deduce the existence and uniqueness of a local minimizer to the problem (4.1) close to Φ^0 which satisfies that (see [8, (4.71)–(4.72)]) for i large enough,

$$\frac{1}{2}\|W_{i,h-P}^0\|_{L^2} \leq \|\Phi_{i,h-P}^0 - \Phi^0\|_{L^2} \leq 2\|W_{i,h-P}^0\|_{L^2}, \tag{4.18}$$

$$\frac{1}{2}\|W_{i,h-P}^0\|_{H_0^1} \leq \|\Phi_{i,h-P}^0 - \Phi^0\|_{H_0^1} \leq 2\|W_{i,h-P}^0\|_{H_0^1}. \tag{4.19}$$

Hence we have proven the following theorem.

Theorem 4.1 *Let Φ^0 be a local minimizer of the Hartree-Fock problem* (1.2), *and assume that it is asymptotically well behaved, i.e., satisfies* (2.3). *Then there exist $r^0>0$ and i^0 such that, for any $i \geq i^0$, the discrete problem* (4.1) *has a unique local minimizer $\Phi_{i,h-P}^0$ in the set*

$$\left\{\Psi_{i,h-P} \in \left(X_{h-P}^i\right)^N \cap \mathcal{M}^{\Phi^0} \mid \|\Psi_{i,h-P} - \Phi^0\|_{H_0^1(\Omega)} \leq r^0\right\}$$

and there exist constants $C>0$ and $c>0$, independent of i such that

$$\|\Phi_{i,h-P}^0 - \Phi^0\|_{H_0^1} \leq C\exp\left(-c\sqrt[4]{N_i}\right).$$

Finally, the discrete solution $\Phi_{i,h-P}^0$ satisfies the Euler equations

$$\left\langle\mathcal{H}_{\rho_{i,h-P}^0}^{\mathrm{HF}} \phi_{j,i,h-P}^0, \psi_{j,i,h-P}\right\rangle_{H^{-1},H_0^1} = \sum_{\nu=1}^{N}\left[\lambda_{i,h-P}^0\right]_{j\nu}\left(\phi_{\nu,i,h-P}^0, \psi_{\nu,i,h-P}\right)_{L^2},$$

$$\forall \Psi_{i,h-P} \in \left[X_{h-P}^i\right]^N,$$

where $\rho^0_{i,h-P} = \rho_{\Phi^0_{i,h-P}}$ and the $N \times N$ matrix $\Lambda^0_{i,h-P}$ is symmetric (but generally not diagonal). Of course, it follows from the invariance property (1.13) that (4.1) has a local minimizer of the form $U\Phi^0_{i,h-P}$ with $U \in \mathcal{U}(N)$ for which the Lagrange multiplier of the orthonormality constraints is a diagonal matrix.

In order to get more precise results on the convergence rate of the eigenvalues, further analysis needs to be performed on the approximation properties in standard Sobolev spaces of the discrete space X^i_{h-P}. As far as we know, these results concerning

(1) inverse inequalities,
(2) convergence properties of the $L^2(\Omega)$ orthogonal projection operator $\Pi^i_{L^2,h-P}$ over X^i_{h-P} for e.g. H^1 functions,
(3) convergence properties of the $H^1_0(\Omega)$ orthogonal projection operator $\Pi^i_{H^1,h-P}$ over X^i_{h-P} for e.g. H^2 functions,

do not exist in optimal form, which is what is required to get the doubling of convergence for the approximation of the eigenvalues with respect to the convergence of $\|\Phi^0_{i,h-P} - \Phi^0\|_{H^1_0}$. These results will be proven in a paper in preparation.

4.2 The Kohn-Sham Problem

Following (3.2), the $h-P$ approximation of the Kohn-Sham problem is

$$\mathcal{I}^{KS,i}_{N,h-P}(V) = \inf\Big\{\mathcal{E}^{KS}(\Phi_{h-P}),\ \Phi_{h-P} = (\phi_{1,h-P},\dots,\phi_{N,h-P})^{\mathrm{T}} \in \left(X^i_{h-P}\right)^N,$$
$$\int_\Omega \phi_{i,h-P}\phi_{j,h-P}\mathrm{d}\mathbf{x} = \delta_{ij},\ 1 \le i,j \le N\Big\}. \tag{4.20}$$

Following the same lines as in the proof of the previous result, and the analysis of the plane wave approximation of the Kohn-Sham problem presented in [8], we can prove the following theorem.

Theorem 4.2 *Let Φ^0 be a local minimizer of the Kohn-Sham problem* (1.5), *and assume that it is asymptotically well behaved, i.e. satisfies* (2.3), *Then there exist $r^0 > 0$ and i^0 such that, for any $i \ge i^0$, the discrete problem* (4.20) *has a unique local minimizer $\Phi^0_{i,h-P}$ in the set*

$$\left\{\Psi_{i,h-p} \in \left(X^i_{h-P}\right)^N \cap \mathcal{M}^{\Phi^0} \,|\, \|\Psi_{i,h-p} - \Phi^0\|_{H^1_0(\Omega)} \le r^0\right\},$$

and there exist constants $C > 0$ and $c > 0$, independent of i such that

$$\|\Phi^0_{i,h-P} - \Phi^0\|_{H^1_0} \le C\exp\left(-c\sqrt[4]{N_i}\right).$$

References

1. Anantharaman, A., Cancès, E.: Existence of minimizers for Kohn-Sham models in quantum chemistry. Ann. Inst. Henri Poincaré **26**, 2425–2455 (2009)
2. Bao, G., Hu, G., Liu, D.: An h-adaptive finite element solver for the calculation of the electronic structures. J. Comput. Phys. **231**, 4967–4979 (2012)
3. Babuška, I., Suri, M.: The hP and $h - P$ versions of the finite element method, an overview. Comput. Methods Appl. Mech. Eng. **80**(1), 5–26 (1990)
4. Bernardi, C., Maday, Y.: Polynomial approximation of some singular functions. Appl. Anal. **42**(1–4), 1–32 (1991)
5. Born, M., Oppenheimer, J.R.: Zur Quantentheorie der Molekeln. Ann. Phys. **84**, 457–484 (1927)
6. Bylaska, E.J., Host, M., Weare, J.H.: Adaptive finite element method for solving the exact Kohn-Sham equation of density functional theory. J. Chem. Theory Comput. **5**, 937–948 (2009)
7. Cancès, E., Chakir, R., Maday, Y.: Numerical analysis of nonlinear eigenvalue problems. J. Sci. Comput. **45**(1–3), 90–117 (2010)
8. Cancès, E., Chakir, R., Maday, Y.: Numerical analysis of the planewave discretization of some orbital-free and Kohn-Sham models. Modél. Math. Anal. Numér. **46**(2), 341–388 (2012)
9. Cancès, E., Defranceschi, M., Kutzelnigg, W., et al.: Computational quantum chemistry: a primer. In: Handbook of Numerical Analysis, vol. X, pp. 3–270. North-Holland, Amsterdam (2003)
10. Cancès, E., Le Bris, C., Maday, Y.: Méthodes Mathématiques en Chimie Quantique. Springer, New York (2006)
11. Cances, E., Le Bris, C., Nguyen, N.C., et al.: Feasibility and competitiveness of a reduced basis approach for rapid electronic structure calculations in quantum chemistry. In: Proceedings of the Workshop for Highdimensional Partial Differential Equations in Science and Engineering, Montreal (2007)
12. Cancès, E., Stoltz, G., Staroverov, V.N., et al.: Local exchange potentials for electronic structure calculations. Maths Act. **2**, 1–42 (2009)
13. Canuto, C., Hussaini, M.Y., Quarteroni, A., Zang, T.A.: Spectral Methods: Fundamentals in Single Domains. Springer, New York (2006)
14. Chen, H., Gong, X., He, L., Zhou, A.: Convergence of adaptive finite element approximations for nonlinear eigenvalue problems. Preprint. arXiv:1001.2344 [math.NA]
15. Chen, H., Gong, X., Zhou, A.: Numerical approximations of a nonlinear eigenvalue problem and applications to a density functional model. Math. Methods Appl. Sci. **33**, 1723–1742 (2010)
16. Chen, H., Schneider, R.: Numerical analysis of augmented plane waves methods for full-potential electronic structure calculations. Preprint 116. dfg-spp1324.de
17. Costabel, M., Dauge, M., Nicaise, S.: Analytic regularity for linear elliptic systems in polygons and polyhedra. Math. Models Methods Appl. Sci. **22**(8) (2012)
18. Dreizler, R.M., Gross, E.K.U.: Density Functional Theory. Springer, Berlin (1990)
19. Edelman, A., Arias, T.A., Smith, S.T.: The geometry of algorithms with orthogonality constraints. SIAM J. Matrix Anal. Appl. **20**, 303–353 (1998)
20. Gavini, V., Knap, J., Bhattacharya, K., Ortiz, M.: Non-periodic finite-element formulation of orbital-free density functional theory. J. Mech. Phys. Solids **55**, 669–696 (2007)
21. Fang, J., Gao, X., Zhou, A.: A Kohn-Sham equation solver based on hexahedral finite elements. J. Comput. Phys. **231**, 3166–3180 (2012)
22. Fournais, S., Hoffmann-Ostenhof, M., Hoffmann-Ostenhof, T., Sørensen, T.Ø.: The electron density is smooth away from the nuclei. Commun. Math. Phys. **228**(3), 401–415 (2002)
23. Fournais, S., Hoffmann-Ostenhof, M., Hoffmann-Ostenhof, T., Sørensen, T.Ø.: Analyticity of the density of electronic wavefunctions. Ark. Mat. **42**(1), 87–106 (2004)

24. Fournais, S., Srensen, T.Ø., Hoffmann-Ostenhof, M., Hoffmann-Ostenhof, T.: Non-isotropic cusp conditions and regularity of the electron density of molecules at the nuclei. Ann. Henri Poincaré **8**(4), 731–748 (2007)
25. Gaussian web site. http://www.gaussian.com
26. Gui, W., Babuška, I.: The h, P and $h-P$ versions of the finite element method in 1 dimension. Numer. Math. **49**(6), 577–683 (1986)
27. Guo, B., Babuška, I.: The $h-P$ version of the finite element method. Comput. Mech. **1**(1), 21–41 (1986)
28. Hermannson, B., Yevick, D.: Finite-element approach to band-structure analysis. Phys. Rev. B **33**, 7241–7242 (1986)
29. Hoffmann-Ostenhof, M., Hoffmann-Ostenhof, T., Sørensen, T.Ø.: Electron wavefunctions and densities for atoms. Ann. Henri Poincaré **2**(1), 77–100 (2001)
30. Hohenberg, P., Kohn, W.: Inhomogeneous electron gas. Phys. Rev. **136**, B864–B871 (1964)
31. Kohn, W., Sham, L.J.: Self-consistent equations including exchange and correlation effects. Phys. Rev. **140**, A1133–A1138 (1965)
32. Langwallner, B., Ortner, C., Süli, E.: Existence and convergence results for the Galerkin approximation of an electronic density functional. Math. Models Methods Appl. Sci. **20**, 2237–2265 (2010)
33. Le Bris, C.: Ph.D. Thesis, Ecole Polytechnique, Paris (1993)
34. Lehtovaara, L., Havu, V., Puska, M.: All-electron density functional theory and time-dependent density functional theory with high-order finite elements. J. Chem. Phys. **131**, 054103 (2009)
35. Lester, W.A. Jr. (ed.): Recent Advances in Quantum Monte Carlo Methods. World Scientific, Singapore (1997)
36. Lester, W.A. Jr., Rothstein, S.M., Tanaka, S.: Recent Advances in Quantum Monte Carlo Methods. Part II. World Scientific, Singapore (2002)
37. Levy, M.: Universal variational functionals of electron densities, first order density matrices, and natural spin-orbitals and solution of the V-representability problem. Proc. Natl. Acad. Sci. USA **76**, 6062–6065 (1979)
38. Lieb, E.H.: Density functional for coulomb systems. Int. J. Quant. Chem. **24**, 243–277 (1983)
39. Lin, L., Lu, J.F., Ying, L.X., Weinan, E.: Adaptive local basis set for KohnSham density functional theory in a discontinuous Galerkin framework. I. Total energy calculation. J. Comput. Phys. **231**(4), 2140–2154 (2012)
40. Maday, Y., Razafison, U.: A reduced basis method applied to the restricted Hartree-Fock equations. C. R. Math. **346**(3), 243–248 (2008)
41. Maday, Y., Turinici, G.: Error bars and quadratically convergent methods for the numerical simulation of the Hartree-Fock equations. Numer. Math. **94**, 739–770 (2003)
42. Martin, R.M.: Electronic Structure: Basic Theory and Practical Methods. Cambridge University Press, Cambridge (2004)
43. Masud, A., Kannan, R.: B-splines and NURBS based finite element methods for Kohn-Sham equations. Comput. Methods Appl. Mech. Eng. **241**, 112–127 (2012)
44. Motamarri, P., Nowak, M.R., Leiter, K., Knap, J., Gavini, V.: Higher-order adaptive finite-element methods for Kohn-Sham density functional theory (2012). Preprint. arXiv:1207.0167
45. Pask, J.E., Klein, B.M., Fong, C.Y., Sterne, P.A.: Real-space local polynomial basis for solid-state electronic-structure calculations: a finite element approach. Phys. Rev. B **59**, 12352–12358 (1999)
46. Pask, J.E., Klein, B.M., Sterne, P.A., Fong, C.Y.: Finite element methods in electronic-structure theory. Comput. Phys. Commun. **135**, 134 (2001)
47. Pask, J.E., Sterne, P.A.: Finite element methods in ab initio electronic structure calculations. Model. Simul. Mater. Sci. Eng. **13**, R71–R96 (2005)
48. Schötzau, D., Schwab, C., Wihler, T.: *hp*-dGFEM for second-order elliptic problems in polyhedra. I: Stability and quasioptimality on geometric meshes. Technical Report 2009-28, SAM-ETH, Zürich (2009)

49. Schötzau, D., Schwab, C., Wihler, T.P.: hp-dGFEM for second-order elliptic problems in polyhedra. II: Exponential convergence. Technical report 2009-29, SAM-ETH, Zürich (2009)
50. Singh, D.J., Nordstrom, L.: Planewaves, Pseudopotentials, and the LAPW Method. Springer, New York (2005)
51. Suryanarayana, P., Gavini, V., Blesgen, T., et al.: Non-periodic finite-element formulation of Kohn-Sham density functional theory. J. Mech. Phys. Solids **58**, 256–280 (2010)
52. Tomasi, J., Persico, M.: Molecular interactions in solution: an overview of methods based on continuous distributions of the solvent. Chem. Rev. **94**(7), 2027–2094 (1994)
53. Tsuchida, E., Tsukada, M.: Electronic-structure calculations based on the finite element method. Phys. Rev. B **52**, 5573–5578 (1995)
54. Tsuchida, E., Tsukada, M.: Adaptive finite-element method for electronic structure calculations. Phys. Rev. B **54**, 7602–7605 (1996)
55. Tsuchida, E., Tsukada, M.: Large-scale electronic-structure calculations based on the adaptive finite element method. J. Phys. Soc. Jpn. **67**, 3844–3858 (1998)
56. Valone, S.: Consequences of extending 1 matrix energy functionals from purestate representable to all ensemble representable 1 matrices. J. Chem. Phys. **73**, 1344–1349 (1980)
57. Vos, P.E.J., Spencer, S., Kirby, R.M.: From h to P efficiently: implementing finite and spectral/$h - P$ element methods to achieve optimal performance for low-and high-order discretisations. J. Comput. Phys. **229**(13), 5161–5181 (2010)
58. White, S.R., Wilkins, J.W., Teter, M.P.: Finite element method for electronic structure. Phys. Rev. B **39**, 5819–5830 (1989)
59. Zhang, D., Shen, L., Zhou, A., Gong, X.: Finite element method for solving Kohn-Sham equations based on self-adaptive tetrahedral mesh. Phys. Lett. A **372**, 5071–5076 (2008)
60. Zhou, A.: Finite dimensional approximations for the electronic ground state solution of a molecular system. Math. Methods Appl. Sci. **30**, 429–447 (2007)

Increasing Powers in a Degenerate Parabolic Logistic Equation

José Francisco Rodrigues and Hugo Tavares

Abstract The purpose of this paper is to study the asymptotic behavior of the positive solutions of the problem

$$\partial_t u - \Delta u = au - b(x)u^p \quad \text{in } \Omega \times \mathbb{R}^+, \quad u(0) = u_0, \quad u(t)|_{\partial\Omega} = 0,$$

as $p \to +\infty$, where Ω is a bounded domain, and $b(x)$ is a nonnegative function. The authors deduce that the limiting configuration solves a parabolic obstacle problem, and afterwards fully describe its long time behavior.

Keywords Parabolic logistic equation · Obstacle problem · Positive solution · Increasing power, subsolution and supersolution

Mathematics Subject Classification 35B40 · 35B09 · 35K91

1 Introduction

In this paper, we are interested in the study of the parabolic problem

$$\begin{cases} \partial_t u - \Delta u = au - b(x)u^p & \text{in } Q := \Omega \times (0, +\infty), \\ u = 0 & \text{on } \partial\Omega \times (0, +\infty), \\ u(0) = u_0 & \text{in } \Omega, \end{cases} \tag{1.1}$$

where $a > 0$, $p > 1$, $b \in L^\infty(\Omega)$ is a nonnegative function, and Ω is a bounded domain with a smooth boundary. Such system arises in population dynamics, where u denotes the population density of given species, subject to a logistic-type law.

Project supported by Fundação para a Ciência e a Tecnologia (FCT) (No. PEst OE/MAT/UI0209/2011). The second author was also supported by an FCT grant (No. SFRH/BPD/69314/201).

J.F. Rodrigues (✉) · H. Tavares
Department of Mathematics and CMAF, Universidade de Lisboa, Avenida Professor Gama Pinto 2, 1649-003 Lisboa, Portugal
e-mail: rodrigue@ptmat.fc.ul.pt

H. Tavares
e-mail: htavares@ptmat.fc.ul.pt

P.G. Ciarlet et al. (eds.), *Partial Differential Equations: Theory, Control and Approximation*, DOI 10.1007/978-3-642-41401-5_15,

It is well-known that under these assumptions and for very general u_0's, Eq. (1.1) admits a unique global positive solution $u_p = u_p(x,t)$. In fact, in order to deduce the existence result, one can make the change of variables $v = \mathrm{e}^{-at}u$, and deduce that v satisfies $\partial_t v - \Delta v + b(x)\mathrm{e}^{pat}v^p = 0$. As $v \mapsto b(x)\mathrm{e}^{pat}|v|^{p-1}v$ is monotone nondecreasing, the theory of monotone operators (see [1, 2]) immediately provides the existence of the solution of the problem in v, and hence also for (1.1).

One of our main interests is the study of the solution u_p as $p \to +\infty$. As we will see, in the limit we will obtain a parabolic obstacle problem, and afterwards fully describe its asymptotic limit as $t \to +\infty$.

This study is mainly inspired by the works of Dancer et al. [3–5], where the stationary version of (1.1) is addressed. Let us describe their results in detail. Consider the elliptic problem

$$-\Delta u = au - b(x)u^p, \quad u \in H_0^1(\Omega). \tag{1.2}$$

For each domain $\omega \subseteq \mathbb{R}^N$, denote by $\lambda_1(\omega)$ the first eigenvalue of $-\Delta$ in $H_0^1(\omega)$. Assuming $b \in C(\overline{\Omega})$, the study is divided into two cases as follows: The so-called nondegenerate case (where $\min_{\overline{\Omega}} b(x) > 0$) and the degenerate one (where $\Omega_0 := \operatorname{int}\{x \in \Omega : b(x) = 0\} \neq \emptyset$ with a smooth boundary).

In the nondegenerate case, it is standard to check that (1.2) has a positive solution if and only if $a > \lambda_1(\Omega)$ (see [6, Lemma 3.1, Theorem 3.5]). For each $a > \lambda_1(\Omega)$ fixed, then in [4] it is shown that $u_p \to w$ in $C^1(\overline{\Omega})$ as $p \to +\infty$, where w is the unique solution of the obstacle-type problem

$$-\Delta w = aw\chi_{\{w<1\}}, \quad w > 0, \quad w|_{\partial\Omega} = 0, \quad \|w\|_\infty = 1. \tag{1.3}$$

It is observed in [3] that u is also the unique positive solution of the variational inequality

$$w \in \mathbb{K} : \int_\Omega \nabla w \cdot \nabla(v - w)\mathrm{d}x \geqslant \int_\Omega aw(v - w)\mathrm{d}x, \quad \forall v \in \mathbb{K}, \tag{1.4}$$

where

$$\mathbb{K} = \{w \in H_0^1(\Omega) : w \leqslant 1 \text{ a.e. in } \Omega\}.$$

In the degenerate case, on the other hand, the problem (1.2) has a positive solution if and only if $a \in (\lambda_1(\Omega), \lambda_1(\Omega_0))$. For such a's, assuming that $\Omega_0 \Subset \Omega$, if we combine the results in [4, 5], we see that $u_p \to w$ in $L^q(\Omega)$ for every $q \geqslant 1$, where w is the unique nontrivial nonnegative solution of

$$w \in \mathbb{K}_0 : \int_\Omega \nabla w \cdot \nabla(v - w)\mathrm{d}x \geqslant \int_\Omega aw(v - w)\mathrm{d}x, \quad \forall v \in \mathbb{K}_0 \tag{1.5}$$

with

$$\mathbb{K}_0 = \{w \in H_0^1(\Omega) : w \leqslant 1 \text{ a.e. in } \Omega \setminus \Omega_0\}.$$

The uniqueness result is the subject of the paper [5]. Therefore, whenever $b(x) \neq 0$, the term $b(x)u^p$ strongly penalizes the points where $u_p > 1$, forcing the limiting solution to be below the obstacle 1 at such points.

Our first aim is to extend these conclusions to the parabolic case (1.1). While doing this, our concern is also to relax some of the assumptions considered in the previous papers, namely, the continuity of b as well as the condition of Ω_0 being in the interior of Ω. In view of that, consider the following conditions for b:

(b1) $b \in L^\infty(\Omega)$.
(b2) There exists Ω_0, an open domain with a smooth boundary, such that

$$b(x) = 0 \quad \text{a.e. on } \Omega_0,$$
$$\forall \Omega' \Subset \Omega \setminus \Omega_0 \text{ open}, \quad \exists\, \underline{b} > 0 \text{ such that } b(x) \geqslant \underline{b} \text{ a.e. in } \Omega'.$$

Observe that in (b2), $\Omega_0 = \emptyset$ is allowed, and $\overline{\Omega_0}$ may intersect $\partial\Omega$. Continuous functions with regular nodal sets or characteristic functions of open smooth domains are typical examples of functions satisfying (b1)–(b2). As for the initial data, we consider

(H1) $u_0 \in H_0^1(\Omega) \cap L^\infty(\Omega)$,
(H2) $0 \leqslant u_0 \leqslant 1$ a.e. in $\Omega \setminus \Omega_0$.

Our first main result is the following.

Theorem 1.1 *Assume that b satisfies* (b1)–(b2), *and u_0 satisfies* (H1)–(H2). *Then there exists a function u such that, given $T > 0$, $u \in L^\infty(0, T; H_0^1(\Omega)) \cap H^1(0, T; L^2(\Omega))$ and*

$$\begin{aligned} u_p &\to u && \text{strongly in } L^2\big(0, T; H_0^1(\Omega)\big), \\ \partial_t u_p &\rightharpoonup \partial_t u && \text{weakly in } L^2(Q_T). \end{aligned}$$

Moreover, u is the unique solution of the following problem:
For a.e. $t > 0$, $u(t) \in \mathbb{K}_0$,

$$\int_\Omega \partial_t u(t)\big(v - u(t)\big)\mathrm{d}x + \int_\Omega \nabla u(t) \cdot \nabla\big(v - u(t)\big)\mathrm{d}x \geqslant \int_\Omega a u(t)\big(v - u(t)\big)\mathrm{d}x \tag{1.6}$$

for every $v \in \mathbb{K}_0$, with the initial condition $u(0) = u_0$.

Next, we turn to the long time behavior of the solution of (1.6).

Theorem 1.2 *Suppose that b satisfies* (b1)–(b2). *Take u_0 verifying* (H1)–(H2). *Fix $a \in (\lambda_1(\Omega), \lambda_1(\Omega_0))$. Let u be the unique positive solution of* (1.6) *and w be the* *unique nontrivial nonnegative solution of* (1.5). *Then* $\|w\|_\infty = 1$ *and*

$$u(t) \to w \quad \textit{strongly in } H_0^1(\Omega), \quad \textit{as } t \to +\infty.$$

Moreover, if $a < \lambda_1(\Omega)$, *then* $\|u(t)\|_{H_0^1(\Omega)} \to 0$; *and if* $a \geqslant \lambda_1(\Omega_0)$, *then both* $\|u(t)\|_\infty$ *and* $\|u(t)\|_{H_0^1(\Omega)}$ *go to* $+\infty$ *as* $t \to +\infty$.

We remark that in the case $\Omega_0 = \emptyset$, we let $\lambda_1(\Omega_0) := +\infty$, and $a \geqslant \lambda_1(\Omega_0)$ is a empty condition. The case $a = \lambda_1(\Omega)$ is the subject of Remark 4.1.

Under some stronger regularity assumptions on b, u_0 and Ω_0, it is known (see [6, Theorem 3.7] or [7, Theorem 2.2]) that $u_p(t,x)$ converges to the unique positive solution of (1.2) whenever $a \in (\lambda_1(\Omega), \lambda_1(\Omega_0))$. Hence in this situation, if we combine all this information together with the results obtained in this paper, then we can conclude that the following diagram commutes:

$$\begin{array}{ccc}
\begin{array}{c} u_p(x,t) \text{ positive solution of} \\ \partial_t u - \Delta u = au - b(x)u^p \end{array} & \xrightarrow{p\to+\infty} & u(x,t) \text{ solution of (1.6)} \\
\Big\downarrow {\scriptstyle t\to+\infty} & & \Big\downarrow {\scriptstyle t\to+\infty} \\
\begin{array}{c} u_p \in H_0^1(\Omega) \text{ positive solution of} \\ -\Delta u = au - b(x)u^p \end{array} & \xrightarrow{p\to+\infty} & \begin{array}{c} w \in H_0^1(\Omega) \text{ nontrivial} \\ \text{nonnegative solution of (1.5)} \end{array}
\end{array}$$

The proof of Theorem 1.1 uses a different approach with respect to the works of Dancer et al. While in [4], the authors use fine properties of functions in Sobolev spaces, here we follow some of the ideas presented in the works [8, 9], and show that a uniform bound on the quantity

$$\iint_{Q_T} b(x)u_p^{p+1}\mathrm{d}x\mathrm{d}t \quad \text{for each } T > 0,$$

implies that $u(t) \in \mathbb{K}_0$ for a.e. $t > 0$ (see the key Lemma 2.4 ahead). As for the proof of Theorem 1.2, the most difficult part is to show that when $a \in (\lambda_1(\Omega), \lambda_1(\Omega_0))$, $u_p(x,t)$ does not go to the trivial solution of (1.5). The key point here is to construct a subsolution of (1.1) independent of p. It turns out that to do this one needs to get a more complete understanding of the nondegenerate case, and to have a stronger convergence of u_p to u as $p \to +\infty$. So we dedicate a part of this paper to the study of this case. To state the results, we start by defining for each $0 < t_1 < t_2$ and $Q_{t_1,t_2} := \Omega \times (t_1,t_2)$, the spaces $C_\alpha^{1,0}(\overline{Q}_{t_1,t_2})$ and $W_q^{2,1}(Q_{t_1,t_2})$. For $q \geqslant 1$, the space $W_q^{2,1}(Q_{t_1,t_2})$ is the set of elements in $L^q(Q_{t_1,t_2})$ with partial derivatives $\partial_t u$, $D_x u$,

$D_x^2 u$ in $L^q(Q_{t_1,t_2})$. It is a Banach space equipped with the norm

$$\|u\|_{2,1;q,Q_{t_1,t_2}} = \|u\|_{L^q(Q_{t_1,t_2})} + \|D_x u\|_{L^q(Q_{t_1,t_2})} + \|D_x^2 u\|_{L^q(Q_{t_1,t_2})} + \|\partial_t u\|_{L^q(Q_{t_1,t_2})}.$$

For each $\alpha \in (0,1)$, $C_\alpha^{1,0}(\overline{Q}_{t_1,t_2})$ is the space of Hölder functions u in $\overline{Q}_{t_1,t_2}$ with exponents α in the x-variable, $\frac{\alpha}{2}$ in the t-variable and with $D_x u$ satisfying the same property. More precisely, defining the Hölder semi-norm

$$[u]_{\alpha,Q_{t_1,t_2}} := \sup\left\{\frac{|u(x,t)-u(x',t')|}{|x-x'|^\alpha + |t-t'|^{\frac{\alpha}{2}}},\ x,x' \in \overline{\Omega},\ t,t' \in [t_1,t_2],\ (x,t) \neq (x',t')\right\},$$

we have that

$$C_\alpha^{1,0}(\overline{Q}_{t_1,t_2}) := \left\{u:\ \|u\|_{C_\alpha^{1,0}(\overline{Q}_{t_1,t_2})} := \|u\|_{L^\infty(Q_{t_1,t_2})} + \|D_x u\|_{L^\infty(Q_{t_1,t_2})} + [u]_{\alpha,Q_{t_1,t_2}} + [D_x u]_{\alpha,Q_{t_1,t_2}} < +\infty\right\}.$$

Recall that we have the following embedding for every $0 \leqslant t_1 < t_2$ (see [10, Lemmas II.3.3, II.3.4]):

$$W_q^{2,1}(Q_{t_1,t_2}) \hookrightarrow C_\alpha^{1,0}(\overline{Q}_{t_1,t_2}), \quad \forall 0 \leqslant \alpha < 1 - \frac{N+2}{q}. \tag{1.7}$$

In the nondegenerate case, we have the following result.

Theorem 1.3 *Suppose that b satisfies* (b1) *and the condition as follows:*

(b2′) *there exists $b_0 > 0$, such that $b(x) \geqslant b_0$ for a.e. $x \in \Omega$.*

Let u_0 satisfy (H1) *and $0 \leqslant u_0 \leqslant 1$ for a.e. $x \in \Omega$. Then, in addition to the conclusions of Theorem* 1.1, *we have that*

$$u_p \to u \quad \textit{strongly in } C_\alpha^{1,0}(\overline{Q}_{t_1,t_2}),\ \textit{weakly in } W_q^{2,1}(Q_{t_1,t_2}),\ \textit{as } p \to +\infty$$

for every $\alpha \in (0,1)$, $q \geqslant 1$ and $0 < t_1 < t_2$. Moreover, u is the unique solution of

$$\partial_t u - \Delta u = a u \chi_{\{u<1\}} \quad \textit{in } Q, \quad u(0) = u_0, \quad \|u\|_\infty \leqslant 1. \tag{1.8}$$

In this case, as $t \to +\infty$, we also obtain a convergence result for the coincidence sets $\{u(x,t) = 1\}$.

Theorem 1.4 *Suppose that* b *satisfies* (b1)–(b2′). *Take* u_0 *satisfying* (H1) *and* $0 \leqslant u_0 \leqslant 1$ *for a.e.* $x \in \Omega$. *Fix* $a > \lambda_1(\Omega)$. *Let* u *be the unique solution of* (1.8), *and* w *be the unique solution of* (1.3). *Then, as* $t \to +\infty$,

$$u(t) \to w \quad \text{strongly in } H_0^1(\Omega) \cap H^2(\Omega)$$

and

$$\chi_{\{u=1\}}(t) \to \chi_{\{w=1\}} \quad \text{strongly in } L^q(\Omega), \quad \forall q \geqslant 1. \tag{1.9}$$

The structure of this paper is as follows. In Sect. 2, we prove Theorem 1.1, while in Sect. 3, Theorem 1.3 is treated. Finally, in Sect. 4, we use the strong convergence up to the boundary of Ω obtained in the latter theorem to prove Theorem 1.4, and afterwards, we use it combined with a subsolution argument to prove Theorem 1.2.

We end this introduction by pointing out some other works concerning this type of asymptotic limit. The generalization of [4] for the p-Laplacian case was performed in [11]. In [8, 9], elliptic problems of the type

$$-\Delta u + f(x,u)|f(x,u)|^p = g(x)$$

were treated, while in the works by Grossi et al. [12, 13], and Bonheure and Serra [14], the authors dealt with the asymptotics study of problems of the type

$$-\Delta u + V(|x|)u = u^p,$$

as $p \to +\infty$ in a ball or an annulus both with Neumann and Dirichlet boundary conditions.

2 The General Case: Proof of Theorem 1.1

To make the presentation more structured, we split our proof into several lemmas. We start by showing a very simple comparison principle which is an easy consequence of the monotonicity of the operator $u \mapsto |u|^{p-1}u$.

Lemma 2.1 *Suppose that* u *is a solution of* (1.1). *Take* v *a supersolution, satisfying*

$$\partial_t v - \Delta v \geqslant av - b(x)v^p \quad \text{in } Q_T,$$
$$v(0) = v_0, \quad v(t)|_{\partial\Omega} = 0$$

with $u_0 \leqslant v_0$. *Then* $u(x,t) \leqslant v(x,t)$ *a.e. On the other hand, if* v *is a subsolution satisfying*

$$\partial_t v - \Delta v \leqslant av - b(x)v^p \quad \text{in } Q_T,$$
$$v(0) = v_0, \quad v(t)|_{\partial\Omega} = 0$$

with $v_0 \leqslant u_0$, *then* $v(x,t) \leqslant u(x,t)$.

Proof The proof is quite standard, but we include it here only for the sake of completeness. In the case v is a supersolution, we have

$$\partial_t(u-v)-\Delta(u-v)+b(x)\big(u^p-v^p\big)\leqslant a(u-v).$$

Multiplying this by $(u(t)-v(t))^+$, we obtain

$$\begin{aligned}&\frac{1}{2}\frac{\mathrm{d}}{\mathrm{d}t}\int_\Omega\big[\big(u(t)-v(t)\big)^+\big]^2\mathrm{d}x+\int_\Omega\big|\nabla\big(u(t)-v(t)\big)^+\big|^2\mathrm{d}x\\&\quad+\int_\Omega b(x)\big(u^p(t)-v^p(t)\big)\big(u(t)-v(t)\big)^+\mathrm{d}x\\&\leqslant a\int_\Omega\big[\big(u(t)-v(t)\big)^+\big]^2\mathrm{d}x.\end{aligned}$$

As $b(x)(u^p-v^p)(u-v)^+\geqslant 0$, we have

$$\frac{\mathrm{d}}{\mathrm{d}t}\int_\Omega\big[\big(u(t)-v(t)\big)^+\big]^2\mathrm{d}x\leqslant 2a\int_\Omega\big[\big(u(t)-v(t)\big)^+\big]^2\mathrm{d}x,$$

whence

$$\int_\Omega\big[\big(u(t)-v(t)\big)^+\big]^2\mathrm{d}x\leqslant \mathrm{e}^{2at}\int_\Omega\big[(u_0-v_0)^+\big]^2\mathrm{d}x=0.$$

The proof of the result for the subsolution case is analogous. □

Next we show some uniform bounds in p.

Lemma 2.2 *Given $T>0$, there exists an $M=M(T)>0$, such that $\|u_p\|_{L^\infty(Q_T)}\leqslant M$ for all $p>1$.*

Proof Take $\psi\geqslant 0$ as the unique solution of

$$\begin{cases}\partial_t\psi-\Delta\psi=a\psi & \text{in } Q_T,\\ \psi(0)=u_0, & u(t)|_{\partial\Omega}=0.\end{cases}$$

Then

$$\partial_t\psi-\Delta\psi-a\psi+b(x)\psi^p\geqslant\partial_t\psi-\Delta\psi-a\psi=0.$$

Hence, ψ is a supersolution, and from Lemma 2.1, we have that $0\leqslant u_p\leqslant\psi$. In particular,

$$\|u_p\|_{L^\infty(Q_T)}\leqslant\|\psi\|_{L^\infty(Q_T)}<+\infty,\qquad\text{as } u_0\in L^\infty(\Omega),$$

which proves the result. □

Lemma 2.3 *Given $T > 0$, the sequence $\{u_p\}_p$ is bounded in $H^1(0,T;L^2(\Omega)) \cap L^\infty(0,T;H_0^1(\Omega))$. Thus, there exists a $u \in H^1(0,T;L^2(\Omega)) \cap L^\infty(0,T;H_0^1(\Omega))$, such that*

$$\begin{aligned} u_p &\to u \quad \text{strongly in } L^2(Q_T), \text{ weakly in } L^2\big(0,T;H_0^1(\Omega)\big), \\ \partial_t u_p &\rightharpoonup \partial_t u \quad \text{weakly in } L^2(Q_T). \end{aligned}$$

Moreover, there exists a $C = C(T) > 0$, such that

$$\iint_{Q_T} b(x) u_p^{p+1} \mathrm{d}x\mathrm{d}t \leqslant C, \quad \forall p > 1. \tag{2.1}$$

Proof Multiplying (1.1) by u_p, and integrating it in Ω, we have

$$\frac{1}{2}\frac{\mathrm{d}}{\mathrm{d}t}\int_\Omega u_p^2(t)\mathrm{d}x + \int_\Omega |\nabla u_p(t)|^2 \mathrm{d}x = a\int_\Omega u_p^2(t)\,\mathrm{d}x - \int_\Omega b(x)u_p^{p+1}(t)\mathrm{d}x.$$

Integrating the above equation between 0 and t, we have

$$\begin{aligned} &\frac{1}{2}\int_\Omega u_p^2(t)\mathrm{d}x + \int_0^t \|\nabla u_p(\xi)\|_2^2 \mathrm{d}\xi + \iint_{Q_t} b(x)u_p^{p+1}\mathrm{d}x\mathrm{d}t \\ &\quad\leqslant \frac{1}{2}\int_\Omega u_0^2 \mathrm{d}x + a\iint_{Q_t} u_p^2 \mathrm{d}x\mathrm{d}t \\ &\quad\leqslant \frac{1}{2}\|u_0\|_2^2 + at|\Omega|\big(M(t)\big)^2. \end{aligned}$$

Hence, for every $T > 0$, $\{u_p\}_p$ is bounded in $L^2(Q_T)$, and (2.1) holds. □

Now using $\partial_t u_p$ as a test function ($u_p = 0$ on $\partial\Omega$ for all $t > 0$, thus $\partial_t u_p(t) \in H_0^1(\Omega)$ for a.e. $t > 0$), we have

$$\int_\Omega (\partial_t u_p)^2 \mathrm{d}x + \frac{1}{2}\frac{\mathrm{d}}{\mathrm{d}t}\int_\Omega |\nabla u_p(t)|^2\mathrm{d}x = \frac{a}{2}\frac{\mathrm{d}}{\mathrm{d}t}\int_\Omega u_p^2(t)\mathrm{d}x - \frac{\mathrm{d}}{\mathrm{d}t}\int_\Omega b(x)\frac{u_p^{p+1}(t)}{p+1}\mathrm{d}x.$$

Again after an integration, we have

$$\begin{aligned} &\iint_{Q_t} (\partial_t u_p)^2 \mathrm{d}x\mathrm{d}t + \frac{1}{2}\int_\Omega |\nabla u_p(t)|^2 \mathrm{d}x + \frac{a}{2}\int_\Omega u_0^2\mathrm{d}x + \int_\Omega b(x)\frac{u_p^{p+1}(t)}{p+1}\mathrm{d}x \\ &= \frac{1}{2}\int_\Omega |\nabla u_0|^2\mathrm{d}x + \int_\Omega b(x)\frac{u_0^{p+1}}{p+1}\mathrm{d}x + \frac{a}{2}\int_\Omega u_p^2(t)\mathrm{d}x \\ &\leqslant \frac{1}{2}\|\nabla u_0\|_2^2 + \frac{b_\infty|\Omega|}{p+1} + \frac{a}{2}M^2|\Omega|, \end{aligned} \tag{2.2}$$

where we have used the fact that $0 \leqslant u_0 \leqslant 1$ whenever $b(x) \neq 0$, together with the previous lemma.

The proofs of the following two results are inspired by similar computations made in [8, 9].

Lemma 2.4 *We have $u(t) \in \mathbb{K}_0$ for a.e. $t > 0$.*

Proof Let $\Omega' \Subset \Omega \setminus \Omega_0$ and take $Q'_T := \Omega' \times (0, T)$. Given $m > 1$, we will show that $|\{(x,t) \in Q'_T : u > m\}| = 0$. Denote by $\underline{b}$ the infimum of $b(x)$ over $\overline{\Omega'}$, which is positive by (b2). Recalling (2.1), we deduce the existence of $C > 0$, such that

$$\begin{aligned} 0 &\leqslant \iint_{\{u_p>m\}\cap Q'_T} \underline{b} u_p \mathrm{d}x\mathrm{d}t \\ &\leqslant \frac{1}{m^p} \iint_{\{u_p>m\}\cap Q'_T} b(x) u_p^{p+1} \mathrm{d}x\mathrm{d}t \\ &\leqslant \frac{1}{m^p} \iint_{Q_T} b(x) u_p^{p+1} \mathrm{d}x\mathrm{d}t \leqslant \frac{C}{m^p}. \end{aligned}$$

Hence, as $m > 1$ and $\underline{b} > 0$,

$$\lim_{p\to+\infty} \iint_{\{u_p>m\}\cap Q'_T} u_p \mathrm{d}x\mathrm{d}t = 0.$$

Now observe that

$$\begin{aligned} 0 &= \lim_{p\to+\infty} \iint_{\{u_p>m\}\cap Q'_T} u_p \mathrm{d}x\mathrm{d}t \\ &= \lim_{p\to+\infty} \left(\int_0^T \int_{\Omega'} u_p \chi_{\{u_p>m\}} \chi_{\{u>m\}} \mathrm{d}x\mathrm{d}t + \int_0^T \int_{\Omega'} u_p \chi_{\{u_p>m\}} \chi_{\{u\leqslant m\}} \mathrm{d}x\mathrm{d}t \right) \\ &\geqslant \lim_{p\to+\infty} \int_0^T \int_{\Omega'} u_p \chi_{\{u_p>m\}} \chi_{\{u>m\}} \mathrm{d}x\mathrm{d}t. \end{aligned}$$

As $u_p \chi_{\{u_p>m\}} \chi_{\{u>m\}} \to u \chi_{\{u>m\}}$ a.e. and $|u_p \chi_{\{u_p>m\}} \chi_{\{u>m\}}| \leqslant L$ on Q_T, then by the Lebesgue's dominated convergence theorem, we have

$$\begin{aligned} &\lim_{p\to+\infty} \int_0^T \int_{\Omega'} u_p \chi_{\{u_p>m\}} \chi_{\{u>m\}} \mathrm{d}x\mathrm{d}t \\ &= \int_0^T \int_{\Omega'} u \chi_{\{u>m\}} \mathrm{d}x\mathrm{d}t \\ &\geqslant m \big| \{(t,x) \in Q'_T : u(t,x) > m\} \big| \geqslant 0. \end{aligned}$$

Hence $|\{(x,t) \in Q'_T : u(x,t) > m\}| = 0$ whenever $m > 1$. □

Lemma 2.5 *Let u be the limit provided by Lemma 2.3. Then, up to a subsequence,*

$$u_p \to u \quad \text{strongly in } L^2\big(0, T; H_0^1(\Omega)\big).$$

Proof Multiply (1.1) by $u_p - u$, and integrate it in Q_T, we have

$$\iint_{Q_T} \partial_t u_p(u_p - u)\mathrm{d}x\mathrm{d}t + \iint_{Q_T} \nabla u_p \cdot \nabla(u_p - u)\mathrm{d}x\mathrm{d}t + \iint_{Q_T} b(x)u_p^p(u_p - u)\mathrm{d}x\mathrm{d}t = \iint_{Q_T} au_p(u_p - u)\mathrm{d}x\mathrm{d}t,$$

which, after adding and subtracting $\iint_{Q_T} \nabla u \cdot \nabla(u_p - u)\mathrm{d}x\mathrm{d}t$, is equivalent to

$$\iint_{Q_T} \partial_t u_p(v - u_p)\mathrm{d}x\mathrm{d}t + \iint_{Q_T} |\nabla(u_p - u)|^2\mathrm{d}x\mathrm{d}t + \iint_{Q_T} \nabla u \cdot \nabla(u_p - u)\mathrm{d}x\mathrm{d}t + \iint_{Q_T} b(x)u_p^p(u_p - u)\mathrm{d}x\mathrm{d}t = \iint_{Q_T} au_p(u_p - u)\,\mathrm{d}x\mathrm{d}t.$$

By the convergence shown in Lemma 2.3, we have that the terms $\iint_{Q_T} \partial_t u_p(u_p - u)\mathrm{d}x\mathrm{d}t$, $\iint_{Q_T} \nabla u \cdot \nabla(u_p - u)\mathrm{d}x\mathrm{d}t$ and $\iint_{Q_T} au_p(u_p - u)\mathrm{d}x\mathrm{d}t$ tend to zero as $p \to +\infty$. Finally, observe that

$$\begin{aligned}\iint_{Q_T} b(x)u_p^p(u_p - u)\mathrm{d}x\mathrm{d}t &= \iint_{\{u_p \leqslant u\}} b(x)u_p^p(u_p - u)\mathrm{d}x\mathrm{d}t + \iint_{\{u < u_p\}} b(x)u_p^p(u_p - u)\mathrm{d}x\mathrm{d}t \\ &\geqslant \iint_{\{0 \leqslant u_p \leqslant u\}} b(x)u_p^p(u_p - u)\mathrm{d}x\mathrm{d}t.\end{aligned}$$

As $u \leqslant 1$ a.e. in $Q'_T = (0, T) \times \Omega \setminus \Omega_0$ (see Lemma 2.4), we have

$$\begin{aligned}\left| \iint_{\{0 \leqslant u_p \leqslant u\}} b(x)u_p^p(u_p - u)\mathrm{d}x\mathrm{d}t \right| &\leqslant \iint_{\{0 \leqslant u_p \leqslant u\} \cap Q'_T} b(x)u^p|u_p - u|\mathrm{d}x\mathrm{d}t \\ &\leqslant \iint_{Q_T} b_\infty|u_p - u|\mathrm{d}x\mathrm{d}t \to 0,\end{aligned}$$

whence $\liminf \iint_{Q_T} b(x)u_p^p(u_p - u)\mathrm{d}x\mathrm{d}t \geqslant 0$. Thus

$$\iint_{Q_T} |\nabla(u_p - u)|^2\mathrm{d}x\mathrm{d}t \to 0, \quad \text{as } p \to +\infty,$$

and the result follows. □

Proof of Theorem 1.1 (1) The convergence of u_p to u are the consequences of Lemmas 2.3 and 2.5. Let us then prove first of all that

$$\iint_{Q_T} \partial_t u(v-u)\mathrm{d}x\mathrm{d}t + \iint_{Q_T} \nabla u \cdot \nabla(v-u)\mathrm{d}x\mathrm{d}t \geqslant \iint_{Q_T} au(v-u)\mathrm{d}x\mathrm{d}t \quad (2.3)$$

for every $v \in \widetilde{\mathbb{K}}_0$, where $\widetilde{\mathbb{K}}_0 := \{v \in L^2(0,T;H_0^1(\Omega)) : v(t) \in \mathbb{K}_0$ for a.e. $t \in (0,T)\}$. Fix $v \in \widetilde{\mathbb{K}}_0$ and take $0 < \theta < 1$. Multiplying (1.1) by $\theta v - u_p$ and integrating it, we have

$$\begin{aligned}\iint_{Q_T} \partial_t u_p(\theta v - u_p)\mathrm{d}x\mathrm{d}t &+ \iint_{Q_T} \nabla u_p \cdot \nabla(\theta v - u_p)\mathrm{d}x\mathrm{d}t \\ &+ \iint_{Q_T} b(x)u_p^p(\theta v - u_p)\mathrm{d}x\mathrm{d}t \\ &= \iint_{Q_T} au_p(\theta v - u_p)\mathrm{d}x\mathrm{d}t.\end{aligned}$$

By Lemmas 2.3 and 2.5, we have

$$\iint_{Q_T} \partial_t u_p(\theta v - u_p)\mathrm{d}x\mathrm{d}t \to \iint_{Q_T} \partial_t u(\theta v - u)\mathrm{d}x\mathrm{d}t,$$

$$\iint_{Q_T} \nabla u_p \cdot \nabla(\theta v - u_p)\mathrm{d}x\mathrm{d}t \to \iint_{Q_T} \nabla u \cdot \nabla(\theta v - u)\mathrm{d}x\mathrm{d}t,$$

$$\iint_{Q_T} u_p(\theta v - u_p)\mathrm{d}x\mathrm{d}t \to \iint_{Q_T} u(\theta v - u)\mathrm{d}x\mathrm{d}t.$$

For the remaining term, as $b(x) = 0$ a.e. in Ω_0 and $v \leqslant 1$ a.e in $\Omega \setminus \Omega_0 \times (0,T)$, we have

$$\begin{aligned}&\iint_{Q_T} b(x)u_p^p(\theta v - u_p)\mathrm{d}x\mathrm{d}t \\ &= \iint_{0\leqslant u_p \leqslant \theta v} b(x)u_p^p(\theta v - u_p)\mathrm{d}x\mathrm{d}t + \iint_{\theta v < u_p} b(x)u_p^p(\theta v - u_p)\mathrm{d}x\mathrm{d}t \\ &\leqslant \iint_{Q_T'} b(x)\theta^p|\theta v - u_p|\mathrm{d}x\mathrm{d}t \to 0,\end{aligned}$$

as $p \to +\infty$, because $\theta < 1$. Thus

$$\iint_{Q_T} \partial_t u(\theta v - u)\mathrm{d}x\mathrm{d}t + \iint_{Q_T} \nabla u \cdot \nabla(\theta v - u)\mathrm{d}x\mathrm{d}t \geqslant \iint_{Q_T} au(\theta v - u)\mathrm{d}x\mathrm{d}t,$$

and now we just have to make $\theta \to 1$.

(2) Given $v \in \mathbb{K}_0$, $\xi \in (0,T)$ and $h > 0$, take

$$\widetilde{v}(t) = \begin{cases} v, & t \in [\xi, \xi+h], \\ u(t), & t \notin [\xi, \xi+h]. \end{cases}$$

Then, $\widetilde{v} \in \widetilde{\mathbb{K}}_0$, and from (2.3), we have

$$\int_\xi^{\xi+h} \int_\Omega \partial_t u(v-u)\mathrm{d}x\mathrm{d}t + \int_\xi^{\xi+h} \int_\Omega \nabla u \cdot \nabla(v-u)\,\mathrm{d}x\mathrm{d}t \geqslant \int_\xi^{\xi+h} \int_\Omega au(v-u)\mathrm{d}x\mathrm{d}t.$$

Multiplying this inequality by $\frac{1}{h}$ and making $h \to 0$, we get (1.6), as required.

(3) Finally, it is easy to show that the problem (1.6) has a unique solution. In fact, taking u_1 and u_2 as solutions to (1.6) with the same initial data, we have

$$\int_\Omega \partial_t\big(u_1(t)-u_2(t)\big)\big(u_2(t)-u_1(t)\big) + \nabla\big(u_1(t)-u_2(t)\big)\cdot\nabla\big(u_2(t)-u_1(t)\big)\mathrm{d}x \geqslant \int_\Omega a\big(u_1(t)-u_2(t)\big)\big(u_2(t)-u_1(t)\big)\mathrm{d}x,$$

which is equivalent to

$$\frac{1}{2}\frac{\mathrm{d}}{\mathrm{d}t}\int_\Omega \big(u_1(t)-u_2(t)\big)^2\mathrm{d}x + \int_\Omega \big|\nabla\big(u_1(t)-u_2(t)\big)\big|^2\mathrm{d}x \leqslant \int_\Omega a\big(u_1(t)-u_2(t)\big)^2\mathrm{d}x.$$

The fact that u_1 and u_2 have the same initial data now implies that

$$\int_\Omega \big(u_1(t)-u_2(t)\big)^2(t)\mathrm{d}x \leqslant \mathrm{e}^{2at}\int_\Omega (u_0-u_0)\mathrm{d}x = 0.$$

Hence, $u_p \to u$ for the whole sequence $\{u_p\}_p$, not only for a subsequence. □

3 The Nondegenerate Case: Proof of Theorem 1.3

As stated, the results of the previous section are true even in the case of $\Omega_0 = \emptyset$. Let us check that in the nondegenerate case (b2′), we have a stronger convergence as well as a more detailed characterization for the limit u (see (1.8)). This is mainly due to the following powerful estimate.

Lemma 3.1 *There exists a constant $M > 0$ (independent of p), such that $\|u_p\|_{L^\infty(Q)}^{p-1} \leqslant M$ for all $p > 1$.*

Proof Let $b_0 = \inf_\Omega b > 0$ and take $M_p > 0$, such that $aM_p - b_0 M_p^p = 0$. Observe that $as - b_0 s^p \leqslant 0$ for $s \geqslant M_p$. Take $N_p := \max\{1, M_p\}$. Multiplying (1.1) by $(u_p(t) - N_p)^+$ (recall that $u_p = 0$ on $\partial\Omega$, whence $(u_p - N_p)^+ = 0$ on the boundary as well), we obtain

$$\begin{aligned}
&\frac{1}{2}\frac{\mathrm{d}}{\mathrm{d}t}\int_\Omega \big((u_p - N_p)^+\big)^2 \mathrm{d}x + \int_\Omega |\nabla(u_p - N_p)^+|^2 \mathrm{d}x \\
&\quad = \int_\Omega \big(au_p - b(x)u_p^p\big)(u_p - N_p)^+ \mathrm{d}x \\
&\quad \leqslant \int_\Omega \big(au_p - b_0 u_p^p\big)(u_p - N_p)^+ \mathrm{d}x \\
&\quad = \int_{u_p \geqslant N_p} \big(au_p - b_0 u_p^p\big)(u_p - N_p) \mathrm{d}x \leqslant 0.
\end{aligned}$$

Thus

$$\frac{\mathrm{d}}{\mathrm{d}t}\int_\Omega \big((u_p - N_p)^+\big)^2 \leqslant 0, \quad \int_\Omega \big((u_p(t) - N_p)^+\big)^2 \mathrm{d}x \leqslant \int_\Omega \big((u_0 - N_p)^+\big)^2 \mathrm{d}x,$$

which is zero because $N_p \geqslant 1$. Then

$$0 \leqslant u_p(t,x) \leqslant \max\{1, M_p\},$$

and the result now follows from the fact that $M_p = (\frac{a}{b_0})^{\frac{1}{p-1}}$. □

Lemma 3.2 *For each $t_2 > t_1 > 0$, $q > 1$ and $\alpha \in (0,1)$, the sequence $\{u_p\}_p$ is bounded in $W_q^{2,1}(Q_{t_1,t_2})$ and $C_\alpha^{1,0}(\overline{Q}_{t_1,t_2})$. Thus*

$$\begin{aligned}
&u_p \rightharpoonup u \quad \text{weakly in } W_q^{2,1}(Q_{t_1,t_2}), \\
&u_p \to u \quad \text{strongly in } C_\alpha^{1,0}(\overline{Q}_{t_1,t_2}), \quad \forall \alpha \in (0,1).
\end{aligned}$$

Proof From Lemma 3.1 we get that

$$\|au_p\|_{L^\infty(Q)} \leqslant C' M^{\frac{1}{p-1}} \leqslant C'', \qquad \|b(x)u_p^p\|_{L^\infty(Q)} \leqslant b_\infty M^{\frac{p}{p-1}} \leqslant C'''.$$

Hence

$$\|\partial_t u_p - \Delta u_p\|_{L^\infty(Q)} \leqslant C, \quad \forall p > 1,$$

which, together with [10, IV. Theorems 9.1 and 10.1] (see also [15, Theorems 7.22 and 7.32]), implies that for every $q > 1$, the sequence $\{u_p\}_p$ is bounded in $W_q^{2,1}(Q_{t_1,t_2})$ independently of p. Thus, we can use the embedding (1.7) to show that $\{u_p\}_p$ is bounded in $C_\alpha^{1,0}(\overline{Q}_{t_1,t_2})$. As the embedding $C_\alpha^{1,0} \hookrightarrow C_{\alpha'}^{1,0}$ is compact for all $\alpha > \alpha'$, we have the conclusion. □

Observe that by Theorem 1.1 the whole sequence u_p already converges to u in some spaces, and hence the convergence obtained in this lemma is also for the whole sequence, not only for a subsequence.

Remark 3.1 It is important to assume Ω smooth (say $\partial\Omega$ of class C^2) to get regularity up to $\partial\Omega$. This will be of crucial importance in the next section. Without such a regularity assumption, we would obtain convergence in each set of the type $\Omega' \times (t_1, t_2)$ with $\Omega' \Subset \Omega, 0 < t_1 < t_2$.

Now, in view of Theorem 1.3, we want to prove that in this case u solves (1.8). By Lemma 3.1, we know that $\|u_p^{p-1}\|_{L^\infty(Q)} \leqslant M$ for all $p > 1$. This implies the existence of $\psi \geqslant 0$, such that, for every $T > 0$,

$$u_p^{p-1} \rightharpoonup \psi \quad \text{weak-}*\ \text{in } L^\infty(Q_T) \text{ and weak in } L^2(Q_T).$$

Thus when we make $p \to +\infty$ in (1.1), we obtain that the limit u satisfies

$$\partial_t u - \Delta u = (a - \psi)u.$$

Moreover,

$$\|u_p\|_\infty \leqslant M^{\frac{1}{p-1}} \to 1, \quad \text{as } p \to \infty,$$

which implies, together with Lemma 3.2, that $0 \leqslant u \leqslant 1$. The proof of Theorem 1.3 will be complete after the following lemmas.

Lemma 3.3 $\psi = 0$ *a.e. in the set* $\{(t,x) \in Q :\ u(x,t) < 1\}$. *In particular, this implies that*

$$\partial_t u - \Delta u = au\chi_{\{u<1\}} \quad a.e.\ (x,t) \in Q.$$

Proof Take (x,t) such that $u(x,t) < 1$. As $u_p \to u$ in $C_\alpha^{1,0}$, we can take $\delta > 0$ such that $u_p \leqslant 1 - \delta$ for large p. Then,

$$0 \leqslant u_p^{p-1} \leqslant (1-\delta)^{p-1} \to 0, \quad \text{as } p \to +\infty,$$

whence $\psi(x,t) = 0$. Thus $\psi = 0$ a.e. on $\{(x,t) : u(t,x) < 1\}$.
Finally, as $u \in W_q^{2,1}$ for every $q \geqslant 1$, we have that

$$\partial_t u - \Delta u = 0 \quad \text{a.e. on } \big\{(x,t) :\ u(x,t) = 1\big\},$$

and the proof is complete. □

Lemma 3.4 *Let w be a solution of* (1.8). *Then w solves* (1.6).

Proof Multiply (1.8) by $v - w$ with $v \in \mathbb{K}$. Then we have

$$\int_\Omega \partial_t w(v-w)\mathrm{d}x + \int_\Omega \nabla w \cdot \nabla(v-w)\mathrm{d}x$$
$$= a\int_\Omega w\chi_{\{w<1\}}(v-w)\mathrm{d}x$$

$$= a \int_\Omega w(v - w)\mathrm{d}x - a \int_\Omega (v - 1)\mathrm{d}x$$

$$\geqslant a \int_\Omega w(v - w)\mathrm{d}x,$$

since $v \leqslant 1$ in Ω. □

Proof of Theorem 1.3 The convergence $u_p \to u$ strongly in $C_\alpha^{1,0}(\overline{Q}_{t_1,t_2})$ and weakly in $W_q^{2,1}(Q_{t_1,t_2})$ for every $T > 0$ is a consequence of Lemma 3.2. By Lemma 3.3, u satisfies (1.8). Finally, Lemma 3.4 and the uniqueness shown for (1.6) imply the uniqueness of solution of (1.8). □

4 Asymptotic Behavior as $t \to \infty$: Proof of Theorem 1.4

In this section, we will study the asymptotic behavior of (1.6) as $t \to +\infty$. First we need to understand what happens in the nondegenerate case (b2′), and prove Theorem 1.4. Then, as we will see, the convergence up to the boundary proved in Lemma 3.2 will be crucial. Only afterwards will we be able to prove Theorem 1.2.

4.1 Proof of Theorem 1.4

We start by showing that the time derivative of u vanishes as $t \to +\infty$.

Proposition 4.1 $\|\partial_t u(t)\|_{L^2(\Omega)} \to 0$ *as* $t \to +\infty$.

In order to prove this proposition, we will show that $\|\partial u_p(t)\|_{L^2(\Omega)} \to 0$ as $t \to +\infty$, uniformly in $p > 1$. To do so, we will use the following result from [2, Lemma 6.2.1].

Lemma 4.1 *Suppose that $y(t)$ and $h(t)$ are nonnegative continuous functions defined on $[0, \infty)$ and satisfy the following conditions:*

$$y'(t) \leqslant A_1 y^2 + A_2 + h(t), \quad \int_0^\infty y(t)\mathrm{d}t \leqslant A_3, \quad \int_0^\infty h(t)\mathrm{d}t \leqslant A_4 \tag{4.1}$$

for some constants $A_1, A_2, A_3, A_4 > 0$. Then

$$\lim_{t\to+\infty} y(t) = 0.$$

Moreover, this convergence is uniform[1] *for all y satisfying* (4.1) *with the same constants A_1, A_2, A_3, A_4.*

With this in mind, we have the following lemma.

Lemma 4.2 *Let u_p be the solution of* (1.1) *and $a > 0$. Then*

$$\|\partial_t u_p(t)\|_2 \to 0, \quad \textit{as } t \to +\infty, \textit{ uniformly in } p > 1.$$

Proof Let us check that $y(t) := \|\partial_t u_p(t)\|_2^2$ satisfies the assumptions of Lemma 4.1. First of all, (2.2) implies that

$$\int_0^\infty \|\partial_t u_p(t)\|_2^2 \mathrm{d}x \leqslant \|\nabla u_0\|_2^2 + \frac{|\Omega|}{2} + \frac{a}{2} M^2 |\Omega|$$

(recall that in the nondegenerate case, $\|u_p\|_{L^\infty(\Omega\times\mathbb{R}^+)}$ is bounded uniformly in p, by Lemma 3.1). Differentiate equation (1.1) with respect to t

$$\partial_t^2 u_p - \Delta \partial_t u_p + p u_p^{p-1} \partial_t u_p = a \partial_t u_p,$$

multiply the above equation by $\partial_t u_p$ and integrate it in Ω at each time t. Then, we obtain

$$\begin{aligned} &\frac{1}{2}\frac{\mathrm{d}}{\mathrm{d}t}\int_\Omega \big(\partial_t u_p(t)\big)^2 \mathrm{d}x + \int_\Omega \big|\nabla\big(\partial_t u_p(t)\big)\big|^2 \mathrm{d}x + p\int_\Omega u_p^{p-1}(t)\big(\partial_t u_p(t)\big)^2 \mathrm{d}x \\ &\quad = a\int_\Omega \big(\partial_t u_p(t)\big)^2 \mathrm{d}x \leqslant \frac{a}{2}\left(\int_\Omega \big(\partial_t u_p(t)\big)^2 \mathrm{d}x\right)^2 + \frac{a}{2}. \end{aligned}$$

Thus

$$\frac{\mathrm{d}}{\mathrm{d}t}\|\partial_t u_p(t)\|_2^2 \leqslant a\|\partial_t u_p\|_2^4 + a.$$

So we can apply the previous lemma with $A_1 = a$, $A_2 = a$, $A_3 = \|\nabla u_0\|_2^2 + \frac{|\Omega|}{2} + \frac{a}{2}M^2|\Omega|$, and $h(t) \equiv 0$, $A_4 = 0$. □

Proof of Proposition 4.1 From the previous lemma, we know that, given $\varepsilon > 0$, there exist $\bar{t}$ and p_0, such that

$$\|\partial_t u_p(t)\|_2^2 \leqslant \varepsilon, \quad \forall t \geqslant \bar{t},\ \forall p > p_0.$$

Thus for every $\bar{t} \leqslant t_1 < t_2$,

$$\int_{t_1}^{t_2} \|\partial_t u_p(t)\|_2^2 \mathrm{d}t \leqslant \varepsilon(t_2 - t_1), \quad \forall t \geqslant \bar{t},\ \forall p > p_0.$$

[1] This uniformity is not stated in the original lemma, but a close look at the proof allows us to easily obtain that conclusion.

As $\partial_t u_p \rightharpoonup \partial_t u$ weakly in $L^2(Q_T)$ for every $T > 0$ (see Theorem 1.1), then taking the lim inf as $p \to +\infty$, we get

$$\int_{t_1}^{t_2} \|\partial_t u(t)\|_2^2 \mathrm{d}t \leqslant \varepsilon(t_2 - t_1).$$

Hence

$$\|\partial_t u(t)\|_2^2 \leqslant \varepsilon, \quad \forall t \geqslant \overline{t},$$

which gives the statement. □

Proof of Theorem 1.4 Fix $a > \lambda_1(\Omega)$. By taking $v = 0$ in (1.6), we obtain

$$\int_\Omega |\nabla u(t)|^2 \mathrm{d}x \leqslant \int_\Omega \big(-\partial_t u(t) u(t) + a u^2\big)\mathrm{d}x,$$

which implies that $\|u(t)\|_{H_0^1(\Omega)}$ is bounded for $t > 0$. Therefore, up to a subsequence, we have $u(t) \rightharpoonup \overline{u}$ in $H_0^1(\Omega)$ as $t \to +\infty$. Given a subsequence $t_n \to +\infty$ such that $u(t_n) \rightharpoonup \overline{u}$, we know that

$$\int_\Omega \partial_t u(t_n)\big(v - u(t_n)\big)\mathrm{d}x + \int_\Omega \nabla u(t_n) \cdot \nabla\big(v - u(t_n)\big)\mathrm{d}x$$
$$\geqslant a \int_\Omega u(t_n)\big(v - u(t_n)\big)\mathrm{d}x$$

for all $v \in \mathbb{K}$, which, together with Proposition 4.1, implies that, as $p \to +\infty$,

$$\int_\Omega \nabla\overline{u} \cdot \nabla(v - \overline{u})\mathrm{d}x \geqslant \int_\Omega a\overline{u}(v - \overline{u})\mathrm{d}x, \quad \forall v \in \mathbb{K}$$

or, equivalently,

$$-\Delta\overline{u} = a\overline{u}\chi_{\{\overline{u}<1\}}$$

(here we are using the equivalence between these two problems, which was shown in [3] and stated in Sect. 1). Since $\|\overline{u}\|_\infty \leqslant 1$ and $a > \lambda_1(\Omega)$, in order to prove that $\overline{u} = w$ (the unique nontrivial solution of (1.3)), the only thing left to prove is that $\overline{u} \not\equiv 0$.

(2) Let us then check that, for $a > \lambda_1$, $\overline{u} \not\equiv 0$. Fix any $\overline{t} > 0$. By the maximum principle, we have that $u(\overline{t}, x) > 0$ in Ω and $\partial_\nu u(\overline{t}, x) < 0$ on $\partial\Omega$. By the convergence in $C_\alpha^{1,0}$-spaces up to the boundary of Ω (see Theorem 1.3), we have that for $p \geqslant \overline{p}$, $u_p(\overline{t}, x) > 0$ in Ω and $\partial_\nu u_p(\overline{t}, x) < 0$ on $\partial\Omega$. Let φ_1 be the first eigenfunction of the Laplacian in $H_0^1(\Omega)$ with $\varphi_1 > 0$ and $\|\varphi_1\|_\infty = 1$. Then

$$c\varphi_1 \leqslant u_p(\overline{t}, x), \quad \forall x \in \Omega,\ \forall p \geqslant \overline{p} \tag{4.2}$$

for sufficiently small c (independent of p). Moreover, observe that

$$\partial_t(c\varphi_1) - \Delta(c\varphi_1) \leqslant a(c\varphi_1) - b(x)(c\varphi_1)^p$$

if and only if

$$b(x)c^{p-1}\varphi_1^{p-1} \leqslant a - \lambda_1. \tag{4.3}$$

Take $\overline{c} > 0$ such that (4.2)–(4.3) hold. Then, $\overline{c}\varphi_1$ is a subsolution of (1.1) for sufficiently small $\overline{c}$ and for each $p \geqslant \overline{p}$. Then, by Lemma 2.1, we have that $u_p(t,x) \geqslant \overline{c}\varphi_1$ for every $t \geqslant \overline{t}$ and $p \geqslant \overline{p}$. Hence as $p \to \infty$, we also have $u(t,x) \geqslant \overline{c}\varphi_1(x)$ for every $x \in \Omega$, $t \geqslant \overline{t}$. Thus $\overline{u} \not\equiv 0$ and $\overline{u} = w$, the unique solution of (1.3). From the uniqueness, we deduce in particular that $u(t) \rightharpoonup w$ in $H_0^1(\Omega)$ as $t \to \infty$, not only for some subsequence. As for the strong convergence, this is now easy to show, since by taking the difference

$$\partial_t u - \Delta\big(u(t) - w\big) = au(t)\chi_{\{u<1\}} - aw\chi_{\{w<1\}}$$

and multiplying it by $u(t) - w$, we get

$$\begin{aligned}\int_\Omega &\big|\nabla\big(u(t) - w\big)\big|^2 \mathrm{d}x\\ &= -\int_\Omega \partial_t u(t)\big(u(t) - w\big)\mathrm{d}x + \big(au(t)\chi_{\{u<1\}} - aw\chi_{\{w<1\}}\big)\big(u(t) - w\big) \to 0,\end{aligned}$$

as $t \to \infty$ (recall that both $u(t)$ and w are less than or equal to 1). Thus $u(t) \to w$ strongly in $H_0^1(\Omega)$.

(3) The convergence of the coincidence sets follows as in [16]. As $0 \leqslant \chi_{\{u=1\}}(t) \leqslant 1$, then there exists a function $0 \leqslant \chi^* \leqslant 1$ such that, up to a subsequence,

$$\chi_{\{u=1\}}(t) \rightharpoonup \chi^* \quad \text{weak-}* \text{ in } L^\infty(\Omega), \quad \text{as } t \to +\infty.$$

Since $\chi_{\{u=1\}}(1-u) = 0$ a.e., we have $\chi^*(1-w) = 0$ a.e. Hence $\chi^* = 0$ whenever $w < 1$. Moreover, from the fact that $\partial_t u - \Delta u = au(1 - \chi_{\{u=1\}})$ a.e. in Q, we deduce that $-\Delta w = aw(1-\chi^*)$. As $\Delta w = 0$ a.e. on $\{w = 1\}$ (in fact, $u \in W^{2,q}(\Omega)$ for every $q \geqslant 1$), we conclude that $\chi^* = 1$ on $\{w = 1\}$, whence $\chi^* = \chi_{\{w=1\}}$. Since in general, the $L^\infty(\Omega)$ weak-$*$ convergence of characteristic functions implies the strong convergence in $L^q(\Omega)$ for every $q \geqslant 1$, we have proved (1.9). As a consequence, actually $u(t) \to w$ in H^2-norm.

(4) For $a < \lambda_1(\Omega)$, the function 0 attracts all the solutions of (1.6) with nonnegative initial data. In fact, by taking $v = 0$ in (1.6), we obtain

$$\int_\Omega |\nabla u(t)|^2 \mathrm{d}x \leqslant a\int_\Omega u(t)^2\,\mathrm{d}x - \int_\Omega \partial_t u(t)u(t)\mathrm{d}x \leqslant \frac{a}{\lambda_1(\Omega)}\int_\Omega |\nabla u(t)|^2\,\mathrm{d}x + \mathrm{o}(1),$$

as $t \to +\infty$. Thus $\|u(t)\|_{H_0^1(\Omega)} \to 0$. □

4.2 Proof of Theorem 1.2

Fix $a \in (\lambda_1(\Omega), \lambda_1(\Omega_0))$. In this case, we have a result stronger than that in Lemma 2.2, with a uniform L^∞ bound in $Q = \Omega \times \mathbb{R}^+$.

Lemma 4.3 *For $a \in (\lambda_1(\Omega), \lambda_1(\Omega_0))$, there exists $C > 0$, such that $\|u_p\|_{L^\infty(Q)} \leqslant C$ for all $p > 1$.*

Proof Here we follow the line of the proof of Claim 1 in [4, p. 224], to which we refer for more details. Define $\Omega_\delta = \{x \in \mathbb{R}^N : \mathrm{dist}(x, \Omega) < \delta\}$. Since $a < \lambda_1(\Omega_0)$, there exists a small δ such that $a < \lambda_1(\Omega_\delta)$ (by continuity of the map $\Omega \mapsto \lambda_1(\Omega)$). Denoting by ϕ_δ the first eigenfunction of $-\Delta$ in $H_0^1(\Omega_\delta)$ and ψ any extension of $\phi|_{\Omega_{\frac{\delta}{2}}}$ to $\overline{\Omega}$ such that $\min_{\overline{\Omega}} \psi > 0$, there exists a $Q > 0$ large enough, such that

$$-\Delta(Q\psi) - aQ\psi + b(x)(Q\psi) \geqslant 0 \quad \text{in } \Omega,$$

and $u_0 \leqslant Q\psi$ in Ω. Thus, $Q\psi$ is a supersolution of (1.1) for all $p > 1$. By Lemma 2.1, we have

$$u_p \leqslant Q\psi \leqslant M \quad \text{for all } (x,t) \in Q. \qquad \square$$

Proof of Theorem 1.2 (1) Fix $a \in (\lambda_1(\Omega), \lambda_1(\Omega_0))$. Having proved Lemma 4.3, we can repeat the proof of Proposition 4.1 word by word and show that

$$\|\partial_t u(t)\|_{L^2(\Omega)} \to 0, \quad \text{as } t \to +\infty.$$

By making $v = 0$ in (1.6), we obtain once again by Lemma 4.3 that $\|u(t)\|_{H_0^1(\Omega)}$ is bounded for $t > 0$. Take $t_n \to +\infty$ such that $u(t_n) \rightharpoonup \overline{u}$ in $H_0^1(\Omega)$ for some $\overline{u} \in H_0^1(\Omega)$. Then $\overline{u} \in \mathbb{K}_0$ and

$$\int_\Omega \nabla\overline{u} \cdot \nabla(v - \overline{u})\mathrm{d}x \geqslant a \int_\Omega \overline{u}(v - \overline{u})\mathrm{d}x, \quad \forall v \in \mathbb{K}_0. \tag{4.4}$$

In [3], Dancer and Du shown that (4.4) has a unique nontrivial nonnegative solution w. In order to prove that $\overline{u} = w$ and conclude the proof for this case, we just have to show that $\overline{u} \not\equiv 0$. This will be a consequence of Theorem 1.4. In fact, considering ϕ_p as the solution of

$$\begin{cases} \partial_t\phi_p - \Delta\phi_p = a\phi_p - \|b\|_\infty \phi_p^p & \text{in } Q_T, \\ \varphi_p(0) = v_0, & \varphi_p(t)|_{\partial\Omega} = 0 \end{cases}$$

with $v_0 := \inf\{u_0, 1\}$, it is straightforward to see that ϕ_p is a subsolution of (1.1), and

$$u_p \geqslant \phi_p \to w, \quad \text{as } p \to +\infty,$$

where $w \neq 0$ is the unique nontrivial solution of (1.3). This last statement is a consequence of Theorem 1.4, as $0 \leqslant v_0 \leqslant 1$ a.e. in Ω. Thus $\overline{u} \geqslant w \not\equiv 0$, which concludes the proof in this case.

(2) If $a < \lambda_1(\Omega)$, the same reasoning as in the proof of Theorem 1.4 yields that $\|u(t)\|_{H_0^1(\Omega)} \to 0$. As for the case $a \geqslant \lambda_1(\Omega_0)$, if either $\|u(t)\|_\infty$ or $\|u(t)\|_{H_0^1(\Omega)}$ bounded, it is clear from the proof of Proposition 4.1 that $\|\partial_t u(t)\|_{L^2(\Omega)} \to 0$. Repeating the reasoning of the previous step, we would obtain a nontrivial solution of (1.6) for $a \geqslant \lambda_1(\Omega_0)$, contradicting [3, Theorem 1.1]. □

Remark 4.1 As for the case $a = \lambda_1(\Omega)$, observe that $c\varphi_1$ is always a steady state solution of (1.8) for all $0 < c < 1$, where φ_1 denotes the first eigenfunction of $(-\Delta, H_0^1(\Omega))$ with $\|\varphi_1\|_\infty = 1$. Hence, the long time limit of (1.6) in this case will depend on the initial condition u_0, and we are only able to conclude that, given $t_n \to +\infty$, there exists a subsequence $\{t_{n_k}\}$, such that $u(t_{n_k})$ converges to $c\varphi_1$ for some $c > 0$.

Acknowledgements The second author would like to thank Jesús Hernández and Pedro Freitas for the useful discussions related with this paper.

References

1. Lions, J.L.: Quelques Méthodes de Résolution des Problémes aux Limites Nonlinéaires. Gauthier-Villars, Paris (1969)
2. Zheng, S.: Nonlinear evolutions equations. Chapman & Hall/CRC Monographs and Surveys in Pure and Applied Mathematics, vol. 133. Chapman & Hall/CRC Press, London/Boca Raton (2004)
3. Dancer, E., Du, Y.: On a free boundary problem arising from population biology. Indiana Univ. Math. J. **52**, 51–67 (2003)
4. Dancer, E., Du, Y., Ma, L.: Asymptotic behavior of positive solutions of some elliptic problems. Pac. J. Math. **210**, 215–228 (2003)
5. Dancer, E., Du, Y., Ma, L.: A uniqueness theorem for a free boundary problem. Proc. Am. Math. Soc. **134**, 3223–3230 (2006)
6. Fraile, J.M., Koch Medina, P., López-Gómez, J., et al.: Elliptic eigenvalue problems and unbounded continua of positive solutions of a semilinear elliptic equation. J. Differ. Equ. **127**, 295–319 (1996)
7. Du, Y., Guo, Z.: The degenerate logistic model and a singularly mixed boundary blow-up problem. Discrete Contin. Dyn. Syst. **14**, 1–29 (2006)
8. Boccardo, L., Murat, F.: Increase of power leads to bilateral problems. In: Dal Maso, G., Dell'Antonio, G.F. (eds.) Composite Media and Homogenization Theory, pp. 113–123. World Scientific, Singapore (1995)
9. Dall'Aglio, A., Orsina, L.: On the limit of some nonlinear elliptic equations involving increasing powers. Asymptot. Anal. **14**, 49–71 (1997)
10. Ladyzenskaja, O., Solonnikov, V., Uralceva, N.: Linear and Quasi-linear Equations of Parabolic Type. AMS, Providence (1988)
11. Guo, Z., Ma, L.: Asymptotic behavior of positive solutions of some quasilinear elliptic problems. J. Lond. Math. Soc. **76**, 419–437 (2007)
12. Grossi, M.: Asymptotic behaviour of the Kazdan-Warner solution in the annulus. J. Differ. Equ. **223**, 96–111 (2006)

13. Grossi, M., Noris, B.: Positive constrained minimizers for supercritical problems in the ball. Proc. Am. Math. Soc. **140**, 2141–2154 (2012)
14. Bonheure, D., Serra, E.: Multiple positive radial solutions on annuli for nonlinear Neumann problems with large growth. Nonlinear Differ. Equ. Appl. **18**, 217–235 (2011)
15. Lieberman, G.: Second Order Parabolic Differential Equations. World Scientific, Singapore (1996)
16. Rodrigues, J.F.: On a class of parabolic unilateral problems. Nonlinear Anal. **10**, 1357–1366 (1986)

Composite Waves for a Cell Population System Modeling Tumor Growth and Invasion

Min Tang, Nicolas Vauchelet, Ibrahim Cheddadi, Irene Vignon-Clementel, Dirk Drasdo, and Benoît Perthame

Abstract In the recent biomechanical theory of cancer growth, solid tumors are considered as liquid-like materials comprising elastic components. In this fluid mechanical view, the expansion ability of a solid tumor into a host tissue is mainly driven by either the cell diffusion constant or the cell division rate, with the latter depending on the local cell density (contact inhibition) or/and on the mechanical stress in the tumor.

For the two by two degenerate parabolic/elliptic reaction-diffusion system that results from this modeling, the authors prove that there are always traveling waves above a minimal speed, and analyse their shapes. They appear to be complex with composite shapes and discontinuities. Several small parameters allow for analytical solutions, and in particular, the incompressible cells limit is very singular and related to the Hele-Shaw equation. These singular traveling waves are recovered numerically.

Keywords Traveling waves · Reaction-diffusion · Tumor growth · Elastic material

Mathematics Subject Classification 35J60 · 35K57 · 74J30 · 92C10

Project supported by the ANR grant PhysiCancer and the BMBF grant LungSys.

M. Tang (✉)
Department of Mathematics, Institute of Natural Sciences and MOE-LSC, Shanghai Jiao Tong University, Shanghai 200240, China
e-mail: tangmin@sjtu.edu.cn

M. Tang · N. Vauchelet · I. Cheddadi · I. Vignon-Clementel · D. Drasdo · B. Perthame
INRIA Paris Rocquencourt, Paris, France

B. Perthame
e-mail: benoit.perthame@ljll.math.upmc.fr

N. Vauchelet · I. Cheddadi · I. Vignon-Clementel · D. Drasdo · B. Perthame
Laboratoire Jacques-Louis Lions, UPMC Univ Paris 06 and CNRS UMR 7598, 75005 Paris, France

P.G. Ciarlet et al. (eds.), *Partial Differential Equations: Theory, Control and Approximation*, DOI 10.1007/978-3-642-41401-5_16,

1 Introduction

Models describing cell multiplication within a tissue are numerous and have been widely studied recently, particularly in relation to cancer invasion. Whereas small-scale phenomena are accurately described by individual-based models (IBM in short, see, e.g., [3, 19, 24]), large scale solid tumors can be described by tools from continuum mechanics (see, e.g., [2, 6, 15–18] and [9] for a comparison between IBM and continuum models). The complexity of the subject has led to a number of different approaches, and many surveys are now available [1, 4, 5, 21, 25, 32]. They show that the mathematical analysis of these continuum models raises several challenging issues. One of them, which has attracted little attention, is the existence and the structure of traveling waves (see [12, 15]). This is our main interest here, particularly in the context of fluid mechanical models that have been advocated recently [29, 31]. Traveling wave solutions are of special interest also from the biological point as the diameter of 2D monolayers, 3D multicellular spheroids and xenografts. 3D tumors emerging from cells injected into animals are found to increase for many cell lines linearly in time indicating a constant growth speed of the tumor border (see [30]).

In this fluid mechanical view, the expansion ability of tumor cells into a host tissue is mainly driven by the cell division rate which depends on the local cell density (contact inhibition) and by the mechanical pressure in the tumor (see [11, 29, 31]). Tumor cells are considered to be of an elastic material, and then respond to pressure by elastic deformation. Denoting by v the velocity field and by ρ the cell population density, we will make use of the following advection-diffusion model:

$$\partial_t \rho + \operatorname{div}(\rho v) - \operatorname{div}(\epsilon \nabla \rho) = \Phi(\rho, \Sigma).$$

In this equation, the third term in the left-hand side describes the active motion of cells that results in their diffusion with a nonnegative diffusion coefficient ϵ. In the right-hand side, $\Phi(\rho, \Sigma)$ is the growth term, which expresses that cells divide freely. Thus it results in an exponential growth, as long as the elastic pressure Σ is less than a threshold pressure denoted by C_p, where the cell division is stopped by contact inhibition (the term "homeostatic pressure" has been used for C_p). This critical threshold is determined by the compression that a cell can experience (see [9]). A simple mathematical representation is

$$\Phi(\rho) = \rho H\big(C_p - \Sigma(\rho)\big),$$

where H denotes the Heaviside function $H(v) = 0$ for $v < 0$ and $H(v) = 1$ for $v > 0$, and $\Sigma(\rho)$ denotes the state equation, linking pressure and local cell density. As long as cells are not in contact, the elastic pressure $\Sigma(\rho)$ vanishes whereas it is an increasing function of the population density for larger value of this contact density. Here, after neglecting cell adhesion, we consider the pressure monotonously depending on cell population, such that

$$\Sigma(\rho) = 0, \quad \rho \in [0, 1), \qquad \Sigma'(\rho) > 0, \quad \rho \geq 1. \tag{1.1}$$

The flat region $\rho \in [0, 1)$ induces a degeneracy that is one of the interests of the model for both mathematics and biophysical effects. This region represents that cells are too far apart and do not touch each other. When elastic deformations are neglected, in the incompressible limit of confined cells, this leads to a jump of the pressure from 0 to $+\infty$ at the reference value $\rho = 1$. This highly singular limit leads to the Hele-Shaw type of models (see [28]). Finally, the balance of forces acting on the cells leads under certain hypotheses to the following relationship between the velocity field v and the elastic pressure (see [14]):

$$-C_S \nabla \Sigma(\rho) = -C_z \Delta v + v.$$

This is Darcy's law which describes the tendency of cells to move down pressure gradients, extended to a Brinkman model by a dissipative force density resulting from internal cell friction due to cell volume changes. C_S and C_z are parameters, relating respectively to the reference elastic and bulk viscosity cell properties with the friction coefficient. The resulting model is then the coupling of this elliptic equation for the velocity field, a conservation equation for the population density of cells and a state equation for the pressure law.

A similar system of equations describing the biomechanical properties of cells has already been suggested as a conclusion in [9] for the radial growth of tumors. That paper proposes to close the system of equations with an elastic fluid model to generalize their derivation for compact tumors that assume a constant density inside the tumor with a surface tension boundary condition. Many other authors have also considered such an approach (see, e.g., [17, 18]). In [8, 10, 13, 15] cell-cell adhesion is also taken into account, in contrast with (1.1). Their linear stability analysis explains instabilities of the tumor front which are also observed numerically in [13, 15]. However, many of these works focus on nutrient-limited growth, whereas we are interested here in stress-regulated growth. Besides, most works deal with a purely elastic fluid model. A viscous fluid model was motivated in [8, 10, 11] and studied numerically in [8]. Here we include this case in our mathematical study and numerical results. Moreover, we propose here a rigorous analysis of traveling waves, which furnishes in some case explicit expressions of the traveling profile and the speed of the wave.

From a mathematical point of view, the description of the invasive ability of cells can be considered as the search of traveling waves. Furthermore, the study in several dimensions is also very challenging, and we will restrict ourselves to the 1-dimensional case. For reaction-diffusion-advection equations arising from biology, several works were devoted to the study of traveling waves (see, for instance, [22, 23, 26, 27, 34] and the book [7]). In particular, our model has some formal similarities with the Keller-Segel system with growth treated in [22, 27], and the main difference is that the effect of pressure is repulsive here while it is attractive for the Keller-Segel system. More generally, the influence of the physical parameters on the traveling speed is an issue of interest for us and is one of the objectives of this work. Also the complexity of the composite waves arising from different physical effects is an interesting feature of the model at hand. In particular, the nonlinear degeneracy of the diffusion term is an interesting part of the complexity of the phenomena

Table 1 The outline of this paper

$C_z = 0$	$\epsilon = 0$	Theorem 3.1
		(Incompressible cell limit) Remark 3.1
	$\epsilon > 0$	Theorem 3.2
		(Incompressible cell limit) Remark 3.2
$C_z > 0$	$\epsilon = 0$	(Incompressible cell limit, $C_S C_p > 2C_z$) Theorem 4.1
		(Incompressible cell limit, $C_S C_p < 2C_z$) Remark 4.1
	$\epsilon > 0$	(Incompressible cell limit, $C_S C_p > 2C_z$) Theorem 4.2

studied here. For instance, as in [33], we construct waves which vanish on the right half-line.

The aim of this paper is to prove the existence of traveling waves above a minimal speed in various situations. For the clarity of the paper, we present our main results in the table below. As mentioned earlier, the incompressible cell limit corresponds to the particular case, where the pressure law (1.1) has a jump from 0 to $+\infty$ when $\rho = 1$.

The outline of this paper is as follows (see Table 1). In the next section, we present some preliminary notations and an a priori estimate resulting in a maximum principle. In Sect. 3, we investigate the existence of traveling waves in the simplified inviscid case $C_z = 0$, for which the model reduces to a single continuity equation for ρ. Finally, Sect. 4 is devoted to the study of the general case $C_z \neq 0$ in the incompressible cells limit. In both parts, some numerical simulations illustrate the theoretical results.

2 Preliminaries

In a 1-dimensional framework, the considerations in the introduction lead to the following set of equations:

$$\begin{cases} \partial_t \rho + \partial_x(\rho v) = \Phi(\rho) + \epsilon \partial_{xx} \rho, \\ -C_S \partial_x \Sigma(\rho) = -C_z \partial_{xx} v + v. \end{cases} \tag{2.1}$$

This system is considered on the whole real line $\mathbb{R}$ and is complemented with Dirichlet boundary conditions at infinity for v and Neumann boundary condition for ρ. Here C_p, C_S, C_z stand for nonnegative rescaled constants. It will be useful for the mathematical analysis to introduce the function W that solves the elliptic problem

$$-C_z \partial_{xx} W + W = \Sigma(\rho), \quad \partial_x W(\pm\infty) = 0.$$

This allows us to set $v = -C_S \partial_x W$ and rewrite the system (2.1) as

$$\begin{cases} \partial_t \rho - C_S \partial_x(\rho \partial_x W) = \Phi(\rho) + \epsilon \partial_{xx} \rho, \\ -C_z \partial_{xx} W + W = \Sigma(\rho). \end{cases} \tag{2.2}$$

We recall that the elastic pressure satisfies (1.1), and the growth function satisfies

$$\Phi(\rho) \geq 0, \quad \Phi(\rho) = 0 \quad \text{for } \Sigma(\rho) \geq C_p > 0. \tag{2.3}$$

2.1 Maximum Principle

The nonlocal aspect of the velocity in terms of ρ makes unobvious the correct way to express the maximum principle. In particular, it does not hold directly on the population density, but on the pressure $\Sigma(\rho)$.

Lemma 2.1 *Assume that Φ satisfies* (2.3) *and that the state equation for Σ satisfies* (1.1). *Then, setting $\Sigma_M^0 = \max_{x\in\mathbb{R}} \Sigma(x,0)$, any classical solution to* (2.2) *satisfies the maximum principle*

$$\Sigma(\rho) \leq \max\left(\Sigma_M^0, C_p\right) \quad \textit{and} \quad \rho \leq \Sigma^{-1}(C_p) =: \rho_M > 1, \quad \textit{if } \Sigma_M^0 \leq C_p. \tag{2.4}$$

However notice that, except in the case when C_z vanishes, this problem is not monotonic, and no BV type estimates are available (see [28] for properties when $C_z = 0$).

Proof Only the values on the intervals such that $\rho > 1$ need to be considered. When $\rho > 1$, multiplying the first equation in (2.2) by $\Sigma'(\rho)$, we find

$$\begin{aligned} &\frac{\partial}{\partial t}\Sigma(\rho) - C_S \partial_x \Sigma(\rho) \partial_x W - C_S \rho \Sigma'(\rho) \partial_{xx} W \\ &\quad = \Sigma'(\rho)\Phi(\rho) + \epsilon \partial_{xx}\Sigma(\rho) - \epsilon \Sigma''(\rho)|\partial_x \rho|^2. \end{aligned}$$

Fix a time t, and consider a point x_0, where $\max_x \Sigma(\rho(x,t)) = \Sigma(\rho(x_0,t))$ (the extension to the case that it is not attained is standard [20]). We have $\partial_x \Sigma(\rho(x_0,t)) = 0$, $\partial_{xx}\Sigma(\rho(x_0,t)) \leq 0$, and thus we obtain that

$$\begin{aligned} \frac{\mathrm{d}}{\mathrm{d}t} \max_x \Sigma\left(\rho(x,t)\right) \leq{}& \Sigma'\left(\rho(x_0,t)\right)\Phi\left(\rho(x_0,t)\right) + C_S \rho \Sigma'\left(\rho(x_0,t)\right)\partial_{xx} W(x_0,t) \\ &- \epsilon \Sigma''\left(\rho(x_0,t)\right)|\partial_x \rho(x_0,t)|^2. \end{aligned}$$

Consider a possible value, such that $\Sigma(\rho(x_0,t)) > C_p$. Then we can treat the three terms in the right-hand side as follows:

(i) From assumption (2.3), we have $\Phi(\rho(x_0,t)) = 0$. Then the first term vanishes.
(ii) Also, by assumption (1.1), since $\Sigma'(\rho(x_0,t)) > 0$ for $\rho(x_0,t) \geq 1$, we have $\partial_x \rho(x_0,t) = 0$. Therefore, the third term vanishes.
(iii) Moreover, since $-C_z \partial_{xx} W(x_0,t) = \max_x \Sigma(\rho(x,t)) - W(x_0,t) \geq 0$ (by the maximum principle $W \leq \max \Sigma$), using (ii), we conclude that the second term is non-positive.

We conclude that

$$\frac{\mathrm{d}}{\mathrm{d}t}\max_x \Sigma\big(\rho(x,t)\big) \le 0,$$

and this proves the result. □

2.2 Traveling Waves

The end of this paper deals with existence of a traveling wave for model (2.2) with the growth term and definition

$$\Phi(\rho) = \rho H\big(C_p - \Sigma(\rho)\big), \quad C_p > 0, \qquad \rho_M := \Sigma^{-1}(C_p) > 1. \tag{2.5}$$

There are two constant steady states $\rho = 0$ and $\rho = \rho_M := \Sigma^{-1}(C_p)$, and we look for traveling waves connecting these two stationary states. From Lemma 2.1, we may assume that the initial data satisfy $\max_x \Sigma(\rho(x,t=0)) = C_p$ and $\max_x \rho(x,t=0) = \rho_M$. Then, it is natural to obtain the following definition.

Definition 2.1 A non-increasing traveling wave solution is a solution to the form $\rho(t,x) = \rho(x - \sigma t)$ for a constant $\sigma \in \mathbb{R}$ called the traveling speed, such that $\rho' \le 0$, $\rho(-\infty) = \rho_M$ and $\rho(+\infty) = 0$.

With this definition, we are led to look for (ρ, W) satisfying

$$-\sigma\partial_x\rho - C_S\partial_x(\rho\partial_x W) = \rho H\big(C_p - \Sigma(\rho)\big) + \epsilon\partial_{xx}\rho, \tag{2.6}$$

$$-C_z\partial_{xx}W + W = \Sigma(\rho), \tag{2.7}$$

$$\rho(-\infty) = \rho_M, \quad \rho(+\infty) = 0, \quad W(-\infty) = C_p, \quad W(+\infty) = 0. \tag{2.8}$$

When $C_z = 0$, (2.6)–(2.7) reduces to one single equation

$$-\sigma\partial_x\rho - C_S\partial_x\big(\rho\partial_x\Sigma(\rho)\big) = \rho H\big(C_p - \Sigma(\rho)\big) + \epsilon\partial_{xx}\rho. \tag{2.9}$$

In the sequel and in order to make the mathematical analysis more tractable, as depicted in Fig. 1, we assume that Σ has the specific form given by

$$\Sigma(\rho) = \begin{cases} 0 & \text{for } \rho \le 1, \\ C_v \ln\rho & \text{for } \rho \ge 1. \end{cases} \tag{2.10}$$

This form represents logarithmic strain assuming cells of the cuboidal shape (see the Appendix). The choice of logarithmic strain conserves the volume of incompressible cells for both small and large deformations. Hence, it is particularly useful as cells, because they are mainly composed of water, and are incompressible on small time scales, such that deformations leave the cell volume invariant.

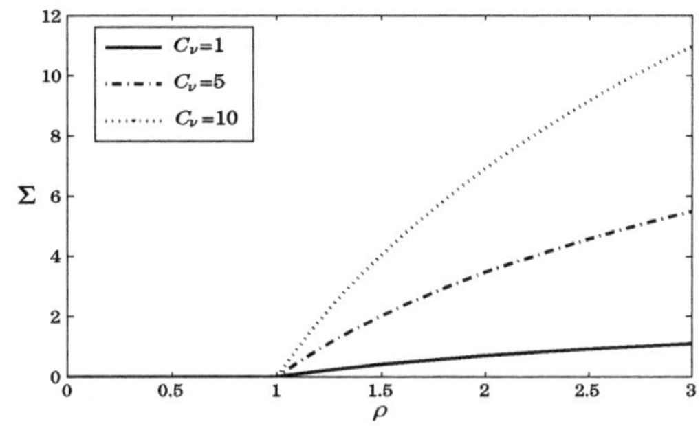

Fig. 1 The equation of state as defined by (2.10) for three different values of C_ν

We will study in particular the case $C_\nu \to +\infty$. We call it the incompressible cell limit, which is both mathematically interesting (see also the derivation of Hele-Shaw equation in [28]) and physically relevant. This limit case boils down to consider the tissue of tumor cells as an incompressible elastic material in a confined environment.

The structure of the problem (2.1) depends deeply on the parameters ϵ and C_z. It is hyperbolic for $\epsilon = C_z = 0$, parabolic when $\epsilon \neq 0$, $C_z = 0$ and coupled parabolic/elliptic in the general case. Therefore, we have to treat the cases separately.

3 Traveling Wave Without Viscosity

When the bulk viscosity is neglected, that is $C_z = 0$, the analysis is much simpler and is closely related to the Fisher/KPP equation (see [7]) with the variant of a complex composite and discontinuous wave. The unknown W can be eliminated. Taking advantage of the state equation for the pression (2.10), we can rewrite (2.9) as a self-contained equation on ρ

$$\begin{cases} -\sigma \partial_x \rho - C_S C_\nu \partial_{xx} Q(\rho) = \rho H(C_p - C_\nu (\ln \rho)_+) + \epsilon \partial_{xx} \rho, \\ \rho(-\infty) = \rho_M, \quad \rho(+\infty) = 0. \end{cases} \tag{3.1}$$

Here f_+ denotes the positive part of f and

$$Q(\rho) = \begin{cases} 0 & \text{for } \rho \leq 1, \\ \rho - 1 & \text{for } \rho \geq 1. \end{cases} \tag{3.2}$$

3.1 Traveling Waves for $\epsilon = 0$

When the cell motility is neglected, we can find the explicit expression for the traveling waves. More precisely, we establish the following result.

Theorem 3.1 *There exists a* $\sigma^* > 0$*, such that for all* $\sigma \geq \sigma^*$, (3.1)–(3.2) *admits a nonnegative, non-increasing and discontinuous solution* ρ. *More precisely, when* $\sigma = \sigma^*$ *and up to translation,* ρ *is given by*

$$\rho(x) = \begin{cases} \rho_M := \exp(\frac{C_p}{C_v}), & x \leq 0, \\ g(x), & x \in (0, x_0),\ x_0 > 0, \\ 0, & x > x_0, \end{cases}$$

where g *is a smooth non-increasing function satisfying* $g(0) = \rho_M$, $g'(0) = 0$ *and* $g(x_0) = 1$. *Its precise expression is given in the proof.*

In other words, when $C_z = 0$ and $\epsilon = 0$, (2.2) admits a nonnegative and non-increasing traveling wave (ρ, W) for $\sigma \geq \sigma^*$.

Notice that, by opposition to the Fisher/KPP equation, we do not have an analytical expression for the minimal speed. Relate that ρ vanishes for large x, a phenomena already known for degenerate diffusion.

Proof Since we are looking for a non-increasing function ρ, we decompose the line to be

$$\mathbb{R} = I_1 \cup I_2 \cup I_3, \quad I_1 = \{\rho(x) = \rho_M\}, \quad I_2 = \{1 < \rho(x) < \rho_M\}, \quad I_3 = \{\rho(x) \leq 1\}.$$

Notice that, equivalently $\Sigma(x) = C_p$ in I_1. To fix the notations, we set

$$I_1 = (-\infty, 0], \qquad I_2 = (0, x_0), \qquad I_3 = [x_0, +\infty).$$

Step 1 (In $I_1 \cup I_2$) ρ satisfies

$$-\sigma \partial_x \rho - C_S C_v \partial_{xx} \rho = \rho H\big(C_p - C_v (\ln \rho)_+\big). \tag{3.3}$$

Therefore, by elliptic regularity, we deduce that the second derivative of ρ is bounded, and therefore $\rho \in C^1(-\infty, x_0)$. On I_1, the function ρ is a constant and by continuity of ρ and $\partial_x \rho$ at $x = 0$, we have the boundary conditions of I_2, such that

$$\rho(0) = \rho_M, \qquad \partial_x \rho(0) = 0. \tag{3.4}$$

In I_2, $H(C_p - C_v (\ln \rho)_+) = 1$. Solving (3.3) with the boundary conditions in (3.4), we find that if $\sigma > 2\sqrt{C_S C_v}$, then

$$\rho(x) = \rho_M e^{-\frac{\sigma x}{2C_S C_v}} \left(A \exp\left(\frac{\sqrt{\sigma^2 - 4C_S C_v}}{2C_S C_v} x \right) + B \exp\left(-\frac{\sqrt{\sigma^2 - 4C_S C_v}}{2C_S C_v} x \right) \right)$$

with

$$A = \frac{\sigma + \sqrt{\sigma^2 - 4C_S C_v}}{2\sqrt{\sigma^2 - 4C_S C_v}}, \qquad B = \frac{-\sigma + \sqrt{\sigma^2 - 4C_S C_v}}{2\sqrt{\sigma^2 - 4C_S C_v}}.$$

In this case, ρ is decreasing for $x > 0$ and vanishes as $x \to +\infty$. Thus there exists a positive x_0, such that $\rho(x_0) = 1$.

When $\sigma < 2\sqrt{C_S C_\nu}$, the solution is

$$\rho(x) = \rho_M \mathrm{e}^{-\frac{\sigma x}{2C_S C_\nu}} \left(A \cos\left(\frac{\sqrt{4C_S C_\nu - \sigma^2}}{2C_S C_\nu} x \right) + B \sin\left(\frac{\sqrt{4C_S C_\nu - \sigma^2}}{2C_S C_\nu} x \right) \right) \tag{3.5}$$

with

$$A = 1, \qquad B = \frac{\sigma}{\sqrt{4C_S C_\nu - \sigma^2}}.$$

By a straightforward computation, we deduce

$$\partial_x \rho(x) = -\frac{2\rho_M}{\sqrt{4C_S C_\nu - \sigma^2}} \mathrm{e}^{-\frac{\sigma x}{2C_S C_\nu}} \sin\left(\frac{\sqrt{4C_S C_\nu - \sigma^2}}{2C_S C_\nu} x \right).$$

Thus ρ is decreasing in $(0, \frac{2C_S C_\nu}{\sqrt{4C_S C_\nu - \sigma^2}} \pi)$, and takes negative values at the largest endpoint. There exists an $x_0 > 0$, such that $\rho(x_0) = 1$.

Finally, when $\sigma = 2\sqrt{C_S C_\nu}$, we reach the same conclusion because

$$\rho(x) = \rho_M \left(\frac{x}{\sqrt{C_S C_\nu}} + 1 \right) \mathrm{e}^{-\frac{x}{\sqrt{C_S C_\nu}}}.$$

Step 2 (On I_3) In $(x_0, +\infty)$, we have $\Sigma = 0$ and $Q(\rho) = 0$ from (3.2). Then Eq. (3.1) is

$$-\sigma \partial_x \rho = \rho. \tag{3.6}$$

We can write the jump condition at x_0 by integrating (3.1) from x_0^- to x_0^+, which is

$$-\sigma[\rho]_{x_0} - C_S C_\nu \big[\partial_x Q(\rho)\big]_{x_0} = 0, \qquad \sigma\big(\rho(x_0^+) - 1\big) = C_S C_\nu \partial_x \rho(x_0^-).$$

Here $\partial_x \rho(x_0^-) < 0$ can be found, due to the expression of ρ on I_2 as computed above. Thus, we get $\rho(x_0^+)$, which is the boundary condition of (3.6). Then the Cauchy problem (3.6) gives

$$\rho(x) = \left(1 + \frac{C_S C_\nu}{\sigma} \partial_x \rho(x_0^-) \right) \mathrm{e}^{-\frac{x}{\sigma}}, \qquad x \in I_3.$$

In summary, when $\epsilon = 0$, a nonnegative solution to (3.1) exists under the necessary and sufficient condition

$$\sigma \geq -C_S C_\nu \partial_x \rho(x_0^-). \tag{3.7}$$

The right-hand side also depends on σ. Therefore, it does not obviously imply $\sigma \geq \sigma^*$. To reach this conclusion, and conclude the proof, we shall use Lemma 3.1. □

Lemma 3.1 *Using the notation in the proof of Theorem* 3.1, *the function* $\sigma \mapsto -C_S C_\nu \partial_x \rho(x_0^-)$ *is nonincreasing. Therefore, there exists a minimal traveling wave velocity* σ^*, *and* (3.7) *is satisfied if and only if* $\sigma \geq \sigma^*$.

Proof We consider (3.3) in $I_2 = (0, x_0)$. We notice that on this interval, $\rho(x)$ is decreasing, and therefore is one to one from $(0, x_0)$ to $(\rho_M, 1)$. We denote by $X(\rho)$ its inverse. Let us define $V = -C_S C_\nu \partial_x \rho$. In I_2, V is nonnegative, and (3.3) can be written as

$$\partial_x V = \sigma \partial_x \rho + \rho = -\frac{V}{C_S C_\nu}\sigma + \rho. \tag{3.8}$$

Setting $\widetilde{V}(\rho) = V(X(\rho))$, by definition of V, we have

$$\partial_\rho \widetilde{V} = \partial_x V \partial_\rho X = \frac{\partial_x V}{\partial_x \rho} = -\partial_x V \frac{C_S C_\nu}{V}.$$

By using (3.8), we finally get the differential equation

$$\begin{cases} \partial_\rho \widetilde{V} = \sigma - \frac{C_S C_\nu \rho}{\widetilde{V}} & \text{for } \rho \in (1, \rho_M), \\ \lim_{\rho \to \rho_M} \widetilde{V}(\rho_M) = -C_S C_\nu \partial_x \rho(0) = 0. \end{cases} \tag{3.9}$$

This differential equation has a singularity at ρ_M. We then introduce $z = \rho_m - \rho$ and $Y(z) = \frac{1}{2}\widetilde{V}^2(\rho_M - z)$ for $z \in (0, \rho_M - 1)$. (3.9) becomes

$$\begin{cases} Y'(z) = -\sigma\sqrt{2Y(z)} + C_S C_\nu(\rho_M - z) & \text{for } z \in (0, \rho_M - 1), \\ Y(0) = 0. \end{cases}$$

This ordinary differential equation belongs to the class $Y' = F(z, Y)$ with F one sided Lipschitz in his second variable and $\partial_Y F(z, Y) \leq 0$. Therefore, we can define a unique solution to the above Cauchy problem. Hence there exists a unique nonnegative solution $\widetilde{V}$ to (3.9).

Define $U(\rho) := \frac{\partial \widetilde{V}}{\partial \sigma}$, and our goal is to determine the sign of $U(1)$. We have

$$\frac{\partial^2 \widetilde{V}}{\partial \rho \partial \sigma} = \frac{\partial}{\partial \sigma}\left(\sigma - \frac{C_S C_\nu}{\widetilde{V}}\rho\right) = 1 + \frac{C_S C_\nu}{\widetilde{V}^2}\rho \frac{\partial \widetilde{V}}{\partial \sigma}.$$

Then $U(\rho)$ solves on $(1, \rho_M)$,

$$\frac{\partial U}{\partial \rho} = 1 + \frac{C_S C_\nu}{\widetilde{V}^2}\rho U. \tag{3.10}$$

Moreover, we have

$$U(\rho_M) = \frac{\partial \widetilde{V}(\rho_M)}{\partial \sigma} = 0. \tag{3.11}$$

Assume $U(1) > 0$. Let us define $\rho_1 = \sup\{\rho_2 \mid \rho \in (1, \rho_2), \text{such that } U(\rho) \geq 0\}$. Then from (3.10), $\frac{\partial U}{\partial \rho}(\rho) \geq 1$ on $(1, \rho_1)$, and thus $U(\rho_1) > U(1) > 0$. By continuity, we should necessarily have $\rho_1 = \rho_M$. However, we then have $\frac{\partial U}{\partial \rho}(\rho) \geq 1$ for all $\rho \in (1, \rho_M)$, which is a contradiction to $U(\rho_M) = 0$. Therefore, $U(1) \leq 0$ and $\widetilde{V}$ is nonincreasing with respect to σ. □

Structural Stability Theorem 3.1 shows that there is an infinity of traveling wave solutions. However, as in the Fisher/KPP equation, most of them are unstable. For instance, we can consider some kind of "ignition temperature" approximation to the system (3.1), such that

$$-\sigma \partial_x \rho_\theta - C_S C_\nu \partial_{xx} Q(\rho_\theta) = \xi_\theta(\rho_\theta) H\big(C_p - C_\nu (\ln \rho_\theta)_+\big), \tag{3.12}$$

where $\theta \in (0, 1)$ is a small positive parameter and

$$\xi_\theta(\rho) = \begin{cases} \rho & \text{for } \rho \in (\theta, \rho_M), \\ 0 & \text{for } \rho \in [0, \theta]. \end{cases} \tag{3.13}$$

Then we have the following result.

Lemma 3.2 *Equations* (3.12)–(3.13) *admit a unique couple of solution* $(\sigma_\theta, \rho_\theta)$ *and* $\sigma_\theta \to \sigma^*$ *as* $\theta \to 0$.

Proof As in Theorem 3.1, we solve (3.12) by using the decomposition $\mathbb{R} = I_1 \cup I_2 \cup I_3$. In $I_1 \cup I_2$, $\rho \geq 1 > \theta$, and therefore, ρ is given by the same formula as computed in the proof of Theorem 3.1. On I_3, (3.12) becomes

$$-\sigma_\theta \partial_x \rho_\theta = \xi_\theta(\rho_\theta). \tag{3.14}$$

By contradiction, if $\rho_\theta(x_0^+) \geq \theta$, then (3.14) implies $\rho_\theta(x) = \rho_\theta(x_0^+) e^{-\frac{x - x_0}{\sigma}}$. Thus, there exists an x_θ, such that $\rho_\theta(x) \leq \theta$ for $x \geq x_\theta$. Then the right-hand side of (3.14) vanishes for $x \geq x_\theta$, and ρ_θ is the constant for $x \geq x_\theta$. This constant has to vanish from the condition at infinity, which contradicts the continuity of ρ_θ. Thus, $\rho_\theta(x_0^+) < \theta$, and (3.14) implies that $\partial_x \rho_\theta = 0$. We conclude that $\rho_\theta = 0$ on I_3. The jump condition at the interface $x = x_0$ gives

$$\sigma_\theta\big(\rho_\theta\big(x_0^+\big) - 1\big) = C_S C_\nu \partial_x \rho_\theta\big(x_0^-\big),$$

which, together with $\rho(x_0^+) = 0$, indicates that

$$\sigma_\theta = -C_S C_\nu \partial_x \rho_\theta\big(x_0^-\big).$$

According to Lemma 3.1, there exists a unique σ_θ^*, satisfying the equality above, so does a unique ρ_θ.

Letting $\theta \to 0$ in this formula, we recover the equality case in (3.7) that defines the minimal speed in Theorem 3.1. By continuity of the unique solution, we find $\sigma_\theta \to \sigma^*$. □

Remark 3.1 (Incompressible Cells Limit) In the incompressible cells limit $C_\nu \to +\infty$, we can obtain an explicit expression of the traveling wave from Theorem 3.1. Since $\rho_M = \exp(\frac{C_p}{C_\nu}) \to 1$, we have $\rho(x) \to 1$ in $I_1 \cup I_2$, but Σ carries more structural information. In the first step of the proof, for large C_ν, by using (3.5), we find

$$\Sigma(x) = C_\nu \ln(\rho) \to C_p - \frac{x^2}{2C_S}.$$

We recall that the point x_0 is such that $\rho(x_0) = 1$ or $\Sigma(x_0) = 0$. Therefore, $x_0 = \sqrt{2C_S C_p}$ and

$$C_\nu \partial_x \rho(x_0^-) = \partial_x \Sigma(x_0^-) \to -\sqrt{\frac{2C_p}{C_S}}, \qquad \text{as } C_\nu \to +\infty.$$

Thus $\sigma^* \to \sqrt{2C_p C_S}$. We conclude that, on $I_3 = [x_0, +\infty)$, $\rho(x) \to (1 - \frac{\sqrt{2C_p C_S}}{\sigma})e^{-\frac{x}{\sigma}}$.

3.2 Traveling Wave when $\epsilon \neq 0$

We can extend Theorem 3.1 to the case $\epsilon \neq 0$.

Theorem 3.2 *There exists a* $\sigma^* > 2\sqrt{\epsilon}$, *such that for all* $\sigma \geq \sigma^*$, (3.1)–(3.2) *admits a nonnegative, non-increasing and continuous solution* ρ.

Thus when $C_z = 0$, *system* (2.2) *admits a nonnegative and non-increasing traveling wave* (ρ, W) *for* $\sigma \geq \sigma^*$.

Proof We follow the proof of Theorem 3.1 and decompose $\mathbb{R} = I_1 \cup I_2 \cup I_3$. Due to the diffusion term in (3.1), $\rho \in C^0(\mathbb{R})$, and we will use the continuity of ρ at the interfaces.

On $I_1 = (-\infty, 0]$, we have $\rho = \rho_M$ and $\Sigma = C_p$.

In $I_2 = (0, x_0)$, Eq. (3.1) implies

$$(C_S C_\nu + \epsilon)\partial_{xx}\rho + \sigma\partial_x\rho + \rho = 0, \qquad \rho(0) = \rho_M, \qquad \partial_x\rho(0) = 0.$$

Therefore, we get the same expressions for ρ on I_2 as in the proof of Theorem 3.1, except that we replace $C_S C_\nu$ by $C_S C_\nu + \epsilon$. Thus, as before, there exists a positive x_0, such that $\rho(x_0) = 1$, and ρ is decreasing in $(0, x_0)$.

On $I_3 = [x_0, +\infty)$, we solve

$$\epsilon\partial_{xx}\rho + \sigma\partial_x\rho + \rho = 0. \tag{3.15}$$

At the interface $x = x_0$, integrating from x_0^- to x_0^+ in (3.1) and using the continuity of ρ, we get

$$C_S C_\nu \big[\partial_x Q(\rho)\big]_{x_0} + \epsilon[\partial_x\rho]_{x_0} = 0,$$

that is,

$$\partial_x \rho\big(x_0^+\big) = \left(1 + \frac{C_S C_\nu}{\epsilon}\right) \partial_x \rho\big(x_0^-\big). \tag{3.16}$$

Solving (3.15) with the boundary conditions $\rho(x_0^+) = 1$ and (3.16), we get that if $\sigma < 2\sqrt{\epsilon}$, then ρ is the sum of the trigonometric functions, and therefore will take negative values. Thus $\sigma \geq 2\sqrt{\epsilon}$. In the case $\sigma > 2\sqrt{\epsilon}$,

$$\rho(x) = A \exp\left(\frac{-\sigma + \sqrt{\sigma^2 - 4\epsilon}}{2\epsilon}(x - x_0)\right) + B \exp\left(\frac{-\sigma - \sqrt{\sigma^2 - 4\epsilon}}{2\epsilon}(x - x_0)\right),$$

where

$$A = \frac{1}{2} + \frac{1}{\sqrt{\sigma^2 - 4\epsilon}}\left(\frac{\sigma}{2} + (\epsilon + C_S C_\nu)\partial_x \rho\big(x_0^-\big)\right),$$

$$B = \frac{1}{2} - \frac{1}{\sqrt{\sigma^2 - 4\epsilon}}\left(\frac{\sigma}{2} + (\epsilon + C_S C_\nu)\partial_x \rho\big(x_0^-\big)\right).$$

After detailed calculation of $\partial_x \rho$ and using $\partial_x \rho(x_0^-) < 0$, we have that ρ is a nonnegative and nonincreasing function if and only if $A \geq 0$, that is,

$$\sqrt{\sigma^2 - 4\epsilon} + \sigma + 2(\epsilon + C_S C_\nu)\partial_x \rho\big(x_0^-\big) \geq 0, \quad \sigma > 2\sqrt{\epsilon}. \tag{3.17}$$

In the case $\sigma = 2\sqrt{\epsilon}$, we have

$$\rho(x) = \left(\left(\frac{1}{\sqrt{\epsilon}} + \left(1 + \frac{C_S C_\nu}{\epsilon}\right)\partial_x \rho\big(x_0^-\big)\right)(x - x_0) + 1\right) e^{-\frac{x - x_0}{\sqrt{\epsilon}}}.$$

Thus ρ is a nonnegative and non-increasing function if and only if

$$\frac{1}{\sqrt{\epsilon}} + \left(1 + \frac{C_S C_\nu}{\epsilon}\right)\partial_x \rho\big(x_0^-\big) \geq 0,$$

which is the same condition as (3.17) by setting $\sigma = 2\sqrt{\epsilon}$. Thus (3.17) is valid for $\sigma \geq 2\sqrt{\epsilon}$. Denoting $U_\epsilon(x) = -(\epsilon + C_S C_\nu)\partial_x \rho(x)$, condition (3.17) can be rewritten as

$$\sigma \geq \mathfrak{F}[\sigma] := \max\left(2\sqrt{\epsilon}, \min\left(2U_\epsilon\big(x_0^-\big), U_\epsilon\big(x_0^-\big) + \frac{\epsilon}{U_\epsilon\big(x_0^-\big)}\right)\right). \tag{3.18}$$

By a straightforward adaptation of Lemma 3.1, we conclude that $\sigma \mapsto U_\epsilon(x_0^-)$ is nonincreasing with respect to σ. When $U_\epsilon(x_0^-) > \sqrt{\epsilon}$, we have

$$\mathfrak{F}[\sigma] = U_\epsilon\big(x_0^-\big) + \frac{\epsilon}{U_\epsilon\big(x_0^-\big)}.$$

Then $\mathfrak{F}[\sigma]$ is an increasing function with respect to $U_\epsilon(x_0^-)$ for $U_\epsilon(x_0^-) > \sqrt{\epsilon}$. Together with $\sigma \to U_\epsilon(x_0^-)$ being nonincreasing, $\mathfrak{F}[\sigma]$ is nonincreasing with respect to σ. For the case $U_\epsilon(x_0^-)^2 < \epsilon$, we have

$$\mathfrak{F}[\sigma] = 2\sqrt{\epsilon}.$$

Therefore, for all $\sigma \in (0, +\infty)$, $\mathfrak{F}[\sigma]$ is a nonincreasing function of σ. Hence, there exists a unique σ^*, such that (3.18) is satisfied for every $\sigma \geq \sigma^*$. □

Structural Stability We can again select a unique traveling wave when we approximate the growth term by $\xi_\theta(\rho)H(C_p - C_\nu(\ln\rho)_+)$. This can be obtained by considering $\epsilon\partial_{xx}\rho_\theta + \sigma\partial_x\rho_\theta + \xi_\theta(\rho_\theta) = 0$ instead of (3.15) and by matching the values of $\partial_x\rho$ on both sides at the point where $\rho = \theta$. Then, the equality in (3.17) holds, and one unique velocity is selected. As for (3.12), we let $\theta \to 0$, and the minimum traveling velocity σ^* is selected. Then the remark below follows.

Remark 3.2 (Incompressible Cells Limit) In the limit $C_\nu \to +\infty$, we have $\rho(x) \to 1$ in $I_2 = (0, x_0)$ and

$$\Sigma(x) = C_\nu \ln(\rho) \to C_p - \frac{x^2}{2C_S}.$$

Therefore, $x_0 = \sqrt{2C_S C_p}$ and

$$C_\nu \partial_x \rho\left(x_0^-\right) = \partial_x \Sigma\left(x_0^-\right) \to -\sqrt{\frac{2C_p}{C_S}}, \qquad \text{when } C_\nu \to +\infty.$$

Thus (3.17) becomes, for $\sigma \geq 2\sqrt{\epsilon}$,

$$\sqrt{\sigma^2 - 4\epsilon} + \sigma \geq 2\sqrt{2C_p C_S},$$

and we conclude, in this incompressible cells limit, that σ^* is defined by

$$\sigma^* := \max\left(2\sqrt{\epsilon}, \min\left(2\sqrt{2C_p C_S}, \sqrt{2C_p C_S} + \frac{\epsilon}{\sqrt{2C_p C_S}}\right)\right). \tag{3.19}$$

The kink induced by this formula is a very typical qualitative feature that is recovered in numerical simulations (see Table 2).

3.3 Numerical Results

In order to perform numerical simulations, we consider a large computational domain $\Omega = [-L, L]$, and we discretize it with a uniform mesh

$$\Delta x = \frac{L}{2M}, \quad x_i = i\Delta x, \quad i = -M, \ldots, 0, \ldots, M.$$

Table 2 Numerical values for the traveling speed σ^* with different parameters for $C_\nu = 17.114$ obtained by solving the evolution equation. We observe that the numerical speeds are close to $\sqrt{2C_pC_S} + \frac{\epsilon}{\sqrt{2C_pC_S}}$ or $2\sqrt{\epsilon}$ as computed in (3.19). In the first four lines $\epsilon < 2C_pC_S$, while in the last two $\epsilon > 2C_pC_S$

C_p	C_S	ϵ	$\sqrt{2C_pC_S} + \frac{\epsilon}{\sqrt{2C_pC_S}}$	$2\sqrt{\epsilon}$	σ^*
0.57	0.001	0.001	0.0634	0.0632	0.0615
0.57	0.01	0.001	0.1161	0.0632	0.1155
1	0.01	0.001	0.1485	0.0632	0.1472
1	0.01	0.01	0.2121	0.200	0.2113
1	0.01	0.1	0.8485	0.632	0.5946
1	0.01	1	7.2125	2.000	1.9069

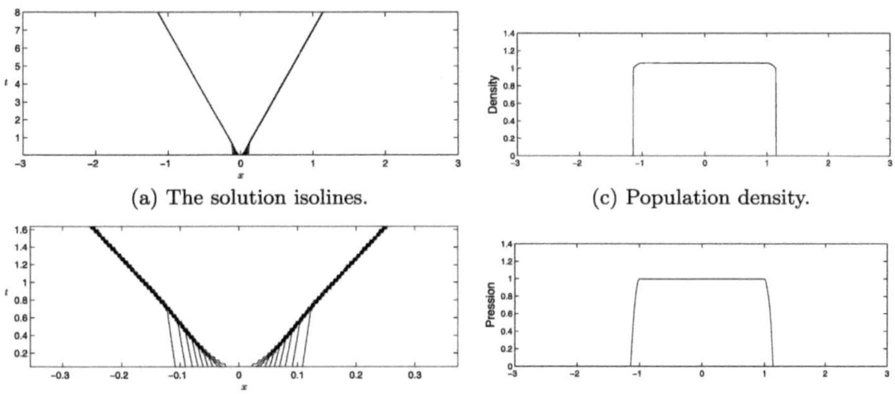

(a) The solution isolines.

(c) Population density.

(b) The zoom in of the subplot in (a).

(d) Pressure.

Fig. 2 The traveling wave solution for $C_z = 0$, $\epsilon = 0$. The parameters are chosen as in (3.20). In (**a**) and (**b**), the horizontal axis is x, and the vertical axis is time. (**c**) and (**d**) show the traveling front at $T = 8$

We simulate the time evolutionary equation (2.2) with $C_z = 0$ and Neumann boundary conditions. Our algorithm is based on a splitting method. Firstly, we discretize $\partial_t \rho - C_S \partial_{xx} Q(\rho) = 0$ by using the explicit Euler method in time and the second-order centered finite differences in space. After updating ρ^n for one time step, we denote the result by $\rho^{n+\frac{1}{2}}$. Secondly, we solve $\partial_t \rho = \rho H(C_p - \Sigma(\rho))$ by the explicit Euler scheme again, using $\rho^{n+\frac{1}{2}}$ as the initial condition. Then we get ρ^{n+1}.

The numerical initial density ρ is a small Gaussian in the center of the computational domain. We take

$$L = 3, \quad C_\nu = 17.114, \quad C_S = 0.01, \quad C_p = 1. \tag{3.20}$$

The numerical traveling wave solution when $C_z = 0$, $\epsilon = 0$ is depicted in Fig. 2. We can see that the two fronts propagate in opposite directions with a constant speed.

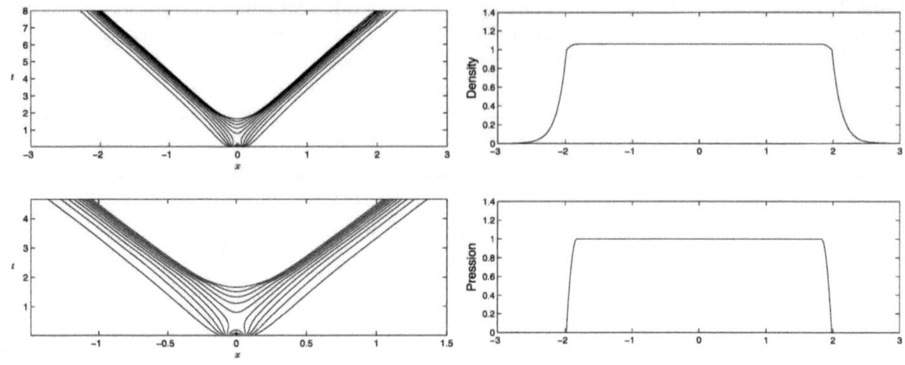

Fig. 3 As Fig. 2 with $C_z = 0, \epsilon = 0.02$

The right propagating front of ρ has a jump from 1 to 0, whereas Σ is continuous, but its derivative $\partial_x \Sigma$ has a jump at the front. Figure 3 presents the numerical results of $C_z = 0, \epsilon = 0.02$, where ρ becomes continuous and the front shape of Σ stays the same as for $\epsilon = 0$. Comparing Figs. 2 and 3, when there is diffusion, the traveling velocity becomes bigger and the density has a tail.

The numerical traveling velocities for different parameters are given in Table 2, where we can compare them with the analytical formula (3.19) in the incompressible cells limit.

4 Traveling Wave with Viscosity

When $C_z \neq 0$, we can not eliminate the unknown, and we have to deal with the whole system

$$\begin{cases} -\sigma \partial_x \rho - C_S \partial_x \rho \partial_x W - C_S \rho \partial_{xx} W = \rho H(C_p - \Sigma) + \epsilon \partial_{xx} \rho, \\ -C_z \partial_{xx} W + W = \Sigma(\rho), \\ \rho(-\infty) = \rho_M, \quad \rho(+\infty) = 0; \quad W(-\infty) = C_p, \quad W(+\infty) = 0 \end{cases} \tag{4.1}$$

still with the equation of state (2.10). In the interval $\{\rho \geq 1\}$, multiplying (2.6) by $\Sigma'(\rho) = \frac{C_\nu}{\rho}$, we get

$$-\sigma \partial_x \Sigma - C_S \partial_x \Sigma \partial_x W - C_S C_\nu \partial_{xx} W = C_\nu H(C_p - \Sigma) + \epsilon \frac{C_\nu}{\rho} \partial_{xx} \rho \quad \text{for } \rho \geq 1. \tag{4.2}$$

Here the situation is much more complicated, and a new phenomenon appears. We need to clarify the meaning of the discontinuous growth term, when $\Sigma = C_p$, which occurs on an interval and is not well defined in the singular incompressible cells limit as we study here (see (4.5)). To do so, we use a linear smoothing of the Heaviside function H, such that

$$H_\eta(u) = \min\left(1, \frac{1}{\eta}u\right) \quad \text{for } \eta \in (0, C_p). \tag{4.3}$$

There are no explicit or semi-explicit solutions for the traveling waves in general due to the non-local aspect of the field W, and we refer to [27] again for a proof of existence in a related case. Thus we will consider the incompressible cells limit. First, we derive formally the limiting system by letting $C_\nu \to +\infty$. From the state equation, we have $1 \le \rho \le \rho_M \to 1$. Therefore, we need to distinguish the two cases, i.e., $\rho = 1$ and $\rho < 1$. Formally when $\rho < 1$, we find that $\Sigma = 0$, and (4.1) reduces to

$$\begin{cases} -\sigma\partial_x\rho - C_S\partial_x\rho\partial_x W - C_S\rho\partial_{xx}W = \rho + \epsilon\partial_{xx}\rho, & \rho < 1, \\ -C_z\partial_{xx}W + W = 0. \end{cases} \tag{4.4}$$

On the interval, where $\rho = 1$, as $C_\nu \to +\infty$, and the function Σ is not defined in terms of ρ and is left unknown, the formal limit of (4.1) implies a coupled system on W and Σ,

$$\begin{cases} -C_S\partial_{xx}W = H_\eta(C_p - \Sigma), & \rho = 1, \\ -C_z\partial_{xx}W + W = \Sigma. \end{cases} \tag{4.5}$$

Then the existence of traveling waves in the asymptotic case $C_\nu \to +\infty$ boils down to studying the asymptotic system (4.4)–(4.5). As in Sect. 3, the structure of the problem invites us to distinguish between the two cases, i.e., $\epsilon = 0$ and $\epsilon \neq 0$.

4.1 Case $\epsilon = 0$

Existence of Traveling Wave in the Limit $C_\nu \to +\infty$ In this case, we can establish the following theorem.

Theorem 4.1 *Assume $C_z \neq 0$, $\epsilon = 0$ and $C_S C_p > 2C_z$. Then there exists a $\sigma^* > 0$, such that for all $\sigma \ge \sigma^*$, the asymptotic system (4.4)–(4.5) admits a nonnegative and non-increasing solution (ρ, Σ). Furthermore, when $\eta \to 0$, we have $\sigma^* = \sqrt{2C_S C_p} - \sqrt{C_z}$, and the solution is given by*

$$\Sigma(x) = \begin{cases} C_p, & x \le 0, \\ -\frac{x^2}{2C_S} - \frac{x}{C_S}\sqrt{C_z} + C_p, & 0 < x \le \sqrt{2C_S C_p} - 2\sqrt{C_z} =: x_0, \\ 0, & x > x_0. \end{cases} \tag{4.6}$$

Therefore, Σ has a jump from $\sqrt{\frac{2C_pC_z}{C_S}}$ to 0 at x_0. The population density satisfies

$$\begin{cases} \rho = 1 & \text{for } x < x_0, \\ \rho = 0 & \text{for } x > x_0, \text{ when } \sigma = \sigma^*, \\ \rho = (\sigma - \sigma^* e^{-\frac{x-x_0}{\sqrt{C_z}}})^{-1-\frac{\sqrt{C_z}}{\sigma}} e^{-\frac{x-x_0}{\sigma}} & \text{for } x > x_0, \text{ when } \sigma > \sigma^*. \end{cases}$$

Proof By the maximum principle in Lemma 2.1, and according to Definition 2.1, Σ is bounded by C_p and is nonnegative. Therefore, due to elliptic regularity, $\partial_{xx} W$ is bounded, and W and $\partial_x W$ are continuous. Following the idea in the proof of Theorem 3.1 or Theorem 3.2, we look for a nonnegative and non-increasing traveling wave defined in $\mathbb{R} = I_1 \cup I_2 \cup I_3$, which has the following form:

(1) On $I_1 = (-\infty, 0]$, we have $\Sigma \in [C_p - \eta, C_p]$, so that the growth term is given by $H_\eta(C_p - \Sigma) = \frac{1}{\eta}(C_p - \Sigma)$.
(2) In $I_2 = (0, x_0)$, we have $\Sigma \in (0, C_p - \eta)$. Thus $H_\eta(C_p - \Sigma) = 1$ and $\rho = 1$.
(3) On $I_3 = [x_0, +\infty)$, we have $\rho < 1$ and $\Sigma = 0$.

On I_1, we have $\rho = 1$, and we solve (4.5). This system can be written as

$$-C_S \partial_{xx} W = \frac{1}{\eta}(C_p - \Sigma), \qquad -C_z \partial_{xx} W + W = \Sigma.$$

Eliminating Σ in this system gives

$$-(\eta C_S + C_z)\partial_{xx} W + W = C_p.$$

Together with the boundary conditions of W at $-\infty$, we have

$$W = C_p + A e^{\frac{x}{\sqrt{\eta C_S + C_z}}} \quad \text{and} \quad \Sigma = C_p + \frac{\eta C_S A}{\eta C_S + C_z} e^{\frac{x}{\sqrt{\eta C_S + C_z}}},$$

which is the bounded solution on $I_1 = (-\infty, 0]$. The constant A can be determined as follows. Since Σ depends continuously on ρ and $\rho = 1$ in $I_1 \cup I_2$, Σ is continuous at x_0. Therefore, A is computed by fixing $\Sigma(0) = C_p - \eta$, which gives $A = -\eta - \frac{C_z}{C_S}$.

In I_2, we still have $\rho = 1$. (4.5) can be written as

$$-C_S \partial_{xx} W = 1, \qquad -C_z \partial_{xx} W + W = \Sigma.$$

At the interface $x = 0$, W and $\partial_x W$ are continuous and given by their values on I_1. Then we can solve the first equation that gives

$$W(x) = -\frac{x^2}{2C_S} - \frac{x}{C_S}\sqrt{\eta C_S + C_z} + C_p - \eta - \frac{C_z}{C_S}. \tag{4.7}$$

Injecting this expression in the second equation implies

$$\Sigma(x) = -\frac{x^2}{2C_S} - \frac{x}{C_S}\sqrt{\eta C_S + C_z} + C_p - \eta.$$

On I_3, since $\rho < 1$, we have to solve (4.4) with $\epsilon = 0$. The second equation in (4.4) can be solved easily, and the only solution which is bounded on $(x_0, +\infty)$ is

$$W(x) = W(x_0)\mathrm{e}^{-\frac{x-x_0}{\sqrt{C_z}}}. \tag{4.8}$$

We fix the value of x_0 by using the continuity of W and the derivative of W at x_0. From (4.8), we have $-\frac{W(x_0)}{\sqrt{C_z}} = \partial_x W(x_0)$. From (4.7), this equality can be rewritten as

$$\frac{1}{\sqrt{C_z}}\left(\frac{x_0^2}{2C_S} + \frac{x_0}{C_S}\sqrt{\eta C_S + C_z} - C_p + \eta + \frac{C_z}{C_S}\right) = -\frac{x_0}{C_S} - \frac{1}{C_S}\sqrt{\eta C_S + C_z}.$$

This is a second order equation for x_0, whose only nonnegative solution (for η small enough) is

$$x_0 = \sqrt{2C_pC_S - \eta C_S} - \sqrt{C_z} - \sqrt{C_z + \eta C_S}. \tag{4.9}$$

Now we determine the expression for ρ on I_3. The jump condition of (4.4) at x_0 in the case $\epsilon = 0$ can be written as $\sigma[\rho]_{x_0} + C_S[\rho\partial_x W]_{x_0} = 0$. The continuity of $\partial_x W$ implies

$$[\rho]_{x_0} = 0 \quad \text{or}$$
$$\sigma = \sigma^* := -C_S\partial_x W(x_0) = x_0 + \sqrt{\eta C_S + C_z} = \sqrt{2C_pC_S - \eta C_S} - \sqrt{C_z}.$$

From the expression (4.8), the first equation in (4.4) with $\epsilon = 0$ gives

$$\left(\sigma - \sigma^*\mathrm{e}^{-\frac{x-x_0}{\sqrt{C_z}}}\right)\partial_x\rho + \left(1 + \frac{\sigma^*}{\sqrt{C_z}}\mathrm{e}^{-\frac{x-x_0}{\sqrt{C_z}}}\right)\rho = 0. \tag{4.10}$$

Looking for a non-increasing and nonnegative ρ implies that we should have $\sigma \geq \sigma^*$. After straightforward computation, we get that

$$\partial_x\rho = -\partial_x\left(\frac{x - x_0}{\sigma} + \left(1 + \frac{\sqrt{C_z}}{\sigma}\right)\ln\left(\sigma - \sigma^*\mathrm{e}^{-\frac{x-x_0}{\sqrt{C_z}}}\right)\right)\rho. \tag{4.11}$$

If $[\rho]_{x_0} = 0$ and $\sigma > \sigma^*$, the Cauchy problem (4.11) with $\rho(x_0) = 1$ admits a unique solution, which is given by

$$\rho(x) = \left(\sigma - \sigma^*\mathrm{e}^{-\frac{x-x_0}{\sqrt{C_z}}}\right)^{-1-\frac{\sqrt{C_z}}{\sigma}}\mathrm{e}^{-\frac{x-x_0}{\sigma}}.$$

When $\sigma = \sigma^*$, the factor of ρ on the right-hand side of (4.11) has a singularity at $x = x_0$. Therefore, the only solution which does not blow up in $x = x_0$ is $\rho = 0$. □

Remark 4.1 When $\sqrt{2C_pC_S} < 2\sqrt{C_z}$, Σ becomes a step function with a jump from C_p to 0 at the point x_0. The corresponding traveling speed is

$$\sigma = -C_S\partial_x W(x_0) = \frac{C_pC_S}{2\sqrt{C_z}}$$

with

$$W(x) = \begin{cases} \frac{C_p}{2}e^{-\frac{1}{\sqrt{C_z}}(x-x_0)}, & x > x_0, \\ C_p - \frac{C_p}{2}e^{\frac{1}{\sqrt{C_z}}(x-x_0)}, & x < x_0. \end{cases}$$

The calculations are similar, but simpler than those in Theorem 4.1.

Remark 4.2 (Comparison with the Case $C_z = 0$) In the asymptotic $\eta \to 0$, and when $C_z \to 0$, the expression for σ^* in Theorem 4.1 converges to that obtained for $C_z = 0$. However, we notice that, contrary to the case $C_z = 0$, the growth term does not vanish on I_1 whereas $\Sigma = C_p$. In fact, if the growth term was zero on I_1, then since $\Sigma = C_p$, we would have $\partial_x\Sigma = 0$ and (4.2) gives

$$-C_SC_\nu\partial_{xx}W = 0.$$

Thus $\partial_{xx}W = 0$ and $W = \Sigma$ on I_1, which can not hold true. That is why we can not use the Heaviside function in the growth term when $\Sigma = C_p$, and the linear approximation in (4.3) allows us to make explicit calculations.

Numerical Results We present some numerical simulations of the full model (2.2) with the growth term $\Phi = \rho H(C_p - \Sigma(\rho))$ and $\epsilon = 0$. As in the previous section, we consider a computational domain $\Omega = [-L, L]$ discretized by a uniform mesh, and use Neumann boundary conditions. System (2.2) is now a coupling of a transport equation for ρ and an elliptic equation for W. We use the following schemes:

(1) The centered three point finite difference method is used to discretize the equation for W.
(2) A splitting method is implemented to update ρ. Firstly, we use a first order upwind discretization of the term $-C_S\partial_x(\rho\partial_x W)$ (i.e., without the right-hand side). Secondly, we solve the growth term $\partial_t\rho = \rho H(C_p - \Sigma(\rho))$ with an explicit Euler scheme.

As before, starting from a Gaussian at the middle of the computational domain, Fig. 4 shows the numerical traveling wave solutions for $C_z = 0.01$ and $\epsilon = 0$. We can observe that, at the traveling front, ρ has a jump from 1 to 0, and Σ has a layer and then jumps to zero. These observations are in accordance with our analytical results, and in particular with (4.6) for Σ.

When $C_z = 0.4$, the relation $C_SC_p > 2C_z$ is no longer satisfied. However, we can perform numerical simulations, and the results are presented in Fig. 5. The proof

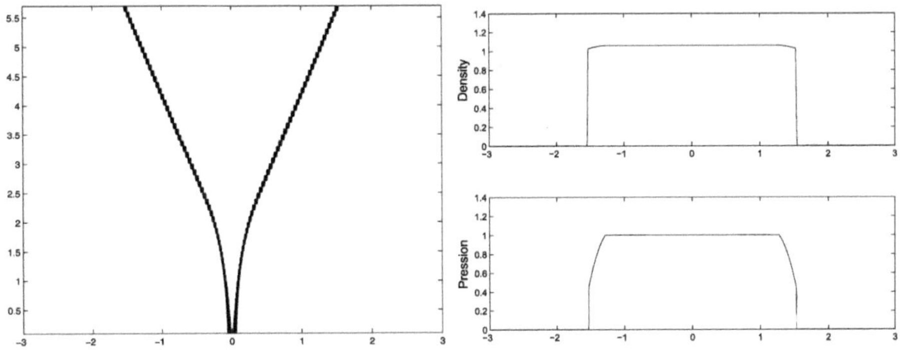

(a) the isolines of the traveling front with respect to time.

(b) the front shapes of the density and pressure.

Fig. 4 Numerical results when $C_p = 1$, $C_S = 0.1$, $C_v = 17.114$, $C_z = 0.01$ and $\epsilon = 0$

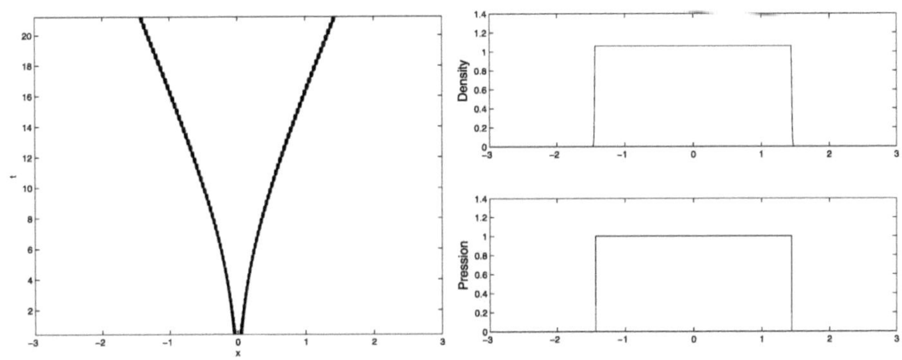

Fig. 5 As in Fig. 4, but the results violate the condition $C_S C_p > 2C_z$ using $C_p = 1$, $C_S = 0.1$, $C_v = 17.114$, $C_z = 0.4$ and $\epsilon = 0$

of Theorem 4.1 shows that we can not have a traveling wave which satisfies the continuity relation for W and $\partial_x W$ at the point x_0. In fact, in Fig. 5, we notice that the pressure Σ seems to have a jump directly from 1 to 0 at the front position, which is in accordance with Remark 4.1.

With different choices of parameters, the numerical values for the traveling velocities σ and the front jumps of Σ at x_0 are given in Table 3, where we can verify the analytical formula in Theorem 4.1.

4.2 Case $\epsilon \neq 0$

Existence of Traveling Waves The case with diffusion such that $\epsilon \neq 0$, can be handled by the same method as above. We have the following result.

Table 3 The traveling speed σ^* for different parameter values satisfying $2C_z < C_pC_S$. The numerical speeds are close to $\sqrt{2C_pC_S} - \sqrt{C_z}$, and the jump of Σ is not far from $\sqrt{\frac{2C_pC_z}{C_S}}$ as calculated in Theorem 4.1

C_p	C_S	C_z	$\sqrt{2C_pC_S} - \sqrt{C_z}$	$\frac{C_pC_S}{2\sqrt{C_z}}$	σ^*	$\sqrt{\frac{2C_pC_z}{C_S}}$	$\Sigma(x_0)$
0.57	1	0.1	0.7515	0.9012	0.7616	0.3376	0.3342
0.57	1	0.01	0.9677	2.8500	0.9686	0.1068	0.1052
0.57	0.1	0.01	0.2376	0.2850	0.2438	0.3376	0.3362
1	0.1	0.01	0.3472	0.500	0.3507	0.4472	0.4129
1	0.1	0.0	0.4472	–	0.4424	0	0

Theorem 4.2 *Assume $\epsilon \neq 0$, $C_z \neq 0$ and $C_SC_p > 2C_z$. Then there exists a $\sigma^* > 0$, such that for all $\sigma \geq \sigma^*$, the asymptotic model (4.4)–(4.5) admits a nonnegative and non-increasing solution (ρ, Σ). As $\eta \to 0$, the following bound on the minimal speed holds*:

$$\max\{2\sqrt{\epsilon}, \sqrt{2C_SC_p} - \sqrt{C_z}\} \leq \sigma^* \leq (\sqrt{2C_SC_p} - \sqrt{C_z}) + 2\sqrt{\epsilon\sqrt{\frac{2C_SC_p}{C_z}}},$$

The solution is given by

$$\Sigma(x) = \begin{cases} C_p, & x \leq 0, \\ -\frac{x^2}{2C_S} - \frac{x}{C_S}\sqrt{C_z} + C_p, & 0 < x \leq \sqrt{2C_SC_p} - 2\sqrt{C_z}, \\ 0, & x > \sqrt{2C_SC_p} - 2\sqrt{C_z}. \end{cases} \tag{4.12}$$

The cell density ρ is a positive, non-increasing $C^1(\mathbb{R})$ function, such that

$$\rho = 1 \quad \text{for } x < \sqrt{2C_SC_p} - 2\sqrt{C_z} \quad \text{and} \quad \rho < 1 \quad \text{for } x > 2\sqrt{2C_SC_p} - 2\sqrt{C_z}.$$

Proof As above, W and $\partial_x W$ are continuous on $\mathbb{R}$. Moreover, due to the diffusion term, ρ is continuous. Using the same decomposition $\mathbb{R} = I_1 \cup I_2 \cup I_3$ as before, we notice that, in $I_1 \cup I_2$, the problem is independent of ϵ. Thus we have the same conclusion as in Theorem 4.1.

(1) On I_1, we have $\rho = 1$, $\Sigma = C_p - \eta e^{\frac{x}{\sqrt{\eta C_S + C_z}}}$ and $W = C_p - (\eta + \frac{C_z}{C_S})e^{\frac{x}{\sqrt{\eta C_S + C_z}}}$.

(2) In I_2, we have $\rho = 1$, $\Sigma = C_p - \eta - \frac{x}{C_S}\sqrt{\eta C_S + C_z} - \frac{x^2}{2C_S}$ and $W = C_p - \eta - \frac{C_z}{C_S} - \frac{x}{C_S}\sqrt{\eta C_S + C_z} - \frac{x^2}{2C_S}$.

(3) On I_3, still from the second equation of (4.4) and the continuity of W and $\partial_x W$, we have

$$\begin{cases} W(x) = \frac{\sqrt{C_z}}{C_S}(\sqrt{C_z + \eta C_S} + x_0)\mathrm{e}^{-\frac{x-x_0}{\sqrt{C_z}}}, \\ x_0 = \sqrt{2C_S C_p - \eta C_s} - \sqrt{C_z} - \sqrt{C_z + \eta C_S}. \end{cases} \tag{4.13}$$

The jump condition at x_0 for the first equation of (4.4) is

$$-\sigma[\rho]_{x_0} - C_S[\rho \partial_x W]_{x_0} = \epsilon[\partial_x \rho]_{x_0},$$

which implies $[\partial_x \rho]_{x_0} = 0$ thanks to the continuity of ρ and $\partial_x W$. Then, from (4.4), when $\rho < 1$, the density satisfies

$$\epsilon \partial_{xx}\rho + \left(\sigma - \frac{C_S}{\sqrt{C_z}} W\right)\partial_x \rho + \left(1 + \frac{C_S}{C_z} W\right)\rho = 0, \tag{4.14}$$

where W is as in (4.13). This equation is completed with the boundary conditions

$$\rho(x_0) = 1 \quad \text{and} \quad \partial_x \rho(x_0) = 0. \tag{4.15}$$

The Cauchy problem (4.14)–(4.15) admits a unique solution. Moreover, at the point x_0, we deduce from (4.14) that

$$\epsilon \partial_{xx}\rho(x_0) = -1 - \frac{C_S}{C_z} W(x_0) < 0.$$

Therefore, $\partial_x \rho$ is decreasing in the vicinity of x_0. We deduce that $\partial_x \rho \leq 0$ for $x \geq x_0$ in the vicinity of x_0. Then if ρ does not have a minimum on $(x_0, +\infty)$, it is a nonincreasing function, which necessarily tends to 0 at infinity from (4.14). If ρ admits a minimum at the point $x_m > x_0$, then we have $\partial_{xx}\rho(x_m) > 0$ and $\partial_x \rho(x_m) = 0$. We deduce from (4.14) that

$$\rho(x_m)\left(1 + \frac{C_S}{C_z} W(x_m)\right) = -\epsilon \partial_{xx}\rho(x_m) < 0.$$

We conclude that $\rho(x_m) < 0$. Thus there exists a point x_c, such that $\rho(x_c) = 0$. Then on $[x_0, x_c)$, we have that $\rho > 0$, and it is nonincreasing. The question is then to know whether there exists a value of σ for which $x_c = +\infty$. In order to do so, we will compare ρ with $\widetilde{\rho}$ that satisfies

$$\epsilon \partial_{xx}\widetilde{\rho} + \left(\sigma - \frac{C_S}{\sqrt{C_z}} K\right)\partial_x \widetilde{\rho} + \left(1 + \frac{C_S}{C_z} K\right)\widetilde{\rho} = 0, \quad x \in (x_0, +\infty) \tag{4.16}$$

with the boundary conditions

$$\widetilde{\rho}(x_0) = 1, \qquad \partial_x \widetilde{\rho}(x_0) = 0. \tag{4.17}$$

Here K is a given constant which will be defined later.

Lower Bound on σ^* Integrating (4.14) from x_0 to $+\infty$, and using $\partial_x W = -\frac{W}{\sqrt{C_z}}$ and the boundary conditions in (4.15), we have

$$\sigma = \sqrt{C_z + \eta C_S} + x_0 + \int_{x_0}^{+\infty} \rho(x)\mathrm{d}x.$$

We deduce that if we had a nonnegative solution ρ, then

$$\sigma \geq \sqrt{C_z + \eta C_S} + x_0 = \sqrt{2C_S C_p - \eta C_s} - \sqrt{C_z}. \tag{4.18}$$

Moreover, from (4.14), we have

$$\epsilon \partial_{xx}\rho + \sigma\partial_x\rho + \rho = \frac{C_S}{\sqrt{C_z}} W\partial_x\rho - \frac{C_S}{C_z} W\rho \leq 0.$$

Using the second assertion of Lemma 4.1, we can compare ρ with $\widetilde{\rho}$ that is the solution to (4.16)–(4.17) with $K = 0$. We deduce that $\rho \leq \widetilde{\rho}$, since when $\sigma < 2\sqrt{\epsilon}$, $\widetilde{\rho}$ takes negative values on I_3. Thus, ρ is no longer nonnegative, which is a contradiction. Therefore,

$$\sigma \geq 2\sqrt{\epsilon}. \tag{4.19}$$

Upper Bound on σ^* We use the bound $W \leq W(x_0)$ to get

$$\epsilon\partial_{xx}\rho + \left(\sigma - \frac{C_S}{\sqrt{C_z}} W(x_0)\right)\partial_x\rho + \left(1 + \frac{C_S}{C_z} W(x_0)\right)\rho \geq 0. \tag{4.20}$$

Using the assertion (1) of Lemma 4.1, wc deduce that ρ is positive on I_3 provided that

$$\sigma \geq \sqrt{2C_S C_p - \eta C_S} - \sqrt{C_z} + 2\sqrt{\epsilon\sqrt{\frac{2C_S C_p}{C_z}}}. \tag{4.21}$$

Thus for all σ satisfying (4.21), there exists a non-increasing and nonnegative solution ρ to (4.14)–(4.15).

However, the bound (4.21) is not satisfactory for small C_z. This is mainly due to the fact that the bound $W(x) \leq W(x_0)$ on I_3 is not sharp when C_z is small. We can improve this bound by using the remark that for any $x_z > x_0$, we have $W(x) \leq K := W(x_z)$. Let us define $x_z = x_0 + \sqrt{C_z}\,\xi(\sqrt{C_z})$ with a continuous function $\xi : (0, +\infty) \to (0, +\infty)$, such that $\lim_{x\to 0} x\xi(x) = 0$. Let us call $\widehat{\rho}$ a solution to (4.16) on $(x_z, +\infty)$ with $K = W(x_z)$ and the boundary conditions $\widehat{\rho}(x_z) = \rho(x_z) > 0$, $\partial_x\widehat{\rho}(x_z) = \partial_x\rho(x_z) \leq 0$. Using the assertion (1) of Lemma 4.1, we deduce that $\rho \geq \widehat{\rho}$, and $\widehat{\rho}$ is positive provided that

$$\sigma \geq \frac{C_S}{\sqrt{C_z}} W(x_z) + 2\sqrt{\epsilon\left(1 + \frac{C_S}{C_z} W(x_z)\right)} \tag{4.22}$$

and

$$\alpha + \sqrt{\alpha^2 - 4\beta} \geq -\frac{2\partial_x \rho(x_z)}{\rho(x_z)}, \tag{4.23}$$

where $\epsilon\alpha = \sigma - \frac{C_S W(x_z)}{\sqrt{C_z}}$ and $\epsilon\beta = 1 + \frac{C_S W(x_z)}{C_z}$. When $x_z \to x_0$, we have $\partial_x \rho(x_z) \to 0$, whereas $\alpha > \frac{2}{\sqrt{\epsilon}}$ from (4.22). Thus for $\sqrt{C_z}$ small enough, (4.23) is satisfied provided that (4.22) is satisfied, i.e.,

$$\sigma \geq (\sqrt{2C_S C_p - \eta C_S} - \sqrt{C_z})e^{-\xi(\sqrt{C_z})} + 2\sqrt{\epsilon}\sqrt{1 + \left(\sqrt{\frac{2C_S C_p - \eta C_s}{C_z}} - 1\right)e^{-\xi(\sqrt{C_z})}}. \tag{4.24}$$

Therefore, choosing the function ξ, such that $\lim_{x\to 0} \frac{e^{-\xi(x)}}{x} = 0$, we deduce that when $C_z \to 0$, (4.24) becomes $\sigma \geq 2\sqrt{\epsilon}$. One possible choice is $\xi(x) = \ln x^2$. □

The proof of Theorem 4.2 uses the following lemma.

Lemma 4.1 *Let α, β, a be positive and $b \leq 0$. For $g \in C(\mathbb{R}_+)$, let f and $\widetilde{f}$ be the solutions to the following Cauchy problems on $\mathbb{R}_+$:*

$$f'' + \alpha f' + \beta f = g, \quad f(0) = a, \quad f'(0) = b \tag{4.25}$$

and

$$\widetilde{f}'' + \alpha \widetilde{f}' + \beta \widetilde{f} = 0, \quad \widetilde{f}(0) = a, \quad \widetilde{f}'(0) = b, \tag{4.26}$$

respectively. Then we have

(1) Assume $g \geq 0$ on $\mathbb{R}^+$. If $\alpha^2 \geq 4\beta$ and $\alpha + \sqrt{\alpha^2 - 4\beta} \geq -\frac{2b}{a}$, then $f(x) \geq \widetilde{f}(x) > 0$ for $x \in \mathbb{R}_+$. Or else, there exists an $x_c > 0$, such that $\widetilde{f}(x_c) = 0$ and $\widetilde{f} \geq 0$ on $[0, x_c]$. Moreover, if $\alpha^2 < 4\beta$, we have $f(x) \geq \widetilde{f}(x)$ for $x \in [0, \frac{2\pi}{\sqrt{4\beta - \alpha^2}}]$; if $\alpha^2 \geq 4\beta$ and $\alpha + \sqrt{\alpha^2 - 4\beta} < \frac{2b}{a}$, we have $f(x) \geq \widetilde{f}(x)$ for $x \in [0, x_c]$.

(2) Assume $g \leq 0$ on $\mathbb{R}^+$. If $\alpha^2 \geq 4\beta$, then $f(x) \leq \widetilde{f}(x)$ for $x \geq 0$. If moreover $\alpha + \sqrt{\alpha^2 - 4\beta} < -\frac{2b}{a}$, then f takes negative values on $\mathbb{R}_+$. If $\alpha^2 < 4\beta$, then we have $f(x) \leq \widetilde{f}(x)$ for $x \in [0, \frac{2\pi}{\sqrt{4\beta - \alpha^2}}]$ and f takes negative values on $[0, \frac{2\pi}{\sqrt{4\beta - \alpha^2}}]$.

Proof Denote by r_1 and r_2 the roots of the characteristic equation $r^2+\alpha r+\beta=0$. Then, if $r_1\neq r_2$, by solving (4.25)–(4.26), we have

$$\begin{aligned}\widetilde{f}(x)&=\frac{r_2a-b}{r_2-r_1}\mathrm{e}^{r_1x}+\frac{r_1a-b}{r_1-r_2}\mathrm{e}^{r_2x},\\ f(x)&=\widetilde{f}(x)+\int_0^x g(y)\left(\frac{\mathrm{e}^{r_1(x-y)}}{r_1-r_2}+\frac{\mathrm{e}^{r_2(x-y)}}{r_2-r_1}\right)\mathrm{d}y.\end{aligned}\tag{4.27}$$

First, we assume that $g\geq 0$ on $\mathbb{R}_+$. If $\alpha^2>4\beta$, then r_1 and r_2 are real negative. We deduce that

$$\frac{\mathrm{e}^{r_1x}}{r_1-r_2}+\frac{\mathrm{e}^{r_2x}}{r_2-r_1}>0,$$

and then $f(x)>\widetilde{f}(x)$ for $x\geq 0$. Moreover, $\widetilde{f}$ vanishes on $\mathbb{R}_+$ if and only if $\min\{r_1,r_2\}\geq\frac{b}{a}$.

If $\alpha^2<4\beta$, r_1 and r_2 are complex and $\overline{r}_1=r_2$. We denote $r_1=R-\mathrm{i}I$, where $2R=-\alpha$ and $2I=\sqrt{4\beta-\alpha^2}$. We can rewrite then

$$\widetilde{f}(x)=\left(\frac{R-b}{I}\sin(Ix)+a\cos(Ix)\right)\mathrm{e}^{Rx}.\tag{4.28}$$

We deduce that there exists an x_c, such that $\widetilde{f}(x_c)=0$ and $\widetilde{f}\geq 0$ on $[0,x_c]$. Moreover,

$$\frac{\mathrm{e}^{r_1x}}{r_1-r_2}+\frac{\mathrm{e}^{r_2x}}{r_2-r_1}=\frac{\mathrm{e}^{Rx}}{I}\sin(Ix)\geq 0\quad\text{for } x\in\left[0,\frac{\pi}{I}\right].\tag{4.29}$$

Thus $f(x)\geq\widetilde{f}(x)$ if $x\in[0,\frac{\pi}{I}]$.

If $\alpha^2=4\beta$, we have $r_1=r_2=-\frac{\alpha}{2}$. By straightforward computation, we have $\widetilde{f}(x)=((b-ar_1)x+a)\mathrm{e}^{rx}$, and

$$f(x)=\widetilde{f}(x)+\int_0^x(x-y)\mathrm{e}^{r_1(x-y)}g(y)\mathrm{d}y.\tag{4.30}$$

For $g\geq 0$, we deduce $f\geq\widetilde{f}$. This concludes the proof of the first point.

Let us consider that $g\leq 0$ on $\mathbb{R}_+$. We deduce the first assertion from (4.27) and (4.30). If $\alpha^2<4\beta$, we deduce $f\leq\widetilde{f}$ on $[0,\frac{\pi}{I}]$ from (4.27) and (4.29). And we have from (4.28) $\widetilde{f}(\frac{\pi}{I})=-a\mathrm{e}^{\frac{\pi R}{I}}<0$, and thus f vanishes on $[0,\frac{\pi}{I}]$. □

Numerical Results We perform numerical simulations of the full system (2.2) by using the same algorithm as in Sect. 4.1 and a centered finite difference scheme for the diffusion term $\epsilon\partial_{xx}\rho$.

We present in Fig. 6 the numerical results still with parameters in (3.20) and $C_z=0.01$, $\epsilon=0.01$. Comparing Figs. 4 and 6, we notice that the profile of ρ has a tail in the latter case.

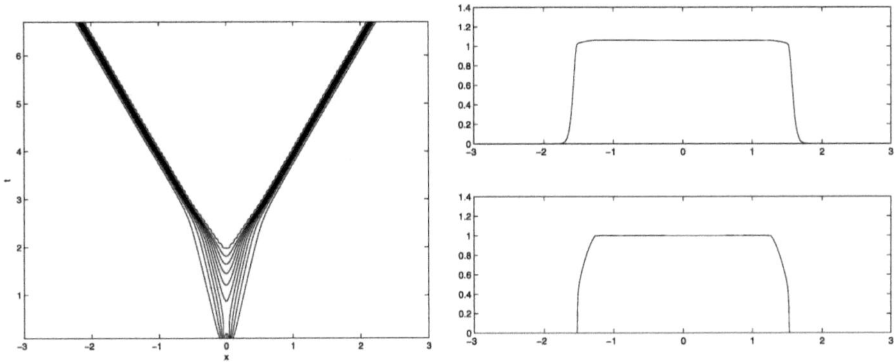

(a) The isolines of the traveling front with respect to time.

(b) The front shapes of the density and pressure.

Fig. 6 The numerical solution when $C_p = 1$, $C_S = 0.01$, $C_v = 17.114$, $C_z = 0.01$ and $\epsilon = 0.01$

Table 4 The traveling speed σ^* for (2.2) with different parameter values

C_p	C_S	C_z	ϵ	$\sqrt{2C_pC_S} - \sqrt{C_z}$	$2\sqrt{\epsilon}$	σ^*
0.57	0.01	0.001	0.01	0.07515	0.20	0.197
0.57	0.1	0.01	0.01	0.2376	0.20	0.321
0.57	1	0.1	0.001	0.7514	0.0632	0.780
0.57	1	0.1	0.01	0.7514	0.2	0.828
0.57	1	0.1	0.1	0.7514	0.632	1.015
0.57	1	0.1	1	0.7514	2	1.974

Table 4 gives numerical values of the traveling velocity for different parameters. We illustrate numerically the bound on σ^* obtained in the proof of Theorem 4.2.

Appendix: Derivation of the Cuboid State Equation

Cells are modelled as cuboidal elastic bodies of dimensions at rest $L_0 \times l_0 \times h_0$ in x, y, z directions aligned in a row in x direction. At rest, the lineic mass density of the row of cells, in contact but not deformed, is $\rho_0 = \frac{M_{\text{cell}}}{L_0}$. We consider the case that the cells are confined in a tube of section $l_0 \times h_0$, where the only possible deformation is along the x axis. This situation can be tested in a direct in-vitro experiment. Moreover, this limit would be expected in case a tumor composed of elastic cells is sufficiently large, such that for the ratio of the cell size L and the radius of curvature R, $\frac{L}{R} \ll 1$ holds, and the cell division is mainly oriented in radial direction as well as the cell-cell tangential friction is sufficiently small, such that a fingering or buckling instability does not occur.

When cells are deformed, we assume that stress and deformation are uniformly distributed, and that the displacements are small. Let L be the size of the cells. The lineic mass density is $\rho = \frac{\rho_0 L_0}{L}$. For $\rho < \rho_0$, the cells are not in contact and $\Sigma(\rho) = 0$; for $\rho \geq \rho_0$, a variation $\mathrm{d}L$ of the size L of the cell corresponds to an infinitesimal strain $\mathrm{d}u = \frac{\mathrm{d}L}{L}$. Therefore, the strain for a cell of size L is $u = \ln(\frac{L}{L_0})$. Assuming that a cell is a linear elastic body with Young modulus E and Poisson ratio ν, one finds that the component σ_{xx} of the stress tensor can be written as

$$\sigma_{xx} = -\frac{1-\nu}{(1-2\nu)(1+\nu)} E \ln\left(\frac{\rho}{\rho_0}\right).$$

The state equation is given by

$$\Sigma(\rho) = \begin{cases} 0, & \text{if } \rho \leq \rho_0, \\ \frac{1-\nu}{(1-2\nu)(1+\nu)} E \ln(\frac{\rho}{\rho_0}), & \text{otherwise.} \end{cases}$$

Here, $\Sigma(\rho) = -\sigma_{xx}$ is the pressure. Let $\overline{\rho} = \frac{\rho}{\rho_0}$, $\overline{\Sigma} = \frac{\Sigma}{E_0}$ and $\overline{E} = \frac{E}{E_0}$ be the dimensionless density, pressure and Young modulus respectively, with E_0 a reference Young modulus. Then the state equation can be written as

$$\overline{\Sigma}(\overline{\rho}) = \begin{cases} 0, & \text{if } \overline{\rho} \leq 1, \\ C_\nu \ln(\overline{\rho}), & \text{otherwise,} \end{cases}$$

where $C_\nu = \frac{\overline{E}(1-\nu)}{(1-2\nu)(1+\nu)}$. In the article, equations are written in the dimensionless form, and the bars above dimensionless quantities are removed.

References

1. Adam, J., Bellomo, N.: A Survey of Models for Tumor-Immune System Dynamics. Birkhäuser, Boston (1997)
2. Ambrosi, D., Preziosi, L.: On the closure of mass balance models for tumor growth. Math. Models Methods Appl. Sci. **12**(5), 737–754 (2002)
3. Anderson, A., Chaplain, M.A.J., Rejniak, K.: Single-Cell-Based Models in Biology and Medicine. Birkhauser, Basel (2007)
4. Araujo, R., McElwain, D.: A history of the study of solid tumor growth: the contribution of mathematical models. Bull. Math. Biol. **66**, 1039–1091 (2004)
5. Bellomo, N., Li, N.K., Maini, P.K.: On the foundations of cancer modelling: selected topics, speculations, and perspectives. Math. Models Methods Appl. Sci. **4**, 593–646 (2008)
6. Bellomo, N., Preziosi, L.: Modelling and mathematical problems related to tumor evolution and its interaction with the immune system. Math. Comput. Model. **32**, 413–452 (2000)
7. Berestycki, H., Hamel, F.: Reaction-Diffusion Equations and Propagation Phenomena. Springer, New York (2012)
8. Breward, C.J.W., Byrne, H.M., Lewis, C.E.: The role of cell-cell interactions in a two-phase model for avascular tumor growth. J. Math. Biol. **45**(2), 125–152 (2002)
9. Byrne, H., Drasdo, D.: Individual-based and continuum models of growing cell populations: a comparison. J. Math. Biol. **58**, 657–687 (2009)

10. Byrne, H.M., King, J.R., McElwain, D.L.S., Preziosi, L.: A two-phase model of solid tumor growth. Appl. Math. Lett. **16**, 567–573 (2003)
11. Byrne, H., Preziosi, L.: Modelling solid tumor growth using the theory of mixtures. Math. Med. Biol. **20**, 341–366 (2003)
12. Chaplain, M.A.J., Graziano, L., Preziosi, L.: Mathematical modeling of the loss of tissue compression responsiveness and its role in solid tumor development. Math. Med. Biol. **23**, 197–229 (2006)
13. Chatelain, C., Balois, T., Ciarletta, P., Amar, M.: Emergence of microstructural patterns in skin cancer: a phase separation analysis in a binary mixture. New J. Phys. **13**, 115013+21 (2011)
14. Chedaddi, I., Vignon-Clementel, I.E., Hoehme, S., et al.: On constructing discrete and continuous models for cell population growth with quantitatively equal dynamics. (2013, in preparation)
15. Ciarletta, P., Foret, L., Amar, M.B.: The radial growth phase of malignant melanoma: multiphase modelling, numerical simulations and linear stability analysis. J. R. Soc. Interface **8**(56), 345–368 (2011)
16. Colin, T., Bresch, D., Grenier, E., et al.: Computational modeling of solid tumor growth: the avascular stage. SIAM J. Sci. Comput. **32**(4), 2321–2344 (2010)
17. Cristini, V., Lowengrub, J., Nie, Q.: Nonlinear simulations of tumor growth. J. Math. Biol. **46**, 191–224 (2003)
18. De Angelis, E., Preziosi, L.: Advection-diffusion models for solid tumor evolution in vivo and related free boundary problem. Math. Models Methods Appl. Sci. **10**(3), 379–407 (2000)
19. Drasdo, D.: In: Alt, W., Chaplain, M., Griebel, M. (eds.) On Selected Individual-Based Approaches to the Dynamics of Multicellular Systems, Multiscale Modeling. Birkhauser, Basel (2003)
20. Evans, L.C.: Partial Differential Equations. Graduate Studies in Mathematics, vol. 19. AMS, Providence (1998)
21. Friedman, A.: A hierarchy of cancer models and their mathematical challenges. Discrete Contin. Dyn. Syst., Ser. B **4**(1), 147–159 (2004)
22. Funaki, M., Mimura, M., Tsujikawa, A.: Traveling front solutions in a chemotaxis-growth model. Interfaces Free Bound. **8**, 223–245 (2006)
23. Gardner, R.A.: Existence of travelling wave solution of predator-prey systems via the connection index. SIAM J. Appl. Math. **44**, 56–76 (1984)
24. Hoehme, S., Drasdo, D.: A cell-based simulation software for multi-cellular systems. Bioinformatics **26**(20), 2641–2642 (2010)
25. Lowengrub, J.S., Frieboes, H.B., Jin, F., et al.: Nonlinear modelling of cancer: bridging the gap between cells and tumors. Nonlinearity **23**, R1–R91 (2010)
26. Murray, J.D.: Mathematical Biology. Springer, New York (1989)
27. Nadin, G., Perthame, B., Ryzhik, L.: Traveling waves for the Keller-Segel system with Fisher birth terms. Interfaces Free Bound. **10**, 517–538 (2008)
28. Perthame, B., Quirós, F., Vázquez, J.L.: The Hele-Shaw asymptotics for mechanical models of tumor growth. (2013, in preparation)
29. Preziosi, L., Tosin, A.: Multiphase modeling of tumor growth and extracellular matrix interaction: mathematical tools and applications. J. Math. Biol. **58**, 625–656 (2009)
30. Radszuweit, M., Block, M., Hengstler, J.G., et al.: Comparing the growth kinetics of cell populations in two and three dimensions. Phys. Rev. E **79**, 051907 (2009)
31. Ranft, J., Basan, M., Elgeti, J., et al.: Fluidization of tissues by cell division and apoptosis. Proc. Natl. Acad. Sci. USA **107**(49), 20863–20868 (2010)
32. Roose, T., Chapman, S., Maini, P.: Mathematical models of avascular tumor growth: a review. SIAM Rev. **49**(2), 179–208 (2007)
33. Sánchez-Garduño, F., Maini, P.K.: Travelling wave phenomena in some degenerate reaction-diffusion equations. J. Differ. Equ. **117**(2), 281–319 (1995)
34. Weinberger, H.F., Lewis, M.A., Li, B.: Analysis of linear determinacy for spread in cooperative models. J. Math. Biol. **45**, 183–218 (2002)

MIX
Papier aus verantwortungsvollen Quellen
Paper from responsible sources
FSC® C105338
www.fsc.org

If you have any concerns about our products,
you can contact us on
ProductSafety@springernature.com

In case Publisher is established outside the EU,
the EU authorized representative is:
Springer Nature Customer Service Center GmbH
Europaplatz 3, 69115 Heidelberg, Germany

Printed by Libri Plureos GmbH
in Hamburg, Germany